Process Safety Spotlights 2016

Topical Conference at the 2016 AIChE Spring Meeting and 12th Global Congress on Process Safety

Houston, Texas, USA
10-14 April 2016

ISBN: 978-1-5108-2480-5

Printed from e-media with permission by:

Curran Associates, Inc.
57 Morehouse Lane
Red Hook, NY 12571

Some format issues inherent in the e-media version may also appear in this print version.

Printed by Curran Associates, Inc. (2016)

AIChE
120 Wall Street, FL 23
New York, NY 10005-4020

Phone: (800) 242-4363
Fax: (203) 775-5177

www.aiche.org

Additional copies of this publication are available from:

Curran Associates, Inc.
57 Morehouse Lane
Red Hook, NY 12571 USA
Phone: 845-758-0400
Fax: 845-758-2633
Email: curran@proceedings.com
Web: www.proceedings.com

TABLE OF CONTENTS

444849 Beyond Big Data Analytics Integrating with Cognitive Computing to Enable Smarter Manufacturing

Abstract:

The oil & gas, downstream and petrochemicals industries are facing challenges such as aging workforce with wide gaps in experience, aging asset base with more complex operations, market variability and speed of change requiring faster and better decisions at real-time, production and energy optimization, safe, secure and reliable operations for compliance.

We will discuss how the winning companies can step up to meet the above challenges by moving beyond the capabilities of Big Data Analytics by integrating Analytics with Cognitive Computing to enable real-time Cognitive Manufacturing. This will enable real-time decision management leveraging the insights and benefits from Analytics using cognitive systems such as IBM Watson that can:

- Ingest unstructured and structured data in all its forms.
- Understand unstructured data, through sensing and interaction;
- Reason with Bid Data by generating hypotheses, considering arguments, and recommendations; and
- Continuously learn from training by experts, from every interaction and from growing data.

IBM Watson has evolved from a single capability of natural language Q&A powered by few technologies such as machine learning and natural language processing (as seen in Jeopardy! in 2011) to more than 28 capabilities such as relationship extraction, personality analysis, tone analysis, concept expansion and trade-off analytics powered by 50 technologies. These capabilities have all been turned into digital services platform offered via APIs. These capabilities are provided as cognitive building blocks in in a cloud based ecosystem which anyone could build into their applications to scale and accelerate innovation.

Decision management platforms utilizing IBM Watson will meet the need for greater consistency in optimal decision making and knowledge retention by providing the capabilities such as the following: 1. Deeper human engagement, 2. Elevated expertise, 3. Cognitive products and services, 4. Cognitive processes and operations, 5. Intelligent exploration and discovery.

444732 How to Start Your Company on the Big Data Journey

Abstract:

The promise of Big Data in the chemical manufacturing industry appears to be coming to fruition. In this presentation we tackle the vexing problem of convincing your organization when and how to take the journey into Big Data. In this paper we will offer our experiences, reveal some of the challenges we met in gaining buy-in and how we overcame these obstacles.

A Rank Order Methodology for Assessing HFE Construction Audit Non-Conformances

Kurt Clowers, Ph.D.
GENESIS Senior Staff Specialist
11750 Katy Freeway Suite 100
Houston, TX 77079
Kurt.Clowers@genesisoilandgas.com

Genebelin Valbuena, Ph.D.
GENESIS Department Manager - Risk and Reliability
11750 Katy Freeway Suite 100
Houston, TX 77079
Genebelin.Valbuena@genesisoilandgas.com

Prepared for Presentation at
American Institute of Chemical Engineers
2016 Spring Meeting
12th Global Congress on Process Safety
Houston, Texas
April 11-23, 2016

A Rank Order Methodology for Assessing HFE Construction Audit Non-Conformances

Kurt Clowers, Ph.D.
GENESIS Senior Staff Specialist
11750 Katy Freeway Suite 100
Houston, TX 77079
Kurt.Clowers@genesisoilandgas.com

Genebelin Valbuena, Ph.D.
GENESIS Department Manager - Risk and Reliability
11750 Katy Freeway Suite 100
Houston, TX 77079
Genebelin.Valbuena@genesisoilandgas.com

Keywords: Construction; Human Factors, Risk Assessment

Abstract

Human factors are those things that enhance or improve human performance and Human Factors Engineering (HFE) applies that knowledge to the design and construction of socio-technical systems. The construction industry is one of the largest industries in the U.S. and today's construction of oil and gas facilities tends to be extensive and elaborate while constrained by schedule and resources. Owner/operators and engineering design firms face similar challenges while having to adhere to HFE guidelines and standards during design of these oil and gas facilities, the result of which can lead to oversight in ensuring HFE principles are properly implemented in the final product. Therefore to address this issue, a growing number of owner/operator and engineering design firms are conducting HFE construction audits. These audits can produce numerous HFE non-conformances such that the remediation of these non-conformances can threaten the feasibility of the project. A method for prioritizing these non-conformances was developed and the results can be applied to form a basis for allocating resources by the project and construction management teams. The approach follows a format similar to a hazard and operability study in that the HFE non-conformance is treated as a node and select HFE screening factors are used as guidewords to develop a consequence from the documented HFE non-conformance. A single-criteria evaluation model is used to rank order the HFE non-conformances and action thresholds are developed to further filter the HFE non-conformances. Using this model not only serves to provide greater focus to high-risk issues and reduce scope, it can also serve as a means to evaluate remediation options and while maintaining the least impact on cost and schedule.

1 Introduction

It is challenging to provide a simple definition or short catch phrase that can adequately characterize the scope of the human factors field. Nevertheless, the recommended practice; OGP 454, suggests that human factors are all those things that enhance or improve human performance in the workplace and states that Human Factors Engineering (HFE) focuses on the application of human factors knowledge to the design and construction of socio-technical systems [1].

Expanding on the definition above, HFE focuses on human beings and their interaction with equipment, facilities, procedures, and products used in everyday life. The objectives for HFE include enhancing the effectiveness and efficiency with which work and activities are completed and improving quality of life by reducing fatigue and stress, improving safety and increasing comfort and job satisfaction. HFE involves the systematic application of information about human capabilities, limitations, characteristics, and behavior to the design of things and the environment in which they are used [2]. Since the birth of HFE as a profession, shortly after World War II, human factors has become more involved and recognized for its contribution to the quality of life and work in numerous industries including construction.

The construction industry is one of the largest industries in the United States. In 2014, the construction industry accounted for 3.8% (or $664 billion) of the US gross domestic product [3]. Despite the recent collapse of oil prices and drilling, in 2015 the greater Houston Texas Metropolitan Area alone saw a major boom in building downstream petrochemical, refining and liquefied natural gas plants with over $50 billion in oil-industry construction underway. There was an enormous expansion seen in Gulf Coast petrochemical plants, accounting for $32.7 billion in new construction in the Houston Metro Area and even more construction spread up and down the Gulf Coast. Approximately $6.7 billion in local construction for in the liquefaction of natural gas for export also began in 2015 [4].

Considering the importance of the construction industry and the available economic resources, HFE application within engineering and construction of petrochemical facilities in the United States tends to be limited and highly specialized. This may be partly a result of the construction project itself. Construction work is generally regarded as temporary commitment with respect to a particular job. Human factors for skilled laborer positions is not normally within their scope and the work for a specific job is short-term, as a result employers may not feel it beneficial to invest in the education of human factors for skilled laborer positions. Furthermore, construction work has a long tradition of working in small groups with autonomous decision making which may reduce workers' tendencies to recognize or identify human factors types of problems [5].

Fully integrating an interdisciplinary practice such as HFE into the construction product in the oil and gas industry is onerous, namely because these projects have an increasing trend toward large scale operations with greater complexity. These large scale projects are often constrained by schedule and resources. Construction work is an important phase in the project execution process yet given the time constraints and complexities (e.g. logistics, technical issues, resources, etc.) associated with this phase it is challenging for construction companies to properly address human factors especially since the source of the problems arise from the design. This is to be expected since their primary service is to construct designs which are fit for purpose after being subjected to various checks during the engineering design process. Compounding the difficulty of assimilating HFE into the construction product, HFE requirements vary based on the maturity of the client's objectives. Regardless of the effort (3D model reviews, design reviews, etc.) put in during the design phase to identify HFE issues, not all of the problems are identified. During the execution phase it is the first time the design is "materialized" and issues that have not been identified during the design phase are realized which imposes additional pressure to projects that are often fast tracked. Therefore the need to develop new approaches or improve on existing ones and consider mechanisms to handle unexpected results is even more important.

Likewise, Owner/Operators and engineering design firms of petrochemical facilities share the pressures of schedule and resources during the construction of a facility though often being governed by HFE design requirements that are based on industry codes and Owner standards and guidelines to safe facility design. These standards and guidelines do not capture trade-offs, considerations of specific applications, and educated opinions which are all important for human use. Even if these were captured during the design phase, the intent of the HFE requirement may not be fully translated to complete fabrication details and as a result the HFE requirements may not be completely implemented in the final product.

Considering that not all potential HFE problems can be identified during the design and construction of the facility, a growing number of Owner/Operators and engineering design firms conduct HFE inspection audits during the fabrication and construction of their facilities and equipment. The purpose of the audits is to ensure HFE principles are integrated into the site construction of the project. A common result of any audit is the identification of non-conformances. It is common for hundreds or thousands of non-compliances to be identified during site construction and equipment fabrication audits. The non-conformances identified range from critical safety issues to limitations in equipment user operability or a simple specification non-compliance with no significant outcome. The challenge for all involved in the construction of the facility is how to address these non-conformances while maintaining the least impact on schedule and resources without compromising the quality.

A methodology is presented whereby the identified HFE non-conformances for a project can be assessed and prioritized for a given construction project. After the HFE non-conformances have been prioritized this information can be used to form a basis for allocating resources by the project and construction management teams. When well-implemented and managed, the methodology allows for a systematic identification of issues after which the project management team can use the results to make trade-off decisions or find alternative solutions.

2 Methodology

2.1 Overview

Risk Ranking is a common method used in Risk Management. A key aim of risk ranking is to provide a distinct focus to the critical risks within a system and assess the consequences of the risks. Risk ranking involves qualitatively assigning a risk value based on the consequence severity and likelihood of a particular risk scenario. The ranking provides a means to prioritize and filter individual risks for further action. This is particularly applicable in settings where resources are limited or some other constraints exists (e.g. tight schedule, limited budge, etc.). In order to achieve a single risk score, risk ranking requires an agreed-upon set of risk factors and evaluation criteria.

The proposed methodology provides conventional risk factors and evaluation criteria that can be used as a qualitative and structured technique to rank HFE hazards identified from inspections during construction of a facility. The process examines and orders each HFE non-conformance associated with a system by considering how the non-conformance has deviated from the design and the potential resulting undesirable consequence. For each HFE non-conformance the criticality of the unit/equipment and the complexity of task(s) and/or amount of human interaction is considered when determining the impact of the consequence of the issue. The severity of the HFE non-conformance is then determined which provides the basis for determining the prioritization level. Last, an additional filtering method using action thresholds is used to further sort the HFE non-conformances. The process outlined above is discussed in detail in the sections that follow and is also shown in Figure 1.

2.2 Components of HFE Non-Conformance Risk

In any risk management approach, the first step is to define the risk assessment scope. The purpose of this step is to develop the context of the assessment. The final goal of this assessment is to provide a list of prioritized HFE non-conformances for which decisions about the non-conformances that require remediation can be made. For the purposes of this technique the scope of the risk assessment can be defined in the form of the following question:

- How the identified HFE non-conformances should be prioritized for remediation?

After outlining the scope of the risk assessment, the risk criteria for evaluating the significance of the risks needs to be defined. Unit/Equipment criticality and complexity of tasks and/or human intervention were selected as HFE risk factors. These HFE risk factors were considered because they reflect the objective and context of the risk assessment. The appropriate HFE risk factor and level of detail is a matter of judgement dependent on the expertise of resources available and the situation to which it is applied.

These two screening factors were adopted from a commonly applied tool that HFE practitioners use. The HFE Equipment Screening Tool was developed by one of the major oil and gas companies and identifies if there are any significant HFE issues related to facilities under development [1]. The HFE risk factors unit/equipment criticality and complexity of tasks and/or human intervention were modified and applied to form the basis of the consequence as described below.

2.2.1 Criticality of Unit/Equipment

The first HFE risk factor designated in prioritizing HFE non-conformances involves determining whether or not the package unit's operation is critical to the safe operation of the facility or the

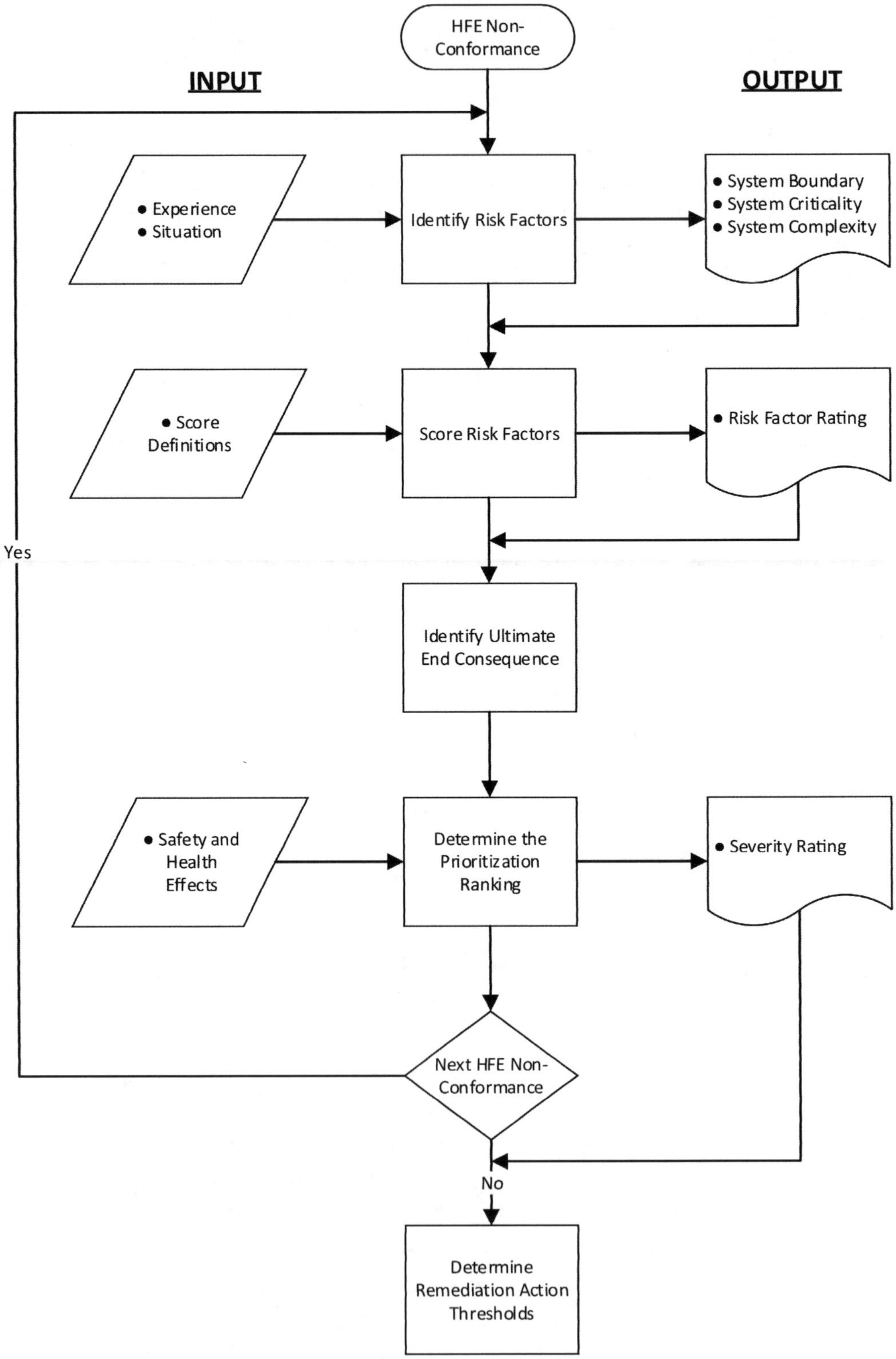

Figure 1. General Methodology Flowchart for Ranking HFE Non-Conformances

ability of the facility to meet its business objectives. Each HFE non-conformance can be evaluated by addressing the following questions:

- Does the HFE non-conformance result in the unit/equipment malfunctioning which could potentially effect the safety of personnel working the facility?
- Does the HFE non-conformance impact the unit/equipment operating correctly which could potentially result in a release?
- Can the facility or unit/package operate if this package is not functioning correctly?

The concern is can the identified HFE non-conformance affect the operation of the equipment and ultimately the health and safety of personnel. These questions should be answered according to the possible ratings provided in Table 2.1 used to develop the consequence of the HFE non-conformance.

Table 2.1. Criticality Ratings

Ratings	**Meaning**
Yes	The unit/equipment is critical or if it does not perform as designed either production would be affected or people would be exposed to hazards.
Possibly	If unit/equipment does not perform as designed there is a chance either production would be affected or people would be exposed to hazards.
No	The unit/equipment does not affect production and if it does not perform as designed people would not be exposed to hazards

2.2.2 Complexity of Tasks and/or Amount of Human Interaction

The Complexity of a unit/equipment is a measure of the amount and complexity of the human interaction(s) during normal operations and maintenance activities. The following question is answered for each HFE non-conformance according to the ratings provided in Table 2.2 in order to achieve a complexity rating.

- What potential impact does the HFE non-conformance have on the manual activities involved in operating and maintaining the unit/equipment?

Table 2.2. Complexity Ratings

Ratings	**Meaning**
Minimal	The manual tasks are inherently simple discrete actions with minimal mental demands (such as pressing Start/Stop buttons, reading gauges, etc.) and the HFE non-conformance minimally impacts these actions.

Ratings	Meaning
Moderate	Neither minimal, nor major.
Major	The manual tasks can be difficult, complex, time consuming and/or require high levels of human reliability and the HFE non-conformance has a significant impact on these factors.
No	No impact.

The answer to the question above are used as input in forming the basis of the consequence of the HFE non-conformance.

2.3 Consequences

The term consequence is defined as the outcome of the HFE non-conformance. Each HFE non-conformance consequence is documented assuming that the identified issue remains uncorrected. The health/safety consequences are recorded and the results of the criticality and complexity assessment are used as inputs for developing the consequence. The consequences are taken to their ultimate end effect, in other words; the consequences are the worst credible. Table 2.3 provides a few examples of a typical HFE non-conformance and possible resulting consequence.

Table 2.3. Typical HFE Non-conformance and possible consequence

HFE Non-Conformance	Consequence
Manual control valve (XV) elevated 80" above working/standing surface exceeding the maximum allowable elevation limit of 70"	The XV is a ***critical*** valve elevated 10" above the working/standing surface, likely resulting in operator stretching to reach the valve during maintenance and/or operation. Although the valve's elevation ***minimally*** impacts its operation, it could potentially result in a musculoskeletal injury (e.g. muscle strain) and ultimately a LTI.
Service oil drain valve is located 1.5" from existing skid structure which is less than the minimum 3" hand clearance required throughout the full range of available travel	The drain valve is 1.5" less than the required 3" minimum hand clearance, likely resulting in operator suffering a minor injury (aka "knuckle-buster"). The operation of the service oil drain valve is ***moderately impacted*** in its current location and could ***possibly*** result in an injury requiring medical treatment.
Ladder and ladder cage extends 40" above landing surface instead of the required 60".	Ladders and ladder cages are required to extend 60" above the landing surface. The ladder is ***critical*** for providing a secondary means of egress and without the ladder and

	ladder cage extended to the full required length its use could be ***moderately*** difficult and potentially result in a fall and ultimately a fatality.

There may be instances where the HFE non-conformance does not result in a significant consequences, as a result, the HFE non-conformance should be prioritized as "No Significant Consequence Identified." Examples of this include but are not limited to a valve that is used in the initial commissioning of a vessel or installation, valves used in drydock, or on equipment where the use of auxiliary access equipment (e.g. mobile platform, scaffolding, etc.) are required/permitted and the use of said equipment is infrequent (e.g. during startups after extended shutdowns).

2.4 Evaluation Criteria: Severity

After the consequences have been established, the evaluation criteria are determined. At a minimum the evaluation criteria should address the severity of the consequences. The prioritization ranking are estimated according to the categories below in Table 2.4. The severity ranking is chosen based on the description which most closely matches the appropriate aspect of the consequence. The consequences are rated assuming the HFE non-conformance were to remain unchanged. The severity reflects the end consequence such that the prioritization ranking is appropriately assessed yielding a single HFE risk "score".

Table 2.4. Severity Ranking Criteria

Severity	Safety and Health Effects
1	Minor injury, very minor irritation or illness
2	Medical treatment, recordable case
3	Lost time or workday injury, significant illness but recoverable
4	Major Injury, possible long term health effects
5	Fatality(s) or multiple long term injuries or illness

Table 2.5 illustrates how the severity ranking criteria is applied to the consequence to achieve a suitable rank order. Because the end consequence drives the severity ranking and the consequence is formed by using HFE risk factors, this method requires the end user to possess a significant level of training and experience in HFE and knowledge about human error when its design induced. Humans can be allowed, encouraged, directed or even coerced into unsafe behaviors by the way equipment is designed. Moreover, if the design of a piece of equipment is considered to be overly difficult or inefficient but cannot be modified, the user will make do with what they have been given (i.e. "get the job done") even if it may be unsafe. Using knowledge

about the principles of human behavior requires an educated judgement to develop the consequence which ultimately decides the severity is imperative. In the first example listed in Table 2.5, the HFE non-conformance specifies that the XV is elevated 10" above the working surface. The XV is a ***critical*** valve where the elevation above the required height ***minimally*** impacts its operation. Applying the principles of human behavior and knowledge of human performance leads to the ultimate end consequence since the valve is elevated just high enough, but not too high, that the user will "get the job done" by most likely stretching to reach the valve which could potentially result in a musculoskeletal injury and a lost time incident (LTI). From Table 2.4 the end consequence of a musculoskeletal injury and a LTI equates to a severity ranking of 3.

Table 2.5. Typical HFE Non-conformance, possible consequence, and severity ranking

HFE Non-Conformance	Consequence	Severity
XV elevated 80" above working/standing surface exceeding the maximum allowable elevation limit of 70"	The XV is a ***critical*** valve elevated 10" above the working/standing surface, likely resulting in operator stretching to reach the valve during maintenance and/or operation. Although the valve's elevation ***minimally*** impacts its operation, it could potentially result in a musculoskeletal injury (e.g. muscle strain) and ultimately a LTI.	3
Service oil drain valve is located 1.5" from existing skid structure which is less than the minimum 3" hand clearance required throughout the full range of available travel	The drain valve is 1.5" less than the required 3" minimum hand clearance, likely resulting in operator suffering a minor injury (aka "knuckle-buster"). The operation of the service oil drain valve is ***moderately impacted*** in its current location and could ***possibly*** result in an injury requiring medical treatment.	2
Ladder and ladder cage extends 40" above landing surface instead of the required 60".	Ladders and ladder cages are required to extend 60" above the landing surface. The ladder is ***critical*** for providing a secondary means of egress and without the ladder and ladder cage extended to the full required length its use could be ***moderately*** difficult and potentially result in a fall and ultimately a fatality.	5

The severity ranking methodology detailed in this paper is a simple model that provides a numerical range where higher values represent higher risk. The severity ranking can be used as a tool to determine where the greatest focus of risk management and/or remediation should be placed. This procedure can also result in reduced scope by elimination of low-risk rankings. For example, thresholds that describe a remediation action can be applied to focus the scope of risk management and/or remediation. A simple set of remediation action thresholds are described in Table 2.6.

Table 2.6. Example of remediation action thresholds

Severity Ranking	Action Threshold
1	Remediation not required
2 - 3	Deviation from the design intent may be submitted; provided evidence supporting risk controls will be used
4 - 5	Remediation required

Employing remediation action thresholds such as the one above not only serves to provide greater focus to high-risk issues and reduce scope, it can also serve as a means to evaluate remediation options. By identifying possible remediation options, the remediation(s) that will have the most beneficial impact to the project's viability can be determined.

3 Discussion

It is inherent in any largescale and complex construction project that non-conformances will be identified, particularly HFE non-conformances, as major oil and gas companies recognize that HFE has a significant contribution to the safety and fitness for purpose of their facilities. The number of HFE non-conformances identified can be copious and as a result create the issue of how to address these non-conformances while maintaining the least impact on schedule and resources. This paper presented a methodology for assessing and prioritizing HFE non-conformances for a given construction project. The technique provided a qualitative and structured procedure to rank order HFE hazards. The process included adapting common and widely accepted risk ranking and HFE screening methods to form a basis for allocating resources by the project and construction management teams.

Risk ranking was chosen as part of the methodology for prioritizing HFE non-conformances primarily because it is one of the most used and widely accepted methods for risk management. The risk ranking technique possess other advantages that made it practical for this application namely it accepts a high degree of complexity, flexible for any type of risk, and it can be used with a variety of quantitative and qualitative evaluation criteria.

HFE is not as simple as executing prescriptive requirements. It is a complex discipline that requires the expertise to recognize the intent of a requirement and the ability to apply educated judgement and trade-offs to meet the intended safety aspect of the requirement. Likewise, the implication of HFE non-conformances from construction audits are often not readily realized. Hence a system that allowed for complexity was needed and risk ranking offered that opportunity. Specifically, it is most appropriate when comparing and managing a collection of complex risks. This is accomplished by the breakdown of a risk into integral risk scenarios with essential components and ultimately generate a consequence that could be estimated using a specific HFE risk ranking.

Another advantage in using a risk based approach is it is flexible for any type of risk. It is used in many industries including the financial sector, engineering, and manufacturing to name a few. A key feature of its flexibility is that, when implemented properly; it can be used to define and narrow the scope of the overall risk question. By carefully applying definitions of boundaries, the scope can be further reduced. The scope for this exercise was "How the identified HFE non-conformances should be prioritized for remediation?" The proposed methodology demonstrated that by applying qualitative filtering techniques such as the use of remediation action thresholds, HFE non-conformances may be narrowed by sifting out non-conformances that do not require remediation. The benefit from using this approach is that it aids in prioritizing risks that can potentially threaten a project's viability due to the cost associated with correcting the non-conformance and achievement of higher standards of production and reliability while not compromising the safety and integrity of personnel and the overall achievement of enterprise objectives and goals.

The result of an HFE construction audit is an HFE non-conformance which is description of how a specific design deviated from an engineering requirement. These non-conformance reports act as the basic data (i.e. non-conformance description) upon which many remediations are based. There can be numerous HFE non-conformances for any given construction project, the issue becomes how to assess the magnitude and scope of a problem before solutions are determined. An advantage of the risk based impact assessment is that it is a tool that can be used with a variety of qualitative evaluation criteria. As described above a risk based approach deconstructs risk into risk scenarios with essential components and ranks them according to a set of evaluation criteria into a single risk score. The technique described in this paper assessed two qualitative risk factors (e.g. unit/equipment criticality and task complexity) to aid in identifying hazards within the HFE non-conformance and ultimately provide a specific risk ranking. It is understood that there are other qualitative HFE factors that could have been selected for use in the model presented in this paper. However, unit/equipment and task complexity were selected because these two screening factors possessed the greatest impact on the risk ranking criteria selected. Other HFE factors such as the novelty of the equipment, design scope (ability to influence HFE design or layout), or known problems (e.g. history of HFE problems associated with equipment or layout) should have been addressed during the detail design phase of the project and would have negligible impact on the evaluation criteria.

It could be reasoned that not including the HFE factor, task frequency (e.g. how frequently operator is likely to interact with the item) could have an impact on the evaluation criteria used in this methodology. This factor was not selected because the use of the HFE factor, unit/equipment criticality, implies some degree of frequency. In the first example after the assessment of the HFE factors the resulting consequence states "The XV is a ***critical*** valve …." An experienced HFE practitioner would be able to ascertain that since the XV is a critical valve the expected operational or maintenance frequency is more than once in a three month period. Thus, inclusion of the HFE factor task frequency may be deemed redundant. Alternatively, the third example given in this methodology declares "The ladder is ***critical*** for providing a secondary means of egress…" Again, a skilled HFE specialist would determine from the text that although the ladder is critical it is a means of secondary egress and its use would be infrequent, reducing the risk ranking. Inclusion of the HFE factor task frequency in this instance may impact the evaluation criteria used and further study is warranted.

Determining the HFE factors to be included and excluded is one disadvantage in using a risk based approach. The use of this approach may require significant effort in establishing HFE risk factors and evaluation criteria. The level of effort is dependent on the complexity of the issue and the expertise of the person(s) involved. The tool presented in this paper is applied to projects in the construction phase. The choice of the risk factors to be included requires skilled judgement with a high level of HFE competence. It is assumed the end user of this tool would be an experienced HFE practitioner with considerable field experience and knowledge of the operation and maintenance of the equipment being audited. When well-implemented and managed, the methodology allows for a systematic identification of issues that can be presented at a level of detail appropriate for various stakeholders.

The use of a single evaluation criteria (i.e. severity) is another limitation in using this method for prioritizing HFE non-conformances. In a traditional risk based approach, such as a HAZOP, a two-criteria evaluation model uses likelihood (i.e. probability) and severity to address a risk scenario. This eliminates some of the conservatism inherent in a single-criteria evaluation model when only using severity to assess the risk. It is understood that the probability or likelihood that the ultimate end HFE consequence would occur provides greater fidelity to the model. However, the probability or likelihood for HFE consequences is not known for facilities under construction because the data does not exist and data from existing facilities is not voluntarily shared by owner/operators. Hence, the single-criteria evaluation model was selected for use in this methodology. Further, the proposed methodology demonstrated how a single-criteria can be implemented to rank order HFE non-conformances and with the selection of the appropriate HFE risk factors some of the conservatism can be reduced.

One of many considerations for any risk assessment is ensuring that the level of detail of evaluation criteria is practical compared to the level of effort required to complete the evaluation. The intended use of the proposed model is a quick and simple technique to rank HFE non-conformances after completing an HFE inspection audit. However, future studies may consider adding supplementary criteria to capture the needs from other stakeholders or reduce conservatism. For example, the probability of occurrence of the HFE non-conformances on a given project could be used as a second evaluation criteria. The challenge in adding a second criteria is “weighting” the factor such that the influence of other evaluation criteria does not adversely impact the risk ranking score. Further study is necessary to determine what additional evaluation criteria could be included in the proposed model that adds value while not sacrificing efficiency and simplicity.

Perhaps the most significant limitation of using this method for prioritizing HFE non-conformances is that there are other factors outside HFE influence. Resource availability, schedule, and cost are huge drivers in today’s large scale construction projects. These are factors that are typically not available to the HFE practitioner conducting the HFE audit and not part of their scope. However, it is believed that the tool provided here can be used by members of the project and construction team along with schedule, cost, and resource availability to make trade-off decisions or find alternative solutions to prioritize risks that can potentially threaten a project’s sustainability.

4 Conclusion

The simple model for prioritizing HFE non-conformances is at an early stage of development compared with many other methods for assessing risks. The qualitative nature of and potential amount of HFE non-conformances for a given construction project presents a major challenge beyond not compromising the safety of personnel, such as remediation of these non-conformances while maintaining the least impact on resources and schedule. With improvements in methods and techniques for prioritizing resolution of HFE non-conformances, such as those described above, construction and project management teams can achieve a more comprehensive basis for allocating resources. Further improvements in the sensitivity and accuracy of the proposed model will enable deeper understanding of the HFE non-conformances and more extensive characterization of resultant ranking. In turn, this will enhance the ability to achieve higher standards of production and reliability and ultimately facilitate attainment of enterprise objectives and goals.

5 Acknowledgments

The authors thank Graeme Cuffy, Ross Cormack and Jeanne Curcio for technical assistance and providing critical comments about the content.

6 References

[1] OGP, "Human Factors Engineering in Projects; Report 454," International Association of Oil & Gas Producers, 2011.

[2] M. S. Sanders and E. J. McCormick, Human Factors in Engineering and Design, New York: McGraw-Hill, Inc., 1993.

[3] B. C. Moyer, "Bureau of Economic Analysis," 2016. [Online]. Available: http://www.bea.gov. [Accessed 7 January 2016].

[4] R. W. Gilmer, "Institute for Regional Forecasting," Bauer College of Buisness University of Houston, 31 December 2015. [Online]. Available: http://www.bauer.uh.edu/centers/irf/houston-updates.php. [Accessed 9 January 2016].

[5] M. Helander, Human Factors/Ergonomics for Building and Constructing, New York: John Wiley & Sons, 1981.

Advanced Procedures: Leveraging a Human Factors Approach for Improving Process Safety Performance

Pranav Kannan[1], Lubna Ahmed[1], Noor Quddus[1], M. Sam Mannan[1],
S. Camille Peres[2], Juan Carlos Batarse[2], Paul Ritchey[2]
[1]Mary Kay O'Connor Process Safety Center, Artie McFerrin Department of Chemical Engineering,
[2]Ergonomic Center, School of Public Health,
Texas A&M University, College Station, TX-77843-3122

Abstract: Procedures form an important part of the industry's operational architecture and are part of the system to ensure that prescribed methods are routinely followed allowing for optimal and safe performance of the facility. The study by the Bullemer *et al.* (2004) has raised concerns over effectiveness of the current state of the art in procedures, with significant incidents being attributed to issues in procedures and deviations from procedures. An example of a process safety incident attributed to procedural deviation is the Texas City refinery explosion of 2005 where deviation from standard operating procedures was amongst the root causes for the incident. The report by the Chemical Safety and Hazard Investigation board pointed out that the outdated and ineffective procedures, combined with poor compliance and execution were major contributing factors to the accident. The objective of the advanced procedures research effort is to create tools that improve procedure design and adherence. The current study focuses primarily on developing a tool, a procedure risk matrix to facilitate the procedure development. This paper illustrates a risk matrix concept and implementation into procedure as a means of enhancing safety that leverages critical components of human factors processes and constructs. Several risk ranks and matrices from different industries and open literature were studied and four parameters have been selected to rank the severity of the failures related to the procedures. Dependency on human performance parameters of these severity ranking parameters and frequency of the procedure related failures will be investigated further.

Keywords: Process safety, procedures, risk matrix, writer's guide, human factor.

1. Introduction

Process safety and performance rely on a number of factors and effective procedures are one of them. However, procedures can come into aid only if they are written well and communicate properly. A significant number of incidents attributed to issues in procedures and deviations from procedures has raised concerns over effectiveness of the current state of the art in procedures [1]. One example of a process safety incident attributed to procedural deviation is the Texas City refinery explosion of 2005 where deviation from standard operating procedures was amongst the root causes for the incident. Report by the Chemical Safety and Hazard Investigation board pointed out that the outdated and ineffective procedures, combined with poor compliance and execution were major contributing factors to the accident [2]. In 2011, Mexico's state-owned oil company Pemex evacuated 638 workers from a semi-submersible platform after water entered into pontoon and it begun to lean on one side. Along with several other root causes contributing to the event, the facility lacked a procedure for cleaning operation of the filter, which was critical to the incident [3].

The objective of the current advanced procedures research effort is to provide a tool that can improve procedure design and performance. As part of this process a writer's guide has been developed and its purpose is to provide procedure writers with guidelines based on peer reviewed scientific literature on how to write procedures in a way that leads to maximum safety performance. Also, the presentation of procedures should be user friendly and accessible. In the first phase of the study, a sample of regulations and standards along with existing procedure writing guidelines was selected from across the world and a thorough analysis of the human performance implications, obtained through systematic categorization of the regulations and guidelines was performed. These human performance implications were expanded by using existing empirical evidence to form guidelines for procedure writing [4]. For further involvement of the stakeholders, mainly the operators; field observations and interviews are being conducted in several sites in various geographies to understand the perceptions and requirements necessitated by the users (*e.g.,* operators). On the basis of these contextual interviews and observation, a survey continues to be developed for wider distribution to reach operators and other personnel, to obtain a more representative picture of the requirements of procedure users. The results and findings obtained from the interviews and surveys will be used to update the procedure writer's guideline document in an ongoing effort to reflect the needs of multiple industries. They have been integrated into process safety paradigms to facilitate understanding about the users.

By ensuring that best practices in process safety and human performance are incorporated, the work presented here provides an insight into how human factors can be leveraged into generating a risk matrix for procedure development. Thus, this paper mainly will present a risk matrix concept and implementation into procedure as a means of enhancing safety. Typically, a risk matrix is a tool for the user to discern the risk presented by a process in the context of the entire system. Risk matrices have been applied in a wide range of qualitative, semi quantitative and more recently quantitative forms in both civil and military applications. Risk matrices are typically used in two steps: the first involving the assessment of the risk of a process or a node of a system under consideration, and the second using the risk matrix to understand the tolerability of the risk given the risk appetite of the user. Risk matrices are especially useful when there is a source of risk which is "rankable" and mitigation/review of the source of risk needs to be done systematically. Risk matrices provide a guide towards the understanding the importance of the source of risk and can be enhanced further to provide guidelines towards effective control and mitigation. The risk ranking system has been extended into the various domains including software development where it is being used as a technique to manage risks associated with projects [5].

2. Objective and methodology

The advanced procedure research initiative is a comprehensive assessment of the procedure life cycle with a dual emphasis on process safety and human factor based procedure design. The study goals include observations and interviews with operators that were conducted with the goal of understanding better the challenges and needs that operators experience when using written procedures and with adhering to standardized procedures. During these interviews, we also inquired about the operators' perceptions of risk in their working environment in general and with procedure use specifically. The observations and interviews are ongoing and thus specific results are not presented here. However, some of the initial findings from operators' regarding risk have been used to develop the risk matrix. The risk matrix that is proposed here can be used as a tool to assist procedure writers by providing information based on relative risk associated with procedures and also as a tool for supervisors for effective management of change and control. Risk matrices currently being used in industry as well as those proposed in literature were reviewed. For the risk ranking and risk matrix tool, various variables were identified and incorporated into a formula to calculate the ranking. Further, information from observations and interviews were factored in during the selection of the variables. From the study, various crucial variables were identified and incorporated into a formula to find relative risk ranking of individual risk matrix. The risk ranking of individual procedures was later used to develop a risk matrix.

3. Results and discussion

Table 1 summarizes several risk matrices and ranks studied under the current scope. Sources of the risk matrices and ranks include both open literature and industry. It also shows the parameters they use to calculate or in some cases rank the risk of a process.

Table 1: Risk matrices and metrics used to risk rank

Risk matrix/rank type	Metrics
Original risk matrix (ORM)	Frequency (probability) of failure, consequence
Surgical risk [6]	Clinical risk, complexity
Extended risk matrix [7]	Severity, detectability, probability, recoverability
Borda method [7]	Borda index is calculated as the sum of the difference between the number of risk items and items which exceed the criteria k in element
Risk matrix supply chain [8]	Probability * severity/ (recoverability*detectability
ARAMIS [9]	Severity = frequency*intensity, Damages = intensity*vulnerability Risk = intensity*vulnerability*frequency
ISO/EN standards [10]	Severity rating, people, assets, environment, reputation
Fuzzy risk matrix [11]	Fuzzification functions
Company A	Job frequency, learning difficulty, business impact, contained hazard
Company B	Task criticality, task familiarity, task complexity

The metrics associated with each of the risk ranking method use various sub-metrics to divide the sources of the risk. A major pitfall however, is that this may introduce subjectivity especially because of biases with risk perception. Existing quantitative procedures such as the Borda method do not adequately address the multilayer interaction commonly associated with the procedures. Further, new techniques such as fuzzy risk matrices may require significant resources to develop on a wide scale basis; hence, a method which can be objective and easily usable is needed.

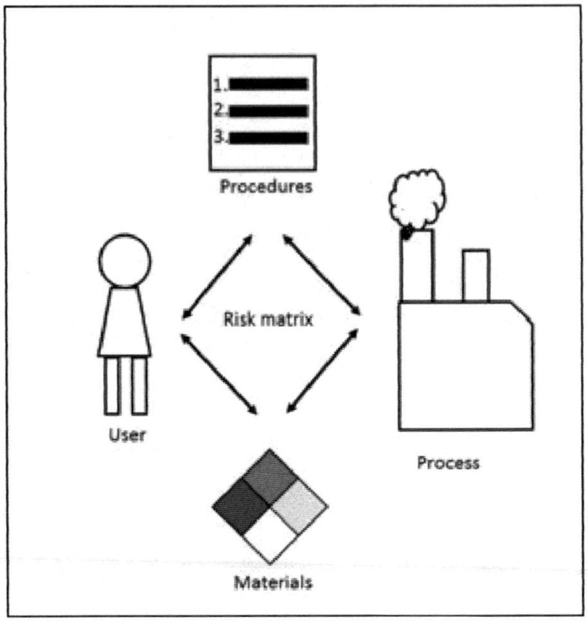

Figure 1: Illustration of "risk matrix" which needs to be captured by the risk ranking process for procedures. The risk ranking metric needs to identify the contributors of the risk from each of these sources in an objective manner to aid the design and management of procedures.

These methods are commonly adapted or a variation is used internally by industries given the availability of data or expert opinion for the accurate development of these metrics. One of the common issues faced in risk matrices which would be amplified for a procedure focused risk matrix is the presence of risk ties. These ties which would require expert opinion to resolve and can be the major obstacle for a large system such as procedures, given the number of procedures in even a small to a medium sized facility. Hence the need for an objective assessment of the risk associated with the procedures that takes into account the various interactions and sources of risk and presenting a measure which can be effectively used as decision support both for operators on the field and procedure writers. Based on the literature review and studies on several risk matrices employed in industry, some crucial elements or variables were identified that can be incorporated into building of a risk matrix. Specifically four variables were identified, namely: complexity (C), hazard of materials (H_m), hazard of the process (H_p) and frequency of usage (f_u) of the procedure. The interaction between the various factors that contribute to the risk associated with procedure usage is illustrated in Figure 1. The usage of procedures places the operator (the user) in a complex interactive environment with the process, the materials he or she is working with, and the content of the procedure itself (the task). The risk matrix aims to capture the combinatorial effects of the risks presented by these factors to provide a system level view of procedures and procedure usage.

Complexity (C). One of the unique approaches to this model is how information from the interviews and observations have been leveraged to better understand complexity. This approach to understanding complexity involves more than just counting the number of actions required to complete the task or the number of steps within a procedure.

Evaluating the complexity of a process requires careful consideration of the user (worker), environment, and task (process). Complexity, as we use it, is a best understood as a quantification of the demands placed on a person by a specific process—the less demand placed on the worker, the less complex the process can be said to be. This conceptualization of complexity is human-centered and is a result of our multidisciplinary approach that includes both process safety and human factors methodologies. By leveraging human factors methodologies and theories, a more complete understanding of a process's complexity can be used to inform the risk matrix presented here. The Interactive Behavior Triad (IBT) is a framework that was originally developed to test computer simulated cognitive models of human-computer interactions [12] but can be applied to physical work, like proceduralized tasks, as well. The IBT framework focuses on the three factors previously mentioned (task, worker, and environment) and the interaction between those factors. Understanding the independent effects of each factor (user, environment, and task) on process complexity is an important first step. Complexity can be increased if the intended user, or worker, is someone that will be suffering from fatigue or drowsiness due to the time at which their work takes place. Complexity could similarly be considered higher if the environment where the work is completed is not well-lit or subject to harsh weather. Tasks that involve difficult and detailed steps, either physically or mentally (recalling settings or thresholds) complicated, would also increase complexity. Each of these factors then need to be addressed in terms of how they may interact to produce greater demand, or complexity. The combination of a difficult task with adverse weather conditions would increase complexity more than just the simple addition of the increased complexity from adverse weather and task difficulty if they were to be evaluated independently.

There are multiple methods for collecting information regarding the user, environment, and task. As part of the advanced procedures project, we collected these data by conducting observations of workers while they completed work with written procedures. After each observation, we interviewed operators in order to better understand the problems and constraints imposed by the tasks themselves, the environment in which these task were completed, and the work-related needs of worker. Primary focus was given to the experiences of the workers and the effect of these task-related factors on them. It is an approach focused on the needs of the human—a human-centered approach. These observations and interviews provided us with rich information regarding the complexities that are involved with each job we observed. It is this information that informed our conceptualization of the complexity component for the risk matrix presented here, which is still in the process of being developed. By not including this human-centered approach when calculating risk, the complexity of a process could be underestimated and by extension the total risk of a process could be underestimated. We have yet to finalize the precise method by which complexity will be calculated.
Hazards of materials (H_m). This pertains to inherent properties of the materials associated with the task being done, the amount being used during the task, *etc.* and this information can be calculated using various hazard indices available.

Hazard of the process (H_p). Hazard of the process includes process variables (*e.g.,* temperature, pressure) as well as equipment that are associated with the task.

Frequency of usage (f_u). The last variable is based on the frequency with which a certain procedure is needed or being used (*e.g.,* once a week, once a month).

After they are established, the four variables are then combined to define the Procedure Risk Rank as following:

$$Procedure\ Risk\ Rank = C \times H_m \times H_p \times f_u \quad (1)$$

After calculating individual risk ranking for procedures, this information can be used to plot against frequency of failure related to procedures to obtain a risk matrix. While traditional a risk matrix is based on frequency of occurrence of a wide number of failures, the current risk matrix (see Figure 2) focuses on procedure related failures only.

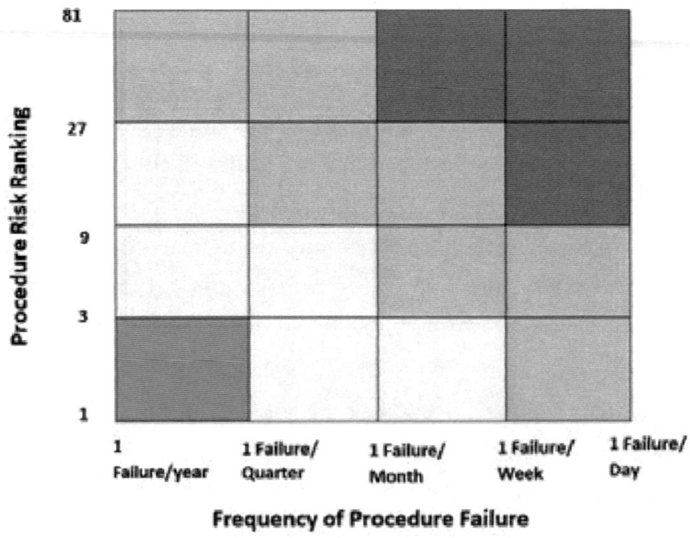

Figure 2: Illustration of risk matrix for procedure risk ranking versus frequency of failure related to procedures for a hypothetical company

4. Conclusions

The research team developing the risk ranking formula and risk matrix presented here is comprised of both process safety and human factors researchers. Process safety approaches alone cannot sufficiently incorporate and accommodate the needs and capabilities of workers. An understanding of both the process and the person interacting with that process yields more appropriate conclusions and guidance for written procedure development. Our team is leveraging the methodologies of both process safety and human factors for the development of tools like this risk matrix. However, the fundamental human factors approach—seeking to understand the effect of the interaction between humans, their tasks, and the environments they work in—can be folded into any existing safety program or industrial development process. Interviews and field observations do not require extensive time or financial investments and can reveal aspects of processes, which would not have been otherwise accounted for, that may place workers at a higher chance of encountering a hazardous situation.

The risk matrix presented here is still in the development phase, once completed it will be a useful and important risk-ranking tool for procedure writers' and review panels. Future work on this risk matrix includes an elaboration on the process of defining procedure failure, finalizing the method for calculating complexity as we have conceptualized it, obtaining data necessary for quantifying procedure failure, and validation of the method by which final risk rankings are calculated (*e.g.*, see Figure 2).The data obtained from the survey will be used to validate the proposed risk ranking formula and risk matrix. Finally, we will conduct extensive testing with procedures writers to determine the most effective method for incorporating this risk matrix into our existing procedure writer's guide.

References

1. Bullemer, P., J. Hajdukiewicz (2004) "A study of effective procedural practices in refining and chemical operations" Proceedings of the Human Factors and Ergonomics Society Annual Meeting, vol. 48 no. 20 2401-2405
2. US Chemical Safety and Hazard Investigation Board (2007) "Investigation report: refinery explosion and fire: BP Texas City" Report no. 2005-04-I-TX
3. Incident Photos - Pemex's Flotel Jupiter platform partially sinks in Gulf of Mexico - gCaptain. (2011, April 14). Retrieved February 16, 2016, from http://gcaptain.com/incident-photos-flotel-jupiter/#.Vdd90flVhBc
4. Peres, S.C., N. Quddus, S. Rahmani, P. Kannan, L. Ahmed, P. Ritchey, W. Johnson, M.S. Mannan (2015) "Development of an effective procedure writer's guide using a Human Factors and regulatory compliance approach" Mary Kay O'Connor Process Safety Center International Symposium, College Station, TX, USA.
5. Xiaosong, L., L. Shushi, C. Wenjun and F. Songjiang (2009) "The application of risk matrix to software project risk management" Information Technology and Applications, 2009. IFITA'09. International Forum on, IEEE.
6. Kneebone, R., D. Nestel, C. Vincent and A. Darzi (2007) "Complexity, risk and simulation in learning procedural skills." Medical education 41(8): 808-814.
7. Ni, H., A. Chen and N. Chen (2010) "Some extensions on risk matrix approach." Safety Science 48(10): 1269-1278.
8. Li, Z., Q. Yee, P. Tan and S.G. Lee (2013) "An extended risk matrix approach for supply chain risk assessment" Industrial Engineering and Engineering Management (IEEM), 2013 IEEE International Conference on, IEEE.
9. Salvi, O. and B. Debray (2006) "A global view on ARAMIS, a risk assessment methodology for industries in the framework of the SEVESO II directive." Journal of hazardous materials 130(3): 187-199.
10. International Organization for Standardization, (2002) "ISO 17776" Petroleum and Natural Gas Industries-Offshore Production Installations-Guidelines on Tools and Techniques for Hazard Identification and Risk Assessment, Geneva.

11. Markowski, A.S. and M. S. Mannan (2008) "Fuzzy risk matrix." Journal of hazardous materials 159(1): 152-157.
12. Byrne, M.D., Kirlik, A. and Fick, C.S. (2006) "Kilograms matter: Rational analysis, ecological rationality, and close-loop modeloing of interactive cognitive systems." Human-Technology Interaction. Methods and Models for Cognitive Engineering and Human-Computer Interaction, A. Kirlik, Ed., ed Oxford/New York: Oxford Press, 2006, pp. 267-286.

442385 Human Factors Considerations in 3D Model Reviews

Abstract:

It is now a common practice to generate 3D models of processing units and facilities during the detailed design and engineering phase of a project. Formal 3D model reviews are conducted at various stages for optimization of resources and for safe construction, operation, and maintenance of the processing units and facilities. Many organizations have internal standards that are referenced with respect to the ergonomic aspects associated with design, construction or modification of processing units and facilities. Some organizations rely upon a "common sense" engineering approach relying upon the experience of the reviewers. However, some human factors considerations may not be reflected in the either the design standards that are utilized during 3D model reviews or identified through engineering experience alone. This paper discusses the human factors considerations such as operator capabilities, limitations, and the concept of cost of compliance that may be critical to the design of a processing unit or facility. We will provide a hierarchy of considerations and a flowchart for decision-making in 3D model reviews. Understanding and incorporating such factors during the 3D model reviews can help address some potential safety concerns in a proactive manner.

444903 Harnessing Big Data for Smart Manufacturing: A Novel Optimization Framework

Abstract:

Manufacturing planning and management practices traditionally feature vertically optimized operational units, where enterprise segments are optimized in a vacuum, planning and scheduling rely on conservative forecasting, and reactive decision making is slow and hindered by incomplete information and inherently static infrastructure. Advanced smart manufacturing frameworks are characterized by enterprises which are fully-integrated and able to harness and optimize vital operational, business, and supply chain information in real time. Current and developing capabilities in the rapid and decentralized acquisition of data both enables and complicates the development of smart manufacturing methods. Existing and new methods that are to be applied to Big Data for smart manufacturing must be able to accept and prioritize highly variable data sources, including factory sensors, historical and real-time information, images, market and demand data, and customer-side product performance collected via the cloud. Applicable methods must also be efficient to draw rapid conclusions such as for fault detection and root cause analysis while still capturing the highly nonlinear and interconnected nature of the data inputs.

In this plenary talk, we will present and discuss optimization- and statistics-based frameworks for data-driven decision making which can identify the key components of an input high-dimensional dataset. The methodology can be applied to a variety of situations under the domain of smart manufacturing by modularly incorporating machine learning approaches, including those for nonlinear regression, classification, outlier detection, or clustering. Optimization models were formulated which cast the high-dimensional feature selection problem into implicit nonlinear feature space, resulting in minimum-maximum mixed-integer nonlinear optimization problems.. Extensive computational studies over numerous benchmark datasets show that new algorithms outperform existing state-of-the-art methods in the machine learning literature and show great promise for accurate, real-time decision making from Big Data in smart manufacturing operations.

444831 Challenges and Research Directions in Big Data

Abstract:

Big Data has opportunities for diagnostic, prognostics, and decision making. This presentation describes challenges associated with Big Data that need to be resolved for these opportunities to be fully realized. The challenges include how to best

(1) extract information from high-order tensor data structures, which arise from spatially resolved photoluminescence, spectral imaging, and color video

(2) integrate information from a variety of types, such as process flowsheet structure, causal relationships between variables, and real-time sensor data

(3) manage the effects of uncertainties, disturbances, faulty sensors, variation in operator practices, and machine drift

(4) address nonlinear dynamic operating conditions of simultaneous discrete and continuous nature such as startup, shutdown, and product changeovers

Several examples are provided where wrong conclusions or poor performance were obtained during the seemingly reasonable application of data analytics, which illustrate some common misconceptions and the potential pitfalls of applying methods naively. Some specific research directions motivated by the challenges are outlined.

12TH GLOBAL CONGRESS ON PROCESS SAFETY

How to Design Human Factors into a PSM System to Ensure Effectiveness and Deliver Risk Reduction Performance

Minu Shikha Mathur, PhD, AEP
JCL Risk Services LLC
16225 Park Ten Place, Suite 500
Houston, Texas 77084
mmathur@jclsc.com

David A. Jones
JCL Risk Services LLC
16225 Park Ten Place, Suite 500
Houston, Texas 77084
djones@jclsc.com

Prepared for Presentation at
American Institute of Chemical Engineers
2016 Spring Meeting
12th Global Congress on Process Safety
Houston, Texas
April 11-13, 2016

How to Design Human Factors into a PSM System to Ensure Effectiveness and Deliver Risk Reduction Performance

Minu Shikha Mathur, PhD, AEP
JCL Risk Services LLC
16225 Park Ten Place, Suite 500
Houston, Texas 77084
mmathur@jclsc.com

David A. Jones
JCL Risk Services LLC
16225 Park Ten Place, Suite 500
Houston, Texas 77084
djones@jclsc.com

Keywords: human factors, human error, process safety management system, risk reduction performance, continual improvement

Abstract

This paper will present a practical application of a design process for developing and implementing effective process safety management (PSM) systems. Management system deficiencies are often the root causes leading to process safety incidents. Such deficiencies can be attributed to inadequately designed management systems, primarily since standards and regulations do not provide detailed specifications for their design. A methodology based on proven quality management principles will demonstrate how human factors can be incorporated directly into the design specification for a PSM system to assure effectiveness and deliver sustainable risk reduction performance. As such, they can have a dramatic impact on how individual management processes can be used to ensure that design, operational and maintenance activities are conducted in a safe and environmentally responsible manner.

Applicable to both onshore and offshore facilities, this design methodology takes process safety beyond regulatory compliance and enables organizations to sustain performance for managing risk, reliability and resilience as part of an integrated operational risk management framework. It applies to both new and modified designs for engineered process systems, as well as individual management processes. Application of human factors to the management system design also allows for continual improvement in response to changes to regulations, technology, and operating practices.

This paper will present a practical step-by-step procedure on "how to" incorporate human factors into the PSM system design, including: leadership, safety culture, human factors engineering and human-machine interfaces, employee engagement and empowerment, communications, etc. Examples will show how quality aspects for human factors are addressed in key management processes within the PSM system in order to improve risk reduction performance.

Introduction

A well-established design concept for management systems is the continual improvement life cycle involving the steps: Plan-Do-Check-Act. Based on standards for managing quality, environment, occupational health and safety, and security [2,22,25,27], design models for effective process safety management (PSM) systems also utilize the Plan-Do-Check-Act life cycle model [28,29]. Even though the implementation of PSM has seen a significant improvement in the recent years, it can be taken a step further by using the knowledge of how humans interact with the facilities, equipment, and management systems within their work environment and within a culture to improve PSM systems. This synergistic integration of human factors and PSM may result in fewer incidents along with higher efficiency and a more productive workforce.

This paper aims to show the human factors deficiencies within the design or implementation of current management processes that can have a cumulative degrading effect on performance results. Failing to properly address these deficiencies leads to higher probabilities for human error. The U.S. Occupational Safety and Health Administration (OSHA) PSM Standard requires that human factors be addressed when conducting process hazard analyses (PHAs). Further, the U.S. Bureau of Safety and Environmental Enforcement (BSEE) notes for its Workplace Safety Rule, or Safety and Environmental Management System (SEMS) regulation, that: "A SEMS program is a comprehensive system to reduce human error and organizational failure" [4]. Since PSM encompasses controlling risks by minimizing human error, applying human factors principles to PSM systems should be included in the design, procurement activities, risk assessments, incident investigations, and virtually all aspects of PSM.

Thus, key processes for managing plans (policies, programs, procedures), managing changes (organizational, procedural, technical), managing performance (conformance, compliance, results) and managing actions (corrections, corrective actions, preventive actions), can each benefit from increased human interaction and input. Engaging the entire workforce through leadership and employee participation is seen as a key driver for integrating human factors in a PSM system.

Human factors implementation to any process requires an understanding of human capabilities and limitations as it applies to facilities, equipment, and management systems [37]. The effect of working environment and culture is well documented and is the key factor in determining how effective implementation of human factors will be to a process or management system [31].

The concepts and examples in this paper will address aspects of the five tenets of "Vision 20/20" from the Center for Chemical Process Safety (CCPS): *committed culture, vibrant management systems, disciplined adherence to standards, intentional competency development, and enhanced applications of lessons learned*, which were developed to demonstrate what perfect process safety will look like in the near future when it is championed by industry [9]. This integration of human factors directly into the design of the PSM system can take the practice of managing process safety to the next level of performance for the industry [9,35].

This paper will draw upon the authors' extensive practical experience in applying the principles and practices of human factors, designing and developing comprehensive PSM systems, conducting PSM system audits, and leading incident investigations.

1. Human Error – The Need for Integrating Human Factors

1.1 Who are the Humans?

Humans are already fully integrated into the life cycle of the engineered processes that produces chemical products. The entire workforce is comprised of humans. Managers who make decisions are humans. Engineers who design and improve the processes are humans. Contractors who construct the processes are humans. Operators who operate the processes are humans. Maintenance personnel who maintain the processes are humans. Support personnel, including those involved with contracts, purchasing, receiving, stores, laboratories, planning, scheduling, logistics, and security, are all humans. In all cases, these humans can affect process safety performance, either negatively or positively. That is, humans are responsible for making the business and the processes run successfully since facilities do not run on their own. Humans are a source of process safety risks through human errors that are possible from their activities. Conversely, they are also the key source of risk control, through their risk management activities.

1.2 What do these Humans do?

All management processes center around human activities; that is, they involve people who manage, perform, and verify the work, and then take action to continually improve quality and performance [22-24]. For example, some of the words we encounter for the management of process safety using the Plan-Do-Check-Act continual improvement life cycle model include:

- Plan (planning & organizing) – a wide variety of humans *manage* tasks for planning and organizing the resources and activities for all processes, in order to prepare for the effective and efficient execution of work. Example human activities include: allocating, acquiring, compiling, defining, determining, developing, encouraging, establishing, gaining, identifying, initiating, obtaining, planning, preparing, selecting, etc.

- Do (implementing & performing) – humans *perform* numerous job tasks on a daily basis, in order to achieve organizational objectives related to their work and the overall business. Example human activities include: accessing, advising, conducting, consulting, communicating, completing, demonstrating, deploying, documenting, employing, executing, explaining, implementing, informing, instructing, issuing, obtaining, performing, responding to, retaining, training, utilizing, etc.

- Check (measuring & evaluating) – humans often engage in specialized tasks to *verify* that actual activities and results are achieved against planned arrangements and objectives, in order to identify any gaps or deficiencies that need to be addressed. Example human activities include: analyzing, ascertaining, assessing, checking, confirming, evaluating, investigating, measuring, monitoring, revalidating, reviewing, verifying, etc.

- Act (reviewing & improving) – humans also *act* to correct identified gaps or deficiencies, in order to improve conditions, behaviors, processes, and performance results. Example human activities include: assuring, certifying, controlling, resolving, correcting, improving, etc.

It is these activities within the PSM system that provide an opportunity to integrate human factors.

1.3 What is Human Factors?

CCPS defines *human factors* as: "A discipline concerned with designing machines, operations, and work environments to match human capabilities, limitations, and needs. Among human factors specialists, this general term includes any technical work (e.g. engineering, procedure writing, worker training, worker selection) related to the person in the man-machine systems." [7].

The UK Health & Safety Executive (UK HSE) defines *human factors* as: "Human factors refer to environmental, organisational and job factors, and human and individual characteristics, which influence behaviour at work in a way which can affect health and safety" [20]. Thus, integrating human factors into the PSM System requires an understanding of: who these humans are, the work activities in which humans are engaged, and the factors needed to be addressed to improve human performance at work [36].

1.4 What are some Human Factors?

Human factors can be conveniently grouped into three interacting groups of performance influencing factors (PIFs) [8,20]:

- <u>Individual Factors (who is doing it?)</u> – e.g. competence, skills, personality, fatigue, stress, attitude, risk perception, etc.
- <u>Job Factors (what are people asked to do and where?)</u> – e.g. task, workload, complexity of task, clarity of procedures, system interfaces, work environment, displays and controls, procedures, etc.
- <u>Organizational Factors (how are they doing it?)</u> – e.g. culture, leadership, resources, work patterns, manning levels, communications, etc.

Human activities and individual behaviors on the job are influenced by each of the PIF groups. Combined with the human tendency to make errors, PIFs are the factors responsible for creating error-likely situations. While humans do not usually make intentional mistakes, misinterpreting situations are far too common and often result in errors. Human decision making is also questionable if they are overworked or are put in stressful situations. On the other hand, with proper experience and training human intervention under certain levels of stress during process upsets and emergency response can be beneficial [6].

Analysis of PIFs can help to determine how effective human performance will be. For example, if an operational procedure for a job provides complete clarity for what needs to be done, how it needs to be done, what may go wrong or require special attention, and how to correct a deviation if it happens, the chances of higher performance is more likely. PIFs can also be utilized for designing a system to increase usability and decrease error probability. For instance, designing a Human Machine Interface (HMI) for the control room operator that displays alarms that have undergone an alarm prioritization to only show those alarms that allow operators time to act upon, would help the operator make better decisions. Alarm prioritization also allows the system to produce fewer alarms and results in operators trusting the alarms to be genuine and provides them with sufficient time to process them and provide corrective action without experiencing a clutter of nuisance alarms that need to be continuously silenced.

1.5 Why integrate Human Factors into the PSM System?

Existing deficiencies within either the engineered process system or the management system create process safety risks through ineffective risk controls intended to prevent or mitigate process hazards. Addressing human factors to reduce incidents is not a novel concept since it has been shown that nearly all incidents (including near misses) can be traced back to deficiencies from a human cause, since humans are responsible for the design, construction, operation and maintenance of the process as well as the development and implementation of the PSM system. However, just stating that human error from human factors is the cause of an incident does not provide any real method to prevent the recurrence of such an event [15,31].

These deficiencies, once identified, represent areas of opportunity for improving the integration of human factors, particularly in designing features of the PSM system to reduce human error, or at least in providing for the recovery of errors through mitigation measures [31]. In fact, being able to trace the human error back to a PSM system deficiency and improving the system can be beneficial on several levels. It can improve communications within the facility, eliminate or at least reduce human error and hence future incidents, and improve human performance and overall PSM system efficiency [6].

A human factors approach can be proactive or reactive leading to differences in cultures observed for implementing changes to reduce risks. Even though a proactive approach can come with limitations of being prescriptive, it is an improvement over a reactive approach where all actions are a direct result of past events. Taking a proactive human factors approach involves a paradigm shift and needs a strong foundation of checklists, auditable improvements, and a structured plan to consistently implement these changes by integrating them with the existing PSM system.

The Baker Panel report following the Texas City incident in 2005 mentions that a majority of incidents in the process industry are a result of failure to minimize or prevent *process* as well as *personal safety* risks. It is a combination of process and personal safety risk reduction that ultimately determines how effectively all the risks are to be managed [5]. The report also cautions against having checklists that groups the risk factors into too tight of categories to really understand the root causes needed for effective risk reduction. In addition, inadequate tracking and implementation of human factors deficiencies identified by PHAs led to inconsistencies in addressing these issues [5].

Existing regulatory and standards references such as the Contra Costa County (California) Human Factors Program [13], UK HSE Human Factors guidelines [16-20], Norsok standards developed by the Norwegian petroleum industry (e.g. NORSOK S-001, S-002) [33,34], and international standards (e.g. ISO 17776) [26], to name a few, provide good guidance for implementing human factors as a means to manage and reduce risks associated with the process industry. These documents refer to human reliability, human error, safety culture, training, operating procedures, organizational factors, process hazard analysis, and incident investigation as important topics to be covered to ensure a complete approach towards understanding the capabilities and limitations of the human element interacting with systems and the need to design these systems accordingly.

Human factors in PSM can be identified by analyzing each element of the PSM system to recognize the opportunities for human interaction with the system and to forecast where these interactions have a probability of producing human error [6].

1.6 Where can Human Factors be integrated into the PSM System?

Human activities already associated with typical PSM system elements (see Table 1) demonstrate the potential for integrating the principles of human factors for all PIF groups to be straightforward, cost-effective, and beneficial in closing a gaps in an existing PSM system where deficiencies exist.

Table 1: Human Activities and Example PSM System Deficiencies by PSM Element

PSM Element	Human Activities	Example Deficiencies	HF PIFs
Employee Participation	Safety culture survey	Inadequate communications	Organizational
Process Safety Information	Design reviews P&ID walk-downs	Inaccurate piping & instrument diagrams (P&IDs), missing data	Job
Process Hazard Analysis	Evaluation of process hazards and safeguards	Inadequate safeguards, incomplete design basis, and design deficiencies	Job
Operating Procedures	Periodic review for currency and accuracy	Outdated operating procedures	Job
Operator Training	Testing / re-testing, on-the-job evaluation	Inadequately trained operators	Individual
Contractors	Pre-job/periodic performance evaluation	Unqualified or under-performing contractors	Individual
Pre-startup Safety Review	Review of design, procedures, training, etc.	Construction and equipment not in accordance with design specs	Job
Mechanical Integrity	Equipment tests and inspections	Equipment deficiencies outside of acceptable limits	Job
Management of Change	Technical review for proposed change	Impact of change on safety and health not assessed or not acceptable	Organizational
Incident Investigation	Causal factor analysis and root cause analysis	Various management system deficiencies as findings	Organizational
Emergency Planning/Response	Post-drill/incident response evaluation	Inadequate planning, lack of resources, inadequate equipment	Organizational
Compliance Audits	Interviews, records reviews, inspections	Various management system deficiencies as findings	Organizational

Some PSM elements already incorporate human factors to some extent, including Operating Procedures, Operator Training, and Employee Participation. Previously referenced regulations, industry standards, and guidelines can be used to provide specific design requirements to implement human factors in other parts of the PSM system. These encompass good industry practices originating as a part of lessons learnt over several decades giving this approach a substantial foundation to build upon for improving the existing PSM system [40].

Additional areas such as Safety Culture Surveys, Communications, Safety Critical Task Analysis, Human Reliability Analysis, Fatigue and Staffing Level Assessment, and Fitness for Duty, can be added to enhance the PSM system and improve productivity and risk reduction performance [6]. For example, Fatigue Assessment studies of staffing levels and workloads provide key information to understand the risks associated with mental and physical workload of a control room operator.

2. PSM System Design – The Solution for Integrating Human Factors

2.1 PSM System Background

A popular definition for the term *process safety management (PSM) system* is a "comprehensive set of policies, procedures, and practices *designed* [emphasis added] to ensure that barriers to episodic incidents are in place, in use, and effective" [7]. This definition implies that a PSM system needs to be adequately designed in order to achieve its objectives of managing process safety risks. Quality management systems have traditionally involved the practice of "say what you do, and do what you say". That is, by documenting what you plan to do, then implementing that plan to achieve results, one can "design in quality" as the basis for a proactive management system aimed at achieving performance, including process safety risk reduction performance.

Borrowing from international standards for quality management systems, the design process for delivering a product (e.g. a PSM system) requires as a starting point the development of design inputs, documented as a design specification [22]. A design specification is a set of requirements, related quality characteristics and performance criteria, translated from identified customer needs and performance objectives; with a procedure being a special type of specification [21]. A design specification is critical for integrating requirements into a management system in order to achieve the overall quality objective of meeting specified requirements [21]. Without defining all of the requirements needed for effective management, we cannot hope to achieve the desired performance results of the management system.

Unfortunately, most international standards for management systems related to quality, health, safety, environment, and security, as well as regulations governing process safety management, do not define detailed specifications for the management system itself, nor do they state specific criteria for its performance [13]. For example, an international standard for occupational health and safety management system states: "… It does not state specific OH&S performance criteria, nor does it give detailed specifications for the *design of a management system* [emphasis added]" [2]. Rather, these "performance-based" standards and regulations define "what" but typically not the "how to" requirements for effective management. These apparent gaps however allow flexibility for determining how best to design the management system in order to meet the general requirements and achieve overall performance.

Further, some PSM regulations (with industry standards incorporated by reference) require that human factors be "addressed", but provide no specifics or guidance on how to meet this requirement [1,3,4,38].

Therefore, in order to benefit from the application of human factors to the management of process safety, we must integrate the principles of human factors directly into the design of the PSM system by defining the requirements in the form of a specification.

This section will address a practical step-by-step procedure on "how to" incorporate human factors into the PSM system design. Examples for a few system components will show how quality aspects for human factors can be addressed in key management processes within the PSM system.

2.2 PSM Design Basis

Borrowing from two of the eight total quality management principles which form the foundation of several international standards for management systems for quality, health and safety, environment, and security, an effective PSM system design basis utilizes [21]:

- Process Approach – A desired result is achieved more efficiently when activities and related resources are managed as a process.
- System Approach to Management – Identifying, understanding and managing interrelated processes as a system contributes to the organization's effectiveness and efficiency in achieving its objectives.

At a higher level, the PSM system is structured with a dozen or so individual management processes (or elements), each one designed to achieve certain objectives which when integrated into a comprehensive system serve as a management tool for use by humans aimed at reducing or controlling process safety risks. At a lower level, each management process is designed as an interrelated set of resources and activities that transforms inputs into outputs (results). Many of these processes are interrelated with their inputs, outputs, and informational resources comprising documented policies, procedures, plans, programs, work instructions, standards, specifications, records, etc.

For each of the management processes within the PSM system, the aim is for humans to perform specific activities on a continual basis that improves the quality of the PSM system resources and outputs and that contributes to its overall effectiveness in reducing and controlling process safety risks. Therefore, the design intents for each PSM element should support the overall performance objective for the PSM system (see Table 2). For people to effectively perform, the PSM system design should define why (the intents or objectives), who (persons responsible), what (the tasks), when (the schedules), where (the applications), and how (the methods).

Table 2: Example PSM Performance Objectives

PSM Element	Performance Objectives to Reduce & Control Risks
Employee Participation	To improve participation and ownership
Process Safety Information	To improve information quality
Process Hazard Analysis	To improve the process design & safeguard effectiveness
Operating Procedures	To improve operational control
Operator Training	To improve competence and performance
Contractors	To improve qualifications and performance
Safe Work Practices (Hot Work Permit)	To improve control of work
Pre-startup Safety Review	To improve the adequacy of PSM elements before startup
Mechanical Integrity	To improve the asset integrity
Management of Change	To improve the control of changes
Incident Investigation	To improve the system and prevent incident recurrence
Emergency Planning & Response	To improve response capability
Compliance Audits	To improve the effectiveness of the PSM system

2.3 PSM Design Process

A proven procedure for designing management processes includes the following steps [28]:

1. Identify Output(s) – define the products or results to be produced.
2. Identify Customer(s) – define who will use or benefit from the outputs.
3. Identify Customer Requirements – define stated and implied requirements.
4. Translate Requirements into Specifications – define specifications for inputs/outputs.
5. Identify Process Steps – define the steps for Plan-Do-Check-Act activities.
6. Select Measurement Criteria – define measures of success and performance criteria.
7. Determine Process Capability – define the resources and requirements to produce results.
8. Evaluate Results – assess conformance and performance results, and evaluate gaps.
9. Improve Results – review and improve on results.

Each of the above steps are described in further detail below, with an emphasis placed on integrating human factors into the PSM system design.

Step 1 – Identify Outputs

All too often, standards and regulations are written based on describing certain activities (e.g. incident investigation, PHAs, compliance audits, etc.) without identifying the purpose of the process or the output expected. However, to design quality (and human factors) into the PSM system to achieve desired results, the output or product of each process activity should be explicitly identified.

For the example activities above, the outputs from the management processes would include: lessons learned with root causes as findings and recommendations for corrections and corrective/preventive actions, documented in an investigation report; process hazards/risks as findings and recommendations to further reduce risks, documented in an PHA report; and management system deficiencies as findings and recommendations, documented in an audit report. Properly defining the process outputs helps to identify their relationships to human factors as both causes of deficiencies and solutions for improvement.

Step 2 – Identify Customers

All processes ultimately exist to deliver products or results to a customer who has expressed needs. Internal and external stakeholders may include: those who are accountable for the results, a third-party such as a regulatory authority or the community, a contract employee or visitor, or for the most part, the end user of the product or the results who needs to be satisfied that they get what is needed in order to subsequently utilize the outputs. For example, an Emergency Response Plan is intended for multiple customers who have defined roles and responsibilities for executing the plan in case of emergency. Such customers may include: the incident commander, any one of several emergency responders, internal and external emergency services providers, operators of process units requiring specific steps for emergency operations, security personnel, etc.

Recognizing that individual humans fill all of the roles of "the customer" establishes a basis for identifying specific needs for job factors within the plan for coordinating, communicating, and controlling the emergency. In addition, identification of all customers helps to define organizational factors interfaces for communications, lines of authority for decision making, and requirements for on-scene and post-incident reporting.

Step 3 – Identify Customer Requirements

Customer needs are expressed in a variety of ways, but mostly fall into two categories: stated and implied requirements, typically expressed by someone other than the customers themselves [28]. *Stated requirements* are often viewed as prescriptive and answer specific questions about the product or results to be delivered or the method of delivery. For example, regulatory requirements for the process of Management of Change may provide detailed requirements on what constitutes a change, what types of changes are addressed, and general steps for managing and documenting the change, including records and communications to those affected by a change. *Implied requirements* on the other hand, are related to specific expectations expressed by a more general requirement. For example, regulations for Management of Change may state that written procedures shall assure that the certain considerations (e.g. authorization) be addressed prior to implementing any change.

However, due to the performance-based nature of some regulations, they may not define how the procedures are to provide such assurance, either in form or content. Rather, it is left up to the operating company (i.e. employer) to determine their own specific requirements as a means to achieving the end result. Thus, requirements play a crucial role in determining the success of a management process since the fulfilment of identified requirements is essentially the definition of *quality* and this fulfilment is the basis for measuring the degree to which desired results are achieved [21].

Step 4 – Translate Requirements into Specifications

Even when customer requirements have been defined, simply having a list may not be that useful. Therefore, requirements are often translated into documented specifications used for communicating information and controlling the update of such information. This translation of requirements may be accomplished by specifying design criteria, by defining qualification expectations, by defining steps in an operating procedure, by providing a checklist for field verification, by devising a flowchart which incorporates decision points, or by providing other documented means to effectively communicate specific needs to other humans. Thus, translating customer needs/requirements into documented specifications or within a procedure or other PSM system component is essential for establishing what humans do, when they do it, where they do it, how they do it, and to what level of quality or performance they need to do it.

Without documented requirements to define specific job factors, there is no hope that humans will be able to consistently achieve the desired results from a given management process, in order to design, construct, operate or maintain a safe and environmentally sound engineered process. Rather, poorly specified requirements may lead to human error that could lead to immediate consequences when the process is implemented, or could remain latent until such time that critical information is called upon to prevent an incident.

Step 5 – Identify Process Steps

Once the process outputs and the customer requirements have been fully defined, the process steps for producing the outputs, products or results should be determined and documented. These steps can be written or expressed in flowcharts or other means to communicate how the process should work. More detailed steps (e.g. work instructions) may be required for more complex management processes.

In some cases, prescriptive requirements of the outputs may dictate the process steps. For example, processes involving the development of safe work practices (e.g. permit-to-work systems) have steps that are more prescriptive in nature since standards or regulations have been well-defined (e.g. confined space entry).

In other cases, the requirements defined by implied needs may offer a wide range of flexibility for carrying out the process, as long as the results are achieved. For example, requirements for Employee Participation may only require that a plan be developed and implemented based on consultation with employees. However, the steps needed to achieve employee participation should also be defined if employees are to have a key role and positive effect in managing process safety. Beyond meeting minimum compliance requirement for a plan, organizational factors such as empowerment and job enrichment can be incorporated into the Employee Participation process as a means to strengthen the safety culture of the organization.

Step 6 – Select Measurement Criteria

Another set of requirements needed to be defined includes criteria for quality characteristics and performance. These can relate to the outputs themselves or the process steps for achieving the outputs; that is, criteria for the product and the process activities and resources. In some cases, an implied need for qualified auditors, PHA facilitators or lead investigators, all of who are expected to possess certain knowledge and/or familiarity, may require operating companies to establish their own criteria for competence, an individual factor. However, without criteria being defined or without even the requirement for defining such criteria, the quality of results for certain critical management processes may fall short of their expectation to employ humans who can positively impact process safety performance.

On the other hand, other areas of process safety have already evolved within industry standards which contain specific criteria. For example, criteria for the Mechanical Integrity management process may need to incorporate an extensive set of acceptance criteria in order to accommodate the tests and inspections for a wide variety of process equipment. Such requirements can be derived from recognized and generally accepted good engineering practices, or otherwise developed as in-house standards for special applications.

Step 7 – Determine Process Capability

Once all of the requirements have been specified for the quality of resources and the outputs, and for the performance of process steps with performance criteria for both the activities and the outputs, the capabilities for each process must be determined. In order to achieve the desired results with expected quality and performance, the organization's leadership has the most influence over commitment of the time and resources necessary to effectively implement the PSM processes.

From a human factors standpoint, all three groups of PIFs (individual, job, and organizational) are relevant, as they can be negatively or positively affected by process capability fulfilment. For example, individual factors such as assigning personnel to a critical operating position before they have been fully qualified puts the individuals and the facility at risk. Hence, the management of organizational change has come to play a more important role in managing process safety performance.

Similarly, failure to allocate sufficient resources for staffing qualified personnel to carry out plans and safely execute procedures has the potential to adversely affect job factors such task and workload. For example, extended work shifts or overtasking of personnel without proper planning during turnarounds results in increased risk of human error. Conversely, leading by example and "walking the talk" can have a profound positive effect on the organizational factors of safety culture and achieving open communications.

Step 8 – Evaluate Results

Based on the specified requirements for quality and performance criteria, measurement of actual results against such criteria offers the opportunity to identify and improve inadequacies, deficiencies, or substandard results, before they manifest themselves as process safety incidents. From a human factors perspective, all three groups of PIFs (individual, job, and organizational) are relevant, as the results from process measurements can be evaluated against defined criteria. For example, individual factors such as competence and refresher training can be measured, and action taken to improve results in order to minimize risks.

Similarly, periodic verification and validation of work products (e.g. currency of operating procedures from reviews, effectiveness of emergency response plans from post-drill evaluations, accuracy of drawings from walk-downs) allow the quality of process safety information to be confirmed and corrected in order to minimize future human errors. Proactive measures, or leading indicators, enable an organization to address organizational factors by monitoring degradations in the PSM system through process design reviews involving identification, evaluation and control of hazard and their associated risks; or through PSM system audits to identify and correct identified deficiencies. Reactive measures, or lagging indicators, can also provide opportunities for improvement, through organizational lessons learned and corrective/preventive actions to prevent recurrence of actual (or near miss) process safety incidents.

Step 9 – Improve Results

From this point forward, continual improvement of the process and overall system is made possible by stating requirements within the process that any identified inadequacies, deficiencies, or substandard results will be improved through responsible actions to close the gaps.

2.4 Examples of Designing Human Factors into the PSM System

A PSM system can be comprised of several documented components which are used to manage the activities within the various processes. Examples of these system components and opportunities where human factors can be integrated include:

- Principles – values and beliefs that address why (e.g. Code of Ethics).
- Policies – management expectations for what to do (e.g. Stop Work Authority policy).
- Plans – written plans on strategy and objectives (e.g. Employee Participation plan)
- Procedures – steps for who, what, when, where, how (e.g. operating procedures).
- Products – outputs of activities that involve people (e.g. reports for PHAs, audits, etc.).
- Processes – inputs, outputs, resources and activities (e.g. incident investigations process).
- Programs – means of achieving objectives, schedules, etc. (e.g. training program)
- Performance – metrics for results of management processes relating to how well objectives are being met (e.g. risks, incidence rates, etc.).

Three examples are presented below (see Table 3) to demonstrate how to design human factors into a PSM system component. Each one uses the Plan-Do-Check-Act continual improvement life cycle model to define the management process, and each one addresses a different human factors group. These examples address three of the "top ten human factors issues facing onshore major hazards sites in the chemical and allied industries, based on research, consultation with industry and intermediaries and inspection experience," as identified by the Energy Institute [14]:

1. Organizational change (and transition management)
2. Staffing arrangements and workload
3. **Training and competence (and supervision)**
4. Fatigue (from shiftwork and overtime)
5. Human factors in design (alarm handling, control rooms, ergonomics)
6. **Procedures (especially safety critical procedures)**
7. **Organizational culture (and development)**
8. Communications and interfaces
9. Integration of human factors into risk assessment and investigations
10. Managing human failure (including maintenance error)

Table 3: Examples for Designing Human Factors into PSM System Components

PSM Element	Component	Example	HF PIF Group
Operator Training	Process	Competence Assurance Process	Individual
Operating Procedures	Procedure	Standard Operating Procedures	Job
Employee Participation	Policy/Program	Stop Work Authority Policy/Program	Organizational

<u>Example #1 – Designing Human Factors into a Process</u>

Human factors can be designed into each management process of the PSM System by specifying the requirements for the key process components: inputs, outputs, resources and activities. This example demonstrates how *individual* human factors requirements can be specified for the process of assuring the competence of personnel, namely those who operate engineered processes.

This management process places emphasis on developing qualified operators as a critical human resource for operating the engineered process. Thus, an unqualified operator is the process input and a qualified operator is the process output with the specified quality characteristic of competence. A competent operator qualified through this process is more likely to reduce the potential for human error and therefore the process safety risk of operating the facility.

The customer in this process is the operator, who has a vested interest in being qualified to fulfill the job; the process unit supervisor or facility management are also customers of this process, as they are accountable for assigning qualified personnel to critical positions. The regulator can also be considered a customer, as they have a stake in the outcome from a compliance standpoint.

Regulatory authorities often only focus on establishing requirements for the training activity of the competence assurance process. For example, the OSHA PSM Standard requires that employees involved in operating an engineered process shall be trained in an overview of the process and in the operating procedures [38]. A process for training is not even required, although it is implied. Further, there are no regulatory requirements established by OSHA to prescribe how an operator should be trained, nor the criteria for determining how they are qualified. The PSM Standard only requires that "the means used to verify that the employee understood the training" is documented as part of the training records [38]. The term *competence* is not even used in this regulation, whereas international health and safety and environmental management system standards require that persons performing tasks shall be competent on the basis of appropriate education, training and experience [2,25]. Quality management system standards define competence as "demonstrated ability to apply knowledge and skills to achieve intended results" [2,21].

Therefore, to assure that qualified operators are competent to perform their assigned tasks of operating the process, a robust competence assurance process should be based on specified requirements, such as: defined competencies with specific criteria for measuring competence, methods for delivering training including qualifications of trainers, and methods for verifying competence and qualifying operators [16]. Such requirements should be specified for each of the process activities for assuring competence. For example, the Plan-Do-Check-Act continual improvement life cycle model can be used to define the steps of the process, as follows:

- <u>Plan</u> – specify requirements for planning and organizing training (e.g. training program, training matrix, training schedule, etc.)
- <u>Do</u> – specify requirements for delivering training (e.g. qualifications of trainers, content and format of training material, training facilities, etc.)
- <u>Check</u> – specify requirements verifying competence (e.g. testing methods and criteria, etc.)
- <u>Act</u> – specify requirements for improving competence (e.g. re-testing, refresher training, management of change, etc.)

<u>Example #2 – Designing Human Factors into a Procedure</u>

Human factors can also be designed into specific procedures for the PSM System by specifying the requirements for how work should be performed. This example demonstrates how *job-related* human factors requirements can be specified in a process for delivering operating procedures.

This management process involves developing, using, and maintaining operating procedures as a critical informational resource for operating the engineered process. Thus, the process output is a set of operating procedures with specified quality characteristics such as accuracy, currency, comprehensiveness, clarity, usability, accessibility, etc. Operators who are trained on operating procedures can only be as effective as the quality of the tools at their disposable. So, having operating procedures with quality designed with the end user in mind is more likely to reduce the potential for human error and therefore the process safety risk of operating the facility.

Besides the process operator as the obvious customer in this process, the regulator can also be considered a customer, as they have a stake in the outcome from a compliance standpoint. As a minimum, the OSHA PSM standard requires that the operating procedures "provide clear instructions for safely conducting activities involved in each covered process consistent with the process safety information", that they address each operating phase (e.g. startup, shutdown, etc.), and that they address several other provisions for the safe operation of the process [38].

One of the ways human factors can be incorporated in writing operating procedures is to conduct a thorough task analysis to understand which job tasks need detailed instructions and which hazards are associated with the tasks before writing the procedures [39]. In order to ensure an effective procedure is drafted, an error analysis may be conducted by experienced personnel to add precautionary notes and warnings immediately prior to a step [13].

In addition, using consistent writing style that explains the purpose of the procedure (e.g., training material, emergency shutdown procedure, normal operation, etc.), references to other supporting documents (data sheets, manuals, etc.), and precautions that must be taken to avoid hazards and undesirable consequences, demonstrate good human factors practices. Avoiding long steps, double negatives, very small font sizes, and steps without any actions associated with them is recommended. The operating procedures should include "who" does "what" and "when" as well as quantitative values and limits for additional clarity [17].

To assure the availability of reliable operating procedures, a robust process should be based on specified requirements, such as: defined quality characteristics with criteria for verification, methods for accessing and using the procedures, and methods for reviewing and maintaining the procedures. For example, the Plan-Do-Check-Act continual improvement life cycle model can be used to design quality into the process and its product, as follows:

- Plan – specify requirements for designing and developing the operating procedures (e.g. task objectives, format, content, quality characteristics, etc.)
- Do – specify requirements for accessing and using the operating procedures (e.g. accessibility, back-up versions, document control, variance authorization, etc.)
- Check – specify requirements verifying and validating the operating procedures (e.g. periodic review frequency and methods to ensure reliability and usability)
- Act – specify requirements for improving operating procedures (e.g. document control for updates and approvals, management of change, etc.)

Example #3 – Designing Human Factors into a Policy/Program

Further, human factors can be designed into specific policies and related programs within the PSM System by specifying the requirements for how policies should be implemented through these programs. Having a safety policy issued by management does not guarantee its success; it must be supported, implemented, evaluated, and continually reinforced to be effective. This example demonstrates how *organizational* human factors requirements can be specified in a management process for a program intended to implement a *stop work authority* policy for preventing incidents.

This management process encompasses developing, implementing, and sustaining a program to support a policy for empowering workers with the authority and responsibility to stop work where work situations may present an imminent risk or danger. Thus, the process output is a program which when successfully implemented not only provides a tool for reducing a process safety risk and potentially preventing a process safety incident, but also affords the opportunity to continually reinforce a workplace safety culture which recognizes process safety as a priority. It allows employees a mechanism to speak up when they see a problem, i.e. "see something, say something", as expressed in the committed culture tenet of the CCPS Vision 20/20 [9].

The customer in this process is the employee or contractor who is equipped with a tool supported by management to be used to protect themselves and others in cases that warrant stopping work. The regulator can also be considered a customer, in that they have a stake in the outcome from a compliance standpoint. Although OSHA's PSM Standard does not address this subject, BSEE's SEMS regulations for offshore operations have recently been revised to require stop work authority to be implemented by operating companies for their facilities [4]. Accordingly, "These procedures must grant all personnel the responsibility and authority, without fear of reprisal, to stop work or decline to perform an assigned task when an imminent risk or danger exists." [4].

Human factors is inherently embedded in this type of policy implementation; starting from a positive change initiation in how safe behavior is encouraged, and leading to being sustained by gaining the trust of every person involved in the policy implementation. This type of policy implementation requires good communication that reinforces a level of credibility that an employee will not be punished when stop work authority is actually utilized [18]. To achieve the credibility and attitude change needed for success of this program, several actions are recommended, such as: regularly stating clear information on safety policies, investigating incidents that lead to a change and informing employees of this change, including employees in a dialogue to improve ownership of the policy, and visibly demonstrating commitment of management to safety [18,32].

To assure that the stop work authority policy is successful, a robust program should include specified requirements, such as: the purpose of the policy, the means to communicate and reinforce the policy, methods for rolling out and using the policy, and methods for monitoring and sustaining its implementation. Such requirements should be specified for each of the process activities for assuring it success. For example, the Plan-Do-Check-Act continual improvement life cycle model can be used to design a program to implement a stop work authority policy, as follows:

- Plan – specify requirements for establishing a stop work authority (e.g. policy, procedures, success criteria, etc.)
- Do – specify requirements for implementing the stop work authority program (e.g. communications and awareness training, stop work authority cards, recordkeeping, etc.)
- Check – specify requirements for monitoring use and effectiveness of stop work authority (e.g. metrics, safety culture surveys, etc.)
- Act – specify requirements for sustaining stop work authority (e.g. reinforcement campaigns, refresher training, etc.)

3. Lessons to be Learned from Process Safety Incidents

Previous sections discussed that human factors can be integrated with key quality aspects into management processes within the PSM System, such as employee participation (e.g. safety culture), operating procedures (e.g. reliability), and training (e.g. competence). A historical perspective can reveal that the lack of human factors integration can contribute to management system deficiencies as the root causes leading to process safety incidents. Examples of three process safety incidents from U.S. Chemical Safety and Hazard Identification Board (CSB) investigation reports show how deficiencies in PSM systems pertaining to human factors led to system failures and how such these situations can be improved to prevent incidents from recurring.

3.1 Plastics Manufacturing Facility Incident (2004) [12]

An explosion and fire resulted in five fatalities and three injuries at the Formosa Plastics Corp. plastics manufacturing plant in Illiopolis, Illinois on April 23, 2004. The CSB reported that a large quantity of highly flammable vinyl chloride monomer (VCM) was released from a reactor.

The CSB noted that this incident resulted from human error by an operator opening the wrong valve that resulted from the inadequate layout of valve controls placed in the facility. The layout of controls made it easier for the operator to make a fatal mistake. The location of valve controls also did not provide feedback of the tank levels for operators to visualize. Further, there were no means for the operators to communicate between the levels where controls were located versus where the tanks were located.

Another cause cited by CSB was a more recent workload change due to staffing reductions. An operator was often required to make decisions without proper supervision since the supervisory positions had been eliminated. Depending on the experience and workload of the operator, as well as the level of stress or fatigue at the time of decision making, not having a supervisor available to check their decisions could lead to a higher probability of committing an error.

3.2 Pesticides Manufacturing Plant Incident (2008) [11]

An explosion at the Bayer CropScience pesticides manufacturing plant in Institute, West Virginia resulted in two fatalities and injured eight persons on August 28, 2008. The CSB reported that "this incident occurred during a lengthy start-up process resulting in a runaway chemical reaction inside a residue treater pressure vessel. The vessel ultimately over pressurized and exploded. The vessel careened into the methomyl pesticide manufacturing unit leaving a huge fireball in its wake."

The CSB noted that several shortcuts were taken during the start-up process that included not completing process hazard analyses (PHAs), bypassing checking valve line-ups, not conducting proper operator training, use of improper operating procedures, and inadequate safety critical equipment. Operator fatigue was also a significant issue leading to poor performance and decision making. In addition, Bayer CropScience management did not communicate properly with the emergency response team leading to exposure of first responders.

3.3 Petroleum Tank Terminal Incident (2009) [10]

A large explosion occurred at the Caribbean Petroleum Corporation (CAPECO) facility in Bayamon, Puerto Rico on October 23, 2009. There were no fatalities but the significant blast pressure generated by this incident injured forty-three (43) persons and caused an estimated $1.5 billion worth of damage to commercial and residential property.

The CSB described that "During an operation to transfer gasoline from the vessel Cape Bruny tanker ship, Caribbean Petroleum Tank 409 overflowed with gasoline, resulting in a vapor cloud that encompassed 107 acres of the CAPECO tank farm." In addition, "Multiple physical causes likely contributed to Tank 409 overfill:

- Malfunctioning of the tank side gauge or the float and tape apparatus during filling operations led to recording of inaccurate tank levels;
- Variations in the gasoline flow rate from the ship may have contributed to the overfill;
- Potential failure of the tank's internal floating roof due to turbulence and other factors may have contributed to the overfill."

The CSB noted that several human factors and safety management issues were at play during this incident. For example, the design of valves made it difficult to determine whether the valves were open or closed. Also, insufficient lighting did not allow the operators to see the overflowing of the tank. Finally, the tank filling procedures were inadequate and the tank gauging equipment was also unreliable, resulting in operators being prone to making errors.

Several similar incidents were also cited for discussion during this investigation. Crucial lessons from these were ignored, pointing towards a poor safety culture and systematic failure to learn from past incidents. Further, management had reduced several positions, leading to operators to being forced to leave their positions during tank filling procedures in order to complete other duties. A lack of emergency preparedness and firefighting equipment on site were also a result of reduced investment in safety by management.

3.4 Summary of Human Factors Causes

The CSB investigation reports from the above process safety incidents point to a singular theme of missing the central driving aspect of how humans interact with their work environment. Risk assessment from a human factors perspective in retrospect during an incident investigation yields a clear path towards enhancing existing PSM system designs by including additional safety measures to allow for recovery from human error and providing a clear feedback of such errors.

In order to allow for minimization of human errors, a conscious effort should be made to ensure good human factors design of equipment, training operators to recognize where to expect higher chances of making errors and how to recover from them, and training operators to use the operating procedures to not only consistently follow a procedure stepwise but also to ask for a revision when the procedure is known to be inaccurate. In addition, management support for improving the safety culture and including human factors in the discussion is crucial to success [36].

Similar lessons to be learned related to the inadequate treatment of human factors within PSM systems can be derived from many other process safety incident investigations [30].

4. Conclusions

This paper has:

- Presented concepts and examples aligned with regulations, standards, and industry guidelines for process safety management and human factors principles and practices.
- Identified the humans and their activities within the PSM system.
- Defined human factors within three key groups of individual, job, and organizational performance influencing factors (PIFs).
- Demonstrated a practical step-by-step design process for "how to" integrate human factors into the PSM system design using the proven quality management design concept of the Plan-Do-Check-Act continual improvement life cycle.
- Shared examples for designing PSM system components which address three of the Top 10 issues for human factors: competence, operating procedures, and organizational culture.
- Presented lessons learned from three process safety incidents which show how human error inherent in management systems can lead to catastrophic failures and how these can be minimized by applying the knowledge of how humans interact with their work environment.

This paper has shown that:

- PSM system deficiencies resulting from human error in design lead to higher probabilities for human error in operation and therefore higher risks of process safety incidents.
- Most process safety incidents can be traced back to management system deficiencies with their root causes related to human factors.
- Identified deficiencies represent areas of opportunity for integrating human factors into the PSM system design for improved human and overall system performance.
- Human activities are already inherent in an existing PSM system, but the application of a structured and systematic approach of human factors principles and practices allows for a successful implementation that can be measured, evaluated, and continually improved.
- Human factors can be integrated directly into the PSM system through detailed design specifications of requirements for quality aspects and performance criteria.
- PSM system processes can be enhanced by defining the humans in the system, their roles, and how their errors can be minimized, then designing the work activities and management tools to accommodate the objective to improve performance.
- Implementation of the PSM system can be used as a means to promote the application of good human factors practices for training and empowering personnel to allow them to recognize avenues for high error probability and how to recover from such errors.
- Safety culture improvement can lead to positive changes in producing policies, procedures, programs and other components that reduce human error potential and lead to the workforce being more engaged in achieving higher human and process safety performance.

5. References

[1] API. API RP 75, *Recommended Practice for Development of a Safety and Environmental Management Program for Offshore Operations and Facilities*, Third Edition, American Petroleum Institute, Washington DC, May 2004 (Reaffirmed May 2013).

[2] BSI. BS OHSAS 18001:2007. "Occupational health and safety management systems – requirements", BSI (British Standards Institution) OHSAS 18001:2007. London: BSI, 2007.

[3] BSEE. Workplace Safety Rule (SEMS), 30 CFR Part 250, "Oil and Gas and Sulphur Operations in the Outer Continental Shelf, Subpart S – Safety and Environmental Management Systems (SEMS)", Bureau of Safety and Environmental Enforcement (BSEE), Federal Register, October 15, 2010, (75 FR 199).

[4] BSEE. Workplace Safety Rule (SEMS II), 30 CFR Part 250, "Oil and Gas and Sulphur Operations in the Outer Continental Shelf, Revisions to the Safety and Environmental Management Systems", Bureau of Safety and Environmental Enforcement (BSEE), Federal Register, April 5, 2013, (78 FR 68).

[5] Baker, James A. , III, et al. *The Report of the BP U.S. Refineries Independent Safety Review Panel*, January 2007.

[6] Bridges, William and Tew Revonda, "Human Factors Element Missing from Process Safety Management (PSM)," 6th Global Congress on Process Safety, AIChE, San Antonio, Texas, March 2010.

[7] CCPS. *Guidelines for Auditing Process Safety Management Systems*, 2nd Edition, ISBN: 978-0-470-28235-9, Center for Chemical Process Safety (CCPS) of the American Institute of Chemical Engineers (AIChE) and John Wiley & Sons, New York, New York, 2011.

[8] CCPS. *Guidelines for Preventing Human Error in Process Safety*, 1st Edition, ISBN: 978-0-8169-0461-7, Center for Chemical Process Safety (CCPS) of the American Institute of Chemical Engineers (AIChE) and John Wiley & Sons, New York, New York, 2004.

[9] CCPS. "Vision 20/20", Center for Chemical Process Safety (CCPS) of the American Institute of Chemical Engineers (AIChE) and John Wiley & Sons, New York, New York. http://www.aiche.org/ccps/resources/vision-2020, Accessed March 15, 2016.

[10] CSB. "Investigation Report – Caribbean Petroleum Tank Terminal Explosion and Multiple Tank Fires" (Caribbean Petroleum Corporation (CPECO)), Bayamon, Puerto Rico, October 23, 2009), No. 2010-02-I-PR, U.S. Chemical Safety and Hazard Identification Board (CSB), June 2015.

[11] CSB. "Investigation Report – Pesticide Chemical Runaway Reaction Pressure Vessel Explosion" (Bayer CropScience., Institute, West Virginia, August 28, 2008), No. 2008-08-I-WV, U.S. Chemical Safety and Hazard Identification Board (CSB), January 2011.

[12] CSB. "Investigation Report – Vinyl Chloride Monomer Explosion" (Formosa Plastics Corp., Illiopolis, Illinois, April 23, 2004), No. 2004-10-I-IL, U.S. Chemical Safety and Hazard Identification Board (CSB), March 2007.

[13] Contra Costa County. Industrial Safety Ordinance Guidance Document, "Human Factors Program, Section B", June 15, 2011.

[14] Energy Institute. "Human factors – Top ten issues", https://www.energyinst.org/technical/human-and-organisational-factors/human-factors-top-ten, Accessed March 15, 2016.

[15] Gambetti, Fabrizio, Casalli, Andrea, and Chisari, Vladimiro, "The Human Factor in Safety Management", Chemical Engineering Transactions, Vol. 26, 274-284, DOI: 10.3303/CET1226047, 2012.

[16] HSE. "Human Factors Briefing Note No. 2 – Competence", UK Health and Safety Executive (HSE), http://www.hse.gov.uk/humanfactors/topics/02competency.pdf, Accessed March 15, 2016.

[17] HSE. "Human Factors Briefing Note No. 4 – Procedures", UK Health and Safety Executive (HSE), http://www.hse.gov.uk/humanfactors/topics/04procedures.pdf, Accessed March 15, 2016.

[18] HSE. "Human Factors Briefing Note No. 7 – Safety Culture", UK Health and Safety Executive (HSE), http://www.hse.gov.uk/humanfactors/topics/07culture.pdf, Accessed March 15, 2016.

[19] HSE. "Procedures Audit Tool", UK Health and Safety Executive (HSE). http://www.hse.gov.uk/humanfactors/topics/procedures-audit-tool.pdf, Accessed March 15, 2016.

[20] HSE. "Reducing Error and Influencing Behaviour", UK Health and Safety Executive (HSE), 1999, http://www.hse.gov.uk/pubns/priced/hsg48.pdf, Accessed March 15, 2016.

[21] ISO. ISO 9000:2015, "Quality management systems –Fundamentals and vocabulary", International Organization for Standardization (ISO), Geneva: ISO, 2015.

[22] ISO. ISO 9001:2015, "Quality management systems – Requirements", International Organization for Standardization (ISO), Geneva: ISO, 2015.

[23] ISO. ISO 9004:2009, "Managing for the sustained success of an organization – A quality management approach", International Organization for Standardization (ISO), Geneva: ISO, 2009.

[24] ISO. ISO 10018:2012, "Quality management – Guidelines on people involvement and competence", International Organization for Standardization (ISO), Geneva: ISO, 2012.

[25] ISO. ISO 14001:2015, "Environmental management systems – Requirements with guidance for use", International Organization for Standardization (ISO), Geneva: ISO, 2015.

[26] ISO. ISO 17776:2000, "Petroleum and natural gas industries – Offshore production installations – Guidelines on tools and techniques for hazard identification and risk assessment", International Organization for Standardization (ISO), Geneva: ISO, 2000.

[27] ISO. ISO 28000:2007, "Specification for security management systems for the supply chain", International Organization for Standardization (ISO), Geneva: ISO, 2007.

[28] Jones, David A. "How to Design Effective EH&S Management Systems for Continual Improvement of Risk Reduction Performance", ASSE Safety 2002, American Society of Safety Engineers (ASSE), Nashville, Tennessee, June 2002.

[29] Jones, David A. "ISO 9000: A Framework for Integrating Safety, Health & Environmental Management Systems," 28th Annual Loss Prevention Symposium, AIChE, Atlanta, Georgia, April 1994.

[30] Jones, David A., Kay Modi, "A Call to Action: Understanding the Key Driver for Continual Process Safety Performance Improvement", American Institute of Chemical Engineers, 12th Global Congress on Process Safety, Houston, Texas, April 10-13, 2016.

[31] Kletz, Trevor. *What Went Wrong? Case Histories of Process Plant Disasters and How They Could Have Been Avoided*, 5th Edition, ISBN-13: 978-1856175319, ISBN-10: 1856175316, Butterworth-Heinemann/IChemE, 2009.

[32] NOPSEMA. "Human Factors: Safety Culture", N-06300-IP1506, Rev 0, National Offshore Petroleum Safety and Environmental Management Authority (NOPSEMA), December 2014.

[33] NORSOK. NORSOK Standard S-001, "Technical safety", Edition 4, Standards Norway, Lysaker, February 2008.

[34] NORSOK. NORSOK Standard S-002, "Working environment", Edition 4, Standards Norway, Lysaker, August 2004.

[35] OGP. Report 368, "Human Factors", International Association of Oil & Gas Producers (OGP), 2005.

[36] OGP. Report 434-5, "Risk Assessment Data Directory", International Association of Oil & Gas Producers (OGP), March 2010.

[37] OGP. Report 454, "Human Factors Engineering in Projects", International Association of Oil & Gas Producers (OGP), August 2011.

[38] OSHA. "Process safety management of highly hazardous chemicals", 29 CFR 1910.119, (e), Process Hazard Analysis, Occupational Safety and Health Administration (OSHA), *Federal Register*, February 24, 1992 (57 FR 6356).

[39] Parker, Christopher W. and Gandhi, Minu Shikha, "Task-Based Approach to Addressing Human Performance in Design," 10th Global Congress on Process Safety, AIChE, New Orleans, Louisiana, April 2014.

[40] Poblete, Benjamin R., Gandhi, Minu, Parker, Christopher W., and Ranasinghe, Sam, "Lessons Learned from the Application of Human Factors Engineering (HFE) in Process Safety Assessments", Mary Kay O'Connor Process Safety Center 17th Annual International Symposium, College Station, Texas, October 2014

439061 Culture, Competence and Compliance

Abstract:

Atlantic LNG is one of the world's largest producers of Liquefied Natural Gas (LNG). From our base in Trinidad, we operate a four-train liquefaction facility with a total capacity of 15 million metric tonnes per annum (tpa)

When Atlantic first embarked on the implementation of a competence management system for our Production Operation and Maintenance Technicians we quickly realized that we needed to focus on compliance just as much if not more than competence. We were grappling to understand why some individuals who after having been deemed competent were observed doing things that their competence should tell them was wrong.

This presentation was our attempt, as the unit responsible for technical competence assurance to help the rest of the organization understand just how culture, competence and compliance interrelate and how focusing on competence alone in a vacuum would not lead to the improved performance the company desired. It will explore the relationship between competence and compliance and how one without the other can be dangerous for an organization. And the importance of understanding people and culture in the design of systems for sustained success.

Human Factors and their Impact on Plant Safety

Luis M. Duran
Global Product Manager Safety Systems
ABB Inc.
3700 West Sam Houston Parkway South, Suite 600
Houston, TX 77042
luis.m.duran@us.abb.com

Jeton Partini
Interaction Designer/ Human Factors
CGM
Borås, Sweden
jeton@cgm.se

Prepared for Presentation at
American Institute of Chemical Engineers
2016 Spring Meeting
12th Global Congress on Process Safety
Houston, Texas
April 11-23, 2016

Human Factors and their Impact on Plant Safety

Luis M. Duran
Global Product Manager Safety Systems
ABB Inc.
3700 West Sam Houston Parkway South, Suite 600
Houston, TX 77042
luis.m.duran@us.abb.com

Jeton Partini
Interaction Designer/ Human Factors
CGM
Borås, Sweden
jeton@cgm.se

Keywords: Human Factors, Safety, Security, Risk Reduction, Safety Systems, Integrated Control and Safety Systems, Independent Protection Layers, Common Cause Failures, Safety Lifecycle, Security Lifecycle, Functional Safety Management System, IEC61508, IEC61511, TÜV, third party certification, control room design, ISO11064, Operator effectiveness

Abstract

History is full of technology breakthroughs, all striving to increase productivity and efficiency, from the steam engine and the telegraph; we've seen technology changing the way we get things done, sometimes in a disruptive way.

Most recently mission critical computing systems have been introduced in manufacturing processes and automated tasks, resulting in increased safety and productivity during normal operation, but can these technologies help keep the plant safe during abnormal process conditions? That's where technology can support but not replace humans. The industry relies on human ability to respond to the unexpected, to handle the odd conditions and ask the right questions to fix the problems at hand.

Today, operators are loaded with numerous activities, is it reasonable to expect they'll be able to respond appropriately to all conditions, what are the human elements that should be taken in to consideration in the design and implementation of modern automation systems? This presentation discusses some of the capabilities available in a modern automation system and how to apply innovative control room planning and technology to support decision making and help humans handle the abnormal situations in a safe and effective manner. Further on we will present results from research performed in-house and centered on operator health, and the impact of an intelligent and ergonomic workspace can both mitigate risk and increase productivity.

1 Introduction

History is full of technology breakthroughs, all striving to increase productivity and efficiency, from the steam engine and the telegraph; we've seen technology changing the way we get things done, sometimes in a disruptive way. Today we are entering the fourth industrial revolution, the era commonly refer to as Industrial Internet of Things, Services and People in America or industry 4.0 in Europe. The first industrial revolution was characterized by the introduction of machine production powered by water and steam energy, which evolved into the second industrial revolution, the technological revolution with the introduction of electricity. The third industrial revolution, the age of information technology (IT) is now giving way to the age of the smart factory, characterized by the keywords big data, the Internet of things, automation, cyber, cloud, connection, analytics. The industrial sector is facing a paradigm shift, how well prepared is the industrial world and how do we address the role of the human?

The introduction of mission critical computing systems in manufacturing processes and automated tasks, resulted in increased safety and productivity during normal operation, but now facing this new industrial revolution, can these technologies help keep the plant safe during abnormal process conditions? This is an area where the authors believe there are technologies available to support humans and to enhance their decision making, not to replace it. The industry relies on human ability to respond to the unexpected, to handle the odd conditions and ask the right questions to fix the problems at hand.

2 Human Factors, Safety Culture and Learning Opportunities

Humans need to be at the center of any safety discussion for many reasons, one of which is the impact of human behaviors as a cause of safety incidents.

Safety Culture is normally found among the top causes of incidents and accidents in the industry and their consequences can range from frequent minor injuries to catastrophes making the headlines of newscast worldwide.

Safety Culture is part of the Corporate Vision. It is frequently presented as an explicit statement by senior management as part of the ethos of the company. It should transpire to every level of the organization so that every individual understands the overarching reasons for the vision, and that the organizational structure, management processes, technology and human resources align together as the fabric of the safety behavior of the company.

The vision and values must be communicated and the effectiveness of the communication must be verified. This to confirm that safety has become a personal commitment of everyone in the company.

There is also a need for a clear and explicit risk management strategy.

3 Understanding and managing risk

As a starting point, plant operators require that a hazard and risk assessment is carried out to identify the overall safety requirements, and then they must focus their attention on proactive measures to ensure that failures do not occur and that a potential catastrophe can be avoided when failures happen.

Individuals, communities, industries and organizations could learn a lot from incidents and system failures. There are opportunities to learn, by reviewing incident investigation reports and expert analysis from any incident in any industry as all industries, chemical, oil & gas, mining, automotive or sports; offers extensive resources and research and in general a body of knowledge that could be used more effectively.

According to research in the area, to learn from a situation or from experiences, those trying to learn should ask a few questions. I'd like to suggest the following questions as a starting point.

- What should you be able to do differently after this experience?
- What can you do to learn more from this experience?
- What would you do differently if you were going to do this experience over again?

The company Safety Culture should provide opportunity for individuals and organizations to learn and motivate positive change and therefore lead to a safer and productive operation.

4 Technology as part of the solution

Automation technology, based on digital programmable systems, is well-known for the positive impact in the productivity of the manufacturing process. However, although it has been used as a risk reduction mechanism for many years, it is constantly perceived as too expensive or negatively affecting productivity.

Traditional design practices allocate the risk reduction across different and independent protection layers. The rationale behind it is simple, "any system that can fail will fail." So the engineering best practice is to distribute the risk reduction tasks across multiple independent functions or systems. One of these systems is a Safety Instrumented System.

Layers of protection

It's common to discuss the concept of Independent Layers of Protection starting with technologies such as Process Control Systems that assist not only in the productivity of the process but also help plant operators to keep the process within the safe boundaries of operation. Today most Process Control Systems will include some degree of Alarm Systems that alert the operator of abnormal conditions and lead to a corrective action by the operator.

It's important to highlight that the operator, the human being, as a form the technology, plays a critical role in process safety. It is critical for an operator to sustain a high level of alertness and understanding of the progress through the production cycle; both during the slow times and, at the same time, have real-time access to critical information in context to be able to make correct decisions immediately when circumstances dictate such as in the case of abnormal conditions in the plant. The technology must be chosen and designed to help operators avoid errors; typically the highest single cause for unscheduled slowdowns and shutdowns.

Risk found in many production processes requires additional Protection Layers. Automatic Safety Instrumented Systems can be one of these protection layers. Designed to detect the dangerous conditions and take immediate action to the process to its safe state and prevent the hazard if the conditions escape the normal behavior in the process, they protect when the operator can't act timely to bring the process back to normal and prevent the hazard.

Numerous project specifications include statements such as "safety instrumented system must be designed following IEC61511/ISA84", or system must be "SIL 3 certified." In a way such statements imply that using a given technology will magically bring safety to the process. Reality is that all these technologies are designed and implemented by human beings and as a result will not be perfect or 100% safe.

International Functional Safety Standards (IEC 61508 and IEC61511) and previously ISA 84, introduced the concept of the Safety Lifecycle, describing the phases that should take place from concept to design to implementation and operation of a Safety Instrumented System. The Safety Lifecycle is a step in the direction of reducing the impact of human factors by establishing the proper design best practices, documentation reviews and validation and verification steps in the execution of a safety project. Additionally the standards introduce two important elements: Competence of Personnel and Functional Safety Management Systems (FSMS). Additionally recent changes to International Functional Safety Standard IEC61508 have turned these requirements (competence and FSMS) into normative clauses of mandatory compliance instead of a recommendation as in the previous versions; therefore giving these processes a significant role in ensure that process are designed, operated and maintained in a safe manner.

If any of these layers (technology or human) fail to prevent the hazard, there are other layers intended to mitigate its consequences such as fire and gas systems or site emergency response or community emergency response systems which will not be covered in this paper.

5 Integrating Control and Safety Systems to deliver consistency for the operator

Integrated Control and Safety Systems provide the enabling technology to drive effective operations and minimize or eliminate some of the sources of Human Error mentioned earlier.

Among the benefits of integrating Control and Safety System we can include

1. Analysis of potential common cause failures and designing those out of the system occurs before the system or technology is released to the market, it's performed by competent experts and tested and certified by independent third party agencies.
2. Consistent and standard access control to prevent unauthorized access to critical facilities and securing the systems is part of the "commercial-off-the-shelf" (COTS) product and accomplished without introducing additional custom programming.
3. Integrated testing occurs at the product test lab rather than during a Factory Acceptance Test, relies on competent personnel with in-depth domain knowledge on the multiple

technologies involved and to ensure that the standard functionality is working as the product was designed, once more prior to releasing the product to the market. Additionally allowing the Factory Acceptance Test (FAT) to be about project specific requirements.

Once the design is complete, all of these systems will have different degrees of interaction with users, and different types of users with different levels of competence depending to their role (i.e. operation, maintenance personnel and engineers). The design, and later the user training, should concentrate on allocating and using functionality relevant to the user function or role.

6 Human-processes for Process Safety

Current Functional Safety standards consider Competence of Personnel and Functional Safety Management Systems (FSMS) as mandatory requirements of a Safety Instrumented Systems. The reason is simple: the most sophisticated automation can't prevent human error. Industrial plants are designed with the highest degree of accuracy in mind, often with several separate safety loops checking the integrity of process systems. However, this approach can't check for human intervention and a small human error could cause an enormous catastrophe.

Results from research conducted by the Health and Safety Executive (HSE) and published in the book "Out of control: Why control systems go wrong and how to prevent failure "[1] shows the impact of the human element in industrial incidents throughout a project lifecycle. Difference sources indicates that 70% of reported incidents in the oil and gas industry worldwide are attributable to human error and account for in excess of 90% of the financial loss to the industry.

Another publication by the Health and Safety Executive mentions the top ten human factors [2] affecting Process Safety are

1. Organizational change and transition management
2. De-manning and staffing levels
3. Safety culture/blame - culture
4. Fatigue from shiftwork and overtime
5. Training, competence and supervision
6. Alarm handling
7. Compliance with safety critical procedures
8. Communications e.g. shift handover
9. Ergonomics: (a) design of interfaces (b) health ergonomics
10. Maintenance error

Different industrial sectors such as oil & gas, energy, marine, aviation, rail and process based industries are really eager to implement new technology in their physical working environment (WE) with the intention of growing efficiency, safety, reducing fatigue and increasing well-being among their workers. Unfortunately, recent research has shown that the physical WE and the automation process do not always provide effective solutions to above-mentioned factors. This article address the questions of matching the psycho-social WE with a physical WE and the impact of this among the operators in Control Rooms (CR) and the role of Human Factors Engineering in reducing financial losses by the use of ergonomic solutions.

7 Human centered design

Designing for humans is one of the most challenging fundamentals during the design process of Control Room or Control Centre working environment. Reducing the Human error by matching the physical and psycho-social WE is one of the most important factors that CGM challenges every day when searching for solutions. Health and Safety Executive in United Kingdom (HSE) has already identified the problem and here is a short definition that HSE has made:

"Physical match includes the design of the whole work- place and working environment. Mental match involves the individual's information and decision-making requirements, as well as their perception of the tasks and risks. Mismatches between job requirements and people's capabilities provide the potential for human error." [3]

Experience has shown that human centered design is the solution for creating the state of the art Control Centre or Control Room working environment. There are plenty of guidelines and standards that tackle the design process of Control Centre or Control Room WE. The offshore industry has established ISO 11064 as the main standard worldwide. CGM has been implementing the five phases of the design process recommended by ISO11064, in order to improve the overall ergonomic standards in CR WE. Those five phases include: Collection of data, Analysis, Conceptual Design, Detail Design and Feedback (iteration process).

Cleary every project is unique in itself, no matter if it's a green or brown field project (innovation or renovation projects), however in pursuing Excellency and the state-of the art we had identified Human-Centered Design best practices that involve the operator in early stages of the design process.

8 Developing the CCR/CR Working environment

From the beginning of our research on Solutions for Control Centre and Control Room, we quickly realized that the industrial sector has been mainly focused in the physical component and automation process of WE rather than in the human as a part of both physical and psycho-social elements in the working environment. The human was ignored regardless of the fact that it is the human that makes most critical decisions when abnormal situations occur in Control Room active failures (as a consequence of operator that is directly involved in decision making) as well as the latent failures (poor design of plant, poor equipment, ineffective communications etc.). Given that many operators usually work in twelve hour shifts therefore improving the health and wellbeing often enhances efficiency, productivity but what is more crucial it increases safety and reduce fatigue.

Health and Safety Executive in United Kingdom also raised this concern in HSG48 where they stated that:

"Proper consideration of 'human factors' is a key ingredient of effective health and safety management. According to Health and Safety Executive (United Kingdom) "It is estimated that up to 80% of accidents may be attributed, at least in part, to the actions or omissions of people. Health and safety failures are still estimated to cost £18 billion annually (only in United Kingdom.)" [4]

The question always remains what happened and why humans (operators) were ignored in the design process during the industrial boom throughout the world? Research has identified two major factors in order to answer this question.

9 Factors pushing the industry toward meaningful working environments

The aging workforce in Sweden or other countries in the Northern Hemisphere is a key factor. According to employment office (Arbetsförmedlingen) in Sweden 40% of industrial process based working force will retire in the next 16 years (www. ams.se 2014). Similar problems are evident in other countries as well. As a result we are losing knowledge and experience that used to be a part of the older generation. Fulfilling that gap and replace this deficit remains a challenge that we must carefully address in order to at- tract the young generation into the industrial world (often called generation Y).

Human Centered Design, understanding the cognitive task performance among the generation Young (Y) is one of the key factors in order to attract them into the industrial world.

"The Generation Y is, perhaps, the most digitally sophisticated generation we have ever seen. They are looking for a sustainable environment offering a social structure within both a physical and virtual environment." [5]

This research has pointed out the significant factors that young generation values the most when it comes to working environment (regardless language or nationality). Among many factors that the research pointed out, life Long Experience and collaboration are very high valued among the younger generation.

"Life Long Learning Experience: The reasons for choosing a company are:
1: Opportunities for Learning
2: Quality of Life
3: Work Colleagues Collaborative: 41% of the Generation Y prefers to have access to a team space and 32% prefers breakout spaces rather than a conventional meeting room." [5]

The Control Room working environment is often perceived by operators as 99% of boredom and 1% of panic. How can we attract the young generation if not by heavily improving Control Room or Control Centre Working Environment? Designs that fulfill the individual requirements of operators is the only way forward to attract young generations into the industrial sector and to welcome them in to the CR Working Environment.

Another important factor is the emergence of Remote Control, due to technological development and high cost of offshore platforms and protected areas in onshore industrial sites etc. the industry is experiencing a trend of industrial migration from Control Room to Control Centres. Centralizing

Control Rooms into Control Centres has its benefits such as controlling different processes in one place with less operators, low living cost for person- al on shore (comparing to offshore) etc. Meanwhile this migration is not completely risk free. Cyber security, far away from the process itself are only some of the risks that are self-evident as a result of the migration.

More alarming risk factors are the psycho-social effects of that migration. Effects such as higher workload for operators, increased stress level, depression, anxiety, burnout, self- esteem, in a combination with some physical parts of WE such as bad ergonomics, bad lightning, bad sound integration results in health complications like headaches, dizziness, muscle pain, and digestive problems), cardiovascular disease, and muscle-skeletal disorders. By aligning the psycho-social and physical WE automatically increase health and wellbeing in CR WE. "Organizations should include the development of stress management policies to identify and eradicate work practices that cause most job dissatisfaction as part of any exercise aimed at improving employee health. Occupational health clinicians should consider counselling employees diagnosed as having psychological problems to critically evaluate their work and help them to explore ways of gaining greater satisfaction from this important aspect of their life." Implementing the future by understanding the past.

> *"A healthy, happy workforce is a vital component of a productive, successful business in the long-term."* [6]

The big question remains in how to identify the needs of the operator instead of implementing technological solution (driven by technology without taking the human into consideration) that often results in more work-flow for the operator, increasing stress, irritation, reducing joy and satisfaction in the working environment and so forth. From our research, we learned that human differs very much in cognitive processes and abilities to solve a problem. For instance some of the operators can be good in multitasking, others in understanding the complexity of work load, some of them analyzing data, others an effective leadership etc. There is one main value that all share in common and that is health. Awareness among the operators to improve their health is one of the main driven factors that pushes us forward to introduce the Operators Activity Cloud (OAC).
OAC will become the operator's best butler by providing feedback to the operator based on the operator's ability to improve their own health and wellbeing in the working environment.

10 A holistic approach to CCR solutions

Improving only the physical part of working environment or only the psycho-social part of working environment does not result in a holistic approach. On the other side matching together the physical and psycho-social elements of the working environment is the way forward in order to achieve good interrelationship in the CR or CCR working environment.
The importance of creating a state of the art platforms for CR working environments, triggered the development of working collaborations with different Swedish Universities such as Chalmers, Lund University, KTH, Skövde university etc.

Research conducted in cooperation with Chalmers performed research comparing the traditional CR with and high end CR working environment (see figure 2.0)

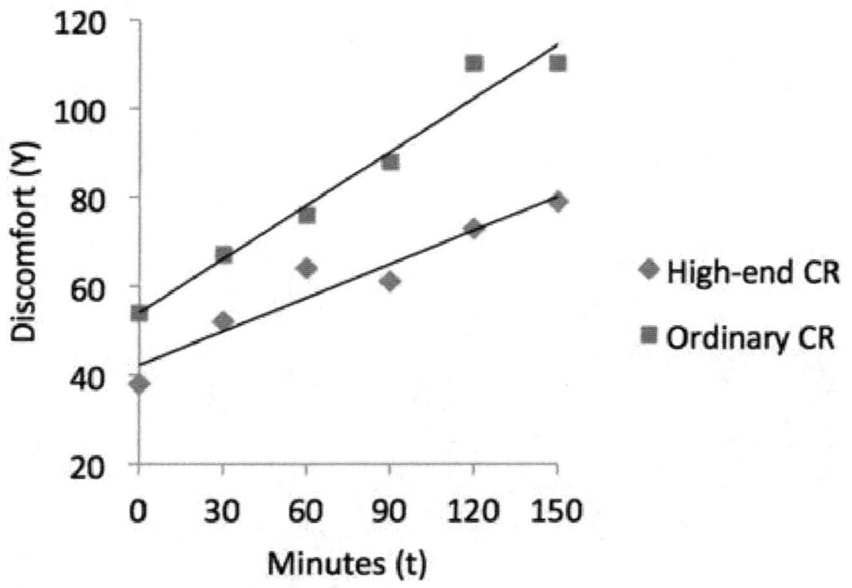

Figure 2.0 Regression curves for perceived discomfort in the ordinary control room (red dots, y=54+0,4t) and the high-end control room (blue dots, y=42+0,2t), where y=discomfort and t=time, t goes from 0 to 150 minutes.

"To summarise, the perceived discomfort increased by time in both control rooms, but the increase was lower in the high-end control room." [7]

Lighting is another field were we has been cooperating with Lund University and other partners in order to provide a human centric lighting platform for operators in CR working environment.

"Light strongly influences employees' performance, making it possible to cope effortlessly with visual tasks and is therefore the first consideration when it comes to lighting design. An adequate lighting level and good glare control are required in order to be able to concentrate on the work at hand and remain motivated. In addition, high illuminance has a positive impact on people's commitment, on error rates and the risk of accidents." [8]

Lighting has a direct impact on the health and well-being since our circadian rhythm is directly connected to the amount of light that humans are exposed to.

"The world rotates on its axis, and as a result, all creatures on Earth are exposed to 24-h cycles of light and dark. Living organisms on this planet have adapted to the daily rotation of Earth by developing biological rhythms that repeat approximately every 24 h." [9]

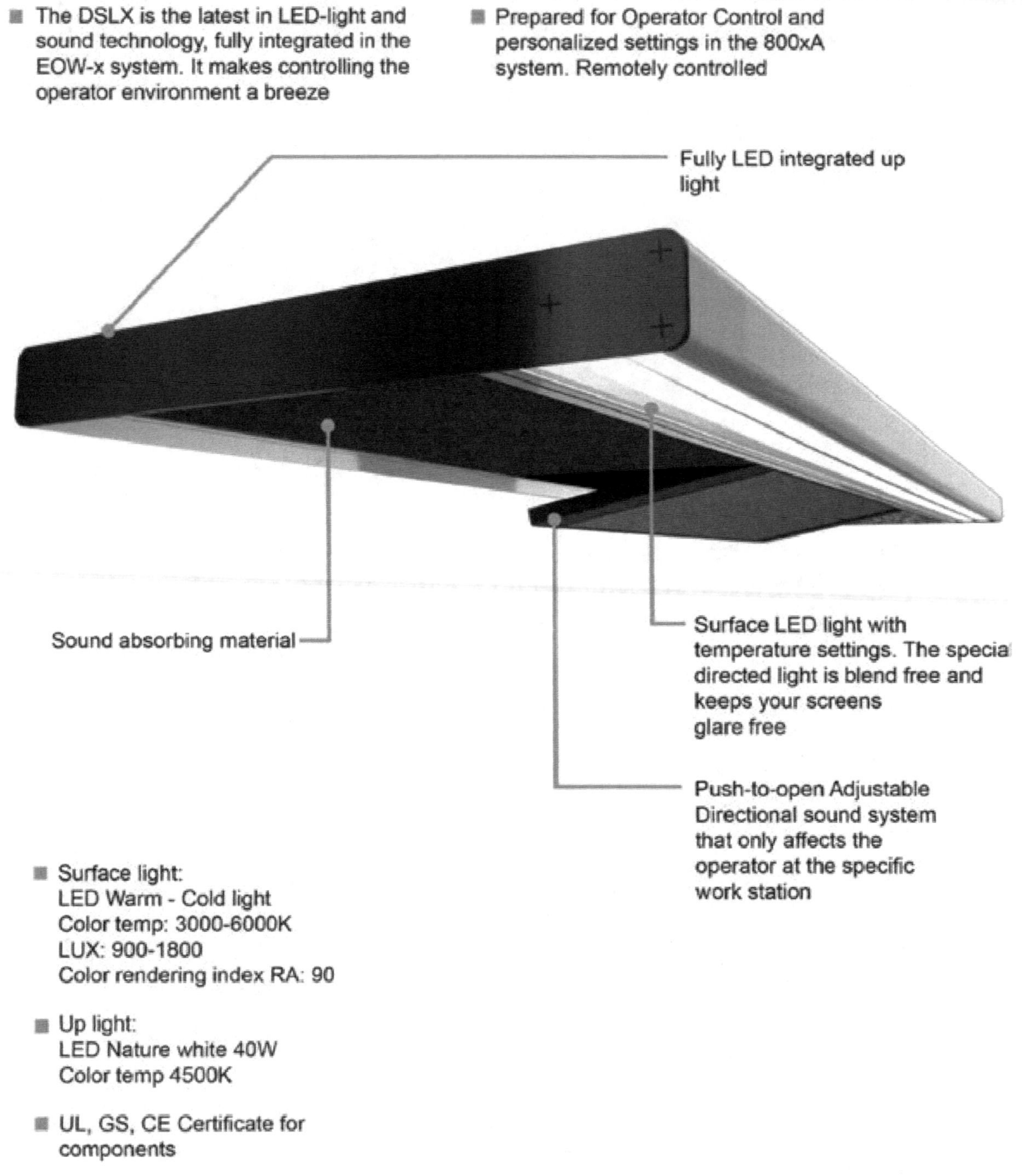

Figure 3.0 CGM adjustable task area lighting up to 1800 lux.

One application of the research was the development of task area lighting for personal adjustments where the operator can freely adjust their own task area lighting by using cold or warm light. The range level of lux on this light is between 900 to 1800 lux and exceeds the minimum 500 lux recommendation of ISO 11064.

Directed sound or sound shower as we called is another function that was implemented as a part of personal adjustment to improve operators working environment. The benefit of the sound shower is enormous for the operator where the operator includes all telecommunication, alarms and other communication throughout sound shower without disturbing the colleague beside.

Integrated communication

Communications inside the Control Room can be considered in a number of categories such as verbal communication inside the Control Room and external communication through electronic system e.g. VHF/UHF radio, telephone, mobile, intercom, public address system etc. the integration of all communication in one device regardless communication type is another solution based on the application of the research- Tacticall (see figure 4.0).

Figure 4.0 Tacticall

11 Operator Activity Cloud (OAC)

Latest research is providing observations and correlations between sitting posture and obesity, type II diabetes, cancer and even early death. [10]

Since 1994 CGM implemented alternate standing and sitting positions into each platform that we provide to CR working environment, now it's time to move forward with Operator Activity Cloud (OAC) unique platform that pro- vides a feedback to the operator in order to improve the operators health and well-being.

Experience have taught us that each individual is unique and each has different requirements. In order to improve health, safety, efficiency, alertness as a significant find of Human Factors we are moving forward by implementing Operators Activity Cloud platform as a new and intelligent solution for future Control Room Working Environment.

In the upcoming platform OAC will measure bio-data such as pulse, stress, alternated posture, walking, running, recovery etc.

Additional data that will be measured is working environment data such as, temperature, light, sound, air quality etc. inside the CR WE (see figure 5.0). All this data will be synchronized in a cloud environment providing feedback to the operator in order to enhance health and wellbeing in CR WE. CGM together with Chalmers University is already participating in a joint project called DIGIN (digitalized wellbeing) to measure real time bio-da- ta as well as working environment data.

Why health and safety?

“We estimate that the total cost to society of health and safety failures could be as high as £18 billion every year.” [11] [12]

The technological developments are making it possible to measure the cost of health and safety in completely new and accurate ways. The importance of good health and safety is evident to anyone who has seen the consequences of health and safety failure, hence the revitalizations of health and safety remains a challenge and must be improved in the industrial sector. The objective of OAC is to optimize human wellbeing and overall system performance in the CR WE.

12 Conclusion

Saying an accident is due to operator error does not lead to constructive action. In manual systems, which require permanent human intervention, the source of error could be the operators and the design should consider that. However, in a manual system, the operator/user receives immediate feedback from failures or deviations, whether caused by technical problems or misoperation, which enables the user to build up experience, to understand system, and to handle all but the most serious eventualities.

As Industries seek increased efficiency and productivity, most facilities will have some degree of automation. In contrast to the case of a manual system, automated system failures are less frequent. However, systems design engineers can induce human error if they have not identified all the operational eventualities and provided a suitable control process or system for those circumstances. These are generally referred to as latent failures that can remain in a system in benign fashion and not be identified until the specific operational eventuality occurs, leading to accident or incident. In this situation, the operator is largely unprepared and unable to cope due to the complexity of the system.

As industries continue to invest in new facilities, or modernize existing facilities (i.e. removing control centers from the blast-zone of the processes) some of the investments could benefit from the adoption of many human-centered design best practices. However it’s an important number of production sites are approaching a stage in which the technologies are close to obsolescence and considering these human elements in the modernization could lead to additional benefits and a safer and more productive environment.

13 References (examples given to show format).

[1] HSE Out of control: Why control systems go wrong and how to prevent failure “ ISBN 978 0 7176 2192 7

[2] HSE Top Ten Human Factors - Definition, Consequences, Resources and Practical Tools

[3] Health and Safety Executive (Reducing error and influencing behaviour ISBN 978 0 7176 2452 2

[4] Johnson Controls, Haworth and DEA (Generation Y and the Workplace Annual Report 2010)

[5] CIPD (Chartered Institute of Personnel and Development, http://www.cipd.co.uk).

[6] Health, Wellbeing & Productivity in Offices, The next chapter for green building (World Green Building Council, sponsored by JLL, Lend Lease and SKANSKA)

[7] The relationship between job satisfaction and health: a meta-analysis (E B Faragher, M Cass, C L Cooper) Occup Environ Med 2005;62:105–112. doi: 10.1136/oem.2002.006734 (Downloaded from http:// oem.bmj.com/ on May 21, 2015 - Published by group.bmj.com)

[8] Disruptive technologies: Advances that will transform life, business, and the global economy (McKinsey Global Institute, May 2013).

[9] Impact of Physical Ergonomics Design of Process Control Rooms on Operator Experience of Comfort, Stress and Emotions (Anna-Lisa Osvalder, Jonas Andersson, Lars-Ola Bligård, Håkan Alm, Chalmers University).

[10] Lightning for Industry and Engineering (http://www.zumtobel.us/PDB/resources/teaser/EN/AWB_Logistics.pdf accessed 1 June 2015).

[11] The impact of watching television on evening melatonin levels (Mariana G. Figueiro Brittany Wood Barbara Plitnick Mark S. Rea).

[12] Role of Low Energy Expenditure and Sitting in Obesity, Metabolic Syndrome, Type 2 Diabetes, and Cardiovascular Disease (Marc T. Hamilton, Deborah G. Hamilton, and Theodore W. Zderic).

Modeling of LNG Spreading and Vaporization Using Smoothed Particle Hydrodynamics Method

Anh Viet Bui
AnhBui@flowsim3d.com

Morgan Reed, Tony Liu, Eric Peterson
MMI Engineering Inc.
11490 Westheimer Road, Suite 150, Houston, Texas 77077
MReed@MMIEngineering.com
TLiu@MMIEngineering.com
EPeterson@MMIEngineering.com

Prepared for Presentation at
American Institute of Chemical Engineers
2016Spring Meeting
12thGlobal Congress on Process Safety
Houston, Texas
April 11-23, 2016

Modeling of LNG Spreading and Vaporization Using Smoothed Particle Hydrodynamics Method

Anh Viet Bui
AnhBui@flowsim3d.com

Morgan Reed, Tony Liu, Eric Peterson
MMI Engineering Inc.
11490 Westheimer Road, Suite 150, Houston, Texas 77077

Keywords: LNG, vaporization, CFD, Smoothed Particle Hydrodynamic (SPH)

Abstract

Smoothed Particle Hydrodynamics (SPH) has been used for numerical simulation of the release of LNG and the subsequent spreading of the liquid. Although the SPH method has been used for years in the astrophysics industry, its application to the LNG industry is quite novel. The spreading and vaporization of LNG creates the potential for a large vapor cloud with the risk of an explosion or flash fire. The SPH method offers excellent mass conservation, especially when modeling small scale changing fluid interfaces (i.e. thin film or small vapor bubbles). Based on LNG consequence hazards, many regulations (i.e. NFPA 59A) restrict vapor cloud dispersion extent to the facility property perimeter; less expensive (time and money) calculation methods are sought.

Vaporization of LNG during its spreading over water or solid substrate involves either evaporation on the air-LNG interface or nucleate/film boiling on the contact of cold LNG with relatively high-temperature substrate surface. For the latter vaporization mechanism to be correctly modeled, surface temperature and its subsequent change due to contact with LNG needs to be described. Consequently, a conjugate heat transfer model which includes wall heat conduction is necessary.

This paper discusses the continued development of a multiphase flow model with heat transfer and phase change based on the SPH model available in the LAMMPS Molecular Dynamics Simulator software. A mechanistic film boiling model is implemented and incorporated into our SPH flow model. Limited validation of the SPH prediction of the LNG spreading will be conducted and examples showing the enhanced boiling and heat transfer capabilities of the proposed LNG spreading model are outlined and discussed.

1 Introduction

LNG release, spreading, and vaporization and the formation of a flammable vapor cloud present a significant fire and explosion hazard to oil & gas facilities. NFPA 59A Standard for the Production, Storage and Handling of Liquefied Natural Gas, requires that the periphery of a vapor cloud (where fuel gas concentration becomes 50% of LFL) formed as a result of LNG release be within the property limits of LNG facilities.

Gravity-driven spread of LNG is a non-trivial process due to the intensive heat exchange of cryogenic LNG with the surrounding environment (soil, water, wind) which involves a rapid thermal exchange and evaporation of LNG. Various boiling regimes, such as nucleate, film, and/or explosive, may happen depending on the thermal and mechanical interaction of LNG and the substrate. In a scenario of LNG release to water, for instance, mechanical mixing of very cold LNG and relatively "hot" water may occur leading to a chain reaction of explosive increase of LNG-water interphase area and fiery LNG boiling dynamics (i.e. rapid phase transition).

The presence of a large free surface of spreading LNG and possible thermal and mechanical interactions of LNG with complex terrain and structural geometries pose significant difficulty to the modeling of LNG spreading using traditional grid-based CFD approaches. Such CFD approaches usually require a refined grid to resolve the physics happening inside an increasingly thinning layer of LNG when it spreads further away from the release origin, which is computationally expensive. Complex geometry of and abrupt change of material properties across the gas-liquid interfaces may also pose a significant problem to numerical methods used to solve the system of transport equations.

The Smoothed Particle Hydrodynamics (SPH) method has been proposed to model the spreading of LNG and the mechanical and thermal interactions with the surrounding environment and structures [1]. In comparison to the traditional grid-based CFD method, SPH can naturally and efficiently deal with complex (moving) fluid interfaces and terrains. This paper discusses the continued development of the SPH LNG spreading model with heat transfer and phase change [1] which is based on a SPH model available in the LAMMPS Molecular Dynamics Simulator software [2]. The model has been revised to enable the simulation of LNG spreading over complex solid boundaries represented by the STL (STereoLithography) geometries. The model of thermal interaction of LNG with such complex boundaries has been implemented which may include a mechanistic LNG boiling model. Limited testing and validation of the model will be conducted and examples showing the enhanced capabilities of the proposed LNG spreading model will be outlined and discussed.

2 Methodology

2.1 A brief description of the SPH method

Smoothed Particle Hydrodynamics method is based on a Lagrangian description of a fluid domain in which a limited number of particles with their own individual masses, momenta, and energies are employed for the modeling of fluid motion. The governing equations of fluid motion, i.e. Navier-Stokes equations, are solved with use of discrete (particle) approximations of field variables, i.e. density, velocity and energy, as well as their derivatives and integrals.

In the SPH method, the system of Navier-Stokes equations is expressed using particle approximations of the field variables and their derivatives as follows

$$\frac{D\rho_i}{Dt} = -\sum_{j=1}^{N} m_j W_{ij},$$

$$m_i \frac{D\mathbf{v}_i}{Dt} = -\sum_{j=1}^{N} m_i m_j \left(\frac{P_i}{\rho_i^2} + \frac{P_j}{\rho_j^2}\right) \nabla_j W_{ij} + \sum_{j=1}^{N} \frac{m_i m_j}{\rho_i \rho_j} (\mu_i + \mu_j) \mathbf{v}_{ij} \left(\frac{1}{r_{ij}} \frac{\partial W_{ij}}{\partial r_{ij}}\right) + \boldsymbol{F}_i,$$

[Eq.1]

$$m_i \frac{De}{Dt} = -\frac{1}{2} \sum_{j=1}^{N} m_i m_j \left(\frac{P_i}{\rho_i^2} + \frac{P_j}{\rho_j^2}\right) : \mathbf{v}_{ij} \nabla_j W_{ij} - \sum_{j=1}^{N} \frac{m_i m_j}{\rho_i \rho_j} \frac{(k_i + k_j)(T_i - T_j)}{r_{ij}^2} \boldsymbol{r}_{ij} \cdot \nabla_j W_{ij}.$$

where $\boldsymbol{r}_{ij} = \boldsymbol{x}_i - \boldsymbol{x}_j$, $r_{ij} = |\boldsymbol{r}_{ij}|$, $\mathbf{v}_{ij} = \mathbf{v}_i - \mathbf{v}_j$; e, P, μ and k are the particle energy, pressure, dynamic viscosity and thermal conductivity, respectively; and m_i, ρ_I and N denote the particle i mass, density, and the total number of particles located within the influence area. W_{ij} or $W(\boldsymbol{x}_i - \boldsymbol{x}_j, h)$ is a smoothing function which defines the influence or support area around each particle. This function determines the accuracy of the SPH approximations and efficiency of the computation. A *cubic spline smoothing function* is used in this work.

A weakly-compressible approach to simulate incompressible fluids is applied which is based on the following equation of state (EOS):

$$p(\rho) = \frac{\rho_0 c^2}{\gamma} \left[\left(\frac{\rho}{\rho_0}\right)^{\gamma} - 1\right] \quad \text{[Eq.2]}$$

where c is the sound speed, ρ_0 is the reference density, and γ is a factor defining fluid stiffness. The time integration of Eqn.2 is conducted using a Velocity-Verlet scheme [3].

2.2 Treatment of complex wall boundary conditions

Most SPH models in the literature employ the so-called ghost (or dummy) particles technique to represent solid wall boundaries. In this approach, two or more layers of ghost particles are placed around a boundary (Figure 1). These ghost particles interact with the fluid particles confined inside the boundary, but are stationary during simulation. For complex boundary geometry with curved walls, appropriate generation and placement of ghost particles can be a challenging task.

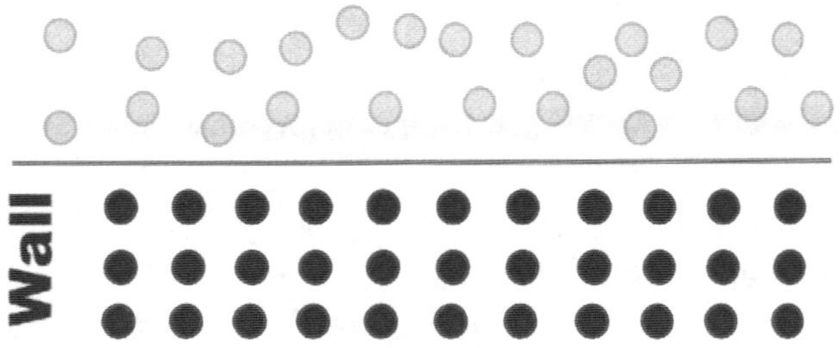

Figure 1: SPH particles – fluid (yellow) and boundary (purple) particles.

In this work, complex wall geometry is treated with use of a repulsive force boundary condition which prevents the penetration of interior fluid particles through the wall [4]. The wall is represented by STL geometries which approximate complex surfaces with triangular meshes. The repulsive force is calculated for every particle which is in the proximity of a boundary triangle. This force acts normal to the wall triangle and is given by

$$F_{rep,i}(d) = \begin{cases} D\left[\left(\frac{r_0}{d}\right)^a - \left(\frac{r_0}{d}\right)^b\right], & d < r_0 \\ 0, & d \geq r_0 \end{cases} \quad \text{[Eq.3]}$$

where d is the distance from a particle to a wall triangle and D, r_0, a, and b ($a > b$) are some regulation parameters.

The viscous forces acting on the directions tangential to the wall surface are accounted for by a modified viscous term which has the following form [4]:

$$\boldsymbol{F}_{vis,i}(d) = \frac{2m_i\mu_i\mathbf{v}_{i,tan}}{\rho_i^2}\left(\frac{1}{d}\frac{\partial W_d}{\partial d}\right) \quad \text{[Eq.4]}$$

with $\mathbf{v}_{i,tan}$ being the projection of the velocity vector of particle $i$on the wall surface plane.

2.3 *Models of surface tension and cohesion forces*

The surface tension force is defined simply by [5]

$$\boldsymbol{F}_{s,i} = -\sigma_s \sum_j m_j \boldsymbol{r}_{ij} W_{ij} \quad \text{[Eq.5]}$$

with σ_s being the surface tension coefficient.

A cohesion force can also be added which attracts nearby small bubbles (vapor particles) to form a large one. This force is given by [5]

$$\boldsymbol{F}_{cohesion,i} = -k_c m_i \sum_j \rho_j \boldsymbol{r}_{ij} \quad \text{[Eq.6]}$$

with k_c being a parameter that controls the strength of the cohesion force.

2.4 *Modeling of wall heat transfer*

Constant temperature or heat flux can be specified on the solid wall boundaries. When a constant wall temperature is set, the heat flux from the wall to the fluid particle $i$is defined by

$$Q_i = h(T_w - T_i) \quad \text{[Eq.7]}$$

where T_w and T_i are the wall and particle temperatures, respectively, and h is the convective heat transfer coefficient.

This wall heat flux can be added to the energy equation using the following term:

$$Q_i = \frac{2m_i^2 h(T_w - T_i)}{\rho_i^2}\left(\frac{\partial W_d}{\partial d}\right). \quad \text{[Eq.8]}$$

Upon a transition from convective heat transfer to nucleate or film boiling heat transfer, an appropriate boiling heat transfer coefficient should be utilized for h.

2.5 *Modeling of LNG-vapor coexistence and evaporation*

The flow of LNG and its vapor can be modeled with use of a *multiphase* or *multi-fluid* SPH model where separate transport equations and equations of state (EOS) are utilized for liquid and vapor LNG [1,6,7]. The mechanical and thermal interactions between phases including phase transition (LNG evaporation) are described by the interphase exchange terms. Different particle types are used to represent LNG and its vapor and the mass source or sink terms define the phase change rates.

Alternatively, the coexisted LNG liquid and vapor phases can be described by a single set of transport equations and a common EOS. The van der Waals equation of state is normally used in such a model, although it may impose some stability constraint on the numerical solutions.

Modeling of the near-wall boiling is complicated due to small scale physics involved. The physics of boiling and the transition from nucleate to film boiling are still not well understood and the split of wall heat flux (to convective and boiling heat transfers) and the wall evaporation rate are still mostly defined based on experimental data.

Phase change, i.e. evaporation or condensation, happens as a result of heat transfer between heating wall and flow or thermal exchange across the fluid-vapor interface. In this work, only evaporation near a heating wall is considered with the wall heat flux (Eq.7) assumed to be decomposed into a fluid heating and an evaporation components [6,7]. The evaporation heat flux Q_{evap} results in an evaporation mass flow rate of $\dot{\boldsymbol{m}}_{\boldsymbol{evap}} = \boldsymbol{Q}_{\boldsymbol{evap}}/h_{evap}$ (h_{evap}– the latent heat of evaporation). A corresponding number of vapor particles needs to be generated and inserted into the computational domain. In the current model, a simple evaporation model based on simple phase/particle type switching is employed.

3 Simulations of Fluid Spreading and Phase Change

Limited testing of the SPH method in modeling the spread of water over dry and wet flat surfaces has been conducted as described in a previous work [1]. In this study, new capabilities of the further developed SPH model are tested.

3.1 Simulation of fluid spreading over STL boundaries

The latest approach tracks the liquid running pool as it transverses a terrain other than a flat surface. Figure 2 shows an example of what can be considered a trench system common at many LNG terminals throughout the United States. The fluid temperature is set at 20 °C, whereas the surface temperature is at 100 °C. While the problem is essentially 2D, the simulation is 3D due to the presence of 2D STL boundary. The initial water column has the dimensions of 0.1m × 0.1m × 0.2m, and is described by approximately 3000 fluid particles.

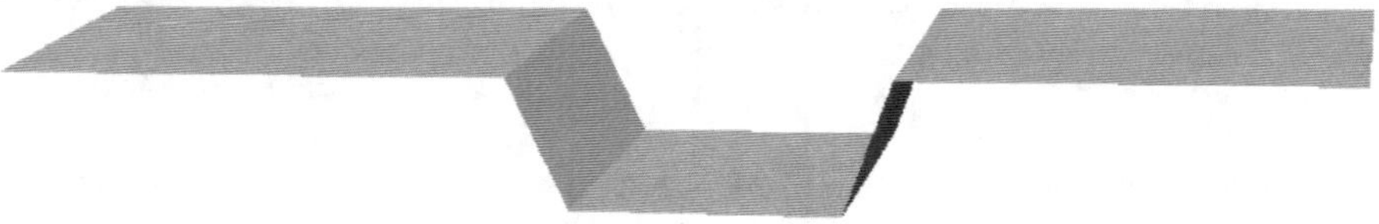

Figure 2: A simple STL surface.

The simulation results are show in Figure 3 which indicates that the SPH model can successfully capture the dynamics of fluid spreading over a relatively complex boundary described by a STL surface meshe. The SPH method is seen to be able to capture the void space developed under the fluid jet in the beginning stage of the spreading (which can hardly be described by a shallow water model relying on a fluid height equation). Heat transfer between the wall and the fluid is also found to be reasonably simulated. It is worth noting that only constant temperature or constant heat flux boundary condition can be specified for the STL walls. If heat conduction inside the wall needs to be considered, the traditional approach of using stationary ghost particles needs to be applied.

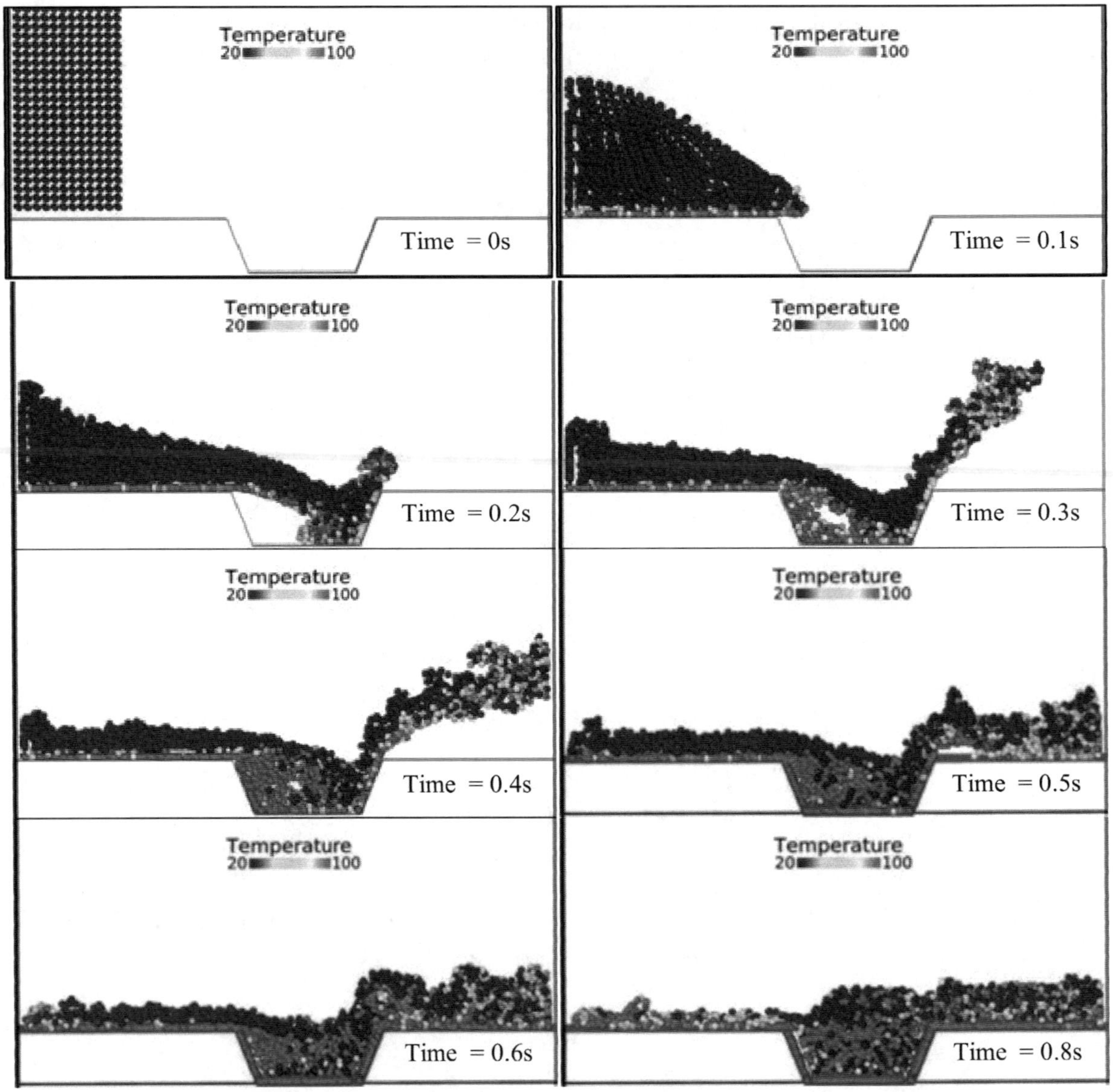

Figure 3: Spreading and heating of fluid over a STL surface.

3.2 *Simulation of fluid evaporation during spreading*

The SPH has been tested for its capability to correctly simulate the evaporation of fluid during spreading and the dynamics of the vapor phase. The fluid spreading is initiated by the collapse of a fluid column, which had the dimensions of 0.5m width by 1m height. The fluid and vapor densities are set at 450 kg/m^3 and 2 kg/m^3, respectively. The temperature difference between the spreading surface and the fluid is set at 900°C to facilitate faster heat transfer and evaporation of fluid.

The simulation results are presented in Figure 4, which shows the creation of an unstable vapor layer along the hot spreading surface. In comparison to previous simulations with similar settings and conditions [1], predictions of the wall evaporation dynamics and the rising of vapor phase across the fluid layer are more accurately predicted in Figure 4. The progressive thinning of the fluid layer is also seen.

To simulate the atmospheric transport and possible combustion of the vapor phase, the open-sourced CFD software OpenFOAM will be used. A coupling of OpenFOAM and this SPH spreading model is currently under development.

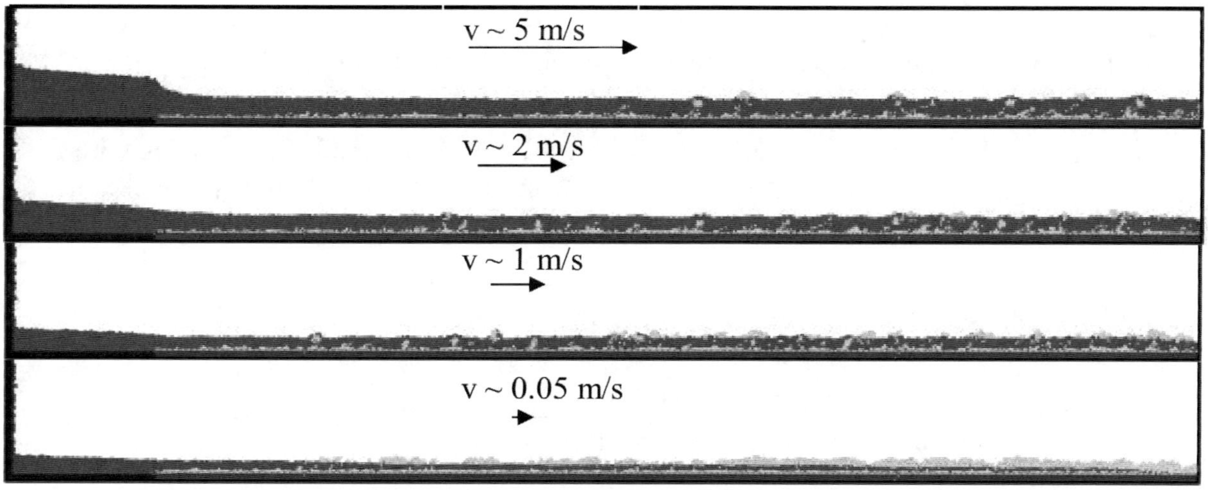

Figure 4: Fluid spreading, evaporation and dryout.

4 Conclusion

A SPH model of LNG spreading and evaporation [1] has been further developed in this work to include the capabilities of

- Modeling the mechanical and thermal interactions of the spreading fluid with terrain and objects of arbitrarily complex 3D geometries represented by STL surface meshes,
- Taking into consideration the effect of interactions between phases, i.e. surface tension and cohesion forces.

The wall evaporation model has also been improved to account for the effect of subcooled boiling.

Limited testing of the model has been conducted, but further model verification and validation are needed before it can be applied to the simulation of realistic industrial scenarios of LNG spreading.

Using CFDEM®coupling simulation engine [9], the proposed LNG spreading model can easily be coupled with the OpenFOAM software to enable 3D CFD modeling of LNG vapor dispersion and fire.

5 References

[1] A.V. Bui, T. Liu, M. Reed and E. Peterson. "CFD Modeling of LNG Spreading and Vaporization," Proceedings of the 11thAIChEGlobal Congress on Process Safety, Austin, Texas, April 2015.

[2] LAMMPS Molecular Dynamics Simulator, Available at http://lammps.sandia.gov, Assessed on 02/15/2016.

[3] G.C. Ganzenmuller, M.O. Steinhauser, and P.V. Liedekerke, The implementation of SPH in LAMMPS – A guide to the SPH-USER package, Germany, 2011.

[4] A. Eitzlmayr, G. Koscher, and J. Khinast, "A novel method for modeling of complex wall geometries in smoothed particle hydrodynamics," Comput. Phys. Comm., 185, pp.2436-2448, 2014.

[5] M. Ihmsen, J. Bader, G. Akinci, and M. Teschner, "Animation of air bubbles with SPH," Proceedings of International Conference on Computer Graphics Theory and Application, 2011.

[6] N. Shirakawa, Y. Yamamoto, H. Horie, and S. Tsunoyama, "Analysis of subcooled boiling with the two-fluid particle interaction method," J NuclSci Tech, 40(3), pp.125-135, 2003.

[7] A.K. Das and P.K. Das, "Modeling of liquid-vapor phase change using smoothed particle hydrodynamics," J Comput Phys, 303, pp.125-145, 2015.

[8] J.L. Woodward and R.M. Pitblado, "LNG Risk Based Safety – Modeling and Consequence Analysis," John Wiley & Sons, Inc., Hoboken, New Jersey, 2010.

[9] CFDEM®project, Available at http://www.cfdem.com, Assessed on 02/15/2016.

LNG Facility Siting Regulations –

Home (USA) and Away (International)

Chaitanya Gottimukkala
RISKbytes Inc
1318 Nance St Houston TX 77002
Chaitanya@riskbytes.com

Prepared for Presentation at
American Institute of Chemical Engineers
2016 Spring Meeting
12th Global Congress on Process Safety
Houston, Texas
April 11-23, 2016

LNG Facility Siting Regulations – Home (USA) and Away (International)

Chaitanya Gottimukkala
RISKbytes Inc
1318 Nance St Houston TX-77002
Chaitanya@riskbytes.com

Keywords: LNG, Facility Siting, Regulations, USA, International, Risk-based, Prescriptive

Abstract

Several LNG facilities are being proposed, designed and built around the world. These facilities will have to be designed, built and operated according to their local and national codes and standards. This paper provides an overview of the various codes, regulations and standards that govern the siting of the onshore LNG Facilities in the U.S.A and around the world. This paper also highlights some of the important siting requirements including risk criteria, vapor dispersion and overpressure criteria, acceptable thermal radiation levels found in these regulations and standards.

The primary siting regulations for U.S. onshore LNG facilities are 49 Code of Federal Regulations (CFR), Part 193 and NFPA 59A, "Standard for the Production, Storage, and Handling of Liquefied Natural Gas (LNG)". The design and siting standard for Europe and some facilities internationally is the European Standard, EN 1473, "Installation and equipment for liquefied natural gas - Design of onshore installations".

1 Introduction

With the demand for cleaner fuels rapidly growing around the world, Liquefied Natural Gas (LNG) is playing a vital role in this market. In order to meet the growing demand, LNG has to be transported from places of availability to places of demand. Therefore, a number of LNG Liquefaction and Regasification facilities are in proposal, design or construction around the world.

These LNG facilities must follow the industry codes, local and national regulations, and standards applicable to their location. The next section of this paper provides an overview of these required codes, regulations and standards for siting of LNG facilities both in the USA (home) and internationally (away). This paper's major focus is on land-based large scale LNG facilities; however, some of the standards for offshore based LNG facilities are also listed.

The major difference between the U.S. and international (for example - Europe & Australia) LNG siting requirements is that the United States (U.S) generally follows a "Prescriptive" approach and majority of the rest of the world follows a "Performance or Risk-based" approach. These approaches are consistent with the general approaches the onshore industries follow in these countries.

Prescriptive approaches in the U.S must follow a defined set of spacing criteria as outlined in NFPA 59A [3] and 49 CFR Part 193 [4]. In the risk-based approach, the proposed facility's spacing needs to demonstrate that it meets the tolerable risk criteria established by the host Nation, company or as outlined in codes such as EN 1473 [5], whichever is most stringent.

The subsequent sections of this paper highlight some of the important requirements with respect to the facility siting process including determination of design spills, risk criteria, allowable vapor dispersion and overpressure criteria, allowable thermal radiation levels.

2 Regulations, Codes, and Standards

The regulations, codes and standards applicable to facility siting in the US and internationally are summarized in the subsections below.

Local regulations, codes, and standards exist for countries such as Australia, United Kingdom, Canada, China, Europe, India, Japan and Mexico. Both the NFPA 59A (Prescriptive) and EN 1473 (Risk-based) have been adopted globally either individually or as a combination of both. Table 1 below summarizes the approaches followed by some of the countries.

Table 1. Summary of Some of the Countries with LNG Facilities and their Approaches

Country	Approach
U.S.A - NFPA 59A 2001 edition	Prescriptive
Canada - CSA Z276	Prescriptive
Japan - JGA-102	Prescriptive
Mexico - NOM-013-SECRE-2004	Combination of Prescriptive and Risk-based
India - OISD-STD-194	Combination of Prescriptive and Risk-based
Europe (U.K, Netherlands, France) - EN 1473	Risk-based
Australia - AS3961	Risk-based
Singapore	Risk-based

For example, the Canadian standard, CSA Z276, adopted NFPA 59A's prescriptive approach and the Australian standard, AS3961, adopted a risk-based approach similar to EN1473.

The primary focus of this paper is to compare the Prescriptive approach in the U.S (home) versus the Risk-based approach followed in Europe and other parts of the world (away). For purposes of this paper, when "U.S" or "Home" is mentioned it refers to the countries following Prescriptive approach and when "International", "EN1473" or "Away" is mentioned it refers to countries following Risk-based approach.

Offshore LNG facility siting process is outside the scope of this paper. However, some of the industry standards applicable to the offshore LNG facilities are listed below:

- DNV-OSS-103, Rules for Classification of Floating LNG/LPG Production, Storage and Loading Units
- ABS Guide for Building and Classing Floating Offshore Liquefied Gas Terminals

2.1 U.S (Home) Regulations, Codes and Standards – Onshore LNG Facility Siting

The following are the facility siting regulations and standards applicable to the U.S land based LNG facilities.

- 49 CFR Part 193, Liquefied Natural Gas Facilities: Federal Safety Standards
- 33 CFR 127 Subpart B, U.S Coast Guard - LNG Facility Inspection
- PHMSA, U.S Department of Transportation Pipeline and Hazardous Materials Administration
- FERC, Federal Energy Regulatory Commission
- NFPA 59A (2001/2009) – Standard for the Production, Storage, and Handling of Liquefied Natural Gas (LNG)
- API RP 752, Management of Hazards Associated with Location of Process Plant Permanent Buildings
- API RP 753, Management of Hazards Associated with Location of Process Plant Portable Buildings
- API RP 500, Recommended Practice for Classification of Locations for Electrical Installations at Petroleum Facilities
- API RP 505, Recommended Practice for Classification of Locations for Electrical Installations at Petroleum Facilities Classified as Class I, Zone 0, Zone 1, and Zone 2
- API STD 2510, Design and Construction of LPG Installations

2.2 International (Away - European Norm) Regulations, Codes and Standards – Onshore LNG Facility Siting

The following are the facility siting regulations and standards applicable to the land based LNG facilities following Risk-based approach, specifically, countries adopting European Standard - EN 1473:

- EN 1473 (2007), Installation and equipment for liquefied natural gas - Design of onshore installations

- EN 1160, Installations and equipment for liquefied natural gas - General characteristics of liquefied natural gas
- GAP 2.5.2, Oil and chemical plant layout and spacing
- GAP 2.5.2 A : Hazard Classification of Process operations for spacing requirements
- IP15, Area Classification code for installation handling flammable fluids Part 15, 2d edition (2002), by the Institute of Petroleum
- Site selection and design for LNG Ports and jetties: SIGTTO

The Host Nations have specific standards and regulations for their LNG facilities in addition to the above standards, which have not been included in this paper.

3 Facility Siting Process

U.S (Home) LNG facilities follow a prescriptive approach to siting as described in US 49CFR193 and NFPA 59A (2001) requiring strict compliance to the prescribed thermal radiation levels, vapor dispersion levels and overpressure levels at the facility's property line that can be built upon. The primary focus of the NFPA 59A and US 49CFR193 regulations is offsite risk to public. Other standards and regulations such as API, Environmental Protection Agency (EPA), Occupational Safety and Health Administration (OSHA) provide guidance for onsite risk criteria.

Some of the International (Away) LNG facilities follow a risk-based approach as described in EN 1473. Three categories of are risk listed in EN 1473 [5]:

- Level 3: Situation which is undesirable and cannot be tolerated. Remedial action required (Not Acceptable);
- Level 2: Situation which shall be improved. A level at which it shall be demonstrated that the risk is made As Low As Reasonably Practical (ALARP);
- Level 1: Normal situation (Acceptable).

These facilities should meet the risk acceptance criteria i.e risks in the "Acceptable" and "As Low As Reasonably Practicable (ALARP)" region for both public and on-site personnel either set by the Host Nation, Company or the EN 1473 standard, whichever is more stringent. If the quantified risk estimate lies in the Acceptable and ALARP regions, then the risk is "tolerable". The acceptability criteria are more stringent for hazards to the public.

The first step for facility siting in both the Home and Away approaches is hazard analysis to determine the maximum design spillage. The criteria for calculation of design spills are different in both approaches. Figure 1 provides a flow chart for facility siting process for both the Prescriptive (US) and Risk-based (International) approaches.

Home (US) Regulations and codes require the determination of design spill, referred to as Single Accidental Leakage Source (SALS) in NFPA 59A based on a maximum 10 minute release. Recent Nominal Failure Rate data published by FERC [2] and PHMSA [1] should be utilized to determine the failure rate criteria of equipment, piping and other appurtenances. Approved modeling tools should be used for evaluating the prescribed criteria for vapor dispersion, thermal radiation and overpressure at the property line that can be built upon. In case the prescribed siting criteria is not satisfied, passive mitigation measures can be specified. Some of the passive mitigation measures are rerouting of piping and relocation of hazardous equipment / tanks; modifying design conditions such as changing flow rates, pipe sizes, and capacities; constructing vapor barriers; or buying more land to extend the property line.

Away (EN 1473) requires the determination of the design spill either following a deterministic approach and/or probabilistic approach as described in Section 4.4.2 of the EN 1473 standard and Figure 1 below. Deterministic approach requires identification of hazards, establishment of credible hazards, determination of consequences, and mitigation measures to reduce consequences if risk is not tolerable. Probabilistic approach requires identification of hazards, determination of consequences, application of failure rate data, determination of frequency of hazards, and quantifying risk level from these data. If risk level is not tolerable, it can be reduced by active or passive mitigation measures including water deluge, fire and gas detection, vapor barriers and thermal shelters. The facility layout can be optimized and finalized following careful evaluation of various "tolerable" options.

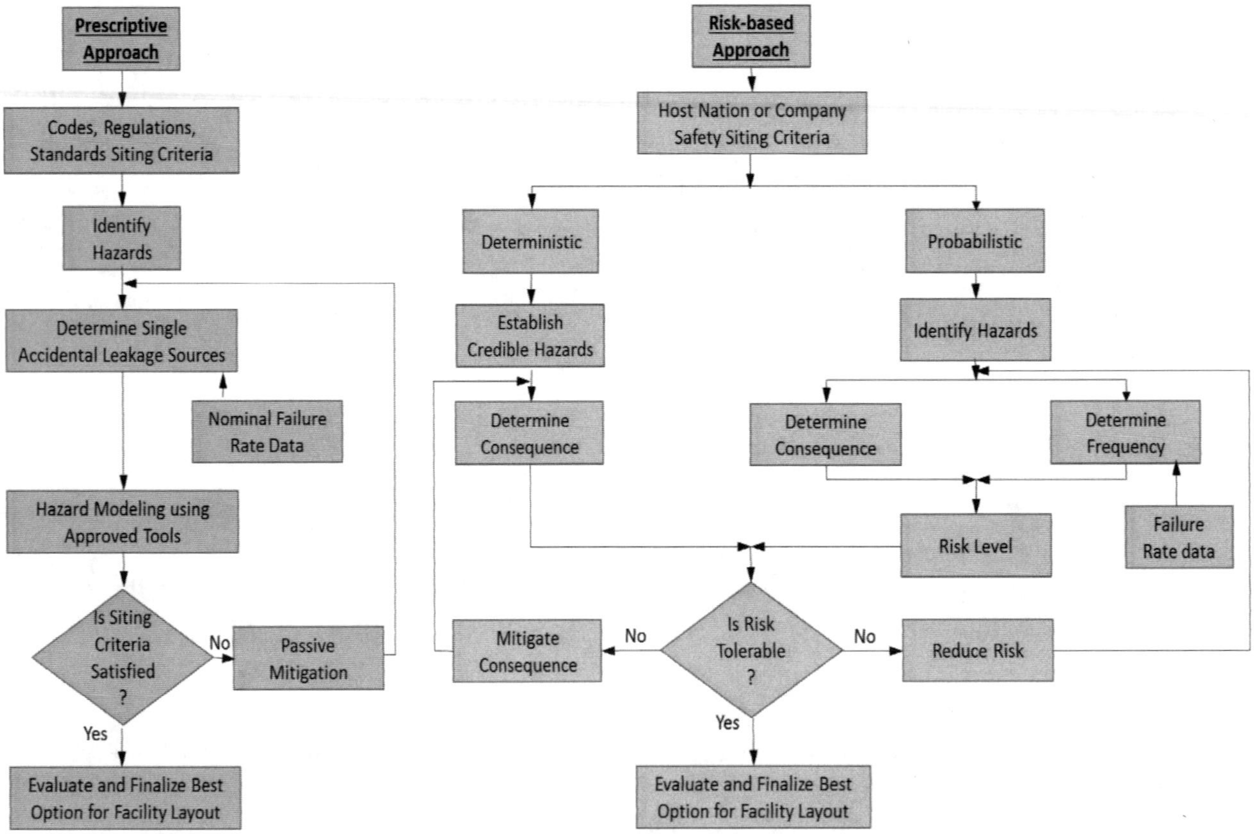

Figure 1. Facility Siting Process – Prescriptive (U.S) vs. Risk-based (International – EN1473) Approaches

The following subsections describe and compare the ambient conditions, required thermal radiation levels, vapor dispersion and overpressure criteria for both the Home (U.S) and Away (European) facilities.

3.1 Ambient Conditions

Table 2 describes the ambient conditions used for both the Prescriptive (U.S) and Risk-based (International) LNG facilities.

Table 2. Ambient Conditions for Home (49CFR193 / NFPA 59A, 2001) and Away (EN 1473, 2007)

Parameter	Home (49CFR193 / NFPA 59A, 2001)		Away (EN 1473, 2007)	
	Thermal Radiation	Vapor Dispersion	Thermal Radiation	Vapor Dispersion
Wind Speed	Wind speed producing the maximum exclusion distances shall be used. [1]	Wind speed producing longer predicted downwind dispersion distances than other weather conditions at the site at least 90 percent of the time. Alternatively, 4.5 miles per hour (2.01 meters/sec)	Combination of wind speed and atmospheric stability that can occur simultaneously and result in the longest predictable downwind dispersion distance that is exceeded less than 10 % of the time. Alternatively, 10 meters/sec	Combination of wind speed and atmospheric stability that can occur simultaneously and result in the longest predictable downwind dispersion distance that is exceeded less than 10 % of the time. Alternatively, 2 meters/sec
Temperature	Ambient temperature that produces the maximum exclusion distances shall be used. [2]	Average ambient temperature in the region	Ambient temperature in the region	Average ambient temperature in the region
Relative Humidity	The ambient relative humidity that produces the maximum exclusion distances shall be used. [2]	50%	50%	50%
Pasquill Stability Class	N/A	F	N/A	Combination of wind speed and atmospheric

Parameter	Home (49CFR193 / NFPA 59A, 2001)		Away (EN 1473, 2007)	
	Thermal Radiation	Vapor Dispersion	Thermal Radiation	Vapor Dispersion
				stability that can occur simultaneously and result in the longest predictable downwind dispersion distance that is exceeded less than 10 % of the time. Alternatively, Stability Class "F" can be used.
Surface roughness	N/A	0.03 m, higher values may be used based on the nature of terrain	N/A	Not specified. Industry standard values can be used.
Average gas concentration in air	N/A	½ of Lower Flammability Level (LFL) or 2.5 %	N/A	Not specified. Industry standard values can be used.

1. Per 49 CFR 193.2057, 'The wind speed producing the maximum exclusion distances shall be used except for wind speeds that occur less than 5 percent of the time based on recorded data for the area."
2. Per 49 CFR 193.2059, "in calculating exclusion distances, the ambient temperature and relative humidity that produce the maximum exclusion distances shall be used except for values that occur less than five percent of the time based on recorded data for the area."

3.2 Allowable Thermal Radiation Levels

The allowable thermal radiation levels are described in both the standards, U.S and International.

3.2.1 U.S (Prescriptive) Thermal Radiation Criteria

U.S 49CFR193 and NFPA 59A specify the thermal radiation flux levels and the exclusion zones for impounding areas containing LNG spills as listed in Table 3. U.S standards do not account for mitigation measures such as water deluge, shelters in reducing the heat flux level to the required limit. This often results in conservative estimates for thermal radiation hazards leading to larger exclusion zones. LNG spills from loading/unloading arms and LNG carriers should also be included when calculating the incident thermal radiation flux. Details on thermal radiation and impounding area design and siting requirements can be found in Section 2.2 of NFPA 59A and 49CFR193.2057 Subpart B. Thermal Radiation calculations should be calculated using the approved computer model LNGFIRE3.

Table 3. Allowable Thermal Radiation Flux Levels per Section 2.2.3 NFPA 59A

Exposure at Exclusion Zone	Thermal Radiation Flux	
	BTU/hr/ft^2	KW/m^2
At a property line that can be built upon upon ignition of a design spill	1600	5
At the nearest point located outside the owner's property line that, at the time of plant siting, used for outdoor assembly by groups of 50 or more persons for a fire over an impounding area containing a volume, V, of LNG	1600	5
At the nearest point of the building or structure outside the owner's property line that is in existence at the time of plant siting and used for occupancies	3000	9
At a property line that can be built upon for a fire over an impounding area containing a volume, V, of LNG	10,000	30

3.2.2 International (Risk-based: EN 1473) Thermal Radiation Criteria

EN 1473 specifies the thermal radiation flux levels excluding solar radiation for equipment inside the boundary and for areas outside the plant boundary as listed in Table 4. There is a provision for reducing the heat flux to the required limit by means of spacing, water sprays, fireproofing and radiation barriers. The main advantage with this approach over prescriptive standards is that overly conservative thermal exclusion zones will not be required after application of mitigation measures to reduce the risk to be in the "Tolerable" risk region.

LNG spills from loading/unloading trucks or LNG carriers should also be included when calculating the incident thermal radiation flux. It also specifies the allowable thermal radiation flux from flares or ignited vent stacks on equipment inside the plant boundary and on areas outside the plant boundary. This standard does not specify a required computer model to estimate thermal radiation and specifies that the radiation flux from an LNG fire shall be calculated using appropriate and validated models.

Table 4. Allowable Thermal Radiation Flux Levels per Table A.1 & A.2 of EN 1473

Exposure at Equipment Inside Boundary or Area Outside Boundary	Maximum Thermal Radiation Flux KW/m^2
Equipment Inside Boundary	
Concrete outer surface of adjacent storage tanks [1]	32
Metal outer surface of adjacent storage tanks	15
The outer surfaces of adjacent pressure storage vessels and process facilities	15
Control rooms, Maintenance workshops, laboratories, warehouses	8
Administrative buildings	5

Exposure at Equipment Inside Boundary or Area Outside Boundary	**Maximum Thermal Radiation Flux** KW/m^2
Area Outside Boundary	
Remote area [2]	8
Critical area [3]	1.5
Other areas [4]	5

1. For pre-stressed concrete tanks, maximum radiation fluxes may be determined by the requirements given in A.1.1 of EN 1473.
2. An area only infrequently occupied by small numbers of persons, e.g. moor land, farmland and desert.
3. This is either an unshielded area of critical importance where people without protective clothing can be required at all times including during emergencies or urban area (defined as an area with more than 20 persons per square kilometre) or a place difficult or dangerous to evacuate at short notice (e.g. hospital, retirement house, sports stadium, school, outdoor theatre).
4. Other areas typically include industrial areas not under control of the operator/occupier of the LNG facilities.

3.3 Vapor Dispersion Criteria

The vapor dispersion criteria have been specified in both the standards, U.S and International.

3.3.1 U.S (Prescriptive) Vapor Dispersion Criteria

In the U.S, the regulation 49CFR193, and code NFPA59A, prescribe the criteria. Clarifications and additional details for these criteria can be found on the PHMSA website [1]. These regulations require assessment of vapor dispersion hazards resulting from design spills of LNG and other flammable substances. Flammable vapor dispersion distances should be calculated at ½ Lower Flammability Limit (LFL) concentrations for both LNG and other flammable fluids.

FERC and PHMSA provide guidance on determining the single accidental leakage source to assess the flammable vapor dispersion hazards. Process and ambient conditions, among other information is used to determine the design spills. The potential release source is determined using nominal failure rate criteria recently published by FERC and PHMSA. However, these failure rates do not account for the consequence part of risk in terms of injury or fatality onsite and/or offsite.

The approved computer models for vapor dispersion exclusion zones calculations are DEGADIS 2.3, FEM3A, FLACS 9.1 Release 2 (Oct. 7, 2011) and PHAST-UDM Version 6.6 and 6. 7 (Oct. 7, 2011).

In addition to the flammable vapor hazard analysis, U.S regulations also require the toxic gas (for example - benzene) hazard analysis, as applicable.

3.3.2 International (Risk-based: EN 1473) Vapor Dispersion Criteria

EN1473 specifies a risk-based approach for vapor hazard analysis. The determination of consequence, probability and associated risk level of a hazardous event is specified in the standard. If the local, country or company criteria are more stringent, then that criteria should be utilized. The criteria for estimation of consequences of LNG and other hazardous fluids is either detailed in EN 1473 or a reference is cited. More information is available in Section 4.4 and Annexures J, K and L of EN 1473.

Risk to personnel and public should be calculated and it should be within the "Tolerable" range of the risk criteria.

3.4 Overpressure Criteria

3.4.1 U.S (Prescriptive) Overpressure Criteria

There is no prescriptive requirement for overpressure and explosion in U.S 49CFR193 but Chapter 2.1.1(d) of NFPA59A requires that all hazards having an impact on public or plant personnel should be evaluated. FERC and PHMSA require the use of the 1.0 psi to determine the offsite impact, and are consistent with the EPA's "Risk Management Program for Chemical Accidental Release Prevention" (RMP, 40 CFR 68).

FLACS 9.1 Release 2 (Oct. 7, 2011) and PHAST-UDM Version 6.6 and 6. 7 (Oct. 7, 2011) are accepted models for use in determination of the overpressure hazards.

3.4.2 International (Risk-based: EN 1473) Overpressure Criteria

EN 1473 describes the methodology for overpressure analysis in Section 4.4.2.5 d. It also specifies a risk-based approach for overpressure analysis. The determination of consequence, probability and associated risk level of a hazardous event is specified in the standard. If the local, country or company criteria are more stringent, then that criteria should be utilized.

EN 1473 refers to EN 1160 for flammability range of mixtures of gas and air. The standard indicates that some of the recognized methods and models such as the multi energy method and/or deflagration at constant speed method can be used to calculate the over-pressure.

Risk to personnel and public should be calculated and it should be within the "Tolerable" range of the risk criteria.

3.5 Other Siting Criteria

3.5.1 U.S (Prescriptive) – Other Siting Criteria

U.S Federal regulations prescribe additional siting criteria for impoundments, equipment, buildings and structures. Spacing criteria within the facility as well as with respect to the property line is described in detail. Details of these prescriptive requirements are in U.S 49 CFR 193 and NFPA 59A.

Consequences from Boiling Liquid Expanding Vapor Explosions (BLEVE) should be considered during siting and design.

Permanent and portable buildings siting shall be in accordance with API 752 and 753.

Escape routes, SIMOPS and construction activities should also be considered during siting.

3.5.2 International (Risk-based: EN 1473) – Other Siting Criteria

EN 1473 specifies a risk-based approach for facility siting based on hazard assessment. Separation distances should include radiation flux levels, LFL contours, noise, blast effects. It also describes the siting of buildings, equipment and tanks. It refers to industry standards such as GAP 2.5.2 and GAP 2.5.2.a for additional guidance on facility siting.

Escape routes, SIMOPS and construction activities should also be considered during facility siting.

4 Prescriptive (US) vs Risk-based (International) Approaches – Pros and Cons

Based on the defined criteria, the siting of facilities using Prescriptive and Risk-based approaches have their share of pros and cons. Table 4 below describes a few pros and cons for both these approaches.

Prescriptive approach warrants the LNG facilities to follow the facility siting requirements without specific consideration for their location with respect to population density in that area, while the Risk-based approach takes into consideration the overall risk to public outside the fence based on the quantitative risk estimates included during modeling. The prescriptive approach does not take into account active mitigation measures for risk reduction while in the risk-based approach risk levels may be reduced using appropriate mitigation measures.

For example, U.S 49 CFR 193 and NFPA 59A require that the LNG facilities should adhere to the thermal radiation flux of 1600 BTU/hr-ft^2 and vapor dispersion exclusion criteria of ½ LFL at the property line that can be built upon. These criteria have to be followed without consideration of the population density in that area which may result in the LNG facility needing to own more acreage than required to satisfy the facility siting requirements. International standard, EN1473, the facility siting requirements are satisfied as long as the thermal radiation, vapor dispersion and other hazards associated with the LNG facility comply with the Company or Country required "Tolerable" risk levels.

Table 5. Prescriptive (Home) vs Risk-based (Away) Approaches – Pros & Cons

Prescriptive (Home - US)	Risk-based (Away - International)
Pros	
Application of Defined criteria is easy to follow	Flexibility in facility siting and layout
Defined criteria provide consistency across LNG industry based on industry knowledge and best practices.	Takes Location's population density into account
Less time consuming during design phases	Active and passive mitigation measure accepted to demonstrate tolerability
Cons	
Strict compliance irrespective of Location's population density	Definition of Tolerability Criteria
Limited mitigation measures (passive)	Requires complex calculations and assumptions
Special cases require longer processing time	Longer Project duration for finalizing Facility layout to demonstrate tolerable risk.

5 Conclusion

The growing demand for LNG requires proposal, design and construction of LNG facilities around the world, U.S (home) and International (away). This paper highlighted some of the regulations, codes and standards distinguishing between U.S (Prescriptive approach) and International (Risk-based approach: EN 1473) requirements.

U.S facilities follow a "Prescriptive" approach (U.S 49CFR Part 193 and NFPA 59A, 2001) and some of the International facilities follow a "Risk-based" approach (EN 1473). This paper highlighted the various requirements between these two approaches with respect to facility siting.

Criteria for ambient conditions, allowable thermal radiation levels, vapor dispersion, overpressure and additional spacing requirements for both approaches has been described.

The pros and cons of the Prescriptive and Risk-based approaches have been discussed.

In the U.S, the newer editions of NFPA 59A such as the 2013 edition include QRA (Quantitative Risk Assessment) methodology for facility siting. However, US 49 CFR 193 and FERC still refer to the 2001 edition and have not adopted the latest edition. Future adoption of the risk-based approach into the U.S regulatory ecosystem such as the U.S 49 CFR 193 and FERC will have its own challenges from public, judicial and regulatory systems.

Acknowledgment:

The author would like to thank his supervisor, Robert McDonnell, for his valuable guidance and review of the manuscript.

6 References

[1] U.S Department Of Transportation Pipeline Transportation Pipeline and Hazardous Materials Administration, Available at http://primis.phmsa.dot.gov/ptr.htm, Accessed on March 15, 2016

[2] Federal Energy Regulatory Commission, Available at http://www.ferc.gov/, Accessed on March 15, 2016

[3] National Fire Protection Agency 59A, "Standard for the Production, Storage, and Handling of Liquefied Natural Gas (LNG)", 2001

[4] Code of Federal Regulations, 49 CFR Part 193, Subpart B, Available at http://www.ecfr.gov/, Accessed on March 15, 2016

[5] European Standard, EN 1473, "Installation and Equipment for Liquefied Natural Gas - Design of Onshore Installations", 2007

[6] Australian Standard, AS 3961, "The Storage and Handling of Liquefied Natural Gas", 2005

[7] Canadian Standard, CSA Z276, "Liquefied Natural Gas (LNG) - Production, Storage, and Handling", 2011

[8] International Association of Oil and Gas Producers, OGP Draft 116901, "Guidance on Performing Risk Assessment in the Design of Onshore LNG Installations including the Ship/Shore Interface", 2013

[9] TNO, "Guidelines for Quantitative Risk Assessment, 'Purple Book,'" CPR 18E, Publication Series on Dangerous Substances (PGS 3), 2005

[10] T.A. Williams and N. Assing, " Risk-Based LNG Facility Siting and Safety Analysis in the U.S.: Recent Developments", LNG 17 Proceedings, Available at http://www.gastechnology.org/, Accessed on March 15, 2016

[11] J.L. Woodward and R.M. Pitblado, "LNG Risk based Safety Modeling and Consequence Analysis", John Wiley & Sons, Hoboken, New Jersey, 2010.

Flare Reduction Strategies for LNG Plant Design

Meghana Cherukuri
Process Engineering Professional
Houston, TX - 77054
Meghana.Cherukuri@yahoo.com

Sumit Sharma
Senior Technical Professional – Process
KBR
601 Jefferson St., Houston, TX - 77002
Sumit.Sharma@kbr.com

Kristina Martinez
Principal Technical Professional – Process
KBR
601 Jefferson St., Houston, TX - 77002
Kristina.Martinez@kbr.com

Prepared for Presentation at
American Institute of Chemical Engineers
2016 Spring Meeting
12th Global Congress on Process Safety
Houston, Texas
April 11-23, 2016

Flare Reduction Strategies for LNG Plant Design

Meghana Cherukuri
Process Engineering Professional
Houston, TX - 77054
Meghana.Cherukuri@yahoo.com

Sumit Sharma
Senior Technical Professional – Process
KBR, 601 Jefferson St., Houston, TX - 77002
Sumit.Sharma@kbr.com

Kristina Martinez
Principal Technical Professional – Process
KBR, 601 Jefferson St., Houston, TX - 77002
Kristina.Martinez@kbr.com

Keywords: Flare system, flare optimization, flare minimization, LNG, Plant design.

Abstract

In liquefied natural gas (LNG) plants, the flare system provides a reliable and safe method for vapor and liquid hydrocarbon disposal. The flare system enables controlled handling of off-spec streams during plant upsets, emergencies, and short term operations such as start-up, shutdown, or equipment maintenance activities. Although the flare system is essential for the safe operation of the plant, proper design and optimization is necessary to control capital cost.

A typical LNG flare system includes Warm (or Wet), Cold (or Dry), and Low Pressure (or LP/Marine) flares. The disposal of streams with potentially freezable components originating from plant inlet and gas treating sections is achieved through the Warm flare. The Cold flare is utilized to dispose of streams with non-freezable components and very low operating temperatures. Low pressure streams from the storage or loading area of the LNG plant are routed through the LP flare for safe disposal. Based on project design criteria, the flare system can require an elevated, ground, enclosed or smokeless flare design.

Addressing flare optimization issues provides opportunities for controlling material loss during plant operation. In this paper, an overview of factors that influence flare type selection and determine flare size are discussed. Significance of flare type and size for overall plot space requirements is also addressed. Based on typical major relieving scenarios, flare load reduction strategies for each of the LNG plant flares are presented.

LNG plant design configurations for multiple trains, refrigerant compressors, and process recycle streams can significantly affect design flare loads. High design flare loads increase flare size and require large plot space due to flare radiation limits. Proper selection of design options for major LNG plant sections, guided by flare reduction strategies, can help in reducing flare design flow rate, flare size, and plot space requirements.

1. Introduction

The flare system in Liquefied Natural Gas (LNG) plants provides a reliable and safe disposal method for hydrocarbon vapor and liquid streams. These streams may result from a deviation from normal plant operation or from intermittent operations such as start-up, shutdown, venting, draining, or purging. Typical plant safety design manages any potential material loss during plant operation by routing the material streams to the flare system for disposal.

Although the flare system is essential for the safe operation of the plant, proper design and optimization is necessary to control capital cost. Flaring design flow rates affect the size of the flare and determine the plot space requirements. [1] High flare loads increase flare stack and header size and require large plot space due to flare radiation limits which altogether can increase the overall project capital cost.

LNG plant design, train, and equipment configurations can significantly affect flare design loads. Addressing flare optimization at the design stage provides opportunities for controlling material loss during plant operation. Proper selection of design options for major LNG plant sections, guided by flare reduction strategies, can help in reducing flare design flow rate, size, and plot space requirements.

2. The Flare System

Disposal of gas and liquid streams in a process plant is achieved through flaring which involves controlled combustion of hydrocarbon streams. Flared hydrocarbons are routed to a remote location and are typically burned in an open flame during which stream components are ideally converted to carbon dioxide and water.

Proper design of the flare system creates an efficient mechanism so that complete combustion of hydrocarbon can be achieved. Without flaring, the alternate options of venting or draining of the process streams can create health, safety, and environmental problems. Local regulations may also make it mandatory to include a flare system for safe disposal of process streams from an operating plant.

A typical flare system (Figure 1) in a process plant consists of three primary components: the flare structure, a plant wide flare header piping system, and plot area for radiation zone. The process streams for disposal are sent to appropriate flares where the streams undergo combustion. Routing of the process streams to the flare is achieved through a plant wide piping system that also includes knockout drums for phase separation of any two phase streams. Liquid content to the flare must be avoided for safe operation.

At the flare tip, the continuous flare flame is maintained using a pilot system, and purge gas is used to prevent oxygen ingress into the flare system. In order to delineate safe work zone conditions, a sterile radiation zone is maintained around the flares.

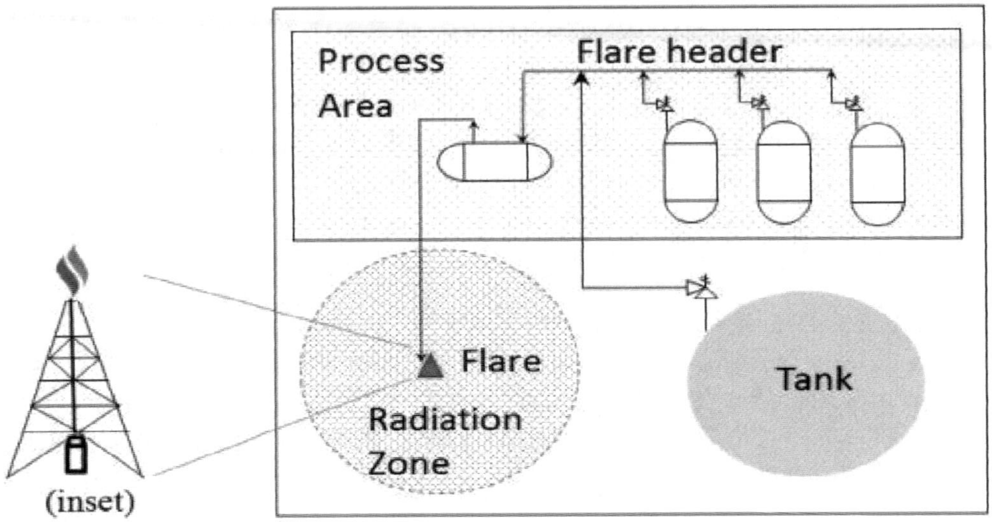

Figure 1. Simplified flare system sketch

2.1. Flare type:

The main component of the flare system is the flare structure that can be classified as elevated or ground flare based on the location of the open flame relative to the flare structure. Selection of the flare type is based on project and process requirements.

- Elevated flares: Elevated flares include a vertical stack with a flare tip that is located at the highest point of the stack. A process stream routed to the flare stack for disposal undergoes combustion at the top of the stack. An elevated combustion point helps in the dispersion of the flare products into the atmosphere.

  The support structure for the elevated flares can be self (Figure 2a), guy wire (Figure 2b), or derrick supported (Figure 2c) based on the height of the flare stack. High elevation of the flare also creates visual and noise impact from the flaring process. Efficient mixing of flaring stream with combustion air is important in achieving smokeless flare design.

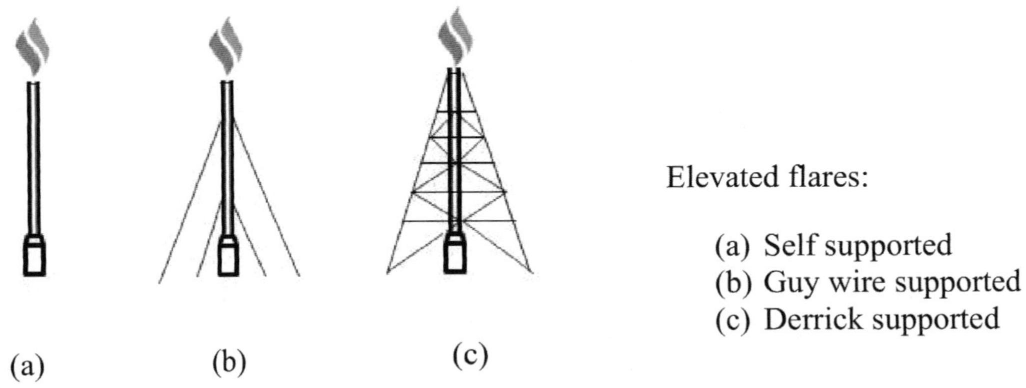

Figure 2. Elevated flares

While there are no regulatory restrictions for the stack height, Federal Aviation Administration (FAA) may require height limits for flare placements in the vicinity of an airport. Flare dispersion modeling during the design phase is essential to ensure that concentrations of toxic emissions do not exceed safe levels at any location where personnel can be present. Flare dispersion is affected by the type of flare tip. In the US, 40 CFR 60.18 (4)(iii) limits flare tip velocity to 400 ft/s with a flared gas lower heating value of at least 1000 Btu/SCF. [3]

- Ground Flares: Open multi-point ground flares (Figure 3a) or enclosed ground flares (Figure 3b) are the preferred option when project requirements prohibit visual and noise disturbances from the flaring process. Ground flares are designed smokeless but dispersion of flare combustion products is typically inefficient. Stream combustion in ground flares is achieved at grade level.

  In an enclosed ground flare the burners are located within a refractory lined physical structure that encloses the flame and contains thermal radiation. Open ground flares are not enclosed but typically are surrounded by a radiation fence which similarly blocks radiation.

  Open ground flares can be designed to handle full flare capacity similar to the elevated flares whereas enclosed ground flares are used for routine small scale flaring. Compared to the elevated flares, the ground flares are costlier and require larger plot area (open ground flares).

(a) Open ground flare

(b) Enclosed ground flare

Figure 3. Ground flares [2]

2.2. Flare header piping system

Plant equipment are protected from overpressure scenarios using relief valves. The outlet flow from the relief valve is collected using a plant wide flare header system and transported to the flare for disposal. Process streams from intermittent plant operations such as start-up, shutdown, and emergency blowdown are also routed to the flare through the flare header piping to dispose of hydrocarbon inventory in a controlled manner to prevent or mitigate escalation of any unsafe event. Flare and flare header piping can be classified based on the operating conditions of the flaring stream:

Table 1. Flare and flare header classification:

• Based on flaring stream operating pressure:	
High Pressure (HP) Flare:	Sonic flare. Handles large high pressure flaring streams.
Low Pressure (LP) Flare:	Sub-sonic flare. Typically used for equipment with low design pressure specifications such as storage tanks.

• Based on flaring stream composition and operating temperature:	
Warm (Wet) Flare:	High temperature flaring streams with components that can potentially freeze at low temperatures
Cold (Dry) Flare:	Low temperature flaring streams with no freezable stream components

Separation of the flares in terms of pressure, temperature, and stream composition is necessary for safe and reliable flare operation. Depending on the operating conditions of the relieving stream, the flare header material can range from carbon steel to stainless steel.

Typically, for the Warm flare, carbon steel piping can be used to handle the flaring stream whereas the Cold flare will require stainless steel piping because of the possibility of very low temperatures that may exist in the flaring streams during relieving scenarios. High design flow flaring rate for the flare results in large flare header size and increases the cost of the system. Stainless steel metallurgy makes the Cold flare relatively more size and cost sensitive compared to the Warm flare.

2.3. Radiation zone

Flaring large quantities of hydrocarbon can result in high intensity thermal radiation and noise around the flare during plant operation. In order to maintain safe working conditions, the area around the flare is demarcated as sterile zone. The permissible levels for flare radiation are summarized in Table-3 below.

Table 3. Radiation and noise limits

• **Radiation: Permissible Radiation Level** [4]	
KW/m^2	
1.58	maximum radiant heat intensity at any location where personnel with appropriate clothing can be continuously exposed
4.73	maximum radiant heat intensity in areas where emergency actions lasting 2 min to 3 min can be required by personnel without shielding but appropriate clothing
6.31	maximum radiant heat intensity in areas where emergency actions lasting 30 seconds can be required by personnel without shielding but appropriate clothing
9.46	maximum radiant heat intensity at any location where urgent emergency action by personnel is required. When personnel enter or work in an area with the potential for radiant heat intensity greater than 6.31 KW/m^2, radiation shielding and/or special protective apparel (e.g. a fire approach suit) should be considered. Safety precaution- It is important to recognize that personnel with appropriate clothing cannot tolerate thermal radiation at 9.46 KW/m^2 for more than few seconds

• **Flare noise limits**		
85 dB (A)	max	continuous exposure
115 dB (A)	max	for short term exposure (up to 15 mins)
115 dB (A)	max	with hearing protection

Heat radiation calculations for flare typically take into consideration the wind velocity, relative humidity, and solar radiation at the flaring location to assess the heat radiation impact of flaring. The impact of flare radiation and flare front can be identified early in the project to determine locations for flare stack and plot area for radiation zone.

3. Flares in LNG Plant

In a typical LNG plant the feed gas to LNG conversion involves four primary process units that include feed gas pre-treatment, refrigeration, liquefaction, and product storage. The feed gas pre-treatment section removes water, acid gas, and mercury from the feed gas stream. The treatment process occurs at high pressure and warm temperatures. Refrigeration and liquefaction units operate at high to low pressure and temperature ranges. The LNG product storage unit operates at close to atmospheric pressure and very low temperature range.

The flare system in an LNG plant is comprised of different flares that are designed to handle streams originating from various units for safe disposal. The hydrocarbon flare streams mainly contain methane, ethane, propane and other heavy components from the natural gas streams or refrigerant streams:

- Warm (Wet) Flare: A high pressure carbon steel flare for warm and wet hydrocarbons, designed to handle streams containing heavy hydrocarbons, contaminated water streams and streams with any freezable components. These streams mainly originate from the pre-treatment units.

- Cold (Dry) Flare: A high pressure stainless steel flare for cold and dry hydrocarbon vapors. These light hydrocarbon streams originate from the refrigeration and liquefaction units of the plant.

- Low Pressure (LP) or Marine Flare: A low pressure stainless steel cold flare for emergency vapor (low pressure) releases from the LNG storage and loading system.

Apart from the Warm, Cold, and LP flares, the LNG plant may also include a Sour Gas Flare, and Thermal Oxidizer or Incinerator for flaring hydrocarbon streams containing a high percentage of acidic gases such as H_2S, CO_2, and volatile hydrocarbons (VOC) originating from pre-treatment units. Some facilities also include dedicated warm and cold liquid collection header systems to separately handle liquid disposal.

4. Flare Design Loads

The design flare loads in LNG plants are determined by identifying major relieving scenarios during process upset, intermittent or normal operations. The Warm and Cold flares are designed to handle the largest gas flows in the LNG plant while the LP or Marine flare is designed based on major flaring incidents from the tank or product storage area.

In the following sections, all discussions related to controlling cases for flare design are based on the Propane Pre-Cooled Mixed Refrigerant Process (C3MR) by Air Products and Chemicals Inc. (APCI).

4.1. Warm Flare Design Load:

The largest flow at ambient or warm temperatures (above 0 °C) in the LNG plant is the feed gas flow into the facility. The inlet pressure to the process units is typically maintained by single or multiple (parallel) pressure letdown valves. Any overpressure scenario in the feed gas line in the inlet or pre-treatment facilities can result in large feed gas flaring to the Warm flare. Another large flow that can determine the Warm flare capacity is the blowdown loads from the inlet and pre-treatment units of the LNG plant. The blowdown operation removes hydrocarbon inventory from the units and routes the hydrocarbon streams to the flare system.

4.2. Cold Flare Design Load:

The Cold flare design flow is typically based on refrigerant flows in the refrigeration and/or liquefaction units of the LNG plant. The primary refrigerants for the APCI process are Propane which is used for pre-cooling the feed gas and a Mixed Refrigerant (typically consists of nitrogen, methane, ethane, and propane) is used for liquefaction process. The Propane (PR) and Mixed Refrigerant (MR) flows are circulated using Propane and MR Compressors respectively. Process upset cases such as PR or MR compressor blocked outlet or local power failure may require routing PR or MR circuit flow to the Cold flare. Blowdown load from the cryogenic heat exchanger and emergency depressuring events are also sent to the Cold flare.

4.3. LP /Marine Flare Design Load:

Low Pressure or Marine flare handles any relieving streams from the LNG storage tanks or product loading operations. The capacity of the LP/Marine flare is typically set by the Boil-Off Gas (BOG) compressor blocked outlets case. The largest flow to the Boil-Off Gas compressor occurs during LNG product loading operations which may utilize multiple BOG compressors operating in parallel.

5. Flare System Optimization

Collectively, a significant quantity of gas is estimated to be flared annually from LNG export terminals. [5] Optimization of the flare system and minimization of flaring incidents can potentially reduce LNG plant's carbon foot print and overall plant cost. Flare design optimization and flare minimization will require consideration of design options that can minimize flaring incidents or reduce the amount of flare load during a flaring incident.

5.1. Warm Flare:

For the Warm flare, the largest flare load is typically from the feed gas pre-treatment section located upstream of the liquefaction unit. An overpressure scenario in which the feed gas forward flow is blocked in the pre-treatment section can result in routing all feed gas flow to the Warm flare through the activated relief valves.

In another scenario the failure of plant inlet control valve can significantly increase flow to the plant inlet. This control valve failure will activate relief valves downstream from the valve and route the relief valve outlet flow to the Warm flare (assuming that there is no additional heating provided and that the upstream pressure is not high enough to cause the downstream gas temperature to be too cold for the Warm flare metallurgy). The flow resulting from the control valve failure is a function of the pressure drop across the control valve. The higher the difference between the inlet and outlet pressures across the valve, the higher the potential flow to the flare subsequent to the control valve failure. High flare load will increase the flare height, the radiation zone, and associated costs.

5.1.1. <u>High Integrity Pressure Protection System (HIPPS) for the feed gas line</u>

An preventative safety system such as High Integrity Pressure Protection System (HIPPS) can be used to eliminate high Warm flare load from any blocked outlet overpressure scenario in the feed gas flow at plant inlet. For the pressure letdown control valve failure case, consideration of a HIPPS downstream of the plant inlet control valve will prevent a flaring incident by shutting down the forward flow to the plant. Implementation of a HIPPS at the plant inlet will trip any dedicated emergency shutdown valves at the plant inlet before reaching a scenario in which the operating pressure exceeds the design conditions. This scheme will significantly reduce the flow to the flare by eliminating full feed gas forward flow as a consideration for the Warm flare sizing.

HIPPS can provide a completely independent safety layer. Inclusion of redundant features such as additional sensors and instrument logic control increases the HIPPS reliability. ASME Sec VIII, Division 1, UG-140 accepts the instrumented protective device in consultation with and on the advice of the owners for certain services.

A HIPPS can provide sufficient reliability to replace a mechanical device such as a pressure relief valve. HIPPS is a particularly attractive option for revamps where limited capacity of an existing flare can prohibit any flare load increase from revamp changes. Implementation of HIPPS for such cases will reduce the occurrence of any flaring scenarios that can impact flare design loads.

HIPPS application requires frequent testing and maintenance to keep the device operative and available when needed. This can be a significant burden for the operator. The use of a HIPPS requires a hazard and risk analysis (per ISA S84.00.01/2/3) to determine the required risk reduction that must be achieved by the overpressure protection. This then sets the required safety integrity level for the instrumented protection. Any HIPPS based design should be thoroughly reviewed in HAZOP, and clearly understood by both the owner management and the plant operation personnel.

5.2. Cold Flare:

The design flare load for the Cold flare usually constitutes the highest flaring rate in LNG plants. Refrigerant compressor blocked outlet cases result in routing the refrigerant system flow to the Cold flare. The refrigerant system flow and inventory capacity affect the size and capacity of the Cold flare and the overall flare system.

5.2.1 Implementation of HIPPS for refrigerant compressor system

Similar to the HIPPS for feed gas inlet control valve, HIPPS can be considered for refrigerant compressor system. A two out of three pressure transmitter voting can be implemented on the refrigerant compressor outlet to trip the machine during a high pressure case. Using HIPPS, the overpressure scenario and high flare load from compressor blocked outlet can be eliminated for the Cold flare.

HIPPS application requires frequent testing and maintenance to keep the device operative and available when needed. This can be a significant burden for the operator. The use of a HIPPS requires a hazard and risk analysis (per ISA S84.00.01/2/3) to determine the required risk reduction that must be achieved by the overpressure protection. This then sets the required safety integrity level for the instrumented protection.. Any HIPPS based design should be thoroughly reviewed in HAZOP, and clearly understood by both the owner management and the plant operation personnel.

5.2.2 Partial condensation at propane condenser by natural convection.

Failure of the propane condenser results in overpressure in the propane refrigerant system. For air cooled LNG plant design, the resulting calculated flare load for the Cold flare sizing can be reduced by considering partial condensation at the propane condenser air coolers.

Power failure affects the propane condenser air cooler fans however partial condensation at elevated pressure will still occur due to natural convection cooling at the air coolers which are usually located high up on the top of pipe racks. Consideration of partial condensation will reduce the relieving load and the flaring requirement.

5.2.3 Recycle relief valve flow to compressor suction [6]

The relief valve downstream of the refrigerant compressor discharge sends the outlet flow to the Cold flare in case of overpressure scenarios in the compressor system. Recycling the relief valve outlet flow to suction of the compressor can significantly reduce the Cold flare load (Figure 4).

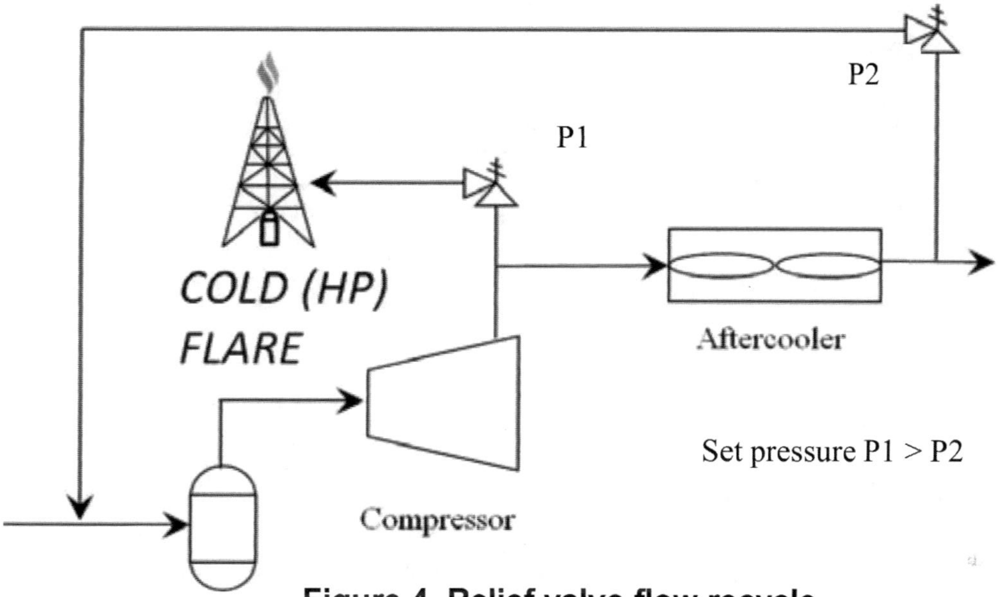

Figure 4. Relief valve flow recycle

The relief valve outlet flow recycle scheme eliminates the flare load from compressor blocked outlet case. The recycle route is parallel to the anti-surge valves that protect the compressors. This scheme can be implemented for both propane and mixed refrigerant compressors.

The proposed relief valve recycle scheme works based on differential pressure which makes it susceptible to unintended relief valve opening. Using pilot operated relief valves can potentially solve this problem. This scheme may require multiple relief valves within the compressor system. Long term operation and heat built up effects may have to be completely analyzed with inputs from compressor supplier and dynamic simulation analysis.

5.2.4 Split scheme for the flare load from the refrigerant compressors

The relief valve flow from refrigerant compressors blocked outlet or power failure cases is conventionally sent to the Cold flare. However the outlet flow from the relief valves may be routed to the warm flare if the stream temperature is at a higher temperature during the compressor system overpressure scenario. A high stream temperature can result in warmer letdown temperature downstream of the relief valves, and may not reach an unacceptable limit.

This scheme (Figure 5) will substantially reduce the size of the cold flare which is all stainless steel material and lower the cost by routing the relieving load to the warm flare which is typically based on carbon steel material.

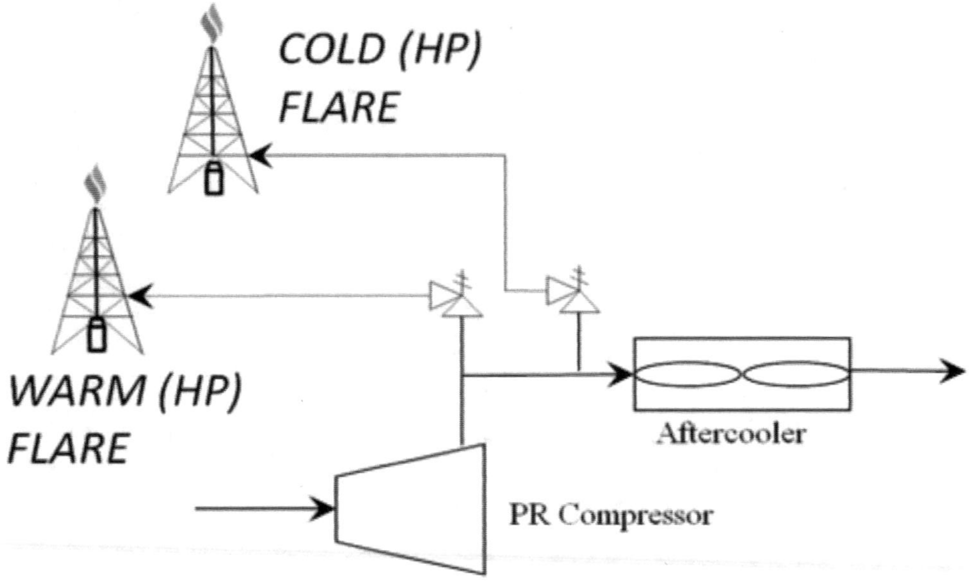

Figure 5: Flare load split scheme

If all of the flare stacks (Warm, Cold, and LP flares) are located at a single location, the heights of the flare stacks are usually the same height so that there is no flame impingement on any flare stack from an adjacent active stack. The choice to route the refrigerant flow to the Warm flare may not provide any benefit if it leads to increasing the Warm flare size beyond the Cold flare size and if other controlling scenarios for the Cold flare maintain the same initial Cold flare size.

The total radiation level from this scheme may stay the same because the total propane system load to flare remains the same based on the overpressure scenario for the compressor system.

5.2.5 Increasing the propane system design pressure

Increasing the propane system design pressure can potentially reduce the flare load to a level so that it no longer controls the Cold flare design. The propane refrigeration system is generally designed in carbon steel materials. The design pressure of the propane system can be increased to a level that is within flange limits of the piping material class. A higher set pressure for the relief valve will reduce flare flow rate from the compressor during overpressure scenarios. Higher relief valve set pressure may also reduce the frequency of overpressure scenarios for the compressor system.

This option will lead to higher design pressure for the air coolers, compressor casings, and other equipment in the propane compressor system. These design modifications can lead to the compressor surging at elevated pressure (if the anti-surge control system fails) which may damage the compressor. Many owners also view the propane compressor discharge relief valve as providing a back-up means of machinery protection in addition to its overpressure protection function.

5.2.6 Dynamic simulation for compressor overpressure scenarios

Dynamic simulation can provide realistic insight into the impact of overpressure scenario on the overall compressor system. Using information on inventory, driver power, compressor curves and other details, Dynamic simulation can predict realistic reliving rates during overpressure incidents. Based on dynamic simulations, the relief flow can be significantly lower compared to a conservative estimate of relieving all compressor refrigerant forward flow and system inventory capacity to the flare during overpressure scenarios. [7]

5.3. Flare system for multiple train design:

In multiple train design each train can have a common or independent pre-treatment, refrigeration, and liquefaction sections. For the independent train design, both Warm and Cold flare systems can be common for the overall plant with each flare sized for their single train controlling cases plus the respective start-up load for one additional train. For independent trains, lacking any common cause failure, double jeopardy considerations can eliminate multiple train failures and limit overall flare loads. Provision of common spare stack can be determined by the project philosophy and safety review of the flare design.

For a single train design with multiple parallel compressor strings (2x50%), the cold flare load may not decrease compared to a same capacity single compressor string (1x100%). In multiple parallel compressor string design, the blocked outlet for the compressor strings can occur at a location where compressor outlet flows from both the strings combine, sending the full combined 100% compressor outlet flow to the Cold flare.

6. Flaring Minimization

Various plant operations can result in flaring in an LNG plant. With increased emphasis on containing plant emissions and operating costs, design modifications can be considered to reduce flaring incidents.

Intermittent planned or unplanned operations such as start-up, shutdown, or plant trips can result in significant flaring as a percentage of feed gas [8]. A full pressure re-start/or partial depressurization for faster re-start of the liquefaction plant after a trip, based on available power, can reduce the amount of gas flared. Using aeroderivative gas turbine driven compressors for refrigeration service can also reduce flaring during start-up activities. Consideration of isolation and maintenance activities during the design phase can provide easy access and means of equipment isolation with minimum flaring.

Design options should also consider recycling or re-routing any gas stream from flare to either front-end of the plant for processing, or to the fuel gas system for consumption.

7. Conclusion

The flare system design in an LNG plant is affected by equipment and system configurations. With increased emphasis on containing plant emissions and costs, design modifications can be considered to reduce flaring incidents during normal or intermittent plant operations. An optimized flare system design will require design considerations for major LNG plant units including the feed gas inlet, pre-treatment, and refrigeration sections.

High flaring rates increase the plant cost by increasing the size of the flares, the flare headers and the required plot area for radiation zone. Flare system optimization for single or multiple train configurations reduces the flare loads by recycling, re-routing, or eliminating the flaring streams from various units of the LNG plant. Implementation of flare load optimization and minimization options during the design phase provides opportunities to limit the overall plant flaring rates, incidents, and related costs.

8. References

[1] L. B. Evans, W. M Vatavuk, D. K. Stone, S. K. Lynch, R. F. Pandullo, W. Koucky, Flares, Available at http://www3.epa.gov/ttncatc1/dir1/cs3-2ch1.pdf. Accessed on February 01, 2016.

[2] TxOGA, Flare Operation Discussion, Available at http://www.tceq.state.tx.us/ assets/public /implementation/air/rules/Flare/Resources_TXOGAFlareOperationDiscussion.pdf, Accessed on February 01, 2016

[3] Electronic code of federal regulations, General control device and work practice requirements, Available at http://www.ecfr.gov/cgi-bin/text-idx?SID=7ae903b94d41527119c 08be5a939da ed&mc= true&node=se40.7.60_118&rgn=div8, Accessed on February 01, 2016.

[4] API Standard 521. Pressure - relieving and Depressuring Systems. American Petroleum Institute, Sixth Edition, January 2014.

[5] C. D. Elvidge, M. Zhizhin, K. Baugh, F. Hsu, and T. Ghosh, Methods for Global Survey of Natural Gas Flaring from Visible Infrared Imaging Radiometer Suite Data, Energies, 9, (2016), 14, Available at http://www.mdpi.com/1996-1073/9/1/14, Accessed on February 01, 2016.

[6] C. E. Lucas, P. A. Brochu, W. C. Rooney, Gas compressor, WO 2005010375 A1, Available at https://patentscope.wipo.int/search/en/detail.jsf?docId=WO2005010375&redirectedID=true, Accessed on February 01, 2016.

[7] Y. N. Liu, C. E. Lucas, R. W. Bower, and J.C. Bronfenbrenner, Reducing LNG Costs by Better Capital Utilization, Available at http://www.airproducts.com/industries/Energy/LNG/ resource-center/whitepapers/reducing-lng-costs-better-capital-utilization.aspx, Accessed on February 01, 2016.

[8] Qatar Gas, Flare reduction programme, Available at https://www.qatargas.com/English /SafetyAndEnvironment/Documents/Qatargas%20Flaring_Fact%20Sheet_20%20Nov%202012. pdf, Accessed on February 01, 2016.

441507 Using Data Analytics to Improve Process Safety

Abstract:

Using Data Analytics to Improve Process Safety

Increasingly, data from multiple sources are compiled into central storage systems for entire facilities and in some cases entire corporations. The low cost of digital memory and the increasing process speed of modern microprocessors and computers continue to make it easier to analyze, manipulate and merge very large databases. Big Data and Data Analytics have been touted as allowing staff at all levels to make real time, data-driven business decisions. These concepts can be used to improve process safety at facilities by proactively analyzing sources of data to identify opportunities for improvement in operating practices, identify and track near misses, and in real time detect incipient instrument and equipment failures that could lead to catastrophic incidents. The aviation industry is one example that is voluntarily implementing similar analytics. Over the last decade the aviation industry has implemented flight operations quality assurance (FOQA) or flight data monitoring (FDM) programs which collect, store, and analyze recorded flight data with the goal of increasing overall safety.

In this presentation we show examples of analytics Exponent performed on data from facilities. These analyses were used to analyze alarm frequencies during specific types of operations, identify near misses, and proactively detect failures of instrumentation and equipment in order to prevent or mitigate catastrophic incidents.

442485 Safety Culture and Leadership- Independent of Language

Abstract:

The existence of safety in the workplace is mandatory for a sustainable business. Strong safety cultures are characterized by human factors, workforce engagement and fully committed safety leadership. With that being said, increased diversity in work environments and international recruitment can oftentimes complicate the process of implementing a common safety culture.

More specifically, those who come from different types of work environments than their current environment may have trouble interpreting the procedures and guidelines that are enforced at their new place of employment. Differences in languages and dialects may also lead to misinterpretations of intended safety guidelines. Lastly, there may be some instances where the workplace follows a non-flexible and hierarchical structure that is not very open to change.

In order to best develop safety practices in the workforce, the senior leaders need to first develop an environment based on the understanding of the different rationales and expectations of a multi-cultural workforce. Once that has been established, communicating with the workforce on safety expectations of the organization and optimizing interpretation of this message is critical.

This paper will continue the discussion from a previous publication; how safety leadership can consider linguistic differences when implementing a safety plan for a diverse workforce. The discussions will emphasize the learnings of the Bradley Curve, showing how important it is to follow safe practices to reach the proximity of an almost 100% incident free area, and will provide real examples from international work assignments and diverse work teams, highlighting this need and offering potential approaches

439897 Indulging in Invincibility (III)- Twelve Steps to Sobriety

Abstract:

<>8/3/2015 <>Indulgence in Invincibility (III)- Twelve Steps to Sobriety <>by Brian P. Kolodji, PE

Could Bhopal, Macondo, or Texas City have been prevented? Directly related examples are presented where PS cultural indulgences contribute to a false sense of invincibility. Light is shed on where the perennial dilemmas of misplaced loyalty, unachievable goals, and belief in corporate higher power contribute to and enable denial of process safety culture failure. Successes and a program are presented where these failed practices are brought out of the shadows and recovery begins.

Case Study of Relief Analysis of Compressor Stations

Kumud Chaurasia
Process Engineer
Lloyd's Register Energy Americas, Inc.
1330 Enclave Parkway, Suite 200, Houston, TX 77077, USA
Kumud.chaurasia@lr.org

Aniket Patankar
Senior Process Engineer
Lloyd's Register Energy Americas, Inc.
1330 Enclave Parkway, Suite 200, Houston, TX 77077, USA
Aniket.patankar@lr.org

Prepared for Presentation at
American Institute of Chemical Engineers
2016 Spring Meeting
12th Global Congress on Process Safety
Houston, Texas
April 11-23, 2016

Case Study of Relief Analysis of Compressor Stations

Kumud Chaurasia, Process Engineer

Aniket Patankar, Senior Process Engineer

Lloyd's Register Energy Americas, Inc.

1330 Enclave Parkway, Suite 200, Houston, TX 77077, USA

kumud.chaurasia@lr.org, aniket.patankar@lr.org

Keywords: Relief analysis, Compressor Station, case study, concern identification

Abstract

A compressor station is a facility which helps the transportation process of natural gas from one location to another. Compressor stations include several key component parts, the primary being the actual compressor unit, either centrifugal or reciprocating. They also typically include scrubbers, strainers or filter separators which remove liquids, dirt, particles, and other impurities from the natural gas. This study will give an overview of the relief analysis conducted for multiple compressor stations. The relief analysis was system based as opposed to relief device based so as to identify any unprotected systems that may need overpressure protection. It will discuss the overpressure scenarios identified for these systems. It will provide some general statistics on the total number of applicable overpressure scenarios. It will give a detailed breakdown of the various concerns found including, but not limited to, undersized devices, unprotected systems, installation concerns, data needs, and documentation discrepancies. The study will discuss possible mitigation solutions for these concerns.

1. Introduction

Compressor stations are typically located along a natural gas pipeline to facilitate the compression of the gas to a specified pressure, which allows the gas to flow along the pipeline to a target location. Since these pipelines typically extend for several hundreds of miles, the gas tends to lose pressure frequently. This necessitates the requirement to compress it frequently during the transportation process.

The size of the station and the number of compressors (pumps) varies, based on the diameter of the pipe, and the volume of gas to be moved. Compressor stations include several key components, the primary being the actual compressor unit, either centrifugal or reciprocating, scrubbers, and strainers or filter separators which remove liquids, dirt, particles, and other impurities from the natural gas.

It is critical to ensure that the natural gas transported to its intended target is free from any contaminants. This is accomplished through four mains steps. The first step is contaminant removal, which occurs upstream of a compressor station near the well head. Here, the raw natural gas, at very high temperature and pressure, is passed through separators in order to remove oil, water, sand etc. It is then passed through a slug catcher, typically located within the compressor station, to remove any leftover condensate. The third contaminant removal step occurs downstream of the compressor unit (after the gas has been pressurized) via a contactor unit that predominantly removes liquid water and carbon dioxide using triethyleneglycol (TEG) as the absorbent. The TEG that separates out at the bottom of the scrubber unit, carrying the water and carbon dioxide with it, is typically regenerated in a glycol regenerator unit before it is recirculated back into the scrubber unit. The final step prior to delivery of the gas to its intended target is to remove any leftover TEG in the gas using a scrubber unit. There are also several tanks within the compressor station for storage of the condensates and TEG. Additionally, the compressor stations contain several components that are part of the glycol regenerator unit including filters, flash tank separators, pumps, heat exchangers, etc. A general process flow diagram of the compression station is shown in Figure 1.

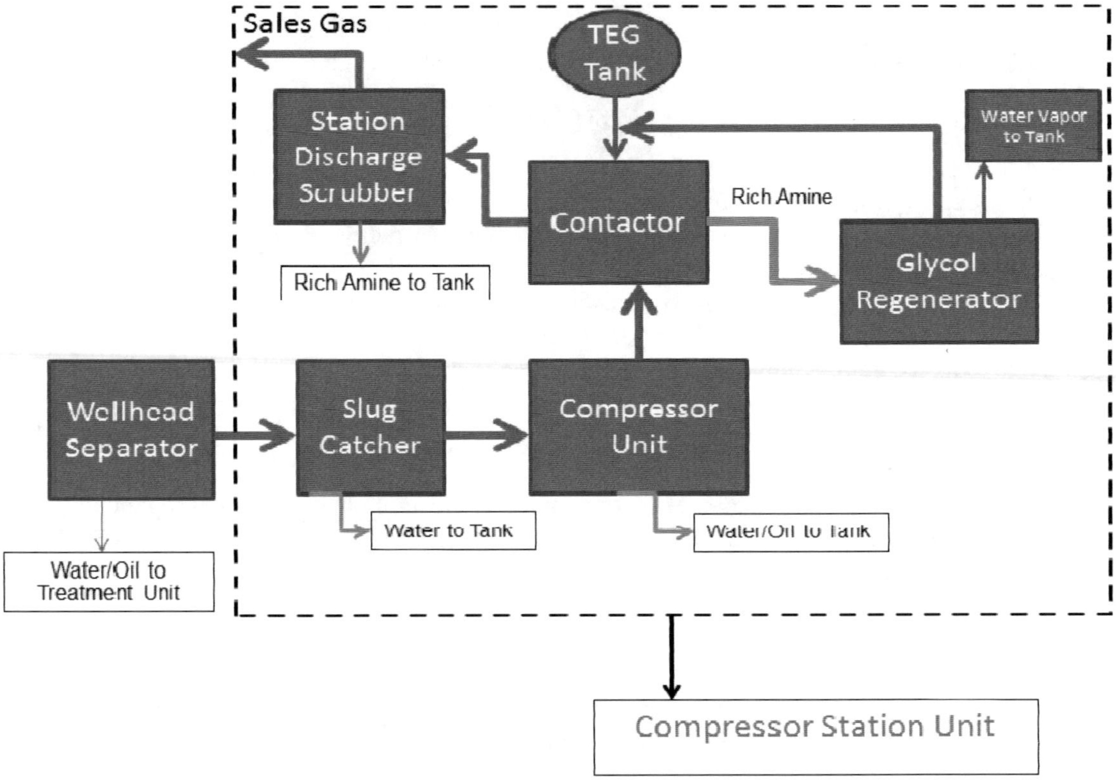

Figure 1: Process flow diagram of Compressor Station

2. Relief Analysis

For this case study, twenty separate compressor stations were analyzed. Specifically, the steps required for performing the analyses, different applicable scenarios, and all the concerns identified with mitigations options are presented. The analysis was conducted using API 520 Part 1 and Part 2 standards, while considering the client's design philosophy as well.

2.1. Steps involved in Relief Analysis

The relief analysis, which is a system based analysis performed for each equipment in the unit, is a methodical four step approach including data and design documentation collection, contingency analysis, pressure relief calculations, concern identification and mitigation. These steps are discussed in detail below.

Data and Design Documentation Collection: Since the client for whom the relief analysis is being performed may not always provide all the information necessary to conduct the analysis, it is imperative that all necessary data and documentation required is collected at the client site prior to commencing the project. Examples of such data collected include isometrics for PSV inlet and outlet piping, PSV and control valve information, equipment and relief device name plate information, U-1A forms, equipment drawings etc. Usually, the piping and instrumentation diagrams (P&IDs) and heat and material (H&M) balance are provided by the client in order to complement the other data collected. All the data collected is then entered into a client project database.

Contingency Analysis: Prior to commencing the contingency analysis, the compressor station is divided into various systems, with each system consisting of several equipment. The interface of systems is defined based on where block valves are situated within the P&ID. Once the systems have been defined, a contingency analysis is performed where all scenarios that could possibly result in an overpressure condition are identified. For the 20 compressor stations that were analyzed in this case study, the following eighteen contingencies were identified: Power Failure, Cooling or Reflux Failure, Steam Failure, Instrument Air Failure, External Fire, Blocked Outlet, Isolated with Heat Input, Equipment Failure, Inadvertent Valve Operations, Failure of Automatic Controls, Gas Breakthrough, Internal Boundary Failures, Reverse Flow, Run Away Reaction, Accidental Mixing, Change in Conditions, Vacuum, Other (applicable scenarios other than listed above).

Calculations: Once the contingency analysis is finished, the required flow rate of the applicable overpressure scenario and the capacity of the relief device (based on the relief device ASME orifice area) are calculated. The inlet and outlet pressure drops, Mach number and sound power level of the piping is also calculated based on the dimensions of the relief device.

Concerns Identification and Mitigation: After the calculations are completed for a system, all the concerns are identified and documented in the database. This is as per the guidelines provided in API 520 Part 1, 2 and the design philosophy provided by client. The concerns are identified based on the calculation results and discrepancies and assumptions made during the analysis. Once all the concerns are identified, recommendations for implementation of mitigation strategies are provided to the client.

3. Results of Case Study

3.1. Relief Device Identification

For the twenty compression stations that were analyzed as part of this case study, the relief devices included all the relief valves, pressure and vacuum vents (PVRVs), goosenecks, and rupture disks. Usually, the relief valves and rupture disks were situated on the pressure vessels while the PVRVs and goosenecks were situated on the low pressure tank. A total of 1099 relief devices were identified for the 20 compressor stations. Depending on the type of relief device, the method of calculation may vary. For example, in the case of PVRVs and goosenecks, it is not necessary to calculate the pressure losses.

The breakdown of the percentage and number of the relief devices, by type, that were analyzed in this study are shown in Table 1 and Figure 2 respectively.

Table 1: Percentage of different relief devices by type

Relief device Type	Percentage
Relief valve	78%
Gooseneck/PVRV's	21%
Rupture Disk	1%

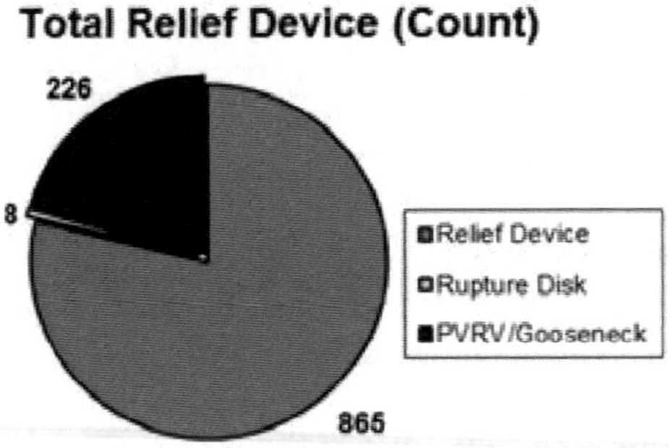

Figure 2: Actual number of different relief devices, by type, for the twenty compressor stations

3.2. Relief Valve Type

There are three types of relief valves that are typically used in the oil and gas industry: conventional, bellows, and pilot valve. Based on the type of valve, the outlet pressure drop limits vary; hence it is important to identify the relief valve type. As shown in Table 2, out of a total of 865 relief valves identified, the majority of the valves in the compression stations analyzed were conventional relief valves,. The breakdown of the number and percentage of the relief valve that were analyzed in this study are shown in Table 2 and Figure 3 respectively.

Table 2: Percentage of different relief valves, by type

Relief Valve Type	Percentage
Conventional	68%
Pilot	23%
Bellows	9%

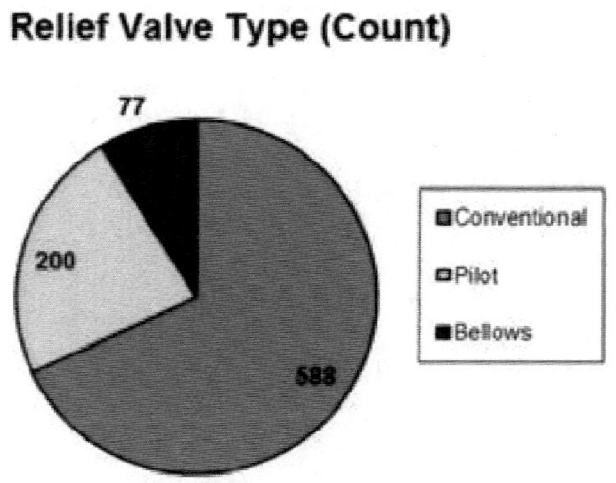

Figure 3: Actual number of relief valves, by type for the twenty compressor stations

3.3. Equipment Type

Based on the equipment type, the overpressure scenario may be applicable or not and the data required to do the analysis will be different. Hence, it is very important to identify the type of the equipment. A total of 2014 equipment were identified in all the compressors stations, of which majority were pressure vessels.

The breakdown of the percentage and number of the equipment type that were analyzed in this study are shown in Table 3 and Figure 4 and respectively. Note that the "Other" type of equipment includes compressors, glycol regenerators etc.

Table 3: Percentage of different equipment, by type

Equipment Type	Percentage
Pressure Vessel	48.5%
Tower	8.1%
Pumps	8.9%
Exchangers	16.6%
Fired Heater	0.1%
Low Pressure Tanks	6.1%
Piping Component	6.4%
Others	5.3%

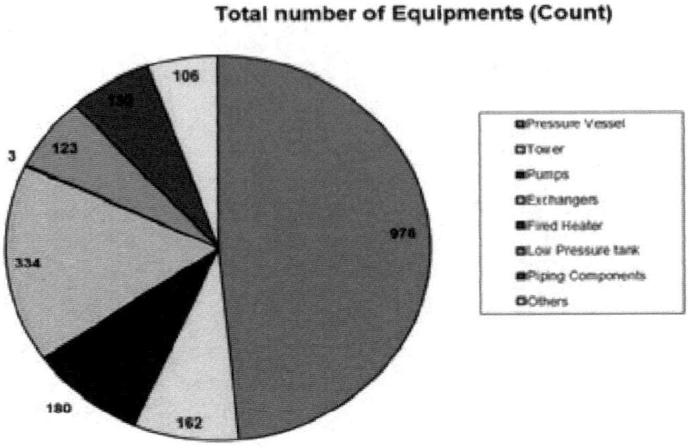

Figure 4: Actual number of equipment, by type, for the twenty compressor stations

3.4. Applicable Overpressure contingencies:

All the applicable scenarios, for each system, were identified after reviewing the P&IDs and other available data. This section shows all the applicable scenarios identified. The breakdown of the percentage and number of the applicable contingencies that were analyzed in this study are shown in Table 4 and Figure 5 respectively. Of a total of 3026 applicable contingencies, the majority of the contingencies in the compression stations analyzed were external or pool fire and blocked outlet.

Table 4: Percentage applicable contingencies by type

Contingency Type	Percentage
External fire	22.8%
Cooling or reflux failure	0.1%
Blocked Outlet	22.6%
Failure of automatic controls	13.3%
Gas breakthrough	8.5%
Inadvertent valve operation	5.6%
Instrument Air failure	2.9%
Internal boundary failure	1.6%
isolated with Heat Input	6.4%
Power failure	0.9%
Reverse flow	3.2%
Vacuum	7.1%
Other	5.0%

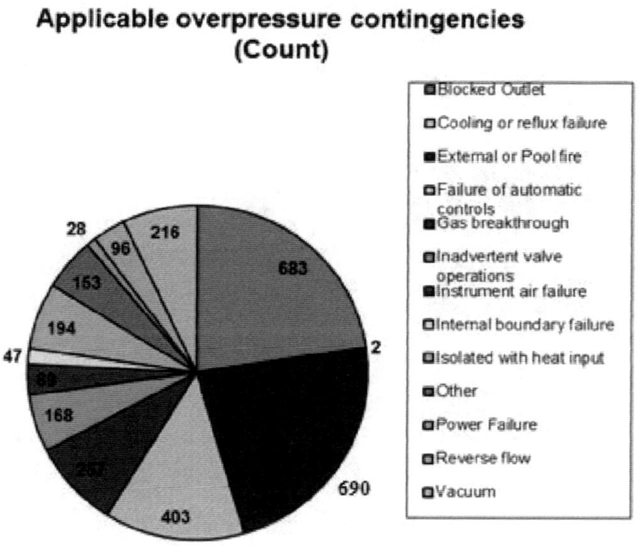

Figure 5: Actual number of applicable overpressure contingencies, by type, for the twenty compressor stations

3.5. Concerns Identification:

This was one of the most important steps in the relief analysis. Based the calculation results, assumptions, discrepancies etc., all the concerns were identified and documented based on the guidelines. The concerns were divided into five major categories: Undersized Concerns, Installation Concerns, Unprotected Systems, Data Requirement, and Documentation Updates. It is important that assumptions made during this stage be documented, verified and approved by the client. The undersized concerns, installation concerns and unprotected system concerns were assigned a higher priority since they could have a direct impact on the operation and failure of the compressor station. Hence, it was recommended that these concerns be mitigated immediately.

The breakdown of the percentage and number of the concerns that were analyzed in this study are shown in Table 5 and Figure 6 respectively. Of a total of 3278 applicable concerns that were identified in this case study, the majority of the concerns in the compression stations were related to Data Requirements.

Table 5: Percentage of concerns identified, by type

Concern Type	Percentage
Undersized	10.4%
Installation	24.6%
Unprotected Systems	9.5%

Data Requirement	50%
Documentation Updates	5.5%

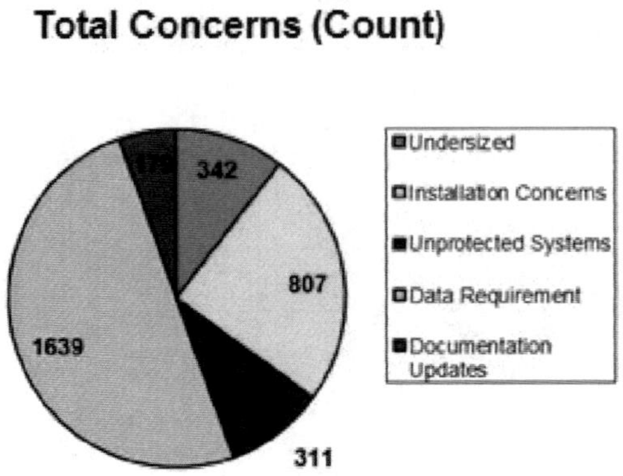

Figure 6: Actual number of different concerns, by type, for the twenty compressor stations

Undersized Concerns:

When the required relief flow rate is more than the capacity of the relief device, the relief device is defined to be undersized. This section shows the overpressure scenarios which were undersized. This includes all the contingencies for which the relief devices were undersized. The most common reasons for relief devices to be undersized were identified as being because of incorrect upstream pressure, incorrect composition of the relief stream, or that the contingency was not identified at all in the previous study.

The breakdown of the number and percentage of the undersized concerns that were analyzed in this study are shown in Table 6 and Figure 7 respectively. Of a total of 342 undersized concerns that were identified in this case study, the majority of the contingencies in the compression stations were Vacuum, Gas Breakthrough, External or Pool Fire and Failure of Automatic Controls.

Table 6: Percentage of type of contingency within undersized concerns

Contingency Type within Undersized Concern	Percentage
Blocked outlet	11.1
External fire	15.2
Failure of Automatic Controls	15.2
Gas breakthrough	15.5
Inadvertent valve operation	1.1
Internal Boundary failure	8.2
Reverse flow	8.5
Vacuum	24

Other	1.2

Undersized Cases (Count)

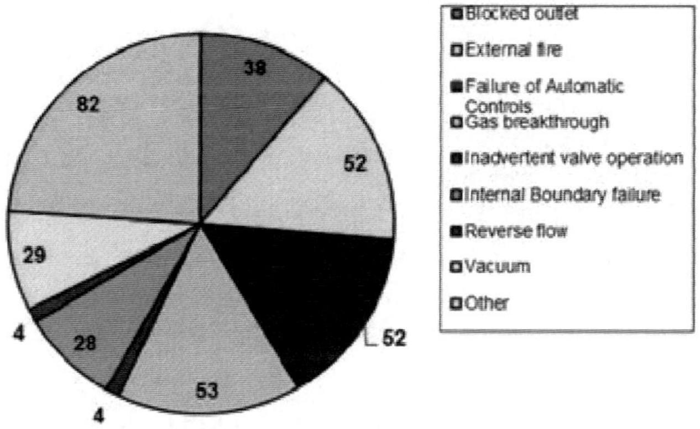

Figure 7: Actual number of different contingency types within the undersized concerns for the twenty compressor stations

<u>Installation Concerns:</u>

The installation concerns (total of 807 were identified) are divided into eleven sub categories as shown below. This is based on the guidelines used to perform the analysis.

Inlet Pressure Drop: The inlet pressure drop is calculated based on the relief device capacity. This includes all the scenarios for which the inlet pressure drop is greater than 3% of set pressure. As per API 520, if the inlet pressure drop exceeds 3%, the potential for capacity reduction and/or mechanical damage may exist for the relief valve. Note that inlet pressure drop concern was not identified for the relief devices which were undersized for one or more overpressure scenarios.

Outlet Pressure Drop: The outlet pressure drop is calculated based on the capacity of the relief device. This includes all the scenarios for which the outlet pressure drop is greater than 10% of set pressure for single conventional valves, or greater than 16% of set pressure for multiple conventional valves except for fire case, greater than 21% of set pressure for conventional valves for external fire, and greater than 50% of set pressure for bellows valves. As per API 520, if the outlet pressure drop exceeds the limit, the potential for capacity reduction and/or mechanical damage may exist for the relief valve. Note that outlet pressure drop concern was not identified for the relief devices which were undersized for one or more overpressure scenarios.

High Exit Velocity: The exit velocity and sound power level of the relief valve piping can affect its stability. All overpressure contingency with Mach number greater than 0.8 and sound power level greater than 155 was documented. Note that the high mach number concern was not identified for the relief devices which were undersized for one or more overpressure scenarios.

High set pressure: The relief device with set pressure greater than maximum allowable working pressure (MAWP) of the equipment was identified since the relief valve can no longer protect the system if this condition exists.

Isolation valve: This includes the block valves that are not chain locked open as required (the most common ones were the block valve on relief device inlet and outlet piping).

Pocketed Piping: In the event of liquid release and if the relief device is pocketed, the liquid can accumulate in the piping and hence the stability of the piping can get affected; hence it is very important to correct this concern.

Restricted line: In the event the inlet segment of the relief device piping is greater than the inlet nozzle as per API 521, these segments needs to be identified and documented.

The breakdown of the number and percentage of the installation concerns that were analyzed in this study are shown in Table 7 and Figure 8 respectively. Of a total of 807 installation concerns that were identified in this case study, the majority was the high outlet pressure drop concern.

Table 7: Percentage of installation concerns by type

Installation Concern	Percentage
High inlet Pressure drop	17.2%
High Outlet pressure drop	24.8%
High Exit Velocity	5.9%
Pocketed Piping	5.3%
Restricted Line	1.7%
Isolation Valves	11.8%
High Set Pressure	0.9%
No Weep Hole	20.8%
Liquid to Atmosphere	1.5%
Valve Stem Rotation	4.0%
Other	6.1%

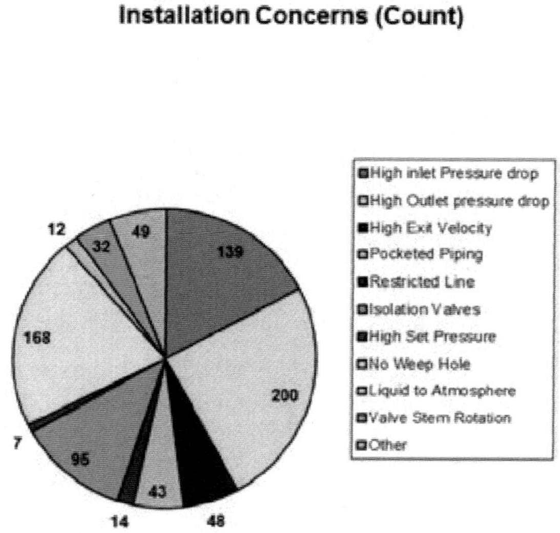

Figure 8: Actual number of installation by type for the twenty compressor stations

Unprotected Systems

These concerns were identified only for those systems that had applicable scenarios without any relief device protecting them. Contingency analysis for all the systems which didn't have any relief device protecting them was performed. As the case study was conducted on system based analysis, it was easier to identify these systems. The systems which had an internal relief device were considered to be unprotected. There were few cases in which there was a block valve present between the system and the relief device; hence these systems were considered unprotected. This list also includes the overpressure contingencies which resulted in over temperature of vessel in the event of external fire. Of a total of 3278 concerns identified, 311 were categorized as unprotected concerns.

Data Requirement

All the assumptions which were made during the analysis were documented under this category. During the analysis, a data need log was created if a proper supporting document was not provided and any follow on assumptions made due to the lack of documentation was approved by the client. Some general assumptions which were made were liquid levels, MAWP of the system (usually was assumed to be same as relief device set pressure), and relief valve information. Of a total of 3278 concerns identified, 1639 was data requirement concerns.

Documentation Updates

During the analysis, some discrepancies were found between different documents such as the U1A-form and the P&ID and these discrepancies were documented, with the correct information needing to be updated on the relevant documents. Documentation updates are usually a low priority concern as they may not impact the operation and safety of plant significantly. Of a total of 3278 concerns identified, 179 were documentation updated.

3.6. Mitigation of Concerns

Once the concerns were identified and documented, different recommendations were provided in order to mitigate the identified concerns. Usually, the clients choose to implement mitigation strategies predominantly for the higher priority concerns such as the unprotected concerns, installation and undersized concerns. An example list (not comprehensive) of the recommended mitigation strategies that were provided to the client for undersized, installation concerns and unprotected systems is detailed below.

Mitigation Strategies for Undersized Concerns

The provided recommendation to resolve undersized concerns depends on the overpressure contingency for which the relief device was undersized.

External Fire: In the event the relief device is undersized for external fire, it was checked if the vessel is insulated and whether credit of insulation could be taken. Usually, in order to overcome this concern, it was recommended that a bigger relief device be installed.

Blocked Outlet: In the event the relief device is undersized for blocked outlet, it was recommended that a bigger relief device be installed.

Isolated with Heat input: In the event the relief device is undersized for "isolated with heat input", it was recommended that a bigger relief device be installed. An alternate recommendation was to see if the undersized concern could be mitigated using administrative controls.

Tank Draining (Vacuum): This was a major concern identified in low pressure tanks. This overpressure scenario was missed in previous studies and hence, a lot of undersized concerns were found. It was recommended that a restriction orifice be installed on the drain line. Another option recommended was to install a bigger PVRV.

Failure of automatic control: For this undersized scenario, it was recommended that a restriction orifice be installed on the line, while still ensuring that it can allow the normal flow through it; however it can regulate the maximum flow such that the relief device is adequately sized. Another option recommended was that a restriction orifice be installed so that the maximum C_v of the control valve can be reduced, which ensures that the relief valve is adequately sized. It was rarely suggested that a bigger relief device be installed for this case.

Inadvertent valve operation: These cases usually included the inadvertent opening of bypass valves around the control valve. For these cases, usually the pipe capacity calculation was reperformed based on the equivalent length of piping from source to destination to check if the concern still existed. If the concern still exists, it was recommended that the bypass valve be chain lock closed or a restriction orifice be installed.

Gas breakthrough: Majority of the time, the failure of automatic controls and inadvertent valve operations resulted in gas breakthrough. For these cases, usually the pipe capacity calculation was reperformed based on the equivalent length of piping from source to destination to ensure if the concern still existed. If the concern still exists, it was recommended that the bypass valve be chain lock closed or a restriction orifice be installed.

Reverse flow: This is one of the most common overpressure contingencies that can be missed while doing the analysis. That is one the reason why the relief device was undersized for this case. In order to overcome this concern, it was recommended that multiple check valves be installed in the direction of flow. An alternate option recommended was the installation of a bigger relief device.

Mitigation Strategies for Installation Concerns

Inlet & Outlet pressure drop: For pressure drop concerns, it was recommended that the concerned piping be modified. The equivalent length of the piping was provided such that it was within the prescribed limit. Alternately, if the relief device was oversized, it was recommended that a smaller relief valve be installed. In this scenario, it is important to still ensure that the relief device is still adequately sized. For inlet pressure drop, it was recommended that a pilot valve with a remote sensing line be installed. For outlet pressure drop, it was recommended that bellows or a pilot valve be installed.

Mitigation Strategies for Unprotected System

If the system is unprotected, the most common recommendation was to install a new relief device. An alternate option recommended was to chain lock open any open block valve to take credit for other relief device. The third option recommended was to mitigate the scenario with administrative controls. This option was common for heat exchangers which have only thermal expansion as the applicable overpressure contingency.

4. Summary

This case study provides an overview of the relief analysis conducted for multiple compressor stations. As stated previously, the relief analysis was system based as opposed to relief device based so as to identify any unprotected systems that may need overpressure protection. It is very important to use the correct operating conditions and fluid composition to perform the relief analysis. The study provided an overview of the various types of overpressure scenarios identified for these systems, with a statistical breakdown of the scenarios by type. It was found that external fire and blocked outlet were the most commonly identified overpressure scenarios. In addition, this case study provided a breakdown of the types of relief devices, a further breakdown of types of relief valves, and the different types of equipment in the compressor stations. A breakdown of the type of concerns identified during the compressor station revalidation was provided along with a general idea of the mitigations that can be implemented to resolve the concerns. When performing relief analysis for compressor stations, it is very important to look out for any contingencies previously not considered, possibly due to changes in facility design or changes in the guidelines.

440451 An Approach to Reducing Risk during Online Isolation, Removal and Maintenance of Relief Devices Using Engineering Assessment

Abstract:

Overpressure protection must be provided for pressurized equipment by means of pressure relief devices, which sometimes require isolation and removal for inspection or repair. It would be advantageous to have a means of isolation without shutting down the process, especially in light of the additional risks associated with shutdown and startup; however, doing so may put the pressurized system at risk of being unprotected in the event of overpressure. Current recognized practices of providing dedicated personnel for monitoring and responding to an overpressure event, or installation of an independent spare, are non-ideal in many instances. An engineering assessment can identify specific ways for providing temporary overpressure protection, including elimination of root causes of overpressure events, provision of alternative pathways to overpressure protection, modification of process operations, and/or establishment of a means of equivalent or better risk reduction for the duration of the isolation. Guidelines and examples are provided to illustrate the analysis, along with recommendations for documentation.

441392 Common "Design Flaws" Identified during Pressure Relief System Analysis

Abstract:

Common "design flaws" identified during a pressure relief system analysis.

It is the goal of a pressure relief system analysis to identify possible gaps in a pressure relief system. In many oil and gas refining and petrochemical plants, it's common to encounter certain "design flaws" which can lead to overpressure scenarios where the system is not adequately protected.

Within these design flaws there are a few which frequently occur, regardless of the process. These can be just simple mistakes or related to consideration not being given to the whole system as part of the design. This paper will focus on five common design flaws:" unusually large bypass valves around a (smaller) control valve"; "liquid head between relief device and equipment being protected"; "how does the suction of the feed pump behaves during the scenario", "tube rupture can affect an entire system and not just the exchanger"; and "common API 650 tank concerns". For each "design flaw" a few common mitigations will also be discussed.

The goal of this presentation is not to discuss the issues in detail, but rather just bring to the industry's attention a few concerns being frequently identified.

Considering Overfilling to be a Non-Applicable Overpressure Scenario

Natalie Doe
Smith & Burgess
190 S. LaSalle St., Ste 550, Chicago, IL 60657
Natalie.doe@smithburgess.com

Dustin Smith, P.E.
Smith & Burgess
7600 W. Tidwell, Ste 600, Houston, TX 77040
Dustin.smith@smithburgess.com

Prepared for Presentation at
American Institute of Chemical Engineers
2016 Spring Meeting
12th Global Congress on Process Safety
Houston, Texas
April 11-23, 2016

Considering Overfilling to be a Non-Applicable Overpressure Scenario

Natalie Doe
Smith & Burgess
190 S. LaSalle St., Ste 550, Chicago, IL 60657
Natalie.doe@smithburgess.com

Dustin Smith, P.E.
Smith & Burgess
7600 W. Tidwell, Ste 600, Houston, TX 77040
Dustin.smith@smithburgess.com

Keywords: PSV, Risk mitigation, liquid overfilling, Safety Instrumented Systems

Abstract

The consequences associated with liquid overfilling are serious for plant safety, as illustrated by many recent incidents (2009, Bayamón, Puerto Rico; 2008, Chesapeake, VA; 2005 Hertfordshire, England; 2005 Texas City, Texas). In many cases, providing overpressure protection solely with pressure relief devices may not protect the facility. Preventing the uncontrolled release of liquid may be the only viable safe option. This paper will present an abbreviated analysis of several incidents that could have been avoided with proper preventive measures. Engineering methods will be presented to eliminate the need to release liquids from a closed system. For instance, the system can be designed such that upstream pressure sources are limited below the allowable downstream pressure. A designer can provide a safety instrumented system with an appropriate Safety Integrity Level (SIL) rating to reduce or eliminate the credibility of overfilling. Alternatively, consideration may be taken for alarms and operator response, with consideration for RAGAGEP, to reduce the credibility of overfilling to an acceptable risk. Systems can be redesigned to be inherently safer by minimizing contained liquids. This paper then gives a few examples of the use of system design, shutdown systems, alarms/operator response combination and inventory controls to eliminate or reduce the potential for liquid release scenarios.

1 Introduction

The consequences of an overfilling overpressure scenario can result in increased risk to the facility when mitigated with pressure relief devices. This could occur due to many different root causes. In order for liquid overfilling to require overpressure protection from a relief device, the following must all be true:

1. The source pressure must be capable of exceeding the system MAWP. Since the relief device is supposed to be set at or below the system MAWP, if the relief device is set lower than the MAWP, a liquid release may occur at a lower pressure.
2. The flow of liquid through the system can be stopped by closing a valve on the system outlet or the loss of a discharge pump (etc). In general, system designers ensure that liquid flows can be stopped, so this is usually a given.
3. There are no acceptable administrative or instrumented controls in place to prevent the scenario.

When reviewing the consequences of an overpressure scenario, the designer must check to ensure that the overfilling event may also not occur during a larger plant upset (power failure, instrument air failure, or other global utility disruption). This would add a simultaneous load to the disposal system, which should be included in the disposal system design.

Overpressure protection

The system designer should consider the root cause of the overfilling scenario to determine how to protect the system or eliminate the scenario. Relief devices are often used to prevent excessive vessel overpressure due to liquid overfilling. For many systems, overfilling is not expected to be the controlling case for relief device sizing; however, a review of incident history shows that many different systems have been overpressured by overfilling with liquid, resulting in relief. Therefore, when relief devices are used to provide protection for overfilling, it is important that proper consideration is given to the effluent handling system.

1. For a system with several pieces of protected equipment, consideration must be taken for the static head throughout the column to ensure equipment at grade is adequately protected. This is particularly an issue for low pressure tanks.
2. When systems overfill, there is the potential for a large amount of liquid to be released to the effluent handling systems. Unless liquid holding drums are sized for the potential liquid rate, there could potentially be an unplanned liquid release with significant consequences.

This paper will outline recent incidents and their consequences due to overfilling, and will overview methods to protect the vessels from overpressure and the system from the consequences of large liquid releases.

2 Overview of recent incidents

The following incidents from the past decade illustrate the consequences of overpressure due to overfilling and the handling of liquid releases.

2009, Bayamón, Puerto Rico;

Per the CSB report, "Tank 409 overflowed with gasoline, resulting in a vapor cloud that encompassed 107 acres of the CAPECO tank farm." [1] The report also states that "Tank 409 overflowed for an estimated 26 minutes before the vapor cloud ignited. The CSB determined nearly 200,000 gallons of gasoline... rushed out of six vents in the tank." These six vents responded to the upset conditions to relieve the excess pressure in the tank; thus, the overpressure protection provided was sufficient to prevent a catastrophic vessel failure. The action of the "engineered" pressure relief devices for this scenario worked and then became the initiating event for this incident. The final result was that the company went bankrupt and 17 of 48 tanks were burned.

2008, Chesapeake, VA;

Per the CSB report, "At about 2:20 pm, as the tank reached a level of 26.72 feet, a vertical split started midway up the shell and rapidly extended to the floor and roof of the tank on the side opposite the workers." [2] In this case overpressure protection to prevent catastrophic failure of the tank would not have worked because the tank was not "full". This incident could not have been protected for by pressure relief devices - only instrumentation, procedures or the like.

2005 Hertfordshire, England;

Per the aforementioned CSB report, "Similar to the CAPECO incident, the vapor cloud explosion and multiple tank fires occurred after a tank was overfilled with gasoline." [1] Thus another incident of overpressure protection for the overfilling scenario working; as such, it operated as designed. The "engineered" overpressure mitigating system was the initiating event for an incident that caused more than a billion dollars' worth of damage to the surrounding commercial and residential property and injured 43 people.

2005 Texas City, Texas

Per the CSB report, "The incident occurred during the startup of an isomerization (ISOM) unit when a raffinate splitter tower was overfilled; pressure relief devices opened, resulting in a flammable liquid geyser from a blowdown stack that was not equipped with a flare. The release of flammables led to an explosion and fire." [3] The pressure relief devices opened, and prevented excessive overpressure of the ISOM tower. As such, the engineered pressure relief devices in this case also worked, preventing a catastrophic overpressure-related loss of containment, but the design for the disposal of the fluid created an additional hazard with significant consequences.

Many of these incidents involved tanks, but that is inconsequential as in each and every case either a relief device was insufficient to prevent overpressure, or the engineered design of the pressure relief device and disposal was the root cause for these scenarios. Not only do the pressure relief devices have to be designed to prevent loss of containment from overpressure, the disposal system has to be designed to ensure the effluent has a safe location to be stored or is adequately destroyed. This can be particularly difficult for liquid releases.

3 Means of overpressure protection

Relief Devices and Disposal

While relief devices are an acceptable method to protect from overpressure due to liquid overfilling, complete risk mitigation requires proper consideration of the ability of the disposal system to handle the liquid effluent. As mentioned previously, additional pressure due to static head should be considered when protecting multiple pieces of equipment. For column systems with relief devices located on the top of the column, this can have significant impact on associated equipment located at grade. It is common practice to perform overfilling relief device calculations at 10% overpressure rather than 16% overpressure when multiple devices are installed, in an attempt to account for the static head. For tall column systems, or for systems with a heavy relief fluid, this may not be sufficient to ensure the allowable accumulation on the equipment located at grade is not exceeded. The relief device set pressure may need to be lowered below the system limiting MAWP to ensure adequate protection for all equipment.

Additionally, the disposal system must be capable of handling the liquid flow from the relief devices. The disposal piping should be evaluated to determine if it can handle the weight of the liquid. A knockout drum must be incorporated as part of the disposal system to ensure liquid is not sent to the flare or vent stack. For large liquid releases, it may not be practical to size one knockout drum to obtain the required particle separation. Multiple knockout drums in parallel can be used to ensure adequate particle separation for these cases. Failure to size the knockout drum(s) correctly can lead to liquid being sent to a system which cannot handle it, which can have serious consequences. The CSB report which reviewed the Texas City incident [3] unduly criticized atmospheric blowdown drums; many of their preferred disposal systems, i.e. flare systems, have had similar incidents [4] with the authors aware of countless near misses from flare systems.

Overpressure protection by system design

The need to design a disposal system for the liquid effluent from relief devices that protect a system of equipment from overpressure by overfilling can be eliminated by designing the system such that relief devices are not needed. The acceptable way to do this is to ensure that the overfilling scenario is not credible. For vessels designed to ASME VIII, UG-140 (formerly known as code case 2211) states "A pressure vessel does not require a pressure relief device if the pressure is self-limiting… and this pressure is less than or equal to the MAWP of the vessel at the coincident temperature…" [5]. This can be achieved in many ways.

Equipment Design

A designer can utilize equipment design to ensure that the disposal system does not need to accommodate effluent from relief devices installed for liquid overpressure protection.

<u>Example 1 – Pump Selection</u>

Instead of feeding a system with a positive displacement pump, a centrifugal pump may be used that is not capable of exceeding the downstream system MAWPs. This is illustrated in Figure 1 below, as P-1 cannot exceed the MAWP of V-1. The defining quality of centrifugal pumps is that their flow will decrease with increased head. Therefore, as the pressure in the downstream system increases, the capacity of a centrifugal pump will decrease. Thus, pump selection can be an effective method of limiting the overfilling rate.

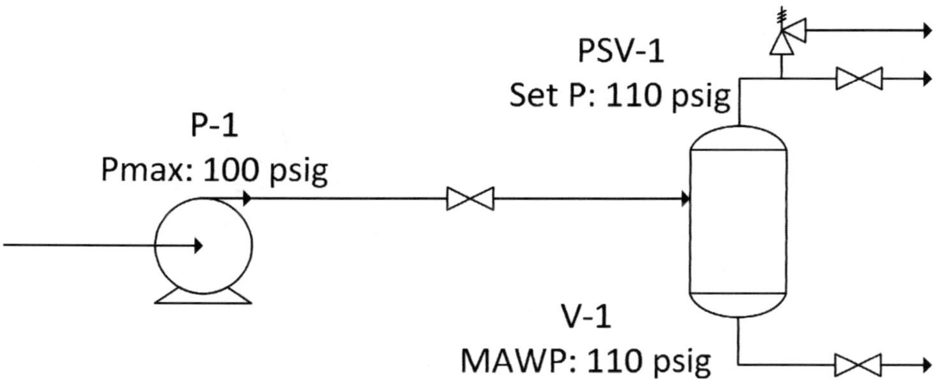

Figure 1: Upstream sources incapable of overpressuring the system

In contrast, positive displacement pumps are constant volume machines that are capable of maintaining high volumetric flow even at increased pressures. Therefore, alternative methods can be applied to protect from overpressure due to positive displacement pumps.

Example 2 – PRV Discharge to Process

If a relief device is installed on the pump instead of the vessel, the discharge piping can be routed to the pump suction. This allows for the liquid relief to be sent to process, rather than to a disposal system. However, the designer is cautioned that this will stop the forward flow of liquid so equipment upstream must be checked for this upset condition. An example is shown in Figure 2 below, as PSV-2A reduces the maximum feed pressure and eliminates the overfilling scenario on V-2.

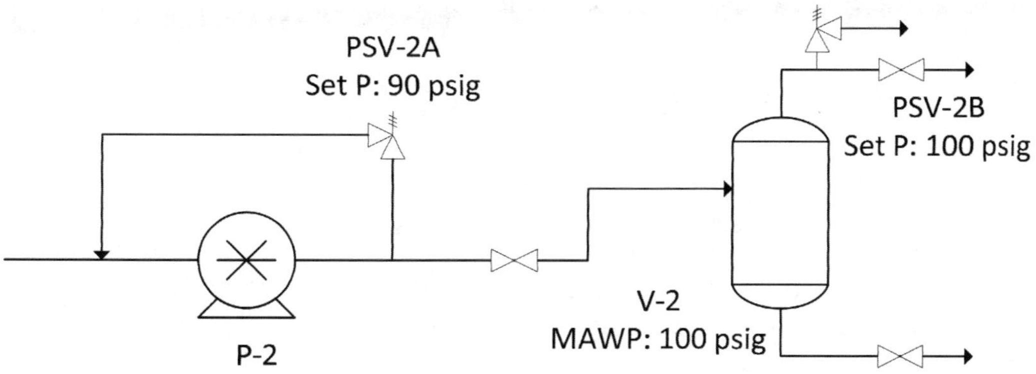

Figure 2: Protecting upstream pressure sources

In other cases, protecting for overfilling by pump selection or upstream protection may not be possible. Inventory controls may be utilized in these situations to ensure overfilling is not credible.

Example 3 - Inventory Controls

If possible, a designer can minimize the liquid inventory in a system. This is also a tactic in inherently safer process design. If the total liquid inventory of a system is insufficient to fill the vessel(s) vapor space then overfilling cannot occur. In this case overpressure does not occur irrespective of the maximum upstream pressure. An example of this is illustrated in Figure 3. The positive displacement pump (P-3) cannot overpressure the vessel (V-3) because the liquid inventory in the upstream tank (T-3) is less than the vessels vapor space.

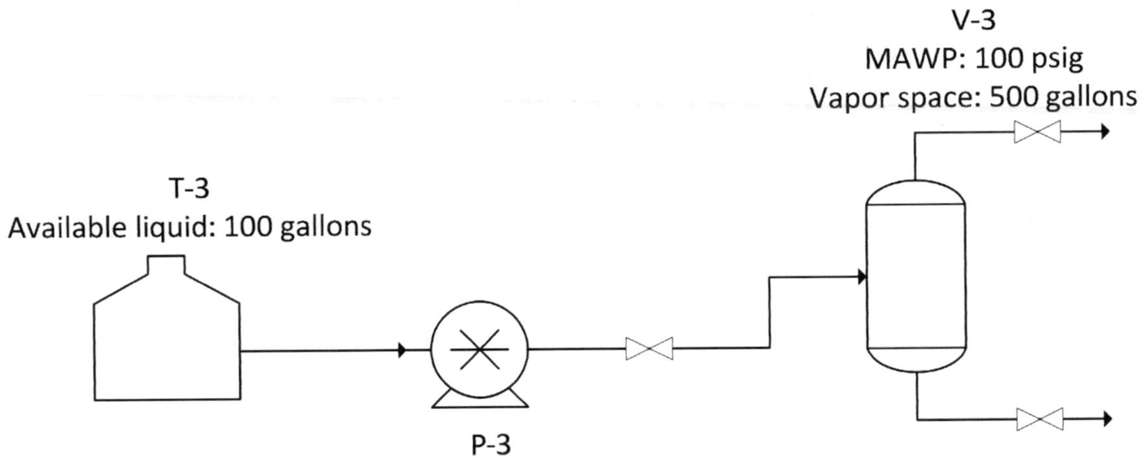

Figure 3: Inventory Controls

Instrumented Solutions

For installations where it is not practical to remove the credibility of the overfilling scenario by equipment design, safety instrumented systems (SIS) can be implemented. There are two ways to protect a facility from the effects of liquid effluent: remove the relief device entirely, or mitigate the consequences of the liquid effluent escaping from the disposal system.

Example 4 – Replace Relief Devices with instrumentation

The system can be designed such that instrumentation is used instead of overpressure protection. A designer can find guidance in API STD 521 which states in Annex E that "High Integrity Protection Systems (HIPS) are designed to avoid overpressure incidents by removing the source of overpressure or by reducing the probability of an overpressure contingency to such a low level that it is no longer considered to be a credible case."[6] ASME VIII UG-140 allows for the use of these systems, provided that detailed description of the safety system is documented, and includes "the identification of all truly independent redundancies and a reliability evaluation... of the overall safety system" [5] The ASME requirements for these systems are further explored elsewhere [7]. Figure 4 below shows an example of a system protected by instrumentation. Three separate level taps are used to determine the liquid level in V-4. Two independent emergency shutdown devices (ESD), each with an independent controller, are used to shut off the feed on high liquid level. The redundancies expressed in this example would be integral to determining the safety integrity level (SIL) of this system.

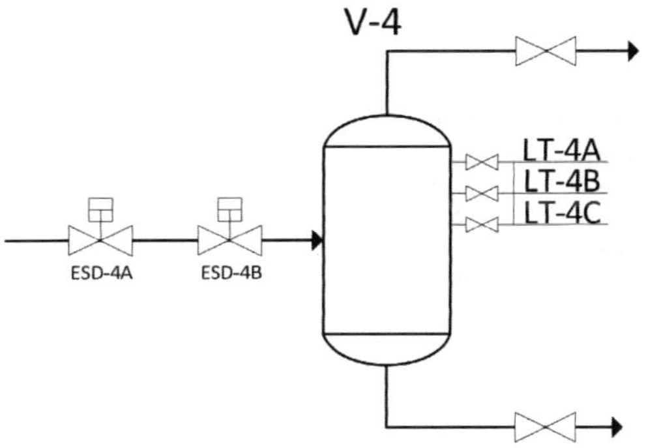

Figure 4: Protection through instrumentation

Example 5 – Add a layer of Protection

Alternatively, the design of a SIS does not require that overfilling is mitigated to a point at which it cannot occur; rather, that the probability is low enough that the total system risk is at an acceptable level. Depending on the existing system, the most cost effective solution may be to mitigate the effluent system release with a SIS but rely on a the relief device to protect the equipment. Due to the uniqueness of each system a risk assessment (such as a LOPA) will be required to determine the reliability of the SIS. The designer may choose to put in a less reliable safety instrumented system in combination with an existing relief device, or the safety instrumented system may replace the relief device. As mentioned above, these redundancies are used to determine the safety integrity level (SIL) of the system. For most cases, a SIL-3 System is needed to remove the credibility of the overfilling scenario. However, a SIL-1 or SIL-2 system may be acceptable when reviewed in a risk analysis [such as Layers of Protection Analysis (LOPA)] to mitigate an incident caused by a release of liquid to the disposal system. Either way, a detailed risk assessment should be performed to determine the SIL rating required to mitigate the hazard [8][1]. An example of a SIS that could provide overfill protection would be level indicators independent of the process control instrumentation from a separate vessel tap feeding a dedicated PLC that triggers a shutdown valve to shut off the feed to the system on a high level alarm.

Operator Response to mitigate overfilling

For many facilities, a combination of instrumentation with an operator response is a reasonable mitigation of risk from overpressure. Research has shown the following factors tend to determine the reliability of operators to take the intended action [9][10]:

1. Trigger – there is a clear indication to the operator that something is wrong triggering the need to take action. It is unreasonable to assume that the operators would have mitigated the overfill scenario at Bayamón PR as they had no indication of level or procedures to know when the tank overfilled [1].
2. Time – Per the work performed by the NRC [9], an individual in a stressful situation with limited time to perform a moderately complex task has a 100 times likelihood of failing compared to the exact same task with sufficient extra time allotted for execution. Others, [10] have put the reliability of an operator completing a clearly understood task on which he/she is trained, with adequate time under stressful conditions, is equivalent to a SIL-III system.
3. Consequences – if the mitigation action causes (or is perceived to cause) a chain reaction that is worse than the failure being mitigated there may be reluctance to perform the action. Additionally, if the operator assumes (or is told) that they will be punished for stopping production, then the likelihood of following the procedure decreases greatly.
4. Training – without the knowledge and training to act, the probability that an action will be taken diminishes greatly, especially if the action is not obvious to the individual operating the facility.

As safety instrumentation systems can be complex and expensive, many facilities choose instead to take credit for operator response to high level alarms. There are several important factors to consider when taking these credits. API 521 4.4.7.3 [6] outlines several factors to consider when relying on level alarms:

a) Whether the level instruments used for safeguards against overfilling are on separate process taps from the process control system
b) Whether level instruments used for safeguards against overfilling are susceptible to the same common mode device failures as those used for the basic process control system.
c) Whether the programmable level transmitter is set to show a low or zero level when the level exceeds the instrument range
d) Whether an open system differential pressure cell is used that can show a false low level due to accumulation of liquid in the upper impulse line if it is overfilled or collects condensed liquid
e) Whether density changes due to temperature effects or composition changes impact readings from differential pressure transmitters
f) Whether the instruments are proven in use for the specific process applications
g) Whether the range of at least one of the level measurement(s) can indicate a valid level reading of the full range between the high critical alarm point and any shutdown or interlock point
h) Whether operating characteristics of the level measurement during off-design, start-up, and shutdown operations are considered in display of level, physical properties, setting alarms, trip points, operator training, and operating procedures
i) Maintenance and testing frequency required for the instrumentation

Even if all of these factors are considered, and the reliability of the operator is evaluated, it is important to analyze the risk associated with human error, and the operator not responding in the allotted time. As with instrumentation systems, relying on level alarms and operator response is not a guarantee overfilling will not occur; rather, it lowers the risk to an acceptable level.

Overfilling of low pressure tanks present a unique challenge as compared to pressure vessels. API has created a specific document to provide further guidance to protect for the overfilling of low pressure storage tanks [11].

Example 6 – Alarms with subsequent operator response

As with Example 4 above, Figure 5 shows a vessel (V-6) with three independent level taps. These level taps are connected to two independent high level alarms (HLA). The designer performed a LOPA, and determined the operator can reliably address the alarm in 30 minutes. This is less than the fill time of V-6 from the time when the alarm sounds (1 hour), so credit may be taken for the operator actuating HV-6A and HV-6B to shut off the feed and prevent overfilling.

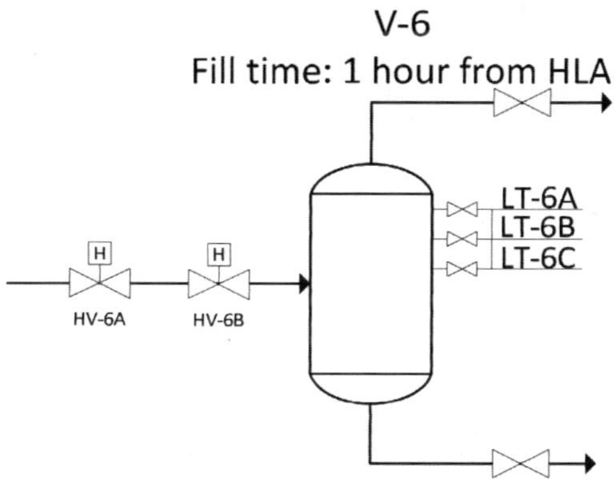

Figure 5: Operator Response

4 Conclusion

Liquid overfill can have serious safety and financial consequences, as seen in the recent incidents outlined in this paper. It is important for the designer to consider both the means to prevent overfilling, as well as methods to respond to an overfilling event. Appropriately designed relief devices and disposal systems are an acceptable method to react to an overfilling event. However, if this is to be the sole means of protection, careful consideration must be given to the large liquid effluent rate sent to the disposal system. Alternatively, the event may be prevented through the proper application of equipment design, instrumented solutions, alarms and operator response, or some combination thereof.

There are a large number of incidents involving overfilling that occurred despite the fact that the installed overfill protection reacted as designed. The protecting relief devices opened and were adequate to prevent overpressure of the overfilled vessel, but the disposal system was not designed for that case. The incidents mentioned in this paper resulted in billions of dollars' worth of damage, and even the bankruptcy of one company. When protecting for overfilling, it is imperative to consider the ramifications of allowing large liquid releases to occur. Overfilling protection design that encompasses consideration of all of the options presented above can prevent an incident with catastrophic consequences.

5 References

[1] U.S. Chemical Safety and Hazard Investigation Board. "Caribbean Petroleum Refining Tank Explosion and Fire" Available at http://www.csb.gov/caribbean-petroleum-refining-tank-explosion-and-fire/. Accessed on 2/5/2016.

[2] U.S. Chemical Safety and Hazard Investigation Board. "Allied Terminals Fertilizer Tank Collapse" Available at http://www.csb.gov/allied-terminals-fertilizer-tank-collapse/. Accessed on 2/5/2016.

[3] U.S. Chemical Safety and Hazard Investigation Board. "BP America Refinery Explosion" Available at http://www.csb.gov/bp-america-refinery-explosion/. Accessed on 2/5/2016.

[4] HSE (Health and Safety Executive) (Eds.), 1997. The explosion and fires at the Texaco Refinery, Milford Haven, 24 July 1994. A report of the investigation by the Health and Safety Executive into the explosion and fires on the Pembroke Cracking Company Plant at the Texaco Refinery, Milford Haven on 24 July 1994, HSE Books, London.

[5] American Society of Mechanical Engineers, 2013 ASME Boiler & Pressure Vessel Code: Section VIII, Division I. s.l.

[6] API Standard 521, "Pressure-Relieving and Depressuring Systems," American Petroleum Institute, 6th ed., 2014.

[7] Smith, Dustin J. "Safety instrumented systems in lieu of pressure relief valves." *Process Safety Progress* 33.4 (2014): 345-349.

[8] A.E. Summers. "Overfill Protective Systems – Complex Problem, Simple Solution," 12th Annual Symposium, Mary Kay O'Connor Process Safety Center, October 2009.

[9] Blackman, Harold S., David I. Gertman, and Ronald L. Boring. "Human error quantification using performance shaping factors in the SPAR-H method." *Proceedings of the Human Factors and Ergonomics Society Annual Meeting*. Vol. 52. No. 21. SAGE Publications, 2008.

[10] Mannan, Sam, ed. Lees' Loss Prevention in the Process Industries: Hazard Identification, Assessment and Control. Vol. 1. Butterworth-Heinemann, 2004.

[11] API Standard 2350, "Overfill Protection for Storage Tanks in Petroleum Facilities," American Petroleum Institute, 4th ed., 2012.

How Reliable Are Your Process Safety Barriers?

Mike Neill
Petrotechnics USA Inc.
10497 Town & Country Way, Suite 700, Houston, TX 77024
mike.neill@petrotechnics.com

Prepared for Presentation at
American Institute of Chemical Engineers
2016 Spring Meeting
12th Global Congress on Process Safety
Houston, Texas
April 11-23, 2016

How Reliable Are Your Process Safety Barriers?

Mike Neill
Petrotechnics USA Inc.
10497 Town & Country Way, Suite 700, Houston, TX 77024
mike.neill@petrotechnics.com

Keywords: Process Safety Management; Process Safety Barriers; Operational Risk Management

Abstract

Using Process Safety Management techniques, we design and implement multiple layers of protection for our hazardous plants. For these barriers to remain effective over the life of the asset is a challenge, particularly with aging assets. We know barriers deteriorate over time without proper, preventative maintenance and inspection. Often a lack of focus on critical barriers can result in activities being deferred or cancelled without recognizing the potential consequence.

There is generally a lack of understanding in our organizations of process safety techniques and approaches. While the criticality of these barriers maybe well-understood in the Process Safety department, this is unlikely true through other organizations, including Operations and Maintenance.

With so much equipment being tagged "safety critical" on our facilities; it is often difficult to ensure safety barriers receive the focus they need to ensure their effectiveness at protecting against high consequences events when needed. Operators need to manage the health of their barriers in near real-time and mitigate the impact arising from any impairments.

By understanding dynamic aspects of cumulative risk, all members of an organization can easily identify barrier impairments. This level of visibility helps make better operational activity management decisions on account of operational activities' potential impact on Major Accident Hazard barriers.

This paper will examine the traditional approach to barrier design and assumptions on reliability, the approach to maintenance, and the reality of operational dynamics. It will introduce the concept of cumulative risk impacts on safety barriers and examine operational dynamics which impact barrier performance.

1 Introduction

Confucius, the Chinese Philosopher from around 500BC is quoted as saying that "a man who lacks reliability is utterly useless." Well the same can be said for our Process Safety barriers, whose sole purpose is to prevent major accidents or mitigate the consequences of such events. These barriers enable us to operate our hazardous facilities, exposing people to potential danger, since we satisfy ourselves that the calculated risk of incident is acceptable, assuming reliable performance. Unfortunately, when we investigate incidents, we see again and again examples where multiple barriers fail to perform. Occasionally we see evidence where the design was at fault and the assumptions made were incorrect, and we see plenty of evidence where Management Systems designed to prevent barrier failure, fail themselves.

All too often the barriers do not work as intended for a variety of reasons ranging from corroded pipe, seized valves, inoperable switches, blocked flowlines to a simple case of responsible people failing to follow operating procedures.

We focus a lot of attention on making sure we design our barriers properly in the first place but do we do enough during the life of the facility to make sure the barriers continually perform per design? Are our organizations set up to optimally manage these barriers and are our frontline workers engaged sufficiently and armed with the right knowledge to play their role in identifying barrier impairments and to make appropriate operating decisions? These are questions this paper will address and will conclude with examining new solutions which are becoming available to help the industry with this challenge.

2 Evidence of Barrier Failure

The failure of barriers designed to protect or mitigate the consequences of Major Accident Hazards (MAH) often results in an incident such as a significant gas release or an over pressure situation. The failure of multiple barriers can escalate into a major accident such as an ignition barrier failure coincident with a gas release which results in an explosion or fire. The term "barrier", refers not just to equipment hardware such as piping, vessels, instruments, actuated valves, etc., but also includes procedural systems, where plant staff follow agreed practices to operate the plant and respond to emergencies.

The latest Marsh and McLennan report (ref1) provides a current update to property damage losses in the hydrocarbon industry. This report, in its 23rd edition, outlines the 100 largest losses across our industry over the past 40 years. A direct quote from that report

> *"Accidents that result in major losses, such as those reported in this document, generally occur because of the failure of a number of barriers within the process-safety management system — all occurring at the same time; typically, none of these losses are the result of the failure of a single barrier or protection measure."*

Furthermore, in the same report, other quotes reflecting on the incidents,.

"N*one of the losses listed in this document should be considered 'black swan' events.*"

("Black Swan" events are those regarded as inconceivable and impossible to consider as a credible threat.)

The implication of this statement is that the scenarios and failure modes were anticipated so either the design of the barriers was inadequate to meet conditions or the design of the barrier was adequate, but its performance did not match expectations. It could also be a combination of both.

In another report, (ref2) Ron Jarvis from Swiss Re looked at a sample of 45 incidents in the Oil, Gas, and Petrochemical fields over the prior 15 years. Despite this being a random sample and not necessarily a statistical analysis, it is worthy of consideration, not least because its aim was to prioritize those risk control elements whose failure resulted in a majority of incidents.

The study concluded 52% of primary incident cause was failure due to corrosion, 27% failure of operating practices and procedures, and 14% was attributed to unsafe maintenance, equipment isolation and permit to work management. Considering secondary and tertiary causes, poor design and/or inadequate hazard identification is another cause which becomes more prominent. The author goes on to identify inadequate piping and equipment inspection programs resulted in 68% of corrosion failures, the remainder being due to incorrect material selection or defective materials when first installed. Regarding the failure of operating practices and procedures, Jarvis highlights poor management of deliberate by-passing of Safety Critical Equipment (SCE) and the management of abnormal situation on the plant. 57% of all incidents occurred during abnormal conditions on the plant, including start-up, shutdown, or maintenance.

The evidence from these two studies is conclusive that Major Accidents occur as the result of multiple barrier failures. You may ask why it is important to recognize this seemingly obvious fact! This Author believes it is having attended many Process Safety Conferences over the years where the emphasis is on barrier design and static risk scenarios with little focus on what happens to barriers over the life of the asset. This, and working in an industry where there seems only to be a tacit recognition of a dynamic component to MAH risk in which a 5 year review of safety management systems seems adequate. The fact is, barriers fail and over time most, if not all, will fail without periodic inspection, testing, and preventative interventions.

The challenge we face is the data on Major Hazard Barriers, which can indicate an impaired status or diminished integrity, is not easily shared with the people who need to know and who routinely make decisions on daily operations based on assumptions that everything is in good health.

3 Barrier Design Versus Barrier Performance

3.1 Design Principles

When designing Oil, Gas, and Petrochemical plants, many of the principles used are determined by regulatory requirements in addition to generally accepted engineering design standards. Wall thicknesses and material specifications for pipes and vessels will be based on the expected fluid properties and operating conditions as will the specifications for valves and rotating equipment selection. HAZOP and HAZID studies look at potential hazard scenarios and failure mechanisms and identify methods for minimizing risk and/or mitigating the outcome. Risk-based studies look at ways to reduce MAH risk to a level which meets the criteria of acceptance. When the risk levels are excessive, additional barriers are designed.

Unfortunately, barriers themselves are not 100% reliable. This can be the result of a material defect in piping or vessel walls, or corrosion of those same walls, a stuck relief valve, a deluge system plugged with scale, or a standard operating procedure which is not followed. A single safety barrier cannot eliminate risk but it can reduce the likelihood of an event happening or mitigate the magnitude of the consequences if it does occur. As such, the traditional design approach is to apply multiple barriers, some preventative and some mitigative. Together they reduce the overall consequential risk.

Modern techniques of Inherently Safe Design, pioneered by Trevor Kletz, seek to build plants where hazards are removed rather than controlled and where volumes of hazardous fluids are minimized to reduce the magnitude of consequences. This is paving a way towards safer plants. At present, the fundamental needs of Oil, Gas, and Petrochemical plants continue to demand processing flammable, explosive, and toxic fluids often at high pressures and high temperatures and in large volumes. As such, the approach of multiple barriers to reduce and mitigate risk is still essential for plants - old and new.

When a hazardous scenario has been identified, the risk of the consequences occurring is calculated using statistical and Quantitative Risk Analysis (QRA) techniques. Risk acceptance criteria is often prescribed or advised by the regulator or set by internal corporate policy for an Operator. For example, the threshold may be a broadly acceptable risk of death of 1×10^{-6}. If a scenario results in the calculated risk of this outcome being higher than the threshold, additional barriers need to be designed to reduce the risk level to one which is acceptable.

Each barrier is credited with a level of risk reduction. This is calculated using performance characteristics of the barrier which, for hardware, is provided by the manufacturer based on empirical data such as "mean time between failure," etc. The assurance from the manufacturers of this performance is founded on a manufacturing quality system designed to provide reproducible results.

For vessels and piping the wall thickness and material specification are the principle drivers of strength and the ability to contain the process fluids. Once again, manufacturers' quality assurances assume manufacturing standards are followed. Mill certificates demonstrate material specification compliance. Hydrostatic testing of vessels and fabricated flowlines / pipelines, at pressures below the material yield points, are part of commissioning new plant and verifying a high level of integrity prior to start-up.

This process of risk reduction through designing additional barriers where necessary is carried out across the facility until the overall level of risk is reduced to an acceptable level.

Unlike hardware barriers, Procedural Barriers typically are less reliable since they require human interventions to be consistent and repetitive and with a discipline to follow standard procedures. Since most Operators are not robotic by nature, reliability is less than for engineered systems. On balance, this is a good thing since, unlike robots, Operators are free-thinking, unconfined by a narrow scope, and are able to identify a wide spectrum of developing situations and respond appropriately to regain control.

Reliability for procedural barriers is improved if the procedures themselves are well-written and workable so have been repeatedly tested and improved over time. The use of checklists provides a simple go/no-go approach, removing ambiguity in the step-by-step procedure. Training, and of course, experience are the other strong determinants of the reliability of procedural barriers.

3.2 Barrier Categories

The diagram below shows an example of fundamental barrier groupings. These are categories of barriers based on their function. While shown as the familiar "Swiss Cheese" model, the layout is similar to a "bow-tie" diagram where the left hand barrier groups are "preventative" and the right hand are "mitigative." It is useful to understand barriers in the context of multiple layers of protection and to appreciate that the term "multiple layers" does not necessarily indicate multiple redundancy.

As an example, if there is a gas leak on a plant than the failure in itself is a serious incident and has the potential for fire and explosion. Another group of barriers would be ignition control, preventing the escaping gas from igniting. Gas detection is another barrier and fire suppression systems a further barrier categorized under Protection. So each of the multiple layers maybe performing different roles to either prevent an event happening or to mitigate the resulting consequences or preventing escalation.

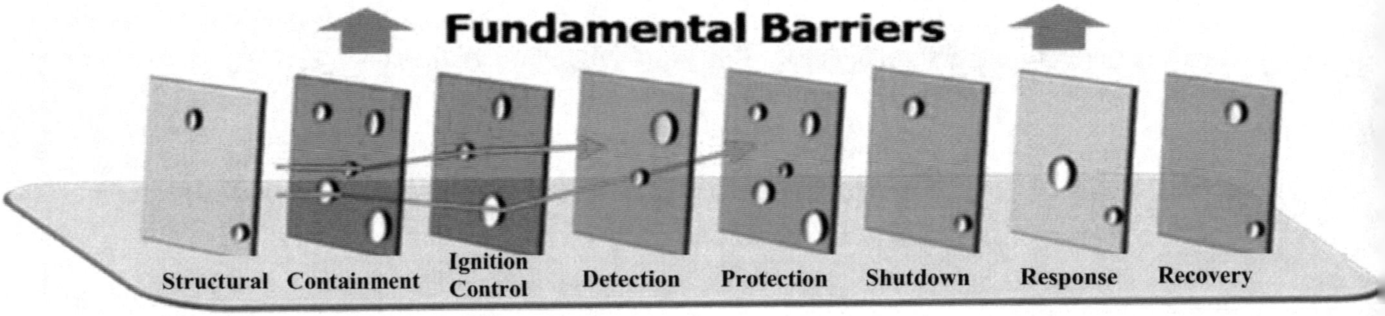

Figure 1. Fundamental Process Safety Barriers

Another example of categorizing is shown in the table below was proposed by Blacklaw, et al. (ref 3), detailing some of the specific barriers for offshore installations:

Table 1. Cumulative Risk Assessment Barrier Model

Barrier System	Example Components
Structural Integrity - (offshore structure)	Jacket, Topsides, Cranes
Process Containment	Vessels, Pipelines, Rotating Equipment, Pressure Systems, Relief Valves
Ignition Control	Ignition Prevention Systems, Flare Systems, Ventilation, Positive Pressure Systems, Gas Seals, etc.
Detection Systems	Fire & Gas Detection and Alarm Systems
Protection Systems	Fire Water Pumps, Deluge & Sprinkler Systems, Safe Refuge, Foam Systems, Blast Walls, etc.
Shutdown Systems	Pressure Trips and Shutdown Valves, ESD Systems, etc.
Emergency Response	Escape Routes, Emergency Lighting, Fire Water Mains & Portable Equipment, etc.
Life-Saving	Life Rafts, Personal Survival Equipment, etc.

3.3 Barrier Failure and Maintaining Barrier Performance

The reliability of engineered barriers will deteriorate over time, eventually to a level where they will just not work. This is simply a factor of material changes due to environmental actions such as corrosion or the rigors of service, erosion, wear, etc. To prevent or slow this deterioration, regular inspection, testing, and preventative maintenance is necessary.

As was explained in section 3.1, the reliability of engineered barrier performance determines the level of risk reduction for which credit is taken resulting in a residual risk which meets the acceptance criteria.

The data used for barrier reliability is based on manufacturer's information but assumes an ongoing schedule of inspection, maintenance, calibration and function testing to maintain that performance. What this means in practice is to take risk reduction credit for a shutdown valve, all the components of the system, from the pressure switch sending the signal to the controller, the actuator, valve stem, valve body and valve seat all have to be periodically inspected, tested, and undergo preventative maintenance. If these activities fail to occur within a specified timeframe, the_only consequence is the risk of major accident increases. Without these sustaining activities, dependability of the system declines, reducing or eliminating the ability to claim a risk reduction credit- thus the consequential risk has to increase.

This simple logic is important because there are often temptations to defer inspection and maintenance of such systems due to other operational priorities. Such deferrals are justified by arguments that it is preventative work and that there have not been reports of any failures…..so there is no reason to assume they will not still work!

The single most important set of barriers on a plant are those classed as "primary containment." Primary Containment includes pipework and vessels, plus a myriad of valves and flanges which connect them all together. It is not only the main flowlines but also includes by-pass lines, impulse lines for pressure switches, temperature gauges, sight glasses, etc. plus small bore pipework for instrument connections, bleed lines, chemical injection, etc.

The performance of these barriers is essential to keep the process fluids contained and under control. They are simpler than instrumented systems, not relying on moveable parts but instead on wall thickness and material specification plus the devices used to couple such as flanges, gaskets, bolt tensions, etc. Corrosion, both internal and external, are the most common failure mechanisms. Erosion, scale deposits, chemical damage and accidental damage will play their part. Inspection programs monitoring for losses in wall thickness are typically in place across the facility. These are supplemented with preventative programs such as protection coatings, corrosion inhibitors, anodic and cathodic protection, etc. Periodic hydraulic testing and recertification of vessels and pipework is also a standard requirement from most regulatory authorities and / or insurance companies.

3.4 Management Systems

Whether specified by the regulator or by internal policy, Management Systems are typically in place to manage the processes described in the sections above. As part of an organization's governance, the suitability of these Management Systems will be periodically reviewed, and their effectiveness tested by audit to ensure they are properly applied. Performance indicators provide a measure of the effectiveness of such systems with API RP 754 (ref. 5) being a standard most commonly used in the O&G & Petrochemical industry for process safety KPI's. In this standard, Tier 1 and 2 measures are known as "lagging indicators" based on historical events such as loss of containment. Tiers 3 and 4 are referred to as "leading indicators" since they are typically based on auditing of management systems in place to prevent or mitigate process safety events. In the UK, the Health & Safety Executive issued guidance, HSG 254, on developing process safety indicators (ref 6) a few years before API published their standard. Here they promoted the identification of MAH scenarios, barrier identification and identifying the "Risk Control System" or RCS which is managing the performance of that barrier. Leading and lagging indicators, similar to API 754 are also recommended in this guide.

While it is essential to the overall safe management of a facility that barrier / risk management systems are in place, both API 754 and HSG254, generate indicators which inform on the system alone and not on specific pieces of equipment or areas of the plant. Furthermore, what are called "leading indicators" are based on data collected months if not years in advance so may not represent what is current. What is clear is that they do not help day to day decisions on risk levels in specific areas of the plant or working on specific equipment.

4 The Reality of Operations and Dynamic Risk

4.1 Deferred Maintenance of Safety Critical Equipment

Maintenance Management Systems are programmed with time-based schedules for inspection and preventative maintenance of all equipment on a facility. Where this equipment forms one of our major accident hazard barriers, it is flagged as "safety critical." A rolling schedule is managed based on different time horizons. As equipment breaks down, repair work is added in based on priority. At two weeks or seven days, the schedule is typically "frozen" and handed over to Operations.

Executing the plan as designed is often a difficult challenge for a variety of reasons and maintenance backlogs, or maintenance activities beyond their planned completion dates, become common as activities slip on the schedule. This becomes more of a concern when the backlog contains safety critical equipment. In such instances, Operators need to justify why the maintenance is not being carried out and a "deferment" review process involving technical and engineering authorities should be conducted.

As we have seen earlier, when inspection and preventative maintenance of safety critical barriers is delayed, the inevitable outcome is that MAH risk levels increase.

Deferred safety critical maintenance is an unfortunate reality in the O&G & Petrochemical industry, especially with old facilities and infrastructure. The magnitude of this issue is difficult to judge since data on the topic is not easy to access as few Operators want to share data which shows them in a bad light; however, there is one exception.

The UK sector of the North Sea is a region where Operators have, for several years, voluntarily submitted data on the maintenance backlog for Safety Critical Equipment (SCE). Oil and Gas UK, the cross-industry body recently published this data in their annual Health & Safety Report in 2015 (ref 4). The diagram below shows how the growth in SCE backlog over the last 4-5 years. The report mentions the regulator has raised concerns the industry has increasingly deferred maintenance backlogs rather than addressing them.

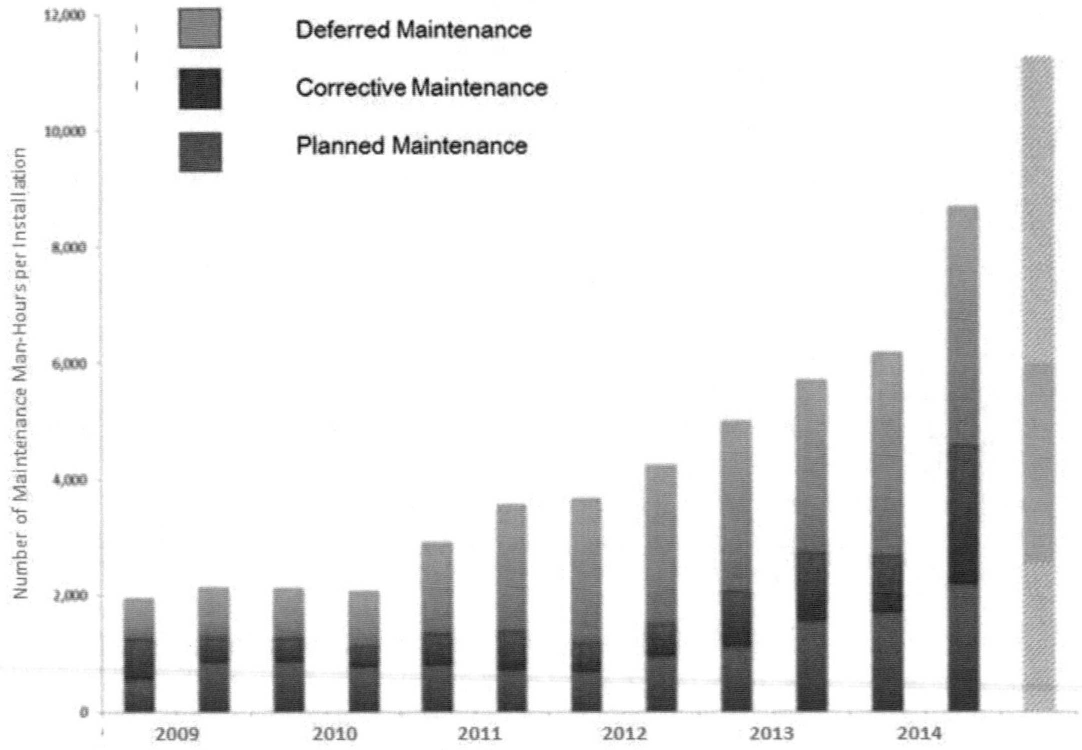

Figure 2. Oil & Gas Health & Safety Report 2015 (ref 4) – Growing Backlog of Safety Critical Maintenance

What this picture says is there are a number of barrier systems on the facility which, with the deferment of inspection and preventative maintenance, are knowingly "impaired. In other words, they are not providing the level of risk reduction that was intended by design. Add to this, the barrier systems which have been inspected but have failed to perform adequately and are therefore also in need of corrective action.

4.2 Deliberate defeat of barrier systems

Another category of compromised barriers must also be considered. Every day, many barriers on a facility are deliberately defeated. This is a necessary on a plant where ongoing maintenance and repair activities have to take place with the plant in a live state or condition. Detection systems, such as gas and fire detectors, have to be deliberately defeated to carry out work in a vicinity which could accidently trip them. Alarms and process trips have to be bypassed for work to be carried out. Ignition sources have to be brought onto the live plant in order for work to be carried out.

The terminology used to describe these situations include: "temporary defeats (of safety system), "instrument override", "interlock by-pass" but they amount to the same thing. A formal, managed process should be followed during this action, often accompanied with a risk assessment. This will often focus on the following critical parameters: **time** – the longer a defeat is in place the greater the risk; **reinstatement** – ensuring the temporary defeat does not get forgotten and becomes permanent; **communication** – ensuring all necessary parties are aware of the status.

Even our most important barrier, primary containment, has to be compromised fairly regularly in order to replace and repair equipment such as valves, pumps, filters, etc. Work involving "breaking containment" is, again, normally managed with a formal process involving energy isolation, draining and purging and gas testing.

All of these situations where safety barriers are deliberately defeated involve an additional level of risk. While the defeats maybe managed with formal process and additional barriers or controls maybe put in place, there will be a residual risk which needs to be considered along with the other sources of risk on the facility.

4.3 Operational Decisions and Corporate Silos

On a daily basis, plant Supervisors make the decision, typically informally and often unconsciously, on whether the plant is safe to operate. They also decide whether it is safe to carry out simultaneously activities - such as maintenance and repair work, project construction work and normal Operational activities. These decisions are often based on incomplete information, specifically with regard to the status of barriers on the plant.

Unfortunately it is quite common within our organizations, that data and information is stored in areas often difficult to access by anyone other than the people who created the data. While this may not be intentional, the result of such "discipline silos" is that updated reports from these discipline groups tend to be too generalized to support specific decision-making.

Given the wide distribution of data, it is not surprising that daily decisions are often made on the basis of exception, i.e. unless told otherwise and in the absence of any visible concerns the assumption is that the barriers are in good health. Visual evidence of barrier impairment cannot be relied upon other than very obvious external corrosion or physical damage to equipment and "local memory" of ongoing status rely on individuals being on shift or very good handover notes and Operations logs. A good supervisor will make great efforts to find the most current information before making decisions but as always when trying to prove a negative….how do you know when to stop looking?

The table below shows some random examples of the types of discrete risk management processes and disconnected data systems which are common in our organizations. This is a short list to demonstrate the principle. The point is that our organizations and discipline groups tend to be very insular and the data they collect is difficult to access from outside the group. This is partly because the work carried out to identify deviations or verify conformance is fragmented and generally covers the whole facility. The output reports tend to be summarized around performance indicators (examples of which are included in the table). These KPI's are useful to judge the effectiveness of the Risk Management Process (RMP) averaged over sizeable time steps (per month , per quarter or per year). They are not necessarily helpful for supporting real time decisions in specific areas of the plant. For that, decision makers need to know the status of specific barriers in particular areas of the plant to identify whether planned activity or plant changes are safe to instigate.

Table 2. Examples of Corporate "Silos" of data and information

Lead Discipline	Risk Management Process (RMP)	Data "silo" system	Performance Indicator
Maintenance	Identified barrier impairment	MMS (Maintenance Management System)	% impaired vs not impaired
Engineering	Corrosion Inspection	Inspection Systems	% systems inspected % failed
Engineering	Safe Operating Limit excursions	DCS / Data Historian	Number of SOL excursions
Instrumentation	Relief Valve testing	Testing Systems	% tests carried out
Instrumentation	Instrument Malfunction	DCS / Data Historian	% malfunctioning instruments
Engineering	Management Of Change	MOC system	% MOC's outstanding
HSE	Incident Reporting	Accident / Incident Database	Tier 1 – 4 reporting statistics
HR	Competency Management	Competency Management / Training database	% workers trained
Process Safety	HAZOP / HAZID	Risk Analysis Database	% Hazops complete
Management	Management Audits	Audit / Action Tracking System	Audit Actions complete

4.4 Cumulative Risk & Dynamic Risk

Two factors of risk management which are gaining some attention in our industry are cumulative risk and dynamic risk. Dynamic risk is perhaps the most simple concept since it is just a recognition that the risk situation is not static in any of our facilities. It changes with a variety of time dependent mechanisms. We have covered extensively the way in which barrier systems fail over time which results in a change in the consequential risk of the considered scenarios. The degree of risk elevation will depend on the length of time the barrier(s) is impaired. Other time dependent components to risk include environmental conditions particularly those at the limit of design such as hurricanes, tornadoes, floods, earthquakes, etc. Ongoing activities on the plant will potentially have an impact on risk either through introduced hazards or introducing potential casualties if an event occurred. Community expansion over time outside the perimeter fence is also a dynamic which changes the risk profile of a facility.

Built into some of our standards and regulatory practice is a 5 year review of hazard and risk scenarios across the facility. This is a suitable time frame for reviewing major equipment items and to ensure that appropriate barriers and risk control systems are in place as well as an opportunity to implement new attitudes towards risk and risk reviews. It does not, however, address the reality of day to day changes in our barriers, temporary situations and the changing hazards of a typical work day and as a result, the changes in overall risk. Instead we rely on informal processes and diligent front line supervisors to recognize when risk levels are reaching unacceptable levels.

Cumulative risk is another component of this dynamic picture which is gaining some recognition in areas, notably, offshore in the North Sea. Oil and Gas UK have recently issued in draft form guidance on the management of Cumulative risk (ref 7). In this document they define Cumulative Risk as follows:

The Cumulative risk on an installation is considered to be the total risk at a particular point in time due to a combination of:

- The risk described in the Safety Case
- Changes to the above caused by one-off operations or other operational changes
- Additional risk from deviations (from a normal state), the interactions between them and the influence of other factors on these risks

The North Sea is a mature basin area with many of the operating facilities having been first installed in the 1970's. Maintaining these facilities is a challenge and, as we saw in section 4.1, a backlog of safety critical maintenance now exists on many facilities. The regulator has a concern on how operators manage the cumulative effect of multiple "deviations", i.e. impaired barriers and how decisions are made.

There is nothing special about the North Sea and it probably represents what Deepwater Gulf of Mexico will look like in 10 to 15 years. Similarly, onshore refineries, chemical plants, storage and distribution facilities suffering the effects of ageing equipment and infrastructure will need to understand and manage the cumulative and dynamic impact of multiple impaired barriers and simultaneous operations. To achieve this successfully, we need to have a better set of tools to help decision makers, in real time, assess overall risk from a number of disparate sources and enable them to more safely and efficiently manage activities and the operation of their plant.

5 A New way to Manage Risk, Barrier Health, and Activity

5.1 New Category of Enterprise Operational Excellence Software

New software platforms are emerging under the category of Enterprise Operational Excellence Management enabling companies to make better decisions in all areas of activity and risk management. These systems are designed to manage activity with the benefit of understanding risk from a number of sources. Barrier impairment, in such systems, is easily tracked and very visible when activities are planned and executed.

These systems help manage some of the following challenges:

- They provide a unified approach to assessing operational risk and governing the status of Process Safety barriers
- They provide opportunities to produce meaningful real-time and predictive risk indicators, to manage the sequence and execution of activities.
- Real time and predictive cumulative risk modeling in time and space dimensions
- Safe execution of frontline activities across the whole planning-to-execution business process.
- Data exchange with other enterprise platforms such as Maintenance Management Systems (CMMS), Inspection databases, Process Data Historians, etc. integrating across many of the data silos in our organizations
- Visualize risk and activity in space and time
- Prioritize work scheduled based on optimizing future reduction of risk
- Easy identification of barrier impairments and management of deviations in risk control systems
- Ability to manage, in real time, all risk and activity in the one place using easy simple graphical representation of the facility.
- Management of all activities including permitted work, energy isolations (LOTO), task risk assessments / Job Safety Analysis, confined space entries, routine activities, shift handover, Operator Duties, etc.
- Efficient and safe execution of work schedule
- Real time and predictive KPI dashboards shwing risk, productivity, plan attainment, time on tools, audit compliance, etc.

The screen capture below shows an example of how graphical views of the plant can indicate where MAH barriers are impaired. Icons show the different categories of impaired barriers. Other icons show ongoing or planned activities. The view can be real-time or predictive depending on the shift date selected. The future position is based on planned activities and estimates for the duration of impacted barriers. The system predicts the overall level of risk for that time period and displays with a simple traffic signal red, orange, green pattern to indicate high, medium, or low level risks. It can also highlight the specific area on the plant where the highest risks are predicted.

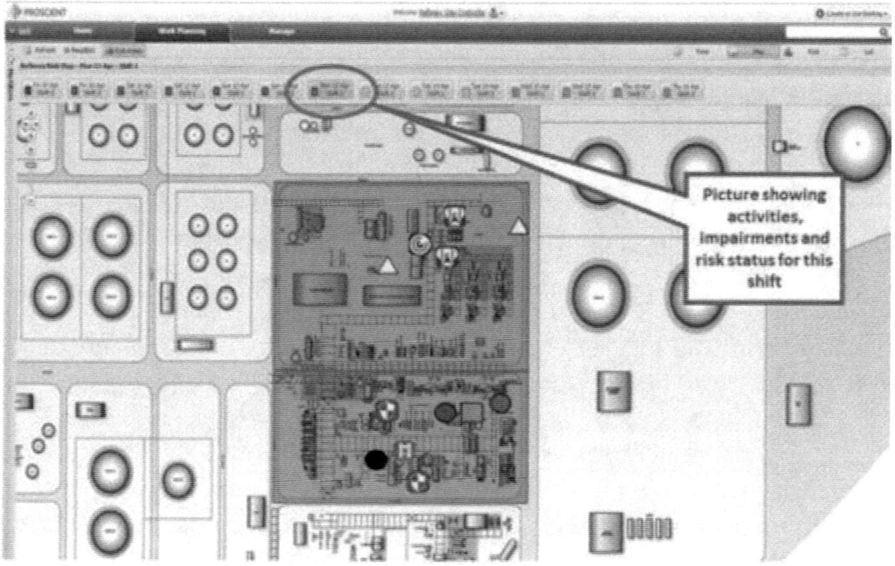

Figure 3. Graphical View of the Plant to Display Operational Risk Impairments and Deviations

Specific risk view screens for each area of the plant allow drill-down into the data to understand which activities or conditions are driving higher risks. This view is also available for current real-time or for future shift periods, with the traffic signals indicating cumulative risk for the facility during that period. This allows an operator to see the current or future state of fundamental safety barriers in each area of the plant.

Figure 4. Risk Views Against Specific Plant Areas for Current or Future Shifts

The impact on risk is driven by specific work activities, plant statuses, or barrier conditions. Each of these is captured as a "line item" and displayed in a Gantt chart style. Much of this data is "sucked in" from the disparate silo stores and is displayed in a single place showing all future work and conditions. This screen automatically shows conflicts depending on configuration rules established. For example, if there was a containment issue in an area and an operator was trying to schedule hot work in the same area, the system would prompt a conflict and shows the activities with a red boundary. Even if it was not a conflict rule, the high risk status for the shift would alert the situation.

Simple drag and drop features allow Frontline Leadership and Planning to simply move conflicts or compounding risk events so that they do not coincide and in doing so, reduce the levels of predicted peak risks.

Digital displays help Operators identify prolonged barrier impairments which are elevating risk levels in specific areas of the plant to unacceptable levels. Simple simulate/fix functionality empowers operators to see the risk reduction impact of fixing an impaired barrier earlier than scheduled.

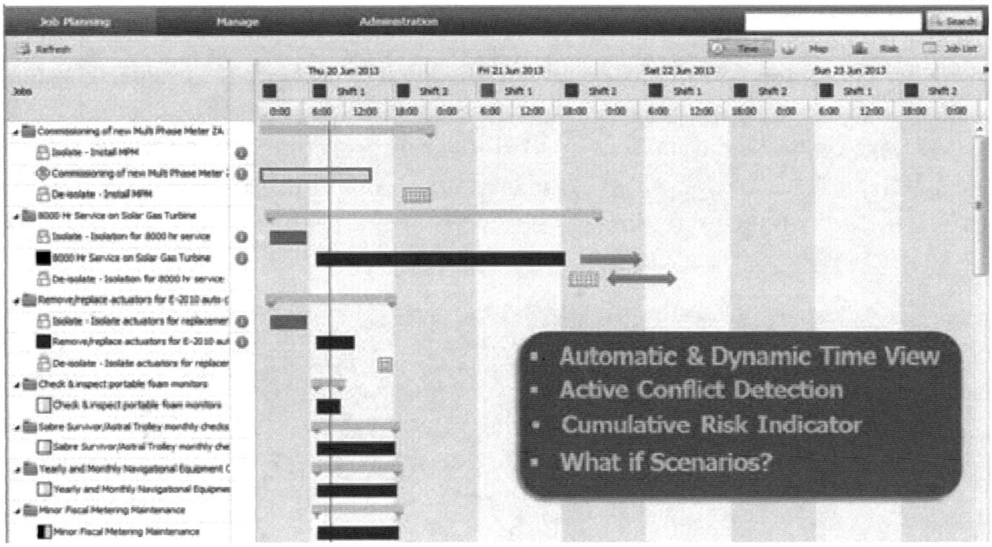

Figure 5. Time View showing Durations of all Planned Activities

A digital risk assessment engine enables Operators to evaluate the risk involved in their planned activities. At the same time, everyone across the plant can see the status of impaired barriers and assess the cumulative risk impact of their activities and those of other departments on all barriers, in all areas, at all times. This means medium-term planning and scheduling (90+ days) can be prioritized using MAH risk criteria. Rather than just blanket labeling of work priorities with the term "safety critical," the true impact on MAH risk can be assessed and activities prioritized accordingly.

With this data being collected centrally and used to assess the overall impact on risks, the quality of analysis and reporting surpasses that which has been achievable to date. The screen below show some of the types of reporting possible such as tracking overdue deviations (i.e. deviations or impairments which are not resolved within an expected timeframe), risk control systems impacts and trends and even trending of MAH risks over time.

These systems are designed for enterprise deployment so reporting can cover a whole organization of assets and identify where barrier impairments and resulting MAH risks are highest.

Figure 6. Examples of Dashboard Type Reports

6 Conclusions

We have seen how major accidents occur due to the failure of several of our protective / mitigative barriers. Simply put, the barriers did not perform as designed. Their reliability is dependent on continued inspection and preventative maintenance which, as we have seen, can often be deferred resulting in an increase of MAH risk. This is because the risk reduction credit for the barrier, taken during design, is no longer applicable. Another problem we face is that the data and information regarding these impaired barriers is often not easily accessible, being stored within corporate silos. Frontline Supervisors are making day to day decisions on the basis of assumed barrier status and cannot easily access information which could contradict some of these assumptions. They also need to take into account the dynamic nature of risk which can change from day to day, hour to hour and the cumulative impact of multiple impaired barriers.

Our frontline supervisors and decision makers have the responsibility to manage a safe plant so they need this information in order to make fundamental decisions such as whether the plant is safe to operate and whether the planned program of work is safe to execute. Currently they do not have easy access the relevant data in order to make these decisions.

Fortunately new technology platforms are emerging where the data on impaired barriers, temporary defeats, equipment isolations, work activities, etc. can be sourced through a single place. In this same place, activities can be planned and executed. Planned maintenance schedules can be imported and work carried out safely. Cumulative risk indicators show the aggregate impact of multiple activities and impaired barriers on risk in specific areas.

These new platforms represent an opportunity for a significant change in the way we manage risk on our facilities and should lead to safer and more efficient operations.

7 References.

[1] "The 100 Largest Losses 1974-2013 – Large Property Damage in the Hydrocarbon Industry", 23rd Edition, Marsh Energy Practice 2014.

[2] "An analysis of common causes of losses in the Oil, Gas & Petrochemical Industries" Presented at IChemE Safety and Loss Prevention Special Interest Group, Ron Garvis Swiss Re Risk Engineering Services, July 7th 2015

[3] The Cumulative Risk Assessment Barrier Model, Alan Blacklaw, Al Ward, Kevin Cassity, BG Group, SPE, 146255, 2011

[4] Health & Safety Report 2015 – Oil & Gas UK

[5] API Recommended Practice 754 - Process Safety Performance Indicators for the Refining and Petrochemical Industries, API, April 2010

[6] HSG 254: Developing Process Safety Indicators, UK Health & Safety Directorate, 2006

[7] Cumulative Risk Guidance (Issue Draft for Comment), Oil & Gas UK, February 2016

445034 Donde, Cuando y Que Enseñar y la nueva certification de seguridad de procesos del CCPS

Abstract:

El primer objetivo de este paper es el de tratar de desmistificar que la Seguridad de Procesos debe enseñarse solamente o fundamentalmente en la carrera de Ingenieria Quimica. Basandonos en el concepto de los 20 elementos propuestos por el libro del Center for Chemical Process Safety "Seguridad de Procesos Basada en el Riesgo " (RBPS por sus siglas en ingles) se puede demostrar que ciertos elementos, como integridad mecanica y planificacion de emergencias, cruza transversalmente a otras disciplinas del conocimento academico como Ingenieria Mecanica, Ingenieria Electrica, etc. y ademas, otras no pertinentes al ambito de la ingeniera, como es el caso de divulgacion a la comunidad. Se presentaran ejemplos de ciertas universidades de los Estados Unidos, Canada, y Colombia. Como segundo objetivo, y conscientes del hecho de que no existe en el mundo actual, un modelo unico y probadamente eficaz para implementar la ensenanza de la Seguridad de Procesos, los autores buscan presentar diferentes modelos y llamar a la accion a las universidades de países hispanoparlantes para desacelerar la brecha entre lo que provee la Universidad y lo que necesita la industria.

Ademas, este paper buscara explicar los requerimientos del nuevo proceso de certificación de seguridad de procesos implementado por el CCPS, que es una nueva herramienta para demostrar competencia en seguridad de procesos, que permitirá a los profesionales agregar a sus credenciales "Certified Process Safety Profesional" or CPSP.

440342 PSM Legal Update

Abstract:

Over the last year, there have been many changes to the PSM Standard. There have been two court cases explaining key terms in the PSM Standard such as "change," "process," and "process equipment" and discussing key issues such as the time to implement PHA and audit findings. In addition, OSHA has issued four memoranda expanding the boundaries of PSM by defining RAGAGEP, creating a new mixture rule for Appendix A substances, completely revising the retail exemption and defining "proximate" for the purpose of establishing the boundaries of a process. OSHA issued a Request for Information to gather information for 14 potential changes to the PSM Standard to modernize it. After reviewing all of the responses to the RFI, OSHA has just initiated the review process under the Small Business Regulatory Enforcement Fairness Act (SBREFA) in which a draft rule with the proposed changes to the PSM Standard will be published and small business will have an opportunity to comment on it. The SBRFA process will provide a framework for additional changes being developed for the PSM Standard. This presentation will provide an overview of the key changes in the PSM Standard that OSHA has implemented and a summary of the changes being considered by OSHA.

440951 Application of Computational Fluid Dynamics (CFD) for Liquefied Natural Gas (LNG) Pool Spreading and Vaporization on Water

Abstract:

A marked increase in LNG utilization is forecast with an increase in LNG marine operations. A ship-ship collision of an LNG marine tanker or a rupture of the LNG loading and unloading lines can result in accidental spillage of LNG on water. Upon release, a spreading liquid can form a pool with rapid vaporization, leading to the formation of a flammable vapor cloud. Safety analyses for the protection of the public and property involve the determination of consequences of such accidental releases. The evaluation of consequences resulting from an accidental spill of LNG on water involves the determination of the rate (vaporization rate/source term) at which flammable hydrocarbon vapor is produced and the dynamics of the spreading pool. A specific offshore scenario of rupture of an offloading line is considered resulting in an LNG spill on water, between the space enclosed by the offloading deck and the LNG ship. This scenario was studied experimentally by releasing LNG in a narrow concrete trench filled with water. The experimental study involves the determination of the vaporization rate for a continuous release of LNG in a trench. Along with the vaporization rate, the spreading rate, pool area and pool height were measured. The computational work involves development of a methodology to model the pool spreading and vaporization behavior in Computational Fluid Dynamics (CFD) for the defined scenario. The overall aim of this research is to improve the understanding of pool spreading and vaporization of LNG releases on water.

12TH GLOBAL CONGRESS ON PROCESS SAFETY

Modeling an Unconfined Liquid Spill on a Sloped Surface

Robert Deshotels
Retired
21492 Pedroso, Mission Viejo, California, USA
R.L.Deshotels@cox.net

Prepared for Presentation at
American Institute of Chemical Engineers
2016 Spring Meeting
12th Global Congress on Process Safety
Houston, Texas
April 11-23, 2016

Modeling an Unconfined Liquid Spill on a Sloped Surface

Robert Deshotels, PE
Retired
21492 Pedroso, Mission Viejo, California, USA
R.L.Deshotels@cox.net

Keywords: liquid spreading, hydraulic radius, pool evaporation, friction factor, liquid spill modeling, simulation of liquid spreading

Abstract

When determining the consequences of a vapor cloud or fire from a liquid spill, the area of the liquid pool is a key variable. If the pool is contained by walls or berms, dispersion or flame parameters can be evaluated directly from the pool area using correlations that are well known to consequence analysts. For a spill onto an unconfined paved area or sloping deck, there is no simple formula for the pool area. For an unconfined pool, area must be calculated from flow modeling. The progressive movement and size of a liquid spill onto a sloped surface can be accurately modeled with well-known fluid equations using an empirical friction factor. The fluid flow formulas for piping can be applied to a sloping flat spill surface by including a correction for the equivalent hydraulic radius, the ratio of flow area to wetted area of the frictional surface. For uncomplicated flow surfaces, spreadsheet calculation is sufficient. Example calculations point out the need for improved knowledge of friction factors in terms of quantitative surface characteristics, and an improved formula for heat transfer during film boiling of cryogenic liquids.

1 Overview

The surface area of a spill of a toxic or flammable liquid spill is crucial to the analysis of hazardous consequences. The evaporation rate from a pool of volatile liquid, and hence the dispersion rate, depends on liquid properties, meteorological conditions, and the surface area of the pool. In the case of a pool fire, the heat flux radiating to nearby structures or people depends on the height and width of the flame, which depend on the size and shape of the pool. Usually, a liquid spill is confined by berms or curbs, so that the worst-case spill area is known from confinement dimensions. For an unconfined spill, there isn't any obvious maximum pool area, even if the source of the spill has a limited volume, because the pool will tend to continue to spread as time progresses. Therefore, unconfined spills require careful analysis to determine whether the spill can affect sensitive areas or lead to accident escalation. As well as environmentally sensitive areas, sensitive areas include escape paths, structures used for the shelter of employees or the public, and areas where fire or explosion could lead to escalation of the consequences.

Equations for the rate of spread of a liquid pool have been published in standard references on hazard evaluation [Ref. 1, 2], but friction loss was not considered in these reference guides. Friction loss is a major factor in the speed of liquid flow, especially for a thin pool depth. For a fixed-volume release, frictional loss can be the most important factor in whether a pool spreads into sensitive regions.

Much of the recent research on liquid spreading across a surface involves the effects of capillary action and surface tension on the motion of a thin film [Ref. 3 - 5]. There is much to be learned about the interaction between spilled liquid and the surface below in terms of capillary action and the hydrophilic/hydrophobic nature of the surface. However, the complexity surrounding surface tension applies only for very thin films of liquid. It will be shown that very thin films are not usually a significant part of industrial-sized releases.

For liquid spills that present a design challenge to industrial-size facilities, equations that are commonly used for design of piping and open channels, such as the Darcy equation, can be adapted to the purpose of modeling the time-dependent flow of an unconfined spill [Ref. 6 - 8].

For volatile liquids, the open channel flow equations give relatively accurate estimates for dispersion consequences. However, the concern may be that the spill will flow into a sensitive area. For non-volatile pools, the following question will be answered: "At what point is the spill thickness too low to produce a significant hazard?"

With a methodology for evaluating the minimum significant pool thickness in place, the basic steps for modeling an unconfined spill are:

1. Determine an appropriate release scenario.
2. Evaluate flow parameters for a generalized frictional flow formula.
3. Estimate the endpoint minimum significant pool thickness.
4. For sequential time increments, calculate pool dimensions, taking evaporation into account.

At each stage of modeling, uncertainties in the models and parameters should be considered to ensure that conservative choices are made. Uncertainty is especially significant when postulating a release scenario, estimating temperature for the liquid and surroundings, and appropriate modeling for heat transfer and evaporation. The largest pool area and evaporation rate produce the worst toxic or flammable release to the atmosphere. When the concern is distance of pool travel, minimum evaporation will create a worse case. There may be several potential worst case scenarios that must be quantified to determine which one is actually the worst.

Two example cases of unconfined liquid spills are provided in this article:
Example A. A leaking tanker truck located on a large paved area;
Example B. A piping leak onto the sloped deck of a floating oil production platform.

2 Release scenario

Generally, the source of the liquid spill will be a vessel or piping system with many potential leak points and spill trajectories. One possible way to address the multitude of possible release scenarios is to choose a worst credible case. If the worst case results in tolerable risk and acceptable mitigation expense, there is no reason to evaluate any less conservative cases. Sometimes, a release scenario may be so conservative as to be speculative. For example, if a release is conjectured to be the total contents of a tank due to accidental impact, there may not be any good basis for determining the speed at which the tank breaks, the size of the break, or the splashing that would occur upon impingement against nearby objects. In this case, simple assumptions should be made, and there should be a discussion of the sensitivity to uncertainties.

Any individual spill case may require a different approach to conservatism, but in general, the most conservative release scenario involves the greatest volume and flowrate. Concerning scenario parameters, pool spread distance is worse when the slope is high, and evaporation is low. Higher liquid temperature decreases viscosity and increases evaporation. When the spill poses multiple hazards or affects multiple possible targets, several potential worst-cases may have to be run to find the actual worst. When risk acceptability or mitigation cost are sensitive to the choice of release scenario, quantitative risk assessment tools such as event tree analysis should be used to pick the appropriate set of release cases. Data from offshore platform leaks [Ref. 9] can be used to estimate the probability of specific leak sizes from common process piping and equipment items.

Standard fluid dynamics methods [Ref. 7, 10] can be used to calculate the flow for a release of a given opening size and specified liquid conditions. The volumetric flow rate through an orifice, q, can be expressed in terms of the following parameters (in order of mention) flow coefficient, flow area of the orifice, gravitational acceleration, and differential pressure across the orifice, expressed in height of liquid:

$$q = C\,A\,(2g\,h)^{1/2}$$

The value of the flow coefficient, C, is a function of the Reynolds number (discussed in the section "Formulas for Flow of a Liquid Pool") and the ratio of the orifice diameter to the diameter of the pipe or vessel that is leaking. Plots of C are readily available [Ref. 7].

For liquids that flash upon release to atmospheric pressure, classical thermodynamics theory [Ref. 10] will accurately predict the vapor-liquid split and liquid properties after flashing. The mass fraction of the leaking fluid that vaporizes after release, f_v, can be expressed in terms of the exponential of the following parameters (in order of mention) liquid specific heat capacity, initial temperature of the liquid, atmospheric boiling point, and latent heat of vaporization. The heat capacity and latent heat should be the mean values over the temperature range T_i to T_b.

$$f_v = 1 - exp\{-C_p(T_i - T_b)\,/\Delta H_v\}$$

Depending on release conditions and liquid physical properties, the liquid release may reach the pool surface as a steady stream or a spray of droplets. If the release is in the form of an undispersed stream, the trajectory can be evaluated using classical Newtonian physics, including the effect of wind pressure on the stream. In some cases, the release will immediately begin breaking into droplets, and in all cases a liquid stream will disperse into droplets if it falls enough distance.

Empirical formulas have been developed to predict the distance that a liquid release will travel before breaking-up into droplets. Droplet formation is characterized by the dimensionless Weber number, which is the ratio of shear force on the liquid surface, $\frac{1}{2}\rho_g v_d^2$, to surface tension pressure on the drop, σ/d. The Weber number is defined in terms of the following parameters (in order of mention) gas density, velocity of the drop relative to the air, drop diameter, and surface tension :

$$W_e = (\tfrac{1}{2}\rho_g v_d^2\, d)\,/\sigma$$

For subcooled liquids, the length between jet release and breakup into droplets, L_b, can be expressed in terms of the following parameters (in order of mention) jet diameter at the narrowest point, Weber number, and Reynolds number [Ref. 11]. This formula applies to leaks large enough to cause a serious hazard, but not to a 2 mm diameter hole or less. The units of measure must be consistent with the dimensionless Weber number.

$$L_b = 540\, D\, W_e^{1/2}\, R_e^{-5/8}$$

For superheated liquids [Ref. 11],

$$L_b = 2.75 \times 10^{10}\, D\, W_e^{1/2}\, R_e^{-2}$$

Based on experimental results for liquids without superheat, the typical droplet size, d, can be expressed in terms of the following parameters (in order of mention) surface tension, gas density, and jet velocity relative to the wind. The drop size depends on the liquid, and could be double the typical value [Ref. 1].

$$d = 24\,\sigma/(\rho_g\, v_d^2)$$

For superheated liquids, the droplet size depends on the difference between the liquid temperature and the atmospheric boiling point [Ref. 1],

$$d = 3 \times 10^4/(10 + 4.0\,\Delta T_{sh})$$

The terminal velocity of a droplet in air is a complicated function of the liquid density and drop diameter, which can be read from a plot in a standard reference [Ref. 12].

Comparison of the terminal velocity to the wind speed will show whether droplets will rain out onto the pool or drift away like a cloud. The direction of release and Newtonian physics determine the trajectory that the droplets will follow from the liquid breakup point to the point where the pool begins growing. Different wind conditions may have to be considered. The probability of the combination of accident scenario, wind and temperature should be consistent with a tolerable risk for the accident.

3 Liquid pool growth and shrinkage

The pool volume will increase as long as the liquid release rate into the pool exceeds the evaporation rate. The release rate changes with time, depending on source inventory, pressure and effectiveness of process shutdown controls. The evaporation rate depends on pool area and changing temperatures.

It is possible to determine a worst-case maximum pool area using the initial leak rate and assuming a minimum evaporation rate. In this case, the maximum pool area equals the maximum pool volume created over the interval of concern (e.g., maximum escape and rescue time) divided by the minimum significant pool thickness. If the hazards evaluated from this bounding case give a tolerable risk, there may be no need to actually model the growth and configuration of the pool. However, even if the worst case results in tolerable risk, dynamic modeling of the pool may provide insight that would be valuable for planning and training.

Dynamic modeling consists of stepwise calculation of pool characteristics at small increments of time. At each increment, the following variables are computed:

1. Based on process conditions of pressure, temperature and composition, determine the flow properties and vapor pressure of the liquid just upstream of the release point. Using classical thermodynamics, determine the fraction of vapor flash after release into the atmosphere.

2. For the release scenario chosen, calculate the liquid flow rate into the pool using the orifice flow equation that is appropriate to the scenario. Calculate the liquid trajectory and dispersion from the jet breakup formula that applies to the orifice size, Weber number, and flashing superheat. It can be assumed that the liquid volume will be in a conical shape, with the target area being circular on a flat substrate, or elliptical on a sloped substrate. While a liquid release is in the momentum-driven region of flow (about 6 to 100 orifice diameters from the release point), a liquid jet will diverge at an angle of 10° from the centerline [Ref. 6]. From the flowrate, trajectory and divergence angle of the release, the target area of the liquid added to the pool is determined, as well as the volume increase in the target area during the time interval.

3. From a predicted Reynolds number, or the Reynolds number from the previous time increment, the applicable flow equation for pool spread is chosen. The volume flow and velocity at the edge of the pool can then be calculated, from which pool edge growth can be calculated. The equations for flow of the pool are given in the section "Formulas for Flow of a Liquid Pool".

4. A conservative calculation of pool evaporation is done, based on the new pool area calculated in the previous step. Depending on the purpose of the modeling, the evaporation calculation should be either a low estimate or a high estimate. An exact calculation is generally not possible because of uncertainty about rates of diffusion and heat transfer. An approach to providing a conservative estimate of evaporation is provided in the section "Evaporation". Evaporation may result in a cooling effect on the pool. The evaporation calculation is expressed as a uniform decrease in pool thickness, which will result in an adjustment in the pool surface area.

5. Knowing the inventory volume and process conditions, new process conditions are calculated from the liquid loss through leakage flowrate calculated in step 2 above. As the process pressure decreases, any evaporative cooling inside the process containment should be considered in the revised process conditions and composition.

Considering the accuracy of the postulated release scenario and air dispersion models, a simple model for pool thickness should suffice. It may be reasonable to model the pool volume as a cone with an elliptical base (because of the slope of the substrate), with the peak of the cone located at the target point of the liquid release trajectory.

4 Minimum significant pool thickness

The driving force for pool flow is the difference in height between different parts of the pool. For a pool moving down a sloped surface, the uphill part will become thinner as the pool spreads downhill. Furthermore, it is common for the leak rate from the process to stop or diminish after a while, causing the rate of evaporation rate from the pool to exceed the fill rate. At some point, part or all of the pool may become too thin to pose a significant hazard. To see whether this is possible in a specific case, it helps to categorize the different hazards posed by a pool, as follows.

Pool liquid that has not evaporated can pose a hazard of toxic exposure by direct contact, or a fire hazard.

For liquids that are not flammable, but are combustible upon ignition by flame, the pool may be too thin to maintain combustion. This happens when heat transfer to the substrate will quench any fire that starts by cooling the liquid below the flash point. This thickness could be calculated, but instead of solving unsteady-state heat transfer calculations using uncertain heat transfer coefficients and uncertain burning rates, the minimum fire hazard thickness for a non-volatile liquid should be determined by a field test. For one rough example, in a backyard test in a one-foot square confined pool of kerosene on concrete sidewalk at 20°C in breezy air, it was found that the minimum liquid thickness that would maintain a flame was between 2 and 3mm.

Where a liquid is highly toxic to people or the environment (e.g., aqueous ammonia) OSHA Standard 1910.119 has established a threshold quantity. A pool below this quantity no longer poses a catastrophic level of hazard, but mitigation is still imperative. Consequently, there is no insignificant pool quantity or thickness in the case of toxic liquids.

Vapor that is evaporating from a pool can create a toxic or flammable cloud leading to toxic exposure, flash fire, or explosion. The severity of exposure of a person to a toxic cloud is measured as the integrated dose of concentration over time. Therefore, for toxics, the quantity of material in the vapor cloud is more important than the peak rate of evaporation. As a result, the thickness of a pool does not affect the severity of a toxic release much, especially for a thin film that will evaporate soon. For a flash fire or explosion, severity is measured as heat flux or blast pressure, which depend on the size and concentration of the flammable cloud. Thus, for a spill of volatile liquid, the pool thickness just prior to complete evaporation does not have much impact on accident consequence.

When there is a concern that a combustible spilled liquid will flow into a sensitive area, the pool can be considered non-hazardous, provided there is no continuous source of ignition, if the entire pool depth is less than the minimum fire hazard thickness, determined from field testing of flame propagation.

5 Formulas for flow of a liquid pool

All practical formulas for the flow of liquids are derived from Bernoulli's equation, an energy balance that includes all of the involved forms of energy: potential energy (elevation difference), pressure, velocity, and frictional loss. The Bernoulli equation for the energy change for a fluid flowing from point 1 to point 2 can be expressed in terms of the following parameters (in order of mention) elevation, pressure, density, velocity, gravitational acceleration, and frictional loss expressed as height of fluid [Ref. 6, 7]. Consistent units of measure must be used. With U.S. customary units of measure, the gravitational acceleration factor of 32.2 ft/s^2 is needed to convert between pounds force and pounds mass.

$$Z_1 + P_1/\rho_1 + v_1^2/2g = Z_2 + P_2/\rho_2 + v_2^2/2g + h_L$$

Because the entire pool is at atmospheric pressure, there is no pressure energy term, P/ρ, to consider for a pool. The change in the elevation term Z is the only driving energy for pool flow, unless the release into the pool has a significant velocity due to the release trajectory at the target point. Further explanation follows for the velocity head term and friction head term.

Flow of a liquid pool on a sloped surface has two components. The first component of pool flow is the downhill motion of the entire pool. The driving force for downhill motion is the gravitational force multiplied by the slope. To determine downhill motion, the downward velocity due to the slope must be added to any velocity head imparted by the trajectory of the release flow into the pool, and the friction head must be subtracted. The downhill momentum increases as time continues. The second component of pool flow is the spreading of the pool outward from the center of mass, which can be modeled from the difference in height between the liquid and the surrounding surface.

Unlike flow through a pipe or in an open channel, liquid spreading on a sloped surface does not have a constant flow depth or cross-sectional flow area. Consequently, velocity varies through the pool. Therefore the value of the velocity used in the velocity head term, $v^2/2g$, and friction head term, h_L, must be considered carefully. For a pool modeled as a cone with a round or elliptical base, spreading begins at the high point as gravity-induced downward flow, then turns outward to be parallel with the base at the outer edge. To determine pool growth, the velocity head due to the elevation difference must be decreased by the friction head to yield the effective elevation change from the high point to the pool edge. From the effective elevation change, a new pool radius can be calculated for a chosen time differential. The new height can be calculated from the mass balance of the pool.

When the substrate is sloped or there is a velocity head imparted by the release stream, the growth in the pool can be evaluated by considering the cone to have four quadrants: uphill, downhill, and two equal sideways quadrants. Each quadrant is broken into differential elements at intervals of radial length along the distance from center of the ellipse outward to a point on the edge in order to evaluate a velocity that applies to each differential element of the quadrant. This allows a calculation of the friction head contribution of each element, from which the radius increase for each quadrant is calculated, and the new pool height determined as discussed in the previous paragraph.

There are different liquid flow formulas (e.g., Darcy, Fanning, Manning) that can be used to estimate the empirical term for frictional loss, h_L.

Regardless of the formula used for friction loss, there are four regimes of flow that can apply to an unconfined liquid spill:

1. Frictionless flow
2. Turbulent flow
3. Laminar flow
4. Thin film flow

5.1 *frictionless flow*

When the flow path has a low elevation gradient and a high pool depth, frictional loss may be insignificant. Also, it is possible for frictional loss to be very low in the case where the pool is heated by film boiling at the substrate interface [Ref. 4]. Otherwise, in actual spill cases that pose a serious risk to a facility, frictional loss is significant. When significant frictional loss is ignored, the simplified formulas for liquid flow rate [Ref. 1, 2] will overestimate the speed of liquid travel. Because friction loss is not difficult to evaluate, there seems to be little incentive to ignore friction.

5.2 *turbulent flow*

Flow turbulence is characterized by the Reynolds number. The Reynolds number is dimensionless, and can be expressed in any consistent units of measure, in terms of the following parameters (in order of mention) pipe diameter, velocity, density, and absolute viscosity.

For pipe, $R_e = D\, v\, \rho / \mu$

Commonly used formulas for liquid flow through a pipe or channel [Ref. 6 - 8] express frictional loss in terms of a friction factor, f, which is determined from experiments. However, formulas for pipe and channels cannot be used directly; the formulas must be modified to account for the different geometry of flow.

In an unconfined liquid spill, the flow often transitions from turbulent flow to laminar flow as the pool spreads thinner. Although the Reynolds number can be calculated exactly, the transition between the turbulent and laminar flow regimes is not consistent from one case to another. The transition is usually gradual, and can depend on flow disturbances and flow history. Turbulent flow is ordinarily exhibited when the Reynolds number is above 4000, with laminar flow exhibited below 2100. Laminar flow has less friction loss than turbulent, resulting in a higher velocity for a given slope. However, laminar flow results in a lower heat transfer rate. In the transition range between flow regimes, the flow should be modeled conservatively as either fully turbulent or laminar. The conservative choice will depend on the situation.

Unconfined liquid flow on a horizontal or sloped surface is analogous to gravity-induced flow in a pipe, except that wall friction only occurs along the bottom of the spill. The concept of hydraulic radius, R_H, [Ref. 6, 7] must be applied.

$$R_H = \text{cross sectional flow area /wetted perimeter}$$

For flow through a round pipe full of fluid, the flow diameter equals $4R_H$. For unconfined pool flow, the hydraulic radius is the pool height, z, at the point where velocity is measured, and the equivalent flow diameter equals $4z$. The Reynolds number for a round pipe full of liquid is expressed in terms of the inside pipe diameter. For full flow through a narrow slot, the Reynolds number is expressed in terms of the height of the slot. For open flow across a flat surface, we typically assume that there is no friction at the top surface of the liquid or the edges, although this is not strictly true due to relative motion between the air and liquid surface. With negligible air friction, the zero-shear level is at the top of the liquid pool, rather than at the centerline as in a round or rectangular pipe. Therefore, the equivalent hydraulic radius used to evaluate the Reynolds number for a round liquid pool, with a flow area (circumference multiplied by height) of $2\pi\ r\ z$ and a perimeter of $2\pi\ r$, should be corrected to twice the usual value, or $2z$. In the Reynolds number and friction head equations, the D used for full-flow round pipe should be replaced by $8z$ for a pool.

For a pool, $R_e = 8z\ v\ \rho\ /\mu$

Wind friction can affect pool motion, but the effect is expected to be minor. Weather stations report wind speed at a reference height of 10 m above grade. If the spill is at grade, the wind is reduced to about 8% of the reference value in the daytime, or 5.5% at night (the difference is due to air stability) [Ref. 4]. It should be noted that wind wave height cannot exceed the pool thickness; therefore, the surface of the liquid will be smooth, even under the influence of wind. Wind drag can be evaluated from low-speed aerodynamics. The drag pressure caused by air friction is the drag coefficient multiplied by the wind velocity head. For wind blowing parallel to a flat surface the drag coefficient, C_D, is 0.04 [Ref. 13]. To convert into liquid head, the wind head is multiplied by the ratio of gas density to liquid density. If the wind is in the opposite direction of flow, the wind head is added to the friction head. If the wind is in the same direction as flow, the wind head is subtracted from the friction head. The wind head, h_{drag}, can be written in terms of the following parameters (in order of mention) drag coefficient, air density, air velocity, and liquid density.

$$h_{drag} = C_D\ \rho_G\ v_G^2\ /2\rho_L$$

Table 1

Formulas for Turbulent Flow Friction Loss

Darcy (Moody)	$h_L = f\,L\,v^2/2D$
Fanning	$h_L = 2f\,L\,v^2/D$
Manning	$h_L = f\,L\,v^2/2R_H$

Note that the Moody friction factor is four times larger than the Fanning friction factor. This can be confusing because plots of friction factor do not always mention which friction factor is given. For the Darcy and Fanning equations, the friction factor is a function of the Reynolds number and relative roughness of the pipe wall, ϵ/D. For a pool having a height of z, D should be replaced by $4R_H$, or 8z. Roughness is determined by experimentation, rather than surface measurement. Based on the limited data available and the difficulty in defining roughness, the friction factor is the major cause of uncertainty in flow rate calculations. The Manning equation is an empirical equation developed from data for water flow in deep open channels. For the Manning equation, a roughness parameter n is used to correlate the friction factor to the hydraulic radius [Ref. 6, 8].

$$\text{Manning } f = 29n^2/R_H^{1/3}$$

Table 2

Roughness Factors for Turbulent Flow

Surface	**ε** ft	**n** ft$^{1/6}$
Clean commercial steel	0.00015	
Galvanized iron	0.0004	0.012
Cast iron	0.00085	0.011
Concrete pipe	0.001 - 0.01	0.015
Riveted steel	0.003 - 0.03	0.014
Earth, straight and uniform		0.019
Brick open channel		0.017
Concrete open channel		0.022
Gravel open channel		0.033

5.3 laminar flow

Unlike turbulent flow, fully laminar flow along an unobstructed path exhibits no eddies in the axial or radial direction. Friction loss is due to viscous effects alone, and the friction factor is a function of Reynolds number alone, not surface roughness.

$$\text{Darcy laminar } f = 64/R_e$$
$$\text{Fanning laminar } f = 16/R_e$$

The value of f for laminar flow is substituted into the Darcy or Fanning equation for friction head. However, the laminar friction factor was developed for round pipe containing full flow, and will be somewhat in error for open flow on a flat surface, due to the parabolic velocity profile that exists between the pipe wall and zero-shear point in laminar flow [Ref. 6]. Although this paper proposes a correction in the hydraulic radius for the location of the zero-shear point, there is no published data to verify that this proposed correction is accurate or conservative.

Like turbulent flow, as long as there is a height difference between the top of the liquid pool and the surface of the substrate, the Bernoulli equation and derived Darcy formula tell us that the pool will continue to spread without limit, although at an ever decreasing rate. For turbulent or laminar flow, the Darcy formula does not consider the effects of surface tension, capillary action, or hydrophilic/hydrophobic nature of the surface under the spill.

5.4 thin film flow

Recently, investigators have been able to explain why liquid pools do not actually continue to spread endlessly [Ref. 5]. When the liquid height is less than around 100 nm (0.01 cm), long-range intermolecular forces (like van der Waals forces) begin affecting the ability of the liquid to wet the surface of the substrate. There is a minimum thickness of liquid film, beyond which the film no longer wets the substrate; instead, the film breaks into individual droplets. In the case of a level substrate, flow stops. The minimum film thickness is affected by substrate surface properties such as roughness, cleanliness, porosity, and surface tension between the solid and liquid. There are many cases for which the proposed model has not been tested (including a sloped surface), but two conclusions can be drawn:

1. For thin pools, the laminar flow model gives a higher spreading rate than any model that considers the effect of surface tension.

The thin film flow model does not apply for liquid pools thicker than about 0.01 cm.

6 Evaporation

Evaporation will occur if the liquid has a significant vapor pressure at the pool temperature, and the atmosphere above the pool is not saturated with the liquid vapor. The temperature of the pool can change because of heat transfer with the substrate or air, which would affect the vapor pressure.

Evaporation rate from the pool is described in terms of classical thermodynamics [Ref. 10]. The driving force for evaporation is the difference between the vapor pressure of the liquid and the partial pressure of the liquid in the air. The mass evaporation rate, W, can be written in terms of the following parameters (in order of mention) molecular weight, mass transfer coefficient, flow area, vapor pressure of the pure liquid, partial pressure of the vapor in the bulk stagnant air above the liquid, gas law constant, and temperature of the liquid. Pressure and temperature are in absolute units.

$$W = M\,K\,A\,(P_{sat} - P_v)\,/(R\,T)$$

At atmospheric pressure and practical operating temperatures, ideal gas can be assumed. Except in the case of a boiling liquid, the rate of evaporation is restricted by the rate of diffusion from the surface of the liquid. Diffusion coefficients have been measured or estimated for common chemicals [Ref. 14]. The resistance to mass transfer from a homogenous liquid to air is mainly due to the resistance of the air. The resistance of the air depends on the transport properties of the air (expressed as the Schmidt number) and flow regime (expressed as the Reynolds number). There are many theoretical and empirical correlations for evaporation, depending on the geometry of the surroundings, flow regime, concentration of the evaporating component in each phase. A great deal of judgement is required in choosing an appropriate correlation. Key issues involve friction modeling for a boiling pool, wind effects, and heat transfer from the substrate [Ref. 5]. It seems that nearly all of the correlations were developed for equipment design and were tested in laboratories, not outdoors in industrial-sized facilities where nearby objects can obstruct or funnel the wind. Therefore, correlation accuracies claimed in literature may not apply to evaporation from a real pool [Ref. 5]. An empirical correlation, rather than theoretical-based, should be applied when the flow is parallel to the interface through which mass transfer occurs (such as pool evaporation) [Ref. 14]. For a calculation that overestimates evaporation, one can assume that there is no resistance to mass flow within the liquid or across the liquid-air interface, which results in an air concentration just above the interface that is equal to the partial pressure of the liquid relative to atmospheric. When the resistance to diffusion is essentially due to air convection alone, the diffusivity is in the range of 10^{-4} to 10^{-6} m^2/s. A simple but crude estimate of diffusivity, K, based on molecular weight, M, may be sufficient [Ref. 10], by assuming the ratio between the diffusivity and cube root of molecular weight is constant for the liquid of interest and some reference liquid with known diffusivity, K_0.

$$K = K_0\,(M_0/M)^{1/3}$$

An evaporation correlation specifically for a pool of volatile liquid has been developed, accounting for the wind speed and stability class of the air [Ref. 2]. This formula does not consider evaporative cooling of the pool. The evaporation rate, W, involves two parameters, λ and n, which depend on the Pasquill-Gifford air stability classification. The equation is written in SI units in terms of the following parameters (in order of mention) λ, wind velocity, vapor density, vapor pressure, ambient pressure, wind velocity squared, pool diameter, and n.

$$W = \lambda\, v_w\, \rho_v\, P_v\,/P\,\{2/(v_w^2\,D\}^n$$

Table 3

Pool Evaporation Parameters

Pasquill Stability Class	**λ**	**n**
A	1.0×10^{-3}	0.078
B	1.2×10^{-3}	0.091
C	1.5×10^{-3}	0.111
D	1.7×10^{-3}	0.130
E	1.8×10^{-3}	0.149
F	1.8×10^{-3}	0.180

To calculate a high estimate of evaporation for non-cryogenic spills, assume the spill occurs on a windy, hot summer day; for a low estimate assume a still, cold winter night.

A separate formula applies to a spill of a cryogenic liquid on land, explicit as a function of time [Ref. 2]. This equation is written in SI units in terms of the following parameters (in order of mention) a surface correction factor, thermal conductivity of the substrate, heat of vaporization, thermal diffusivity of the substrate, substrate temperature, liquid temperature, and time. The surface correction factor, X, is assigned a value of 1 for a smooth substrate, or 3 for a rough substrate such as gravel. Note that this equation crudely assumes that heat transfer is proportional to the temperature difference between the substrate and liquid; this assumption does not apply at all beyond the point of maximum nucleate boiling, when partial or full film boiling occurs.

$$W_{cryo} = 2X\, k / \Delta H_v\, (\pi\, \alpha)^{-1/2}\, (T_S - T_L)\, t^{1/2}$$

If the pool is boiling, the evaporation rate is usually limited by heat input. If the liquid flashes upon release from the process, the liquid and vapor reach equilibrium soon after release, usually before the liquid touches the ground. The liquid will not continue to boil unless heat is added from the air or substrate. The main uncertainty involves heat transfer from the substrate under the pool. Different modes of heat transfer are possible, nucleate boiling, film boiling or transition [Ref. 15, 16]. The transition from nucleate boiling to film boiling occurs at a critical temperature difference between the substrate and liquid, which depends on liquid properties. Heat transfer rates for boiling LNG and NPG have been documented from field experiments with liquefied gas inside level confinement [Ref. 16]. In the case of film boiling, which often occurs in cryogens (especially LNG), the friction factor at the substrate interface is markedly decreased. Equations for the friction factor during film boiling of LNG spilled over a water surface have been published [Ref. 4]. If the pool is supported by the ground, transfer through the substrate depends on several factors including moisture content. One bounding option that overestimates evaporation is to assume that the substrate temperature remains constant. A bounding assumption that underestimates evaporation is that for fully developed film boiling, the heat flux equals the minimum for film boiling for that liquid.

Several references show an "S" curve for typical boiling liquid heat flux as a function of temperature difference between the substrate and liquid. The typical curve shows a low flux at low temperature difference, rising like the front end of a sine wave to reach a maximum flux at the point of fully developed nucleate boiling. When the temperature difference increases beyond this critical point, the heat flux decreases like the back end of a sine wave, reaching a minimum flux point caused by a layer of vapor that separates the substrate and liquid. Beyond the minimum point, the "typical" flux curve curves up again at a slightly lower slope than the initial curve for nucleate boiling. This increase is attributed to the increase in radiative heat transfer to the liquid from the hotter substrate. The data that supports this "S" curve is mainly drawn from data from experiments with silicone oil, which boils at a high temperature. In the typical experiment, the substrate is heated beyond temperatures where radiant heat transfer is significant. This data may apply to film boiling of cryogenic liquids inside pressure vessels, but an atmospheric spill of LNG is different. If the increase in film boiling heat transfer is due to radiative transfer, it should not be expressed as a linear relationship with temperature. Radiative heat transfer is a function of absolute temperature to the fourth power. The equation for transfer between parallel plates, or a flat substrate to a liquid suspended by a flat vapor layer is [Ref. 15]:

$$Q/A = \epsilon\, \sigma\, T_1^4 - \alpha\, \sigma\, T_2^4$$

where Q/A is heat flux, ϵ is emissivity, σ is the Stefan-Boltzmann constant, T_1 is the absolute temperature of the substrate, α is the absorptivity of the liquid, and T_2 is the absolute temperature of the boiling liquid. Units of measure must be consistent with the Stefan-Boltzmann constant, which is 1.713×10^{-9} Btu/(ft^2 h °R^4) or 5.67×10^{-8} W/(m^2 °K^4). The emissivity and absorptivity of common materials is given in standard references [Ref. 15]. It would seem that the radiative effect on film boiling heat flux could be calculated, but there has been no experimental data to validate this conjecture.

7 Conclusions

Use of the Bernoulli equation for liquid flow provides a method for calculating the spread of an unconfined liquid pool on a sloping surface. The Bernoulli equation is an expression of the fluid flow energy balance, so there is no need for the user to characterize the flow as gravity-driven or momentum-driven in order to choose an appropriate model. Friction loss can be calculated the same as for piping or open channel flow after flow geometry is accounted for. Although flow can be calculated exactly if all parameters are known, the surface roughness parameter can be uncertain by an order of magnitude. The most important variable, release flow rate into the pool is generally determined by accident conditions that are difficult to predict. Furthermore, the evaporation rate from the pool can be affected by heat transfer rates and mass diffusion coefficients that are somewhat difficult to predict outside of a laboratory. Therefore, care should be taken to assure conservative assumptions, methods and data.

The calculation accuracy of liquid flow across a sloping surface is limited by roughness data and heat transfer knowledge. Within this current limitation, one-dimensional spreadsheet calculations can provide useful bounding results. However, when accurate data has been developed, computational fluid dynamics will provide more accurate results.

The non-skid checkered deck plating used on the decks of land-based processing facilities and floating production platforms has well known surface dimensions, but no documented friction factor for flow calculations. Rather than relying on field testing of qualitatively defined surfaces, isn't it time that someone develop a method for estimating friction factors based on dimensions and some measurable index of roughness?

Spills of cryogenic fuels such as LNG and LPG pose a high risk of catastrophic accident worldwide. The heat flux during film boiling as a function of differential temperature could be more accurately calculated as function of the difference of the fourth power of the absolute temperatures, like the Stefan-Boltzmann equation, rather than a first-order difference. Experiments with cryogenic liquids such as liquid nitrogen could test this possibility and be used to develop a better correlation for film boiling, especially for liquids with low boiling points. Furthermore, experiments with cryogenic liquids flowing on sloped surfaces would be useful for accident analysis.

8 References

1. J.L. Woodward, *Estimating the Flammable Mass of a Vapor Cloud*, "Discharge Characteristics", p 49-59, Center for Chemical Process Safety of the American Institute of Chemical Engineers, 1999.

2. A. Cox, F. Lees, and M. Ang, "Classification of Hazardous Locations," pp 46-47, Inter-Institutional Group Classification of Hazardous Locations of the Institution of Chemical Engineers, Rugby, UK, 1990.

3. M. Apel-Paz and A. Marmur, "Spreading of liquids on rough surfaces," J Hazard Mater, 146, pp 273-279, 1999.

4. S. Brambilla and D. Manca, "Accidents involving liquids: A step ahead in modeling pool spreading, evaporation and burning," J Hazard Mater, 161, pp 1265-1280, 2009.

5. A. A. Pahlavan, et al., "Thin Films in Partial Wetting: Internal Selection of Contact-Line Dynamics," Phys. Rev. Lett. 115, 034502, July, 2015.

6. R. H. Perry and D. W. Green, *Perry's Chemical Engineers' Handbook,* 7th ed., "Incompressible Flow in Pipes and Channels," pp 6-9 through 6-16, 6-20, 6-21 McGraw-Hill, New York, 1997.

7. Crane Co., *Flow of Fluids, Technical Paper 410*, Ver. 15, pp 1-1, 1-4 through 1-7, www.flowoffluids.com, 2009.

8. S.D. Lin, *Water and Wastewater Calculations Manual,* pp 242-252, McGraw-Hill, 2001.

9. UK Health and Safety Executive, "Hydrocarbon Releases Database," https://www.hse.gov.uk/hcr3/, 2005.

10. D.A. Crowl and J.F. Louvar, *Chemical Process Safety Fundamentals with Applications,* 3rd ed., pp 122-140, Prentice Hall, 2011.

11. J.H. Lienhard and J.B. Day, "The Breakup of Superheated Liquid Jets," J Basic Eng. Trans., ASME, pp 515-521, 1970

12. R. H. Perry and D. W. Green, *Perry's Chemical Engineers' Handbook,* 7th ed., "Particles Dynamics," pp. 6-50 through 6-54, McGraw-Hill, 1997.

13. B. Jones, *Elements of Practical Aerodynamics,* 2nd ed., p 17, Wiley & Sons, 1939.

14. R. H. Perry and D. W. Green, *Perry's Chemical Engineers' Handbook,* 7th ed., "Mass Transfer," pp 5-42 through 5-79, McGraw-Hill, 1997.

15. R. H. Perry and D. W. Green, *Perry's Chemical Engineers' Handbook,* 7th ed., "Heat Transfer," pp 5-8 through 5-29, McGraw-Hill, 1997.

16. J.L. Woodward and R.M. Pitblado, *LNG Risk Based Safety: Modeling and Consequence Analysis,* pp 142-152, Wiley, 2010.

17. Western Regional Climate Center, "Cooperative Climatological Data Summaries," http://www.wrcc.dri.edu/climatedata/climsum/ Feb. 2016.

18. ASHRAE, *Fundamentals Handbook,* ASHRAE, Ch. 22, 1977.

Example A

Leaking LNG tanker truck located on a large paved area

A-1 Accident scenario

The concern is that a tanker truck at a LNG terminal could spill LNG into a storm sewer, with the potential that the natural gas could ignite in a sewer line in a public area. A calculation must be performed to see whether the contents of the truck would evaporate before reaching the nearest storm sewer drain.

At a terminal used to fill city vehicles powered by compressed natural gas, the cryogenic storage tanks are filled by a 10,000 g tanker truck. The tanker is double-walled for insulation and impact protection. The tanker contains LNG at about 70 psig and -210°F. In this example, the LNG is assumed to consist of 100% methane. At an actual facility, the LNG composition would be known. At atmospheric pressure, methane boils at -259°F. The tanker is unloaded through a nominal 3" flexible hose.

The terminal equipment and storage tanks are located inside a protective building. The tanker unloading hose is connected to a double-walled pipe that goes through the building wall. To reach the unloading station, the truck must drive along a concrete driveway that is 59 ft wide. The driveway slopes at a 1% grade toward a storm drain that is located a little more than 400 ft down the driveway.

A-2 Release modeling

The intent is to model a bounding case, to see if the worst-case spill of LNG can reach the storm drain. Leaks from tanker trucks seldom occur, and when they do, it is usually due to accidental impact or failure of the unloading hose or coupling. Failures involve some degree of human error, and therefore are difficult to assign a probability. The failure chosen for modeling is a sudden loss of the total contents of the tanker, which may not be credible, and certainly is a bounding case.

The LNG spill quickly flashes upon release to the atmosphere. The flash can be assumed to be adiabatic. After the initial flash, the LNG continues boiling during the life of the pool. The longest spill travel occurs when there is minimum LNG evaporation or lowest surface friction. Evaporation is minimized by cool air, cool pavement, and poor heat transfer. The lowest evaporation would occur on a cold winter night. However, friction is minimized during film boiling, which is favored by a hot summer day. Calculations that follow show that the summer temperature case is worse for pool travel. At the terminal site, the lowest annual temperature is 39°F (4°C), and the highest annual temperature is 96°F (36°) [Ref. 17]. The maximum temperature typically occurs on a clear, breezy afternoon, consistent with a pavement surface temperature of about 180°F (82°) due to solar gain. Frictionless flow is assumed when the temperature difference between the substrate and liquid indicates the possibility of film boiling. The summer case is average for about 60 days a year, or 16% of the year. Therefore, the summer temperatures could be exceeded 16% of the time. It turns out that the ability of the LNG pool to travel 400 ft is not sensitive to surrounding temperature within the range that occurs at the site.

The modeling covers two phases of pool spreading: unconfined and partially confined. For simplicity, the first phase begins with the volume of the tanker in the shape of an upright cylinder, the same height as the center of the tanker, located at the tanker unloading station. After the cylinder expands to the width of the driveway, which takes about 3 seconds, the pool shape changes due to partial confinement imposed by the curbs on each side of the driveway.

A-3 Pool spread methodology

The boiling regime and heat transfer rate depend on the temperature of the pavement, which will vary with time due to contact with the LNG. A preliminary step of analysis is to correlate pavement surface temperature as a function of contact time with boiling LNG. The heat transfer from the pavement to the LNG depends on the temperature profile of the pavement and sub-pavement as a function of depth. The starting point profile was created from typical day and night temperatures as a function of time, using thermal properties of concrete and aggregate and a plot of air-concrete heat transfer coefficient as a function of wind speed [Ref. 18]. The summer daytime case requires data on solar radiation, which varies with location [Ref. 17]. The profile of temperature versus pavement depth depends on the recent history of air temperature and solar radiation. For the winter night case, there is radiative heat loss on a clear night from the pavement to the sky which is at a temperature of nearly absolute zero. However, the winter night radiation effect is insignificant. For the summer day case, there is no historical data of pavement surface temperatures, so a pavement profile was calculated by solving the one-dimensional differential equations for conductive heat transfer, starting with the temperature profile at sunrise at an average daily minimum summer temperature, allowing for night solar radiation. The air temperature and solar gain varied until six hours later, when air temperature and solar gain reached the average summer maximum. Several trial calculations showed that the daily temperature cycles did not have a significant effect on the profile below a depth of about 12". The summer temperature at 12" depth is assumed to be the average of the summer minimum and summer maximum air temperatures for the month. The equations for one-dimensional heat transfer in to a thin section of substrate at the surface of the pavement are:

$$Q_{radiation}/A = F_R \, \Delta t$$
$$Q_{convection}/A = h\,(T_{air} - T_1)\,\Delta t$$
$$Q_{conduction}/A = -k\,x\,(T_1 - T_2)\,\Delta t$$
$$\Delta T = (Q/A)/m\,C_P$$
$$m = \rho\,A\,x$$

where Q is the heat transfer rate
A is the transverse heat transfer area
F_R is the solar heat flux
Δt is the differential time step
T_1 is the temperature in the 1st section of pavement
k is the thermal conductivity of the pavement
x is the thickness of the section of pavement
ΔT is the temperature change in a section during time dt
m is the mass of the section
C_P is the specific heat capacity of the pavement

ρ is the density of the pavement

By taking a small enough time step, the heat transfer equations can be solved accurately enough by using the temperatures from the previous step. The exception occurs at the first section, where the temperature changes rapidly with time. This section was solved by obtaining the difference between the first and second sections by trial and error. Trial and error is easily done by manually adjusting the temperature difference in the spreadsheet for each time step. This procedure was used to calculate the temperature profiles for the summer case influenced by solar gain, and for the profile when pavement is exposed to LNG.

A summary of the summer temperature profile at different pavement depths is tabulated below. The bottom row shows the profile at the hottest part of a typical summer day:

Table A-1
Pavement Temperature Profile before an LNG Spill
Pavement sections are 1" thick. Section numbering begins with the section on top.
The profile for the hottest hour (hour 6) is used for spill calculations.

Time h	T_{air} °F	**Solar** Btu/h-ft^2	**Section 1**	**Section 2**	**Section 3**	**Section 4**	**Section 5**	**Section 6**	**Section 7**
0	60.4	0	59.1	61.1	63.1	65.1	67.1	69.1	71.1
1	66.4	50	69.9	68.2	68.1	68.5	69.3	70.3	71.4
2	72.5	100	85.4	80.2	77.6	75.9	74.9	74.4	74.4
3	78.5	150	105.1	96.4	91.1	87.0	83.9	81.7	80.1
4	84.5	200	128.6	116.3	108.0	101.2	95.8	91.4	88.0
5	90.6	250	155.3	139.3	127.9	118.3	110.2	103.4	97.7
6	96.6	250	178.6	161.2	147.7	135.9	125.5	116.4	108.4

When LNG is spilled onto pavement in the summer or winter, the pool stays in the film boiling regime until the surface of the pavement cools to within 20°C (36°F) of the LNG boiling point. In the film boiling regime, there is virtually no friction loss. As the tanker spill spreads to fill the area between the curbs of the driveway, the liquid is modeled as an upright cylinder in film boiling, non-frictional flow. For lack of better information, it is assumed that the vapor film under the liquid is of uniform thickness. Energy loss due to internal turbulence is ignored, which overestimates velocity and distance travelled. In the case of frictionless, gravity-induced open flow, the change in elevation equals the resulting velocity head: horizontal velocity squared divided by twice the acceleration of gravity, where the velocity is the average value across the cross-sectional area of flow.

$$z = v^2/2g$$
$$v = (2g\,z)^{1/2}$$

The velocity head equation applies to both components of pool motion – downhill motion of the entire pool, and radial spreading. In the case of downhill motion, z_d is the horizontal distance travelled, Δy, (including the downhill radius of the cylinder) multiplied by the slope of the surface.

$$v\,\Delta t = \Delta y = (2g\, z_d)^{1/2}\,\Delta t$$

In the case of radial spreading, z_r is the height of the pool at the center

$$\Delta r = (2g\, z_r)^{1/2}\,\Delta t$$

Neglecting viscous forces, the initial radial increase equals $(2g\, z)^{1/2}\,\Delta t$, where Δt is the time interval. Prior to considering evaporation loss, the cylinder volume is the same as before, but the cylinder is shorter by Δz. Therefore, the new pool height is the volume of the pool divided by the area calculated from the new radius.

In the beginning, the pool is deep, and the time required to reach the curbs is short, so it is not yet important to model the effect of the driveway slope. However, when the elevation difference between the uphill and downhill portions of the pool becomes significant relative to the pool thickness, the slope should be considered. The cylinder can be partitioned into four quadrants: uphill, downhill, and two sides. Without surface friction, the top of the cylinder is level, so each point on the circumference has a different height above the pavement. The sloped z values for the uphill and downhill quadrants are used to calculate new radii of the uphill and downhill quadrants. The net effect of the longer downhill radius is that the cylinder will spread farther in the downhill direction and become slightly elliptical.

The evaporation loss is the combination of evaporation from the surface of the liquid plus the vapor generated from boiling. The boiling heat transfer rate for LNG can be conservatively calculated by using the minimum heat flux value of 20 kW/m^2 when the temperature difference between the substrate and LNG exceeds 20°C [Ref. 16]. Evaporation from the surface is due to the combined effects of heat input from solar heat and convective heat transfer with the air. Evaporation is assumed to be limited by heat flux, not diffusion into the air. The heat flux terms for solar heat and air convection are the same as used for calculating the pavement temperature profile for a summer day.

As the pool spreads thinner, there is a time when Δz exceeds the pool thickness on the uphill edge of the pool. When this happens, the uphill edge is moved so that the thickness becomes zero at the uphill edge.

A-4 Results

Upon release to atmospheric pressure, 17.3% of the LNG flashes, based on an adiabatic flash, taking data from the methane pressure-enthalpy diagram.

The one-dimensional, time-dependent heat transfer calculations showed that on an example hot summer day, concrete that is in contact with boiling LNG can take almost 10 minutes before the liquid is in frictional contact with the pavement. This is because a film of natural gas vapor insulates the concrete, maintaining a high temperature difference between the boiling LNG and the pavement surface. Film boiling occurs for about 8.7 minutes in the summer example. For the next minute, boiling becomes more nucleate, finally reaching the point of maximum heat flux. The same spill on a cold winter night showed that the duration of film boiling was about half as long at this site, but still greatly exceeding the time required for the pool to evaporate. Other factors, such as difference in evaporation rate between summer and winter are not important. Therefore, because flow will be essentially frictionless during film boiling, the hot summer case is the worse case for distance travelled by the moving pool of LNG.

Example B

Piping Leak onto the Sloped Deck of a Floating Oil Production Platform

B-1 Accident scenario

There is a potential for a large fire on a lower deck of a floating oil production platform. To assess the potential for this fire to cause escalating damage to the platform and possibly interfere with escape, the size of the pool must be estimated. The area of the oil spill is the major factor affecting the height and heat flux of the potential pool fire. This particular spill case is one possible event of an event tree performed for a quantitative risk assessment. The length of the spill path is known, but the width of the pool as it flows down the deck must be estimated by flow calculation.

The spill source is a manifold of 12-inch pipe on a deck made of metal checker plate of the kind typically used in industrial applications. The piping is supported every few meters, so that the piping does not interfere with the flow of oil across the deck. The deck plates are galvanized metal with tread ridges to provide a non-skid surface. The deck has a slope of 2.5% from the keel line down to the edge of the deck, 31.5 m away. The lower deck is shielded from sunlight, and there is little difference in temperature between the air, deck, piping, and contained oil.

The potential leak consists of light crude oil that has been separated from produced water and gas. The crude oil contains enough dissolved gas to be flammable, but the oil loses an insignificant volume of light components when flashed to the atmosphere. The crude oil is contained at a maximum pressure of 8 barg. The transport properties of the oil at flowing temperature are:

Density	889 kg/m^3
Viscosity	11 cp, or 0.00112 kg s/m^2
Mol Wt	250 kg/mol
Surf tension	27 dyn/cm, or 0.027 N/m

Based on the number of potential leak points and the probability of leakage from each type of potential leakage, the probability of a serious leak is on the order of 10^{-3} per year for a leak equivalent to a 50-mm diameter hole. The most important contributor to serious leakage is gasket failure at a flange or valve. Nearly all flanges are arranged vertically, with the opening aimed downhill. Ordinarily, any leak onto the deck will be quickly detected by flammable gas detectors, resulting in a manual shutdown of the piping system. In the accident case for the fault tree branch of interest, the detector alarm system or operator response fails, which is estimated to happen less than 10% of the time, which lowers the probability to 10^{-4} per year. The probability of ignition is about 5% to 10%, which places the accident scenario in the barely credible range. The average wind speed is 4.3 m/s, directed crosswise to the slope of the deck. Extreme weather conditions at the time of the accident would make the scenario beyond credible, and consequently need not be considered.

B-2 Release modeling

The maximum credible leak, equivalent to a 50-mm hole is likely to occur as a narrow gap rather than a round opening, either as a tear in piping due to impact or a gap between flanges due to improper bolt tightening or gasket failure. Round holes often result from corrosion, which is not a problem with crude oil in this case. The gasket thickness is 2 mm, so a 2 mm gap is assumed. A full blow-out of a 2 mm gasket from a 12-inch line results in an opening with an area of 1920 mm^2, which is about equal to the area of a 50-mm hole.

The flow rate through the leak opening can be calculated from the flow coefficient [Ref. 7], which is a function of the Reynolds number for the pipe. The Reynolds number for the pipe is 17,400 and the ratio of orifice diameter to pipe diameter is less than 0.2, which gives a square-edge flow coefficient, *C*, of 0.60. The standard equation for flow through an orfice gives the leak flow rate. Based on the pressure drop, density, and equivalent orifice size, the leakage flow is 0.0500 m^3/s. The liquid velocity at the release point is the volumetric flow rate divided by the flow area, or 26.0 m/s.

For holes less than 2 mm wide, the drop diameter is about 1.9 times the narrow dimension (2 mm in this case) [Ref. 1, 11]. This correlation gives a drop diameter of 3.8 mm. A second method for calculating drop size is based on the Weber number being between 12 and 22 [Ref. 1]. For a Weber number of 12,

$$d = 24\,\sigma /(\rho_g\, v_d^2).$$

For the site air density of 1.204 kg/m^3, surface tension of 0.027 kg/s^2, and drop velocity of 26.0 m/s, the estimated drop diameter is between 0.80 mm and 1.46 mm. The free fall velocity for a drop between 0.80 mm and 3.8 mm is between 3 m/s and 10 m/s [Ref. 12]. The chosen values are 2.3 mm and 2 m/s.

The release will break up into droplets at a distance of L_b, which is a function of the Weber number and Reynolds number. For a drop diameter of 2.3 mm, and an air density of 1.204 kg/m^3, the Weber number is 34.7. The Reynolds number at the release point is 47,500. The distance from release to break-up into drops is

$$L_b = 540\, D\, W_e^{1/2}\, R_e^{-5/8}$$

In this case, L_b = 0.0874 m.

The trajectory of the release begins as a circular sheet 2 mm thick sprayed from a pair of vertical flanges. Before the sheet contacts the deck below, it breaks into droplets and the velocity slows to 2 m/s, relative to the air. For the portion of the spray directed straight down, that liquid lands directly under the leak point. For the portion of the spray directed up, the droplets will be in the air for about 0.36 s, calculated from the free fall drop velocity. The wind force on a droplet F_d, is a function of the drag coefficient C_d, projected drop area facing the wind A_d, air density ρ_a, and relative velocity v between the drop and air [Ref. 12]:

$$F_d = \tfrac{1}{2}\, C_d\, A_d\, \rho_a\, v^2$$

From Newtonian physics, acceleration of the droplet is force divided by mass:

$$a = F_d/(V_d\,\rho_d)$$

Distance travelled by a droplet is:

$$y = \tfrac{1}{2}\, a\, t^2$$

At the free-fall velocity, the droplet Reynolds number is 12,900. For droplet Reynolds numbers exceeding 1,000, C_d has a value of 0.445. The volume of a spherical droplet is:

$$V_d = \pi\, d^3/6$$

Solving the above, the wind will move liquid droplets downwind around 0.2 m, crosswise to the slope of the deck. At the same time, the trajectory of the leak will move droplets up to 2 m downhill, depending on the position on the vertical flange from which the droplet was released.

B-3 Pool spread methodology

The exact direction of leakage (and splashing against objects) cannot be predicted, but the model should be plausible. The release of crude oil is assumed to be continuous and at a constant flow of 0.0500 m^3/s. The starting point of the pool is directly below the leak point, located 31.5 m from the downhill edge of the deck. The initial pool area is 2 m in radius, based on droplet dispersal radius from the assumed leak due to gasket failure. After the release lands on the deck, the effect of wind is insignificant compared to gravity. Evaporation is not significant. The fluid properties of the oil remain constant. Friction loss is calculated from the Darcy equation for laminar flow. The pool flow is in the laminar regime at the edges of the pool, and is a conservative bound on areas that are in the turbulent or transition flow regime.

The pool is modeled as a conical volume with the peak located directly under the leak point. In the case of gravity-induced open flow, the change in elevation equals the resulting velocity head plus head loss from friction. The velocity head is the horizontal velocity squared divided by twice the acceleration of gravity. The friction loss is a linear function of the area of contact and the Reynolds number.

$$z = v^2/2g + h_L$$

$$v = (2g\,(z - h_L))^{1/2}$$

To include the effect of the sloping deck, the conical pool is partitioned into four quadrants: uphill, downhill, and two sides. The height of the peak is the same for all quadrants, but in each quadrant the vertical distance between the peak and the edge can be different. The radius for the uphill quadrant is adjusted as needed so that the uphill edge is never higher than the peak. As pool spreads faster in the downhill direction, the pool model changes from a right cone with a circular base to an angled cone with an elliptical base. The contact area for a quadrant of a cone is $\tfrac{1}{4}\,\pi\, r^2$. In this example, the quadrants are calculated as sections of cones of different radii, which do not match at the edges. The result is a pool area that looks like a wind rose. This mismatch results in an error in pool area calculated as an ellipse. It would have been possible to calculate the area more correctly or increase the number of sections from four if it had seemed that this error was important.

Because the friction factor is a linear function of Reynolds number in laminar flow, the average friction factor occurs at the radius associated with half of the area, r divided by the square root of

2, or approximately $\underline{r} = r/1.414$. The Darcy equation for friction loss, with the hydraulic diameter, $8z$, substituted for pipe diameter, is

$$h_L = f\,L\,v^2/(16\underline{z}\,g)$$

where $\underline{z}$ and v are evaluated at $r/1.414$, and the quadrant radius, r, is substituted for L.
For laminar flow, the Darcy equation for friction factor is

$$f = 64/Re$$

where

$$Re = 8\,\underline{z}\,v\,\rho/\mu$$

giving the following for laminar flow friction head

$$h_L = \mu\,r\,v/(2\underline{z}^2\,\rho\,g)$$

For turbulent flow ($Re > 4{,}000$), the Churchill equation for the Fanning friction factor can be used [Ref. 6]

$$1/f^{0.5} = -4\,\log\{0.27\epsilon/D + (7/Re)^{0.9}$$

where ϵ is the surface roughness. The roughness for galvanized steel is 0.152 mm. The variable D is replaced with the equivalent value of $8z$. The Fanning f must be multiplied by 4 for use in the Darcy equation.

After the average friction loss is calculated at $\underline{r}$, the velocity at that point can be calculated. Then the outward spread at that point can be calculated by

$$.\ \Delta\underline{r} = v\,\Delta t$$

The outward spread at the radial edge of the pool, Δr, is calculated by assuming that the ratio between $\underline{r}$ and r does not change:

$$\Delta r = 1.414\,\Delta\underline{r}$$

The equation for Δr and even the proposal that pool spreading can be modeled with a conical shape has not been validated by computational fluid dynamics or actual experiments.

The deck material is assumed to be flat after installation, and has typical checker plate ridges. The ridges rise from the base at an angle of 45° to a height of 1.5 mm. The ridges occupy about 25% of the surface area. Each ridge has about 1.4 times the surface area of the flat base, which increases the overall area of the plate by 8%. Therefore, the ridges increase the hydraulic radius by a factor of 8% above that of flat plate. The roughness value, ϵ, is estimated to be the same as for galvanized steel pipe.

B-4 Results

Table B-1

Oil Spill Spread on a Sloping Ship Deck from a Constant Continuous Leak

The radii for the side, downhill, and uphill quadrants are labelled ***a***, ***b***, and ***c***, respectively.
The height of the conical peak, **z**, is relative to the deck directly below.
The velocities for the quadrants are calculated at the centroid.

Time s	V m^3	a m	b m	c m	z m	v(a) m/s	v(b) m/s	v(c) m/s
0	0.005	2.00	2.00	2.00	0.001	0.004	0.790	0.088
1	0.055	2.28	3.65	0.33	0.012	0.271	1.136	0.175
2	0.105	2.67	5.80	0.44	0.012	0.269	1.409	0.123
3	0.155	3.03	8.40	0.44	0.011	0.238	1.675	0.000
4	0.205	3.35	11.46	0.44	0.010	0.204	1.944	0.000
5	0.255	3.62	14.98	0.44	0.009	0.175	2.207	0.000
6	0.305	3.85	18.96	0.44	0.008	0.150	2.486	0.000
7	0.355	4.05	23.40	0.44	0.007	0.130	2.759	0.000
8	0.405	4.26	28.30	0.44	0.006	0.141	3.034	0.000
8.6	0.435	4.43	31.46	0.44	0.006	0.152	3.199	0.000

As an overall characterization of a spill on a sloped surface, the spill can be viewed as having a roughly triangular footprint. In this case, with a surface slope of 2.5%, the spill could be viewed as a triangle 31.5 m long and 8.9 m wide at the base. This corresponds to a divergence angle of about 8° from the centerline of flow.

Uncertainties in the results include the question of whether the conical model is representative of the action of an actual liquid spill at an industrial site. There is no published data that could be found containing a friction factor for checker plate decking. The checker plate ridges may have an effect on flow pattern, such as deflecting the flow. In this case, the flow pattern that has the largest area (greatest width before stopping at the edge of the deck) gives the worst consequences. There was no study done for this report to investigate the effect of various conditions affecting the width of the flowing pool. However, results are based on recognized flow equations and should be viewed as useful in the absence of experimental or computational fluid dynamics results.

440778 Apparent Discrepancies in Methods for Predicting the Explosion Energies of Bleves

Abstract:

The authors have used calculation methods for BLEVE blasts as described in the CCPS book *Guidelines for Vapor Cloud Explosion, Pressure Vessel Burst, BLEVE and Flash Fire Hazards*, and most recently similar methods in the PHAST process industry hazard analysis/consequence analysis software tool, developed by DNV-GL. Predictions made using PHAST appeared to be overly conservative, which instigated a comprehensive comparison of PHAST to other methods in the literature. There is not universal agreement concerning the dominant mechanisms driving the air blast pressure wave from a BLEVE, and there are major differences in methods used to predict the explosion energy driving the blast wave. The CCPS guidelines book and PHAST consider the explosion energy driving the blast wave to be from both the vapor and liquid phases present at the time of loss of containment. For each phase the contribution to explosion energy is the difference in internal energy between the thermodynamic states of the liquid and vapor at the time of loss of containment and the final states of the materials when isentropically expanded to atmospheric pressure. Recently, Birk compared various prediction methods to experimental data for propane BLEVEs and concluded that the flash vaporization of the liquid may be too slow to contribute significantly to the blast, therefore the blast energy is primarily from the expansion of the vapor in the vessel at the time of loss of containment. Other authors also base the explosion energy on the gas expansion and others on the liquid superheat. Ogle and coworkers have discussed the available energy of a BLEVE (exergy) and compared this to the BLEVE explosion energy predicted using several methods in the literature. These various approaches can lead to drastically different predictions of explosion energy and the resultant blast pressures and impulses. BLEVE theories and models are discussed in this paper and predictions are compared to limited experimental data. This paper is meant to provide comparisons between the many various approaches, including many more that are not listed above, and summarize their differences.

441089 Automated Optimization of Gas Detector Placement

Abstract:

Chemical process facilities are moving toward the use of performance-based design techniques for fire & gas detector placement. These approaches are displacing methods for layout of gas detection based on uniform "grids" that date back more than 25 years. First generation performance-based design techniques utilized 'gas volume detection' principals that are valid in terms of detecting and mitigating a large explosion event, but can result can result in an excessively large number of detectors and a costly design. This paper presents a superior alternative to the first generation technique of volume detection that uses sophisticated modeling of gas release scenarios with computational fluid dynamics (CFD). The techniques described in this paper align with ISA TR 84.00.07 guidelines for scenario-by-scenario basis. Furthermore, the paper breaks the paradigm that CFD can only be used sparingly on complicated gas dispersion problems. With the proper definition of the process, optimal selection of boundry criteria, and automated scripting of multiple scenarios and data collection from those scenarios, the thousands of CFD runs required to develop results can be efficiently and cost effectively performed. This paper will demonstrate practical application of CFD modeling and probabilistic risk analysis in order to improve on the performance of gas detection systems while at the same time, reducing unnecessary conservatism in the number of detectors. The paper will present methodology for determining scenario-based gas detector coverage and include a worked application example for an offshore gas processing platform.

441414 Quantitative Risk Assessment in the Process Industries: Beyond Numerical Estimates of Individual and Societal Risk

Abstract:

The Chemical Process and Oil and Gas industries (both on and offshore) have largely adopted Quantitative Risk Assessment (QRA) with open arms following its application in the nuclear industry. Currently, the methodology not only forms a core aspect of the risk management programs of numerous organizations who design, build, own, manage or operate high hazard facilities (refineries, LNG terminals etc.). It also occupies a central role in the legislative and regulatory framework in many jurisdictions across the world. It has become a tool of choice for both the regulators and the regulated parties.

As its application has become more widespread, it has been the subject of extensive critical inquiry, much of which has revolved around the primary product of the process, namely numerical estimates of risks to exposed people. Advocates highlight the utility and rationality such numbers offer in the often challenging decision making process associated with risk management. Critics make reference to the wide variability and uncertainty associated with such predictions as well as the implicit value judgments they embody. Less emphasis has been placed on the wealth of other insights (often latent) afforded by a well scoped out, executed and documented QRA.

The aim of the paper is to examine and explore the value of QRA over and beyond the well-known orthodox output of individual or societal risk estimates. It identifies and outlines what these added benefits are and elaborates on their significance - particularly in the context of achieving risk management objectives. It argues that such latent benefits should be considered to be equally or more important to the typical output and that more attention should be paid to these whilst conducting a QRA. A popular adage goes thus "sometimes, it's the journey that teaches you a lot about the destination"; so it is with QRA.

Quantifying the Importance of Accurate Physical Property Data in Dispersion Modeling

Devin Averett
Baker Engineering and Risk Consultants, Inc.
11011 Richmond Avenue, Suite 700
Houston, TX 77042
daverett@BakerRisk.com

Jef Rowley
Baker Engineering and Risk Consultants, Inc.
11011 Richmond Avenue, Suite 700
Houston, TX 77042
jrowley@BakerRisk.com

Prepared for Presentation at
American Institute of Chemical Engineers
2016 Spring Meeting
12th Global Congress on Process Safety
Houston, Texas
April 11-23, 2016

Quantifying the Importance of Accurate Physical Property Data in Dispersion Modeling

Devin Averett
Baker Engineering and Risk Consultants, Inc.
11011 Richmond Avenue, Suite 700
Houston, TX 77042
daverett@BakerRisk.com

Jef Rowley
Baker Engineering and Risk Consultants, Inc.

Keywords: Dispersion Modeling, Physical Properties, DIPPR

Abstract

Modeling discharge and dispersion of materials from processing vessels is an important part of a facility siting study. In this paper, we investigate the dependence of dispersion modeling on physical property data accuracy, including the impact of uncertainty in the lower flammability limit, heat of vaporization, liquid density, and vapor pressure data. Vapor pressure is further analyzed by comparing correlations obtained from the Design Institute for Physical Properties (DIPPR) to the Clausius-Clapeyron equation using the normal boiling point as the reference.

Overall, the predicted dispersion distances are relatively resilient to uncertainty in physical properties. However, errors in dispersion distances can be large under certain conditions. It is important to have accurate data for these cases. To ensure data accuracy, we recommend using a modern, critically reviewed database such as the latest DIPPR database, which contains thermophysical data for a large number of compounds and routinely employs rigorous analysis and consistency checks to assure the reliability of the data.

1 Introduction

Dispersion modeling can be a main driver for many major economic and safety decisions. Modeling of a toxic release may help dictate where and what type of personal protective equipment (PPE) is needed, where to place high-occupancy buildings and work areas, and whether risk to personnel or the surrounding community warrants investment of capital into mitigating the effects or decreasing the likelihood of such a release. Modeling of a flammable vapor cloud may provide similar information regarding safety from the risks posed by flash fires and explosions, but it can also aid in estimating the economic impact from explosion events by determining which buildings and equipment are likely to be damaged. This impact is not just in the cost of replacing equipment, but also in the loss of production during the repair downtime. This information may in turn determine insurance rates and proper amounts of coverage.

With such important decisions hinging on the results of a model, it is vital that the model is reliable and provides as accurately as possible a representation of what will happen during an event, especially when there is already so much inherent uncertainty due in large part to the natural variability in the weather. So selecting a good model that will make the best of a difficult situation is vital. When modeling a dispersion scenario, there are many models available. Typically the most accurate models are based on computational fluid dynamics (CFD), as most other models do not explicitly account for the presence of obstacles; however, use of CFD is computationally expensive and impractical if it is necessary to model numerous release scenarios subject to various weather conditions. Other models have been developed that sacrifice some accuracy in the presence of obstacles to gain computational speed.

Whatever the model, similar inputs are typically required, many of which are scenario-dependent (e.g., atmospheric conditions such as wind direction and speed, weather stability class, etc.). The inputs that remain constant across all different scenarios are the physical properties of the material being released. Unlike the weather, a material's chemical properties do not exhibit large and unpredictable fluctuations. So if there is any input to the model in which one can place a reasonable level of confidence, then surely it is the physical property data.

Of course there is disagreement in the literature pertaining to physical properties. Aside from the more obvious sources of error (human error, bad equipment calibration, etc.), different experimental techniques may yield differing results, and there is even variability in experiments employing the same or similar techniques. The trouble, then, is determining which literature value is likely to be the most correct, which may not be an easy task.

In this paper, we explore whether the effort of finding accurate material properties for consequence modeling is worth the effort and cost. If the physical properties are not exact, does it make a significant difference in dispersion model results? We take a quantitative approach to answer this question by utilizing the dispersion model currently used in SafeSite$_{3G}$®, BakerRisk's in-house software for consequence modeling. By varying a number of different properties, we can determine the impact they have on the final dispersion results. Conclusions can then be drawn on the importance of obtaining the best physical property data in dispersion modeling.

2 Methodology

Dispersion model results can't be summarized in a single number. There are many points of comparison that could have been used in our analysis. In this study, we use for our basis of comparison the horizontal distance at which the centerline of our release reaches a concentration equal to the lower flammability limit of the material.

The dispersion model used is detailed by Rowley [1]. In short, it solves the conservation of mass, momentum, and energy in one dimension, and then concentration profiles in the vertical and cross-wind directions are obtained based on similarity concepts. The model itself actually draws from several other ancillary models for handling phenomena such as droplet trajectory, pool formation, and droplet and pool evaporation. Physical properties for materials used in the model were obtained from the most recent version of the DIPPR® 801 database [2].

Several different physical properties were varied to monitor the effect any variance in the values would have on dispersion distances. These were chosen based on which properties intuitively would seem to have the greatest impact on dispersion results; namely lower flammability limit (LFL), heat of vaporization (HVP), liquid density (LDN), and vapor pressure (VP).

Several chemicals were chosen for the analysis, selected to give a wide range of normal boiling points (NBP) to ensure materials of different volatilities were adequately explored. The chemicals, in order of increasing normal boiling points, are methane, ethylene, propylene, ethylene oxide, methanol, *n*-octane, phenol, 1-decanol, glycerol, and *n*-eicosane. The intent was to select fairly common chemicals that differed in normal boiling point by approximately 50 K.

Dispersions were run at two different weather conditions, both at atmospheric pressure of 1 atm and an ambient temperature of 70 °F. The first weather condition used a wind speed of 1.5 m/s and a Pasquill stability class of F, and the second used a wind speed of 3 m/s with a D stability class. In addition, different dispersion runs were performed for different storage temperatures and pressures. Pressures of 10, 100 and 1000 psig were run at temperatures ranging from -275 to 675 °F in 50 °F intervals. Releases where the material exists as a solid at those storage conditions were ignored.

To establish a baseline set of results, dispersions were run for the different materials with physical properties set equal to the accepted values from DIPPR. Additional dispersion runs were then gathered where one physical property was altered slightly. Each property was altered by 5% in either direction. (Note that if altering the property resulted in a phase change at the storage conditions, these cases were omitted from the analysis, the presumption being that even if a physical property may not be accurately known, it is known in which phase the material is being stored.) We were then able to obtain an estimate as to how much the dispersion distances change in relation to a standard step change in a specific physical property. The vapor pressure analysis was further extended by looking at how dispersion results differed when using the Clausius-Clapeyron equation using the normal boiling point from DIPPR as the reference point. This was done to see how important it is to obtain accurate data across a broad range of temperatures rather than simply at a single point and extrapolating from there numerically with a basis on theory.

3 Results and Discussion

Taking one property at a time, results from the dispersion runs are summarized below, accompanied by some discussion on implications that can be drawn from the data. All results are plotted as:

$$\frac{\%\Delta(Distance\ to\ LFL)}{\%\Delta(Physical\ Property\ of\ Interest)}\ vs.\ \frac{Process\ Temperature\ (K)}{NBP(K)}$$

where $\%\Delta$ refers to a percent change in the pertinent value.

3.1 Lower Flammability Limit (LFL)

The flammability limit has the most straightforward effect on dispersion distances. This is because it has no effect on the dispersion model whatsoever. It merely changes the target concentration we are looking for when calculating the extent of the plume to LFL. Figure 1 shows the results for all runs.

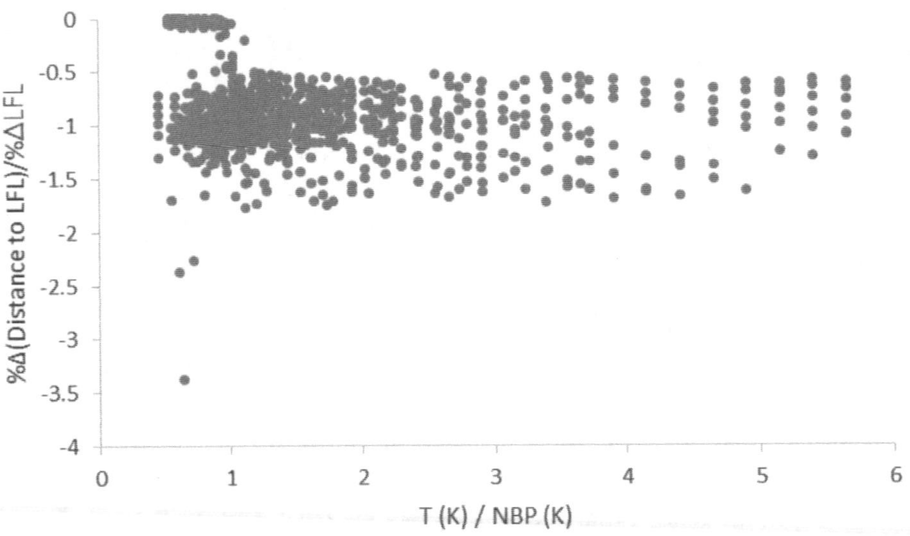

Figure 1. Ratio of change in dispersion distance and change in LFL vs. ratio of process temperature and NBP.

The uppermost cluster showing virtually no change in dispersion distance represents those scenarios that form highly non-volatile pools. Otherwise, all scenarios show a decrease in dispersion distance with an increase in LFL, which intuitively makes sense. The largest deviations occur for materials near saturation (T/NBP = 1). On average, dispersion distances changed only slightly, with an average change of -0.81 % in dispersion distance for a 1 % increase in LFL and a standard deviation of 0.45 %. While this may not appear to be very significant, the trouble with LFL is the inaccuracy in measuring LFL. Literature values may vary widely, and these large discrepancies will result in large differences in dispersion distance. For example, within the DIPPR database, looking at only methane, the accepted value of 5 % vol [3] differs by as much as 12 % from other experimental values (a value of 4.4 % vol [4]). A previous study has shown that large discrepancies can occur simply due to the method of measurement. Data obtained using the US Bureau of Mines tube method can give dispersion distances to LFL that differ by as much as 200 % from data obtained using the EN 1839 tube method [5]. Indeed, the same study discusses the reality that rather than a discrete value, the LFL exists as a range of concentrations that may yield varying degrees of flame propagation, and different methods of measurement may give measurements biased towards one end of this range or the other. Fortunately, in this study we found that small differences in LFL generally didn't have a large impact on dispersion distances, but certainly caution should be taken in cases where the LFL is low or wide ranges of LFL data are available in the literature.

3.2 Heat of Vaporization (HVP)

For heat of vaporization, we examine a much narrower range of conditions. This is because for HVP to play any significant role at all, an appreciable amount of both liquid and vapor phases must be present. Thus high-temperature vapor releases and low-temperature liquid releases would give virtually the same results for large changes in HVP because a negligible amount, if any, of vaporization is occurring. Figure 2 shows the results from the analysis.

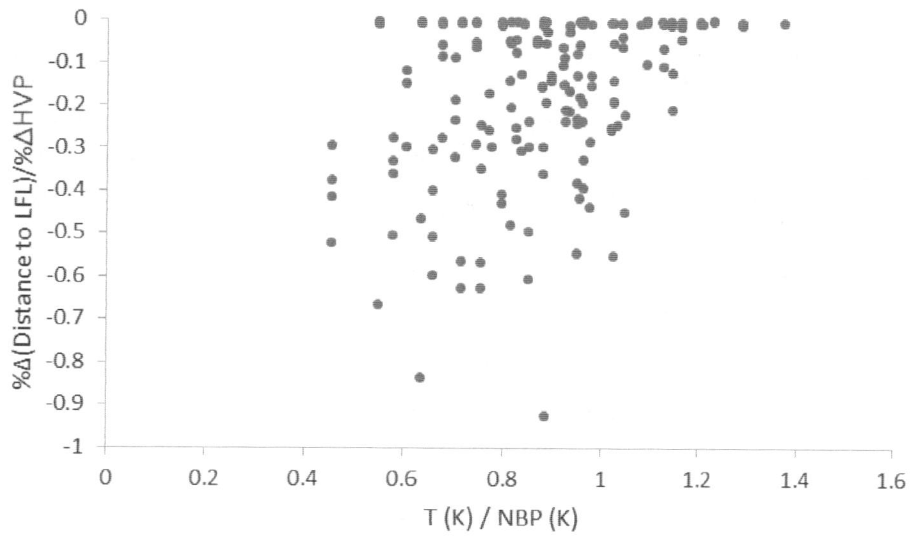

Figure 2. Ratio of change in dispersion distance and change in HVP vs. ratio of process temperature and NBP.

Similar to LFL, we see that small changes in HVP result in approximately the same order of magnitude of change in dispersion distances. For every one percent increase in HVP, the distance to ½ LFL will on average change by -0.17 %, with a standard deviation of 0.19 % and a minimum of -0.93 % occurring in the cases we analyzed. And this, remember, is only for cases where any appreciable phase change is taking place. If at least somewhat reliable HVP data can be obtained, then we can be confident that it will not affect dispersion distances to any significant degree, and especially not if the scenario does not include a phase change.

3.3 Liquid Density (LDN)

The effect of liquid density on dispersion models is much more complex than the effect of LFL or HVP. First of all, liquid density will affect the discharge rate. It would seem logical that a higher density would lead to a material that would discharge at a greater rate from a specific orifice, suggesting greater dispersion distances. However, this would affect other characteristics of the flow as well such as the choke pressure. The change in choke pressure can cause additional issues, particularly in cases where the liquid flashes upon release. Any change in pressure will then have a large effect on the vapor fraction of the discharge, and hence will cause a large swing in the overall density of the initial discharge.

The second issue to consider is of plume buoyancy. For plumes of high liquid fractions, an increase in liquid density will cause a decrease in plume buoyancy, resulting in the liquid raining out at a nearer distance. This would likely result in shorter dispersion distances, which could either supplement or compete with any effects from the change in discharge.

Although there are other issues affected by LDN, the last issue we will mention here is that momentum must be conserved. Given two identical droplet-filled plumes (both plumes start with the same velocity and have the same number of droplets with equal diameter) with the only difference being the density of the material in the droplets, and assuming an equal air entrainment rate between both plumes, conservation of momentum dictates that the denser plume will decelerate at a slower rate than the less dense plume. So with these simplistic plumes, the liquid in the denser plume will rain out at a farther distance, exactly the opposite conclusion that was reached when considering plume buoyancy. We see that with LDN, trying to reason out what effect changing LDN might have on a single scenario without running the model can become very complicated.

We turn to the model, then, for some answers. Figure 3 shows results for LDN dispersion runs. Again, some data was omitted because it was not relevant to the analysis, namely the cases where no liquid was present throughout the scenario. Note that unlike previous properties, both negative and positive deviations in dispersion distances are seen, and the differences become more pronounced the more the temperature is decreased. The point that gives the largest deviation (a surprising 13.1 % increase in dispersion distance per one percent increase in density) is phenol at P = 1000 psig and T = 275 °F for the F-classified weather. Under these storage conditions, phenol exists as a liquid. Looking at this specific case in more detail, we see that the droplets in the denser plume rain out more quickly (less than a second compared to about 11 seconds), but the velocity in the less dense plume is knocked out very quickly, causing rainout to occur at about 28 ft compared to the denser plume at about 60 ft.

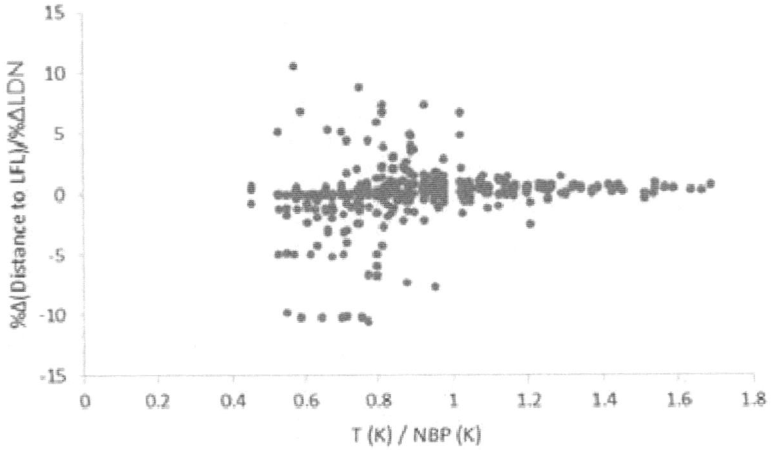

Figure 3. Ratio of change in dispersion distance and change in LDN vs. ratio of process temperature and NBP.

On average, a one percent increase in LDN results in only a 0.08 % increase in dispersion distances, but that has a standard deviation of 2.21 %. Thus, having inaccurate LDN data can cause large differences in dispersion results, and those differences can be difficult to predict. Fortunately LDN is a relatively simple property to measure, but if the scenario deals with a wide range of temperatures it is imperative to have accurate temperature-dependent data for LDN.

3.4 Vapor Pressure (VP)

3.4.1 Deviation in Vapor Pressure

Like HVP, vapor pressure can really only affect releases where two phases are possible. Hence, pure vapor releases are ignored, as well as releases low enough in temperature to form pools with no significant amount of evaporation. Also ignored are cases where the change in VP would result in a phase change at process conditions, the assumption being that the initial phase is known.

Like LDN, VP may have competing effects on dispersion modeling. An increase in VP would cause a flashing liquid upon discharge to have an increased vapor fraction and thus a decreased mass flow, likely suggesting shorter dispersion distances. However, it also means droplets and pools evaporate at higher rates, serving to lengthen dispersion distances.

Figure 4 shows the results from the VP runs. As would be expected, most of the variability occurs when the temperature of the fluid is near the normal boiling point where a two-phase release is much more probable. The most extreme result (an 11 % increase in dispersion distance per 1 % increase in VP) come from *n*-eicosane at P = 100 psig and T = 625 °F for the D-classified weather. In this case, the difference is largely buoyancy driven. For higher VP, the resulting excess vapor is sufficient to increase the buoyancy of the plume and significantly delay rainout, keeping the concentration and mass of material in the plume at higher levels for a longer time.

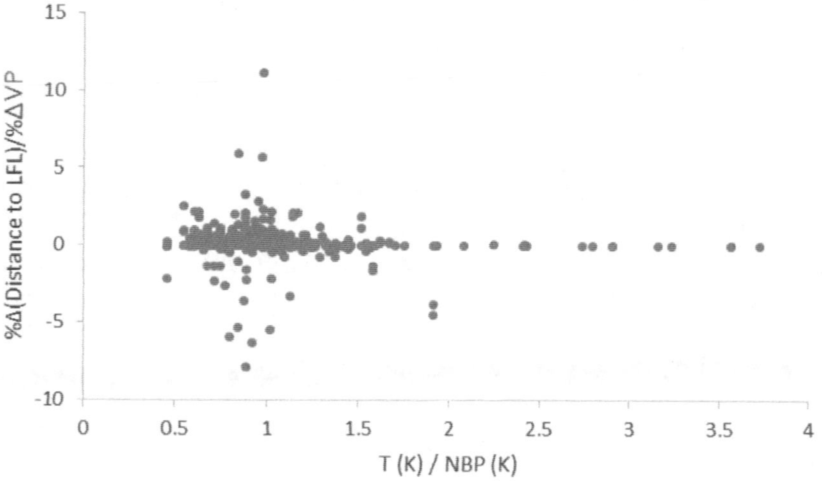

Figure 4. Ratio of change in dispersion distance and change in VP vs. ratio of process temperature and NBP.

The average result is only a 0.18 % increase in dispersion distance per 1 % increase in VP, but with a standard deviation of 1.19 %. While it does not tend to affect dispersion distances to the same degree as LDN, VP does have the potential to significantly alter dispersion results.

3.4.2 Clausius-Clapeyron Equation

It was also of interest to see if one could still get reliable results using the Clausius-Clapeyron equation, or at least the well-known approximation of it, shown here:

$$\ln\left(\frac{P_1}{P_2}\right) = \Delta H_{vap}\left(\frac{1}{T_2} - \frac{1}{T_1}\right) \tag{1}$$

where P_1, T_1 and ΔH_{vap} are the vapor pressure, temperature, and heat of vaporization, respectively, at a reference state, and P_2 and T_2 are the vapor pressure and temperature, respectively, at some new state. Hence, at the reference state, the equation is exact, but errors will develop and worsen as the desired temperature moves farther from the reference temperature. Often, heat of vaporization data is available at the normal boiling point of a material. To establish a basis for this study, then, we select the normal boiling point of the material to be our reference point, using the HVP value obtained from the DIPPR database at this temperature. The DIPPR database base correlations for HVP and VP are based on rigorously evaluated experimental data, so we can be confident that the DIPPR correlation will always result in a more accurate answer than the Clausius-Clapeyron equation, regardless of the temperature. Hence, in the following discussion we assume any deviation from results derived from DIPPR data are unfavorable. Again, pure vapor releases, releases low enough in temperature to form pools with no significant amount of evaporation, and cases where the change in VP would result in a phase change at process conditions are ignored.

Earlier results relating to vapor pressure would suggest that using the Clausius-Clapeyron equation would likely have little effect on dispersion distances since significant deviations seen in Figure 4 are concentrated around the NBP, the area in which the Clausius-Clapeyron (as we have defined it) should be the most accurate. Following the same procedure as the other analyses (testing several materials at several temperatures and pressures), there surprisingly were some significant deviations in dispersion distances. Figure 6 and Figure 6 show the results. Note that because we are not dealing with a strict percentage deviation in vapor pressure, the y-axes in these figures are not presented as ratios, but rather Figure 5 shows straight percent differences in dispersion distance and Figure 6 shows absolute differences in dispersion distances. This makes it difficult to draw a comparison with the results from earlier in this study. Nevertheless, it is clear that particularly at temperatures just below the NBP, significant deviations can occur from the use of Clausius-Clapeyron. Over all runs, there was an average change of 10.0 %, with a standard deviation of 40.4 %.

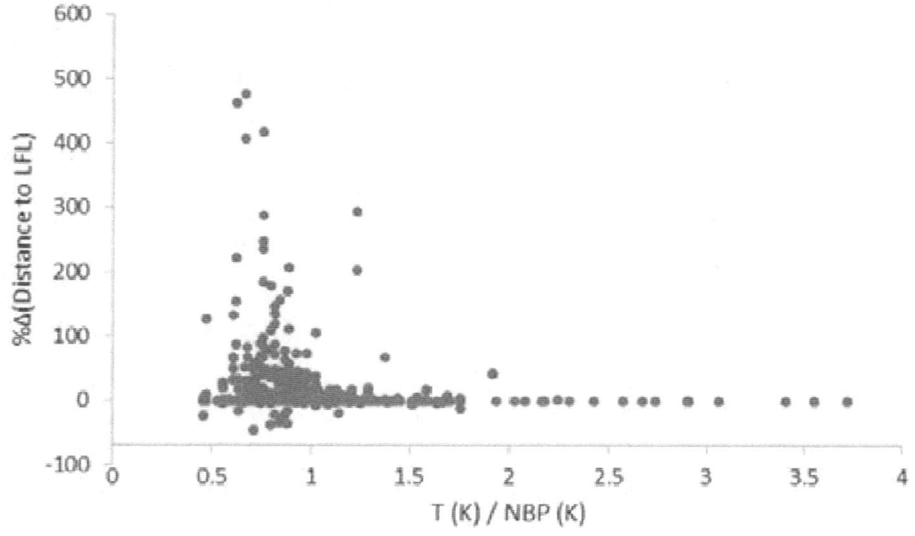

Figure 5. Percent change of dispersion distance vs. ratio of process temperature and NBP.

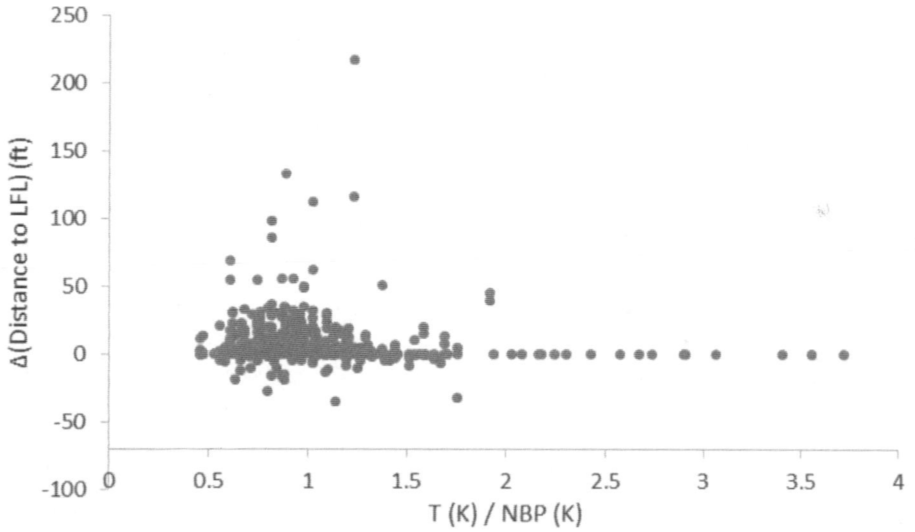

Figure 6. Absolute change in feet of dispersion distance vs. ratio of process temperature and NBP.

Still, the magnitude of these deviations are at times so extreme that it warrants further inspection to determine the reasons behind it. Figure 7 helps make results from Figure 5 and Figure 6 more relatable to results from Figure 4 by showing exactly how much VP as calculated from Clausius Clapeyron differs from VP calculated from the DIPPR correlation, once again plotted against T / NBP. Figure 4 shows deviations per percent change in VP, and Figure 7 shows that using Clausius Clapeyron will frequently result in a deviation much higher. The general trend seen in Figure 7 is that higher-boiling components reach greater deviations at lower T/NBP ratios. (Note that if plotted against T - NBP, the data would look much more uniform across all compounds.) It is the high-boiling compounds that make up the majority of the higher deviations in Figure 5. So not only do we find that Clausius Clapeyron can give very different results from correlations based on temperature-dependent experimental data, but we also see that the magnitude of those differences have a positive correlation with the NBP of the material. For heavier compounds, then, it becomes even more important to use accurate VP data in dispersion modeling.

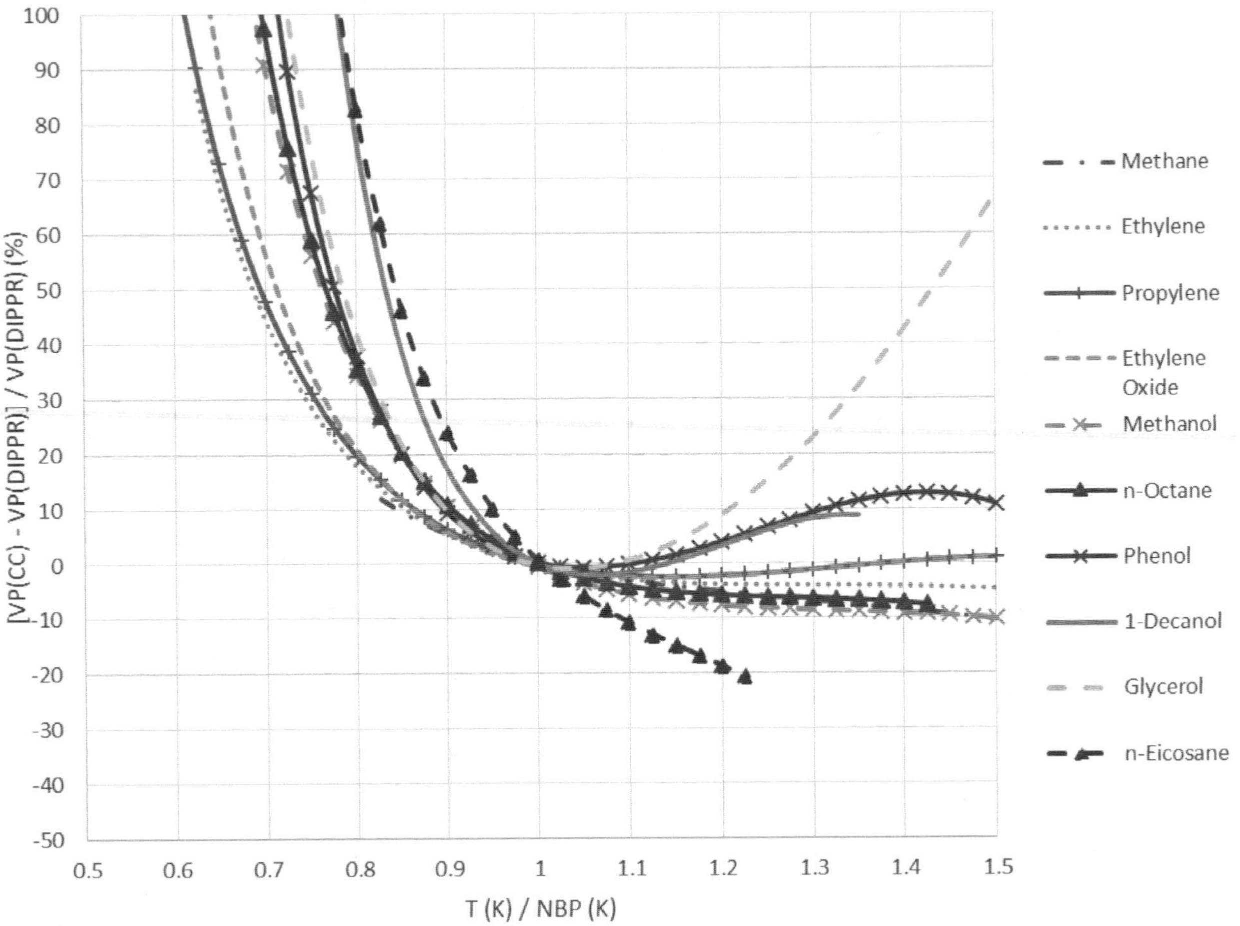

Figure 7. Percent difference between VP as obtained from the DIPPR correlation and the VP as given by Equation 1 plotted against T / NBP.

4 Conclusions

Overall, dispersion distances turned out to be quite resilient to deviations in physical properties. Minor changes in HVP or LFL resulted in only slight deviations in dispersion distances. In many cases, minor deviations in LDN and VP resulted in only minor deviations of dispersion distances. However, under the right conditions (far into the liquid regime for LDN or near the NBP for VP) large deviations in predicted dispersion distances can occur with only minor changes in physical property data. Consequently, great care should be taken as to where one obtains the physical property data used in dispersion modeling.

Furthermore, the Clausius-Clapeyron equation may not work well across the wide range of conditions that may be encountered during even a single dispersion. The processing temperature may be different from the temperature upon discharge, the temperature of the droplets in the plume, or the temperature of any pool formed, and we see in this study that it is vital to have reliable temperature-dependent vapor pressure data rather than rely on an extrapolation from a single point.

Obtaining accurate physical properties may be difficult and time-consuming, and we can often be tempted to settle for any values we can find, regardless of the source. Particularly in the case of LDN and VP, this study has shown that all physical property data should be met with some level of scrutiny when attempting to model dispersions.

We recommend using a critically reviewed database such as the current version of the DIPPR database [2]. The DIPPR database contains thermophysical data for a large number of compounds, and it routinely employs rigorous analysis and consistency checks to assure the reliability of the data. The database is continually updated and reviewed by experts in thermodynamics to ensure the recommended data reflect the latest advancements in thermophysical property measurement and evaluation. For properties where literature values are not available, state-of-the-art prediction methods are utilized to offer a best-guess value for situations where an estimate is needed.

5 References

[1] J. R. Rowley, "A new integral dispersion model based on one-dimensional turbulence theory," in *IChemE's 24th Hazard Symposium*, Edinburgh, UK, 2014.

[2] R. R. Rowley, W. V. Wilding, J. L. Oscarson and N. F. Giles, DIPPR® Data Compilation of Pure Chemical Properties, New York, NY: Design Institute for Physical Properties, AIChE, 2015.

[3] M. G. Zabetakis, "Flammability Characteristics of Combustible Gases and Vapors," in *Bulletin No. 627*, U.S. Bureau of Mines, 1965.

[4] K. Nabert, G. Schön and T. Redeker, Sicherheitstechnische Kenngrößen brennbarer Gase und Dämple, Deutscher Eichverlag, 2004.

[5] J. R. Rowley and J. E. Bruce-Black, Proper application of flammability limit data in consequence studies, Southport, UK: IChemE's 23rd Hazard Symposium, 2012.

[6] G. H. Tryon, Fire Protection Handbook, 12th ed., Boston: National Fire Protection Associations, 1962.

441423 Sustainable Risk Reduction with Dynamic Execution of Procedures for Any Operating Condition

Abstract:

Managing the volume of new information created daily is a challenge. Add to that the complexity of new equipment, employee turnover, and third party contractors working on-site, documenting clearly and concisely is critical to the operation. To truly mitigate risk we need look to operating companies focused on leading indicators of risk management where improved work conditions and profitability of the operation are the by product.

- Reduce human error
- Improve user adoption
- Improve operator efficiency
- Optimize SME time
- Quick resolution incident investigation
- Improve time "on the job"
- Extend equipment life

These objectives, met through the implementation of procedure best-practices to significantly reduce risk and improve the reliability of the operation.

During this presentation we will share the results of our procedure best-practices benchmark report and review the impact of asset level operating procedures for our clients.

As operating companies evolve through the knowledge management maturity model, they not only standardize best-practices but leverage that knowledge to optimize performance with a collaboration hub and dynamically generated procedures to fit any situation.

We will demonstrate how asset level operating procedures, designed for routine operator and facility maintenance, directly relates to improvements in asset life and performance, site-safe incident reduction based on human factors, and improve equipment start-up predictability when returning from planned maintenance.

This is so much more than simply documenting procedures; this is about sustainable adoption to improve operating discipline. Innovatia will introduce a complete procedure management solution to facilitate knowledge transfer in real-time to mitigate risk and improve reliability of the operation.

441155 Process Hazard Analysis of River-Crossing Natural Gas Pipeline

Abstract:

In order to control the process hazard of the river-crossing natural gas pipeline, HAZOP analysis is performed to qualitatively analyze its potential risk. Based on the results of HAZOP analysis, high risk scenarios are identified. So as to calculate the risk caused by pipeline's catastrophic rupture, FLACS software which based on CFD data simulation is used to quantitatively calculate the flame propagation and explosion overpressure of the explosion of natural gas. With the results of HAZOP analysis and quantitative calculation, safety measures are put forward to ensure the long-term safe operation of river-crossing natural gas pipeline.

Safe Depressurization of Large Thickness Vessels Under Pool Fire

Subhash S. Kale and George Maliakkal [(1)], Bechtel Corporation

(1) To whom all correspondence should be addressed

Abstract:

Vessels fail at pressures lower than the Maximum Allowable Working Pressure (MAWP) during fire due to degradation of strength at high temperatures. Analytical solution of the vessel wall temperature profiles as a function of fire exposure time available in API 521 Annex A for thicknesses of 3.2 mm (0.125 in.), 12.7 mm (0.5 in.) and 25.4 mm (1 in.) was extended for vessel thicknesses up to 152.4 mm (6 in.) to include high pressure/large diameter vessels encountered in gas processing. For fire depressuring, plots of ratio of the maximum vessel pressures to avoid failure to MAWP were developed for various wall thicknesses and compared to a plot of the ratio of vessel internal pressure to MAWP, both as a function of time. If the vessel internal pressures throughout fire depressuring were less than the reduced vessel failure pressures, then vessel rupture is avoided. The methodology is illustrated for the case of a high pressure vessel of wall thickness 57.15 mm (2.25 in). in natural gas service.

Introduction:

Large vessels, for e.g. vertical compressor suction KO drums, have a substantial flammable hydrocarbon liquid inventory, which if leaked can form a source of pool fire. The vessels usually have a large diameter 3 - 6 m which requires vessels with thicknesses higher than 25.4 mm (1 in.). API 521 [1] Section 4.4.13.2.4.1 gives information on wall temperatures for plates 3.2 mm (1/8 in.), 12.7 mm (1/2 in.) and 25.4 mm (1 in.). Also, Appendix A of API 521 gives information on maximum pressure as % of Maximum Allowable Working Pressure (MAWP) for plates of thickness up to 25.4 mm (1 in.). No data is available in API 521 for plates of thicknesses greater than 1 in. This work uses the analytical method given in Appendix A of API 521 to extend the calculations for vessels of thickness higher than 25.4 mm (1 in.) - for e.g. 50.8 mm (2 in.) to 152.4 mm (6 in.). thicknesses. The analytical equation given in Appendix A of API 521 is solved to develop wall temperature profiles for various wall thicknesses including those beyond 25.4 mm (1 in.). The temperature versus allowable stress data given in Table A-7 of the Appendix A can then be used to develop temporal profiles of maximum allowable pressure to avoid failure as a function of MAWP. The vessel internal pressure experienced during fire case depressurization is calculated using HYSYS v 7.3 depressurization utility. The internal pressures during depressurization, expressed as a ratio of the internal pressure to MAWP, is overlaid on the failure pressure temporal profiles to ascertain whether the vessel can avoid rupture during a pool fire scenario. The following describes the assumptions used in the calculations, the methodology to compute the temporal profiles of wall temperature and the maximum pressure to avoid failure.

Assumptions:

1. A carbon steel vertical vessel with hemispherical heads is assumed.
2. The surfaces are assumed to fully engulfed in flames hence both convection and radiation terms in equation A-1 of API 521 Appendix A apply.
3. The parameters given in Appendix A are slightly modified to better fit the temperature profiles given in Figure C-9 of Appendix C of API 521.
4. Since the vessels have thicknesses above 25.4 mm (1in.), the depressurization is from the design pressure, which is assumed to be same as MAWP, to half the design pressure in 15 minutes per API 521 guidelines. For these calculations, a design pressure of 80 barg is assumed.
5. The calculations are illustrated for a vessel of controlling thickness 57.2 mm (2.25 in.) for the hemispherical heads. The methodology can be extended to larger thicknesses from 25.4 mm (1 in.) to 152.4 mm (6 in.).

Wall Temperature Calculation by Analytical Method:

The analytical equation given in A-1 of Appendix A of API 521 is stated as:

$q_{absorbed} = \sigma\ (\ \alpha_{surface} * \varepsilon_{fire} * T_{fire}^4 - \varepsilon_{surface} * T_{surface}^4\ ) + h\ (T_{gas} - T_{surface})$ (1)

Also, by a heat balance,

$q_{absorbed} = m\ c_p\ (\text{Delta}\ T_{surface} / \text{Delta time})$ (2)

The following gives a description and the values used for the various parameters in this equation per API 521 Table C-3:

$q_{absorbed}$, Heat absorbed by the wall

σ, Stefan Boltzmann Constant = 5.67 E-8 W/m^2.K^4 (1.713E-09 BTU/H-ft^2-°R$^{4)}$

$\alpha_{surface}$ is the equipment absorptivity = 0.8

ε_{fire} is the fire emissivity = 0.85

$\varepsilon_{surface}$ is equipment emissivity = 0.75

T_{fire} is the fire temperature = 904 °C (1652 °F)

$T_{surface}$ = equipment temperature initially assumed to be 59 °C (122 °F or 582 °R)

The surface temperature increases as the surface heats up due to radiation from the fire

h is the convective heat transfer coefficient 20 W/m^2-K (3.52 BTU/hr-ft^2-R)

Plate area is assumed to be 2.3 m^2 (24 ft^2) (per API Section 4.4.413, Figure 1)

Density of carbon steel is 7850 kg/m^3 (490.1 lb/ft^3)(from internet)

Plate thickness varies from 3.2 mm (1/8 in.), 12.7 mm (1/2 in.) and 25.4 mm (1 in.).

Heat capacity of carbon steel = 502 J/kg°-K (0.12 BTU/lb-°F)

Additionally, larger thicknesses 50.8 mm (2 in) and 76.2 mm (3 in) are also used.

From equation (1), $q_{absorbed}$ is calculated.

The plate volume is calculated for each thickness from the given area.

Using the given carbon steel density, the plate mass is calculated.

From equation (2) Delta $T_{surface}$ is calculated.

Starting with the initial surface temperature, the change in the surface temperature is calculated from equations (1) and (2) for different times in 0.5 second increments. The resulting wall temperatures are plotted and compared to values given in API 521 Appendix C (Figure C-9) as shown in Figure 1.

Figure 1 Calculated wall temperature profiles as a function of time

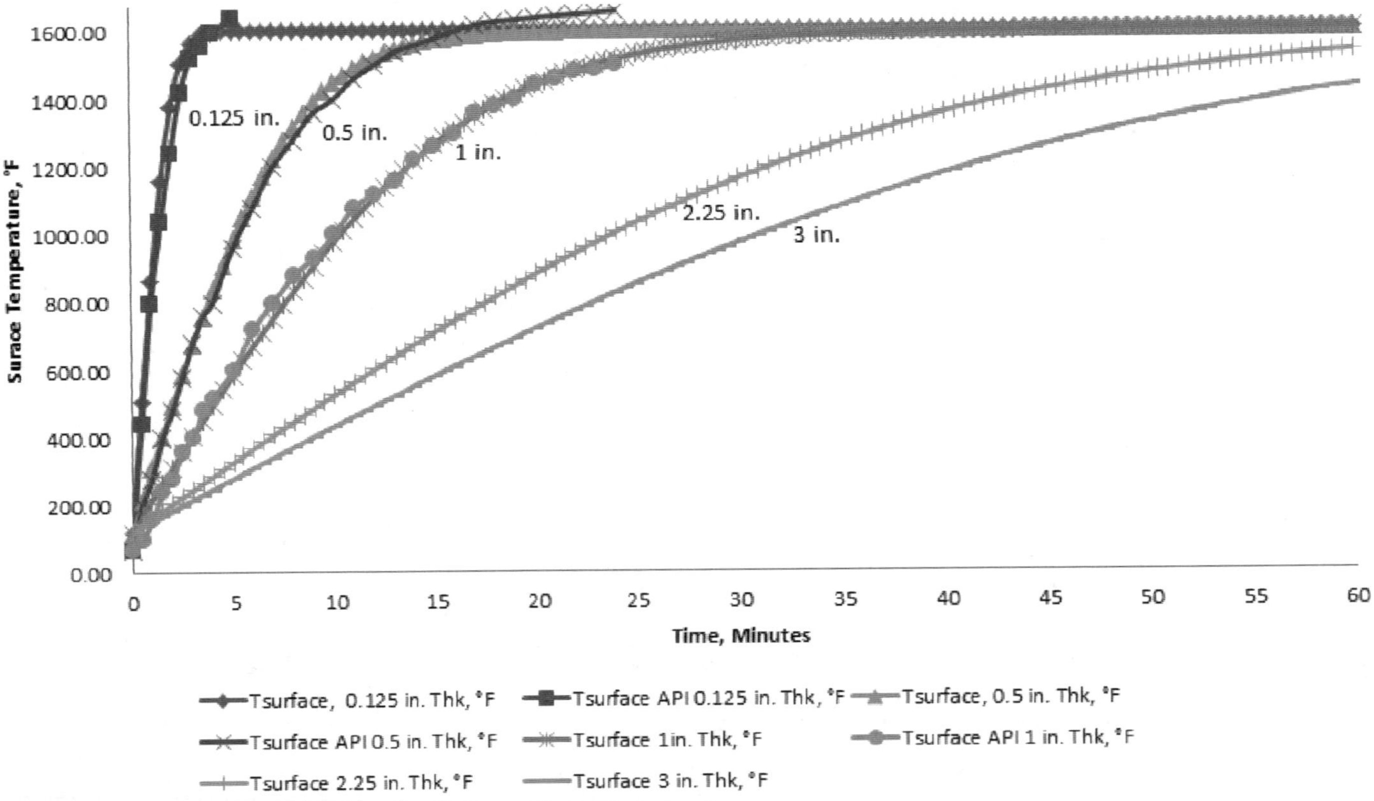

Calculation of the maximum pressure to avoid failure for various wall thicknesses:

As described in Appendix A Section A.3.5.4.6, the wall temperatures can be combined with tensile strength versus temperature data for carbon steel to determine the tensile strength, or the maximum pressure to avoid failure of a vessel, as a function of time in a pool fire scenario.

The tensile strength versus temperature data for carbon steel from Table A-7 in API 521 Appendix A is given in Table 1 for carbon steel.

Table 1 Tensile strength versus temperature for Carbon Steel (SA-515, SA-516)

Temperature (°C)	Temperature (°F)	Tensile Strength (MPa)	Tensile Strength (psi)
482	900	313.7	45,500
538	1000	251.7	36,500
593	1100	187.5	27,200
649	1200	137.9	20,000
704	1300	93.1	13,500
760	1400	62.2	9025

As the wall temperature heats up to 482°C (900°F), the tensile strength or the maximum pressure to avoid failure, decreases. These values are normalized with respect the tensile strength of 313.7 MPa (45,500 psi) and a safety factor of 0.75 is applied to generate Figure A-5 in Appendix A of API 521.

These values are extrapolated to larger thickness greater than 1 in using regression to determine the times to reach failure pressures expressed as a ratio of maximum pressure to avoid failure to MAWP for various thicknesses.

Table 2 Vessel rupture strength as % of MAWP versus tire exposure times

% of MAWP	100%	80%	60%	40%	34%	30%	26%	24%	
Thickness									
in	Time (mins)	Time (mins)	Time (mins)	Time (mins)	Time (mins)	Time (mins)	Time (mins)	Time (mins)	Comment
0.125	1.2	1.5	1.9	2.4	3.1	6.0	12.0	17.5	API 521
0.5	3.6	3.9	4.5	6.1	7.6	10.5	17.0	28.0	API 521
1	12.6	13.1	14.1	17.0	20.0	23.1	31.0	37.0	API 521
2	25.19	25.91	27.58	33.24	38.83	42.15	52.16	59.66	Extrapolated
2.25	28.52	29.28	31.14	37.48	43.74	47.11	57.68	65.18	Extrapolated
3	38.49	39.44	41.81	50.20	58.44	62.00	74.22	81.72	Extrapolated

Figure 2 shows a plot of the maximum allowable pressures to avoid failure for plates of different thicknesses from 25.4 mm (1 in). to 76.2 mm (3 in.)..

Figure 2 Failure pressures for various thicknesses as a function of time

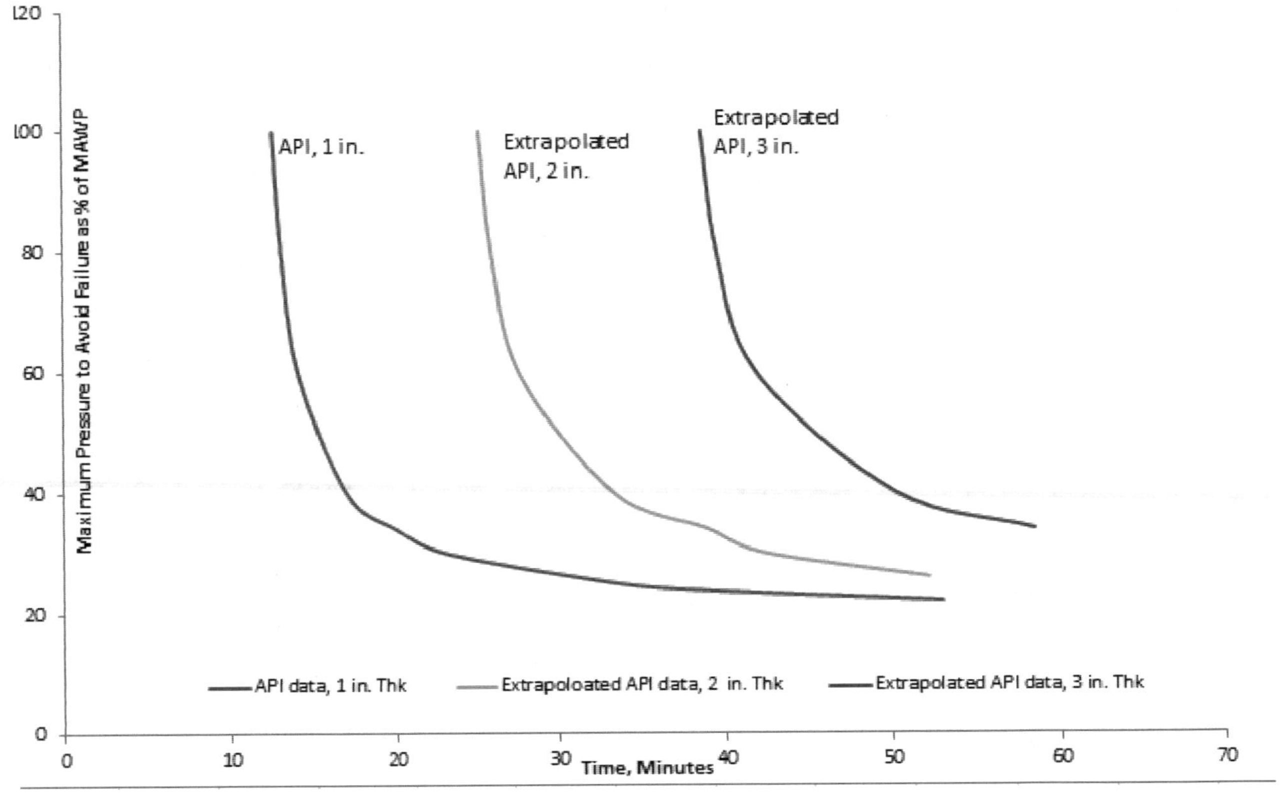

A hypothetical vessel of 4 m diameter and 7 m height with a vessel thickness of 101.6 mm (4 in.) having a hemispherical heads of thickness 57.2 mm (2.25 in.) is depressuerd from a design pressure of 80 barg to a pressure of 40 barg in 15 minutes using HYSYS v 7.3 depressuring utility. The resulting profile of the vessel internal pressure as a function of MAWP is overlaid on the temporal profiles of the maximum allowable pressures to avoid failure for various thicknesses. Data for about 60 minutes is plotted for comparison with the profiles of internal pressures as shown in Figure 3.

Figure 3 Plot of failure pressures for various thicknesses and the vessel internal pressures as a function of time

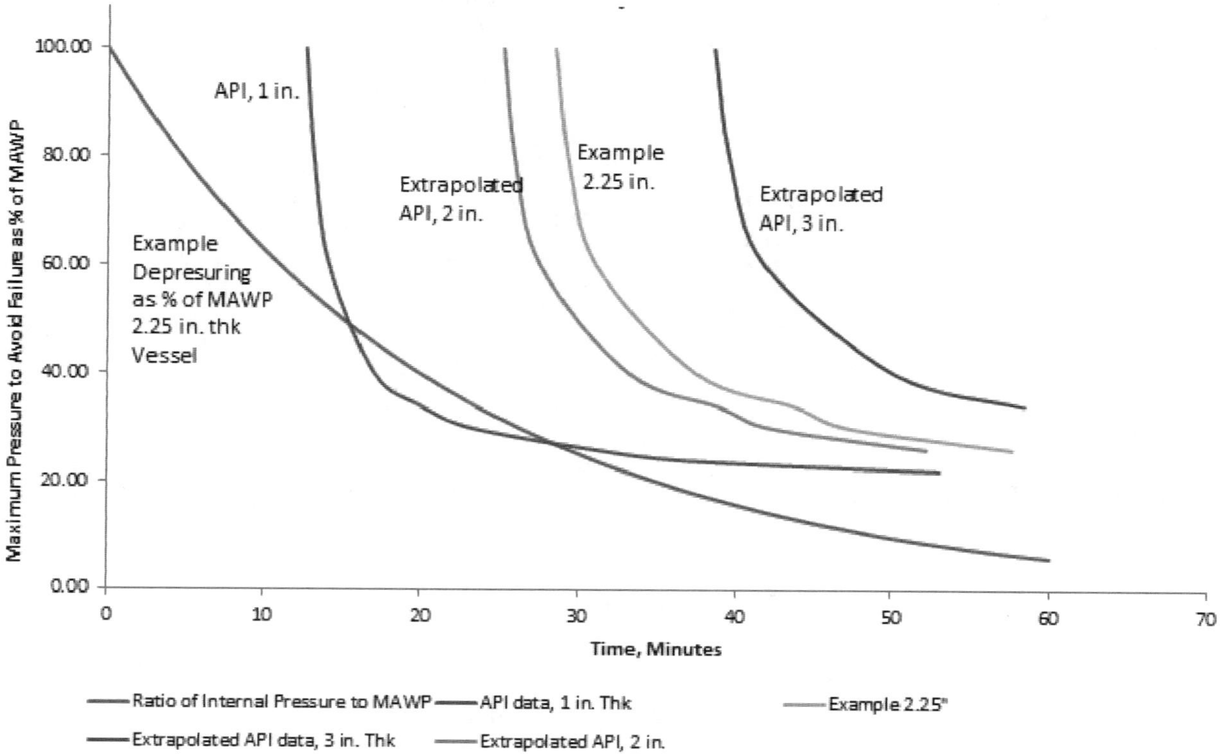

Discussion:

The kinks in the plot are due to round off errors and errors in the statistical regression. However, as seen from Figure 3, the vessel internal pressure for depressuring in a pool fire scenario is less than the maximum allowable pressure to avoid failure for thicknesses greater than 1". Thus, a methodology is developed for calculating wall temperature profiles during pool fire for vessels of various thicknesses and to check if the vessel internal pressures during depressurization exceed the failure pressures for vessel thicknesses greater than 1in.

References:

1. Pressure-relieving and Depressuring Systems, API Standard 521, Sixth Edition, January 2014.

12TH
GLOBAL
CONGRESS
ON PROCESS
SAFETY

Quantifying the Impacts of Human Factors on Functional Safety

Julia V. Bukowski, Ph.D.
Dept. of Electrical & Computer Engineering
Villanova University, Villanova, PA 19085
Julia.Bukowski@villanova.edu

Loren Stewart, CFSP
exida
64 N. Main St., Sellersville, Pa 18960
lstewart@exida.com

Prepared for Presentation at
American Institute of Chemical Engineers
2016 Spring Meeting
12th Global Congress on Process Safety
Houston, Texas
April 11-23, 2016

Quantifying the Impacts of Human Factors on Functional Safety

Julia V. Bukowski, Ph.D.
Dept. of Electrical & Computer Engineering
Villanova University, Villanova, PA 19085
Julia.Bukowski@villanova.edu

Loren Stewart, CFSP
exida

Keywords: systematic failure rate, product failure rate, site-generated failure rate, FMEDA, PFDavg

Abstract

It is not difficult to think of numerous ways in which human factors might impact functional safety. For example, imperfect repair, improper calibration, faulty installation, etc. can all contribute to decreases in functional safety performance. However, if actions are to be taken to mitigate these impacts then it is essential to have some quantitative measure of the impacts so that the actions' effectiveness can be assessed. This paper describes how the Site Safety Index™ (SSI) is used to adjust safety metrics, computed under the assumptions that human factors play no part in safety system performance, to reflect the effects of human factors on safety system performance on a site by site basis. Examples of the application of the method showing the safety performance changes attributable to human factors are provided.

1. Introduction

A safety instrumented function (SIF) has an inherent product failure rate based on its design and implementation and the assumption that the end user will take all necessary actions (i.e., meet all IEC 61508 [1] and/or IEC 61511 [2] requirements) to insure that the inherent product failure rate will be achieved. In practice, end-user activities and practices often fall short of this ideal and, consequently, the safety performance achieved by the SIF is often less than the ideal [3]. The SSI measures, on a scale of 0 to 4, the extent to which end-user actions support the achievement and retention of ideal SIF safety performance. The SSI is described in detail in [4]. This paper explains how end-user actions as measured by SSI affect various parameters that are used to calculate SIF safety performance and show how different levels of SSI impact various safety metrics.

2. Notation

FFD	field failure data
FMEDA	failure modes, effects and diagnostic analysis

IEC	International Electrotechnical Commission
MTTFS	mean time to fail safe
OREDA	Offshore and Onshore Reliability Data
PFDavg	average probability of dangerous failure on demand
RRF	risk reduction factor
SIF	safety instrumented function
SIL	safety integrity level
SRS	Savannah River Site
SSI	Site Safety Index™
1oo1	1-out-of-1 architecture

3. Using SSI to Measure Impacts of End-User Activities and Practices on Safety Performance of SIFs

3.1 Why a Site Safety Index™?

Studies [5] have shown that the failure rates estimated from field failure data (FFD) for identical modules in the same industrial application exhibit a difference in estimated failure rates for the same product from site to site. Typically, when compared to inherent predicted failure rates for the same products, the FFD estimates are approximately 1.2 to 3 times higher depending on product type. Site audits have attributed these differences to dissimilarities in end-user activities and practices which vary from site to site. This variation in end-user activities and practices can be captured by the SSI [4]. Further, these variations in activities and practices directly impact many variables used in models which assess the safety performance of SIFs.

3.2 The Key Variables Impacting PFDavg and Other Metrics

A key metric for measuring the safety performance of SIFs is average probability of failure on demand (PFDavg). Other closely related metrics are risk reduction factor (RRF), safety integrity level (SIL), and mean time to fail safe (MTTFS). Based on studies of FFD and proof test reports, nine key variables impacting PFDavg have been identified [6]. These nine variables are

1. Failure rates of each device including failure modes and any diagnostic coverage from automatic diagnostics (an attribute of devices chosen and of end-user practices);
2. Proof Test Effectiveness (an attribute of the proof test method and, to the extent that there is more than one possible proof test method, an attribute of end-user practices);
3. Proof Test Intervals (assignable by end-user practices);
4. Mean Time to Restore (an attribute of end-user practices);
5. Probability of Initial Failure (PIF) (an attribute of end-user practices);
6. Mission Time (an attribute of end-user practices);
7. Proof Test Duration (an attribute of end-user practices);
8. Site Safety Index™ (an attribute of end-user practices); and
9. Redundancy of Devices including common cause failures (an attribute of SIF design).

3.3 How Key Variables Are Modified to Quantify the Impacts on PFDavg and Other Metrics

Clearly, redundancy of devices is not an attribute of end-user practices or activities. The SSI itself is not modified in safety performance calculations; rather it is used to modify many of the other variables which impact PFDavg and other metrics. Specifically, whenever SSI is less than 4, the following five key variables are modified:

- failure rates of each device
- proof test effectiveness
- proof test intervals
- mean time to restore
- PIF

Failure rates of each individual device in the SIF are increased by a specific multiplier which is determined by the SSI value and the device itself. It is known that final elements are more likely to be negatively impacted by less than ideal end-user practices than are sensors or logic solvers. By increasing device failure rates on an individual device basis, it is possible to more accurately account for the effects of site practices on safety performance. Proof test effectiveness, i.e., the likelihood that all failures are discovered and completely repaired, is also decreased with decreasing SSI.

Proof test intervals, mean time to restore, and PIF are increased by specific multipliers determined by SSI value. It is common to find that less than ideal end-user practices include delaying proof testing, reducing proof test coverage, requiring additional restoration time, and introducing initial failures through imperfect pre- or post-installation testing. Final elements that may develop failures simply due to lack of motion of the moving parts over an extended time (a few months) are especially vulnerable to initial failures in excess of 1% [7, 8, 9].

The two variables not modified are mission time and proof test duration. Mission time is not modified because its impact is generally small unless the mission time is sufficiently beyond the SIF useful life so as to cause the SIF to see failures due to ageing, i.e., failures no longer modeled by a constant failure rate, without renewal during the mission time. Proof test duration has no impact if proof testing is performed with the process off-line. For proof testing performed on-line, proof test duration generally has only a very small effect unless proof testing frequency is very frequent, i.e., greater than or equal to once per month – an impractical situation for most industrial processes.

3.4 Calibrating the Multipliers

As with the development of the Failure Modes, Effect and Diagnostic Analysis (FMEDA) failure rate predictive technique [10] and its associated validated databases [11, 12, 13], it is necessary to calibrate the database of multipliers for the various modifications and to provide on-going validation against actual FFD. The initial calibration phase is still in progress and uses the extensive FFD of the Offshore and Onshore Reliability Data Project (OREDA) [14, 15] as well as the detailed proof testing and root cause failure records of the Savannah River Site (SRS) [8].

4. Examples of Quantifying the Impacts of SSI on Various Metrics

The same simple 1oo1 SIF was analyzed under five different levels of SSI. Figure 1 shows the impacts on PFDavg and SIL of these variations in SSI. For the same 1oo1 system, Figure 2 shows the impacts on RRF and SIL of variations in SSI. Finally, Figure 3 shows the impact on MTTFS of variations in SSI.

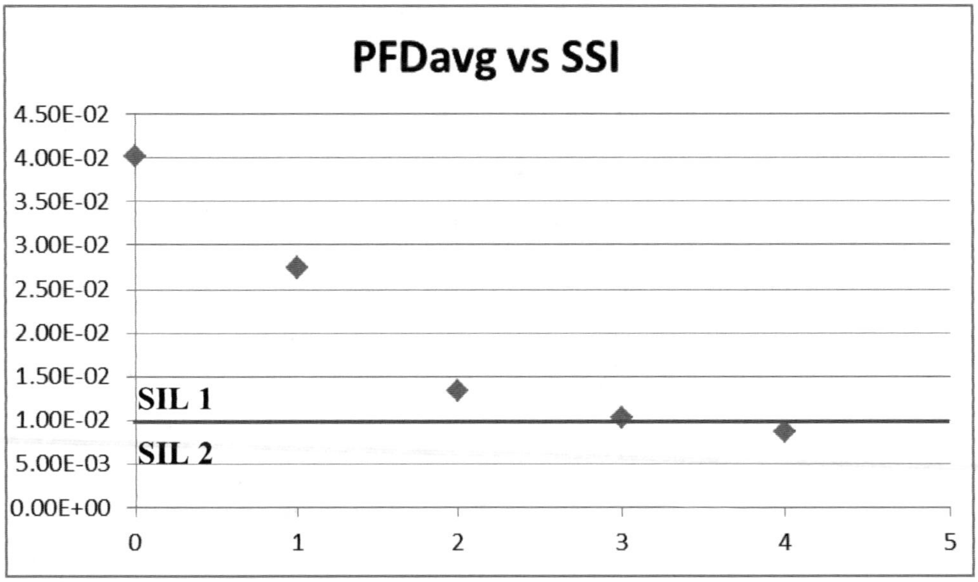

Figure 1. Impacts on PFDavg and SIL of Variations in SSI.

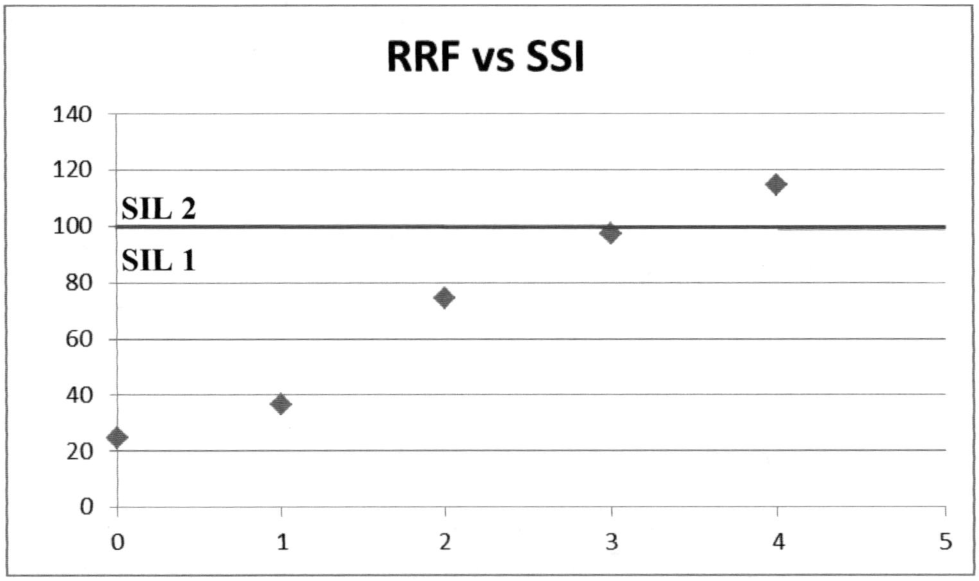

Figure 2. Impacts on RRF of Variations in SSI.

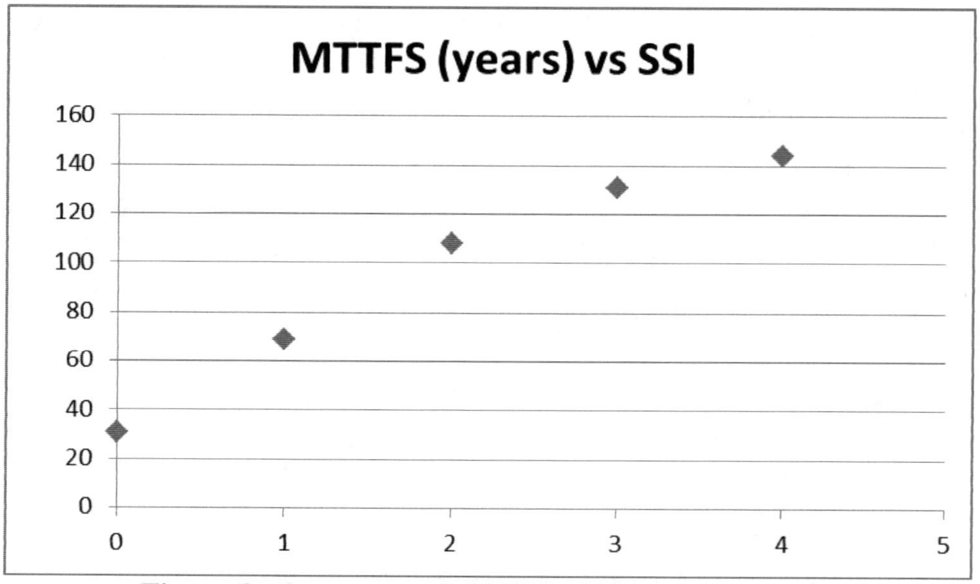

Figure 3. Impacts on MTTFS of Variations in SSI.

5. Summary

This paper clearly shows that the impacts of end-user activities and practices on safety performance of a SIF can be quantified through a simple model like the SSI. It also explains how the SSI modification multipliers are being calibrated and will be updated as increasing amounts of FFD become available. The methodologies presented here and in [4] allow end-users to measure quantitatively their own activities and practices on a site-by-site basis and to measure the impacts of those activities and practices on SIF safety performance.

6. References

[1] IEC 61508, *Functional Safety of electrical / electronic / programmable electronic safety-related systems*, Geneva, Switzerland, 2010.

[2] IEC 61511, *Application of Safety Instrumented Systems for the Process Industries*, Geneva, Switzerland, 2nd edition, 2004.

[3] J. V. Bukowski, W. M. Goble, and I van Beurden, "Product Failure Rates vs Total Failure Rates at Specific Sites: Implications for Safety," *Proceeding AIChE 11th Global Congress on Process Safety*, Austin, TX, April 2015.

[4] J. V. Bukowski and D. Chastain-Knight, "Assessing Safety Culture via the Site Safety Index™," *Proceeding AIChE 12th Global Congress on Process Safety*, Houston, TX, April 2016.

[5] W. M. Goble, "Maintenance capability: Does it matter in the long term?" *Hydrocarbon Processing,* June 1, 2013.

[6] "The Key Variables Needed for PFDavg Calculation," White Paper, exida 2015, http://www.exida.com/Resources/Whitepapers/the-key-variables-needed-for-pfdavg-calculation

[7] J. V. Bukowski, R. E. Gross and W. M. Goble, "Probability of Initial Failure for Spring Operated Relief Valves," *Proceedings of the ASME 2011 Pressure Vessels and Piping Division Conference*, Baltimore, MD, July 2011.

[8] J. V. Bukowski, R. E. Gross and W. M. Goble, "The Adhesion Failure Mode in Stainless Steel Trim Spring Operated Pressure Relief Valves," *ASME Journal of Pressure Vessel Technology*, Vol 135 No 6, December 2013.

[9] L. Stewart, J. V. Bukowski and W. M. Goble, "Improving Reliability & Safety Performance of Solenoid Valves by Stroke Testing," *Proceedings AIChE 9th Annual Global Conf on Process Safety – Process Plant Safety Symposium*, San Antonio, TX, April-May 2013.

[10] W. M. Goble and A. C. Brombacher, "Using a Failure Modes, Effects and Diagnostic Analysis (FMEDA) to Measure Diagnostic Coverage in Programmable Electronic Systems," *Reliability Engineering and System Safety*, Vol. 66, No. 2, Nov 1999, pp. 145-148.

[11] W. M. Goble and J. V. Bukowski, "Development of a Mechanical Component Failure Database," *Proceedings of the 2007 Annual Reliability and Maintainability Symposium*, Orlando, FL, January 2007, pp. 451-455.

[12] J. V. Bukowski and, W. M. Goble, "Validation of a Mechanical Component Constant Failure Rate Database," 2009 *Proceedings Annual Reliability and Maintainability Symposium*, Fort Worth, TX, January 2009, 338-343.

[13] J. V. Bukowski and L. Stewart, "Comparing Failure Rates for Safety Devices: FMEDA Prediction vs OREDA Estimation," exida White Paper available at www.exida.com

[14] SINTEF, *OREDA Offshore and Onshore Reliability Data Handbook, Vol 1. – Topside Equipment and Vol. 2 – Subsea Equipment*, 6th Ed, OREDA Participants, 2015.

[15] SINTEF, *OREDA Offshore Reliability Data Handbook, Vol 1. – Topside Equipment and Vol. 2 – Subsea Equipment*, 5th Ed, OREDA Participants, 2009.

441468 Using HAZOP/LOPA to Create an Effective Mechanical Integrity Program

Abstract:

Many people view the conduct of a HAZOP/LOPA to address regulatory requirements as a chore, and stop there. However, the implementation of a quality HAZOP/LOPA has the potential to provide a framework for addressing numerous safety and operational optimization issues at plants, including the formulation/refinement of the Mechanical Integrity Program. The purpose of this paper is to focus on the mechanical integrity program, illustrate how a quality HAZOP/LOPA can support the effective implementation of some of the new Damage Mechanism Review requirements for California Refineries (e.g., 5189.1(k)), and optimize key elements of an effective Mechanical Integrity Program, e.g.:

- inspection/testing methods
- testing intervals
- maintenance outage periods
- repair prioritization and allowable outage
- identification of low priority equipment

441486 Reactive Chemical Hazards of Azide Compounds: a Case Study and Lessons Learned

Abstract:

While a researcher was servicing a vacuum pump an unintended detonation occurred. This pump had been used in a sulfonyl azide synthesis ($R-SO_2-N_3$). The detonation occurred when he was applying effort to the hose barb from the outlet of the pump to connect it with a rubber hose. Differential Scanning Calorimetry (DSC) and headspace Gas Chromatography/Mass Spectrometry (GC/MS) were employed for the root cause investigation. The DSC showed a large exothermic decomposition (-1388J/g) for the residual sample from the pump fitting. Headspace GC/MS identified the presence of hydrazoic acid in the pump oil. The investigation concluded that either the azide precursor (sodium azide) or hydrolysis of the product (sulfonyl azide) was responsible for generating hydrazoic acid, which escaped from the dry ice trap and appeared in the downstream equipment, and probably reacted with metal parts (brass) to form a highly shock sensitive material (copper azide).

This paper reviews the reactive chemical hazards for azide compounds involved in this case study. Sodium azide is not particularly sensitive to mechanical impact, but its propensity to form hydrazoic acid in water is well-known, even at neutral pH. Likewise, the test data on sulfonyl azide product showed that it is not particularly sensitive to mechanical impact, but it can generate hydrazoic acid upon hydrolysis with water. Metal azides which are formed by the reaction between hydrazoic acid and metal surfaces of equipment parts are highly hazardous due to their extreme impact and friction sensitivity.

Based on lessons learned, this paper also describes further steps for synthesizing organic azides using aqueous sodium azide. These steps include, but are not limited to: (1) Utilizing a dedicated hood and equipments to limit potential HN_3 vapor streams from contacting any reactive metal surfaces. (2) Using a dual caustic trap system designed to scrub the outlet gases swept from the azide reaction step, instead of dry ice trap. (3) Using a 5% aqueous solution of ceric ammonium nitrate as a decontaminating agent for metal surfaces inside the fume hood at the completion of the campaign. (4) Eliminating all brass or copper fittings from equipment used in the synthesis, due to the potential formation of the unstable explosive copper azide that can form from exposure of brass or copper parts to HN_3.

441495 The Application of Frequency Data in Quantitative Risk Assessment

Abstract:

Quantitative Risk Assessment (QRA) is a vital resource for estimating risk in the process industry. There are four primary steps involved in conducting a QRA: hazard identification, frequency analysis, consequence analysis, and risk calculation (the combination of frequency and consequence to determine the risk). Frequency analysis is a key component of the QRA and can have a major impact on the risk profile.

The primary method of estimating event frequency for QRA utilizes historical leak frequency data, of which the Hydrocarbon Release Database (HCRD) is the most widely used leak frequency database. One main issue with applying historical frequency is not having transparency of the causes of the events. Using the historical data approach, a single value is generated for the frequency of an event occurring, with no association to the cause (e.g. corrosion, dropped object, etc.). It has also become widely accepted to apply these offshore databases to onshore processes. There are many concerns in applying these databases to onshore processes, as there are differences in the types of equipment, the hazards the equipment is exposed to, and the safety management systems in place.

When QRAs were first conducted during the 1980s, they were an exhaustive and detailed process. In today's fast track industry, often frequency analysis is conducted without undertaking an evaluation to determine if the frequency being applied is appropriate for the hazard at hand. Leak frequency analysis has become an isolated activity, rather than relying on frequencies established during HAZOP and reliability assessments such as fault tree analysis.

An alternative method to employing historical databases to estimate event frequency is conducting fault tree analysis. Fault tree analysis involves dissecting a hazard into its causes, using a "top down" approach, beginning with hazard identification and working down to a level in which applicable frequency data is available. Fault trees provide a logical relationship between the hazard and its causes.

If historical leak frequency data is employed, which is independent of the cause, the resulting risk is also independent of the cause. Therefore, mitigation measures and the decision making process tend to focus on the back-end consequence of the event, rather than the cause of the event.

Fault tree analysis, although it has had limited application in onshore and offshore risk assessment, provides a major advantage of providing a link between the cause of an event and the associated frequency. This transparency between the event frequency and cause enables us to identify major contributors to the risk and provide risk reduction measures based on what is actually causing the event to occur.

The objective of this paper is to provide an alternative to employing offshore historical frequency data to QRA and demonstrate that the alternative method can provide major advantages in mitigations by linking the estimated risk to the cause and focusing the decision making process on the front-end of the problem rather than solely on the back-end.

441528 Uncertainty in Quantitative Risk Assessment and Decision Making

Abstract:

Including uncertainty in Quantitative Risk Assessment (QRA) is recommended in most published Best Practice Recommendations for QRA, including those published by TNO, CCPS, and Norwegian authorities. Unfortunately typical commercial QRA techniques and models used in Process Risk Analysis do not include the capabilities of uncertainty within the model formulations. Effects on uncertainty may be based on qualitative estimates or ignored, or at most accounted for by sensitivity analysis by "swinging" variables to estimate individual variable sensitivity. By not presenting the uncertainty, the risk assessment results may be misrepresented as possessing an accuracy greater than is warranted.

Failure to account for uncertainty also tends to support centrist thinking and poor quality decision making. The natural tendency of users of risk analysis is to believe in their results, especially if much effort has been put into developing the results. If the result is a single number, that number may be assumed by the analyzer and users as "right". In addition, there is a tendency in people to assume that by picking the most likely values for the inputs into a model, the end result is also the most likely result. This may not be true if the model is complex and non-linear. The most likely results must be determined by including variable distributions and uncertainty modeling in the analysis.

Decision analysis is also deceptively simple when single point criteria or simple curves are used for decision making purposes. A challenge in risk assessment is utilizing results and presenting them in a straight forward way that capture the uncertainty and complexity while allowing clear, high quality decisions by end-users.

This paper presents sources of uncertainty, analysis methods, examples of potential errors introduced by ignoring uncertainty, and illustrates how one can include uncertainties in your quantitative risk assessment (QRA) and benefit from it in your risk assessment decisions.

441529 Analysis of Process Systems during Accidental Fire

Abstract:

In the past 40 years, there have been over 60 major fire accidents reported on offshore platforms. The thermal effects imposed on process equipment and systems (such as pressure vessels, flare scrubbers, etc.) has the likely potential to result in catastrophic consequences such as explosions given that these process systems typically operate at very high pressures. In the event of an accidental fire, blowdown valves are engaged to relieve the excess pressure of the process system to acceptable values in a specific time period. During that time, however, the increased temperature of the blowdown system due to the impinging fire may result in the degradation of its structural capacity. The combination of a degraded vessel shell and a high internal pressure typically leads to sudden ruptures and explosions which could result in the escalation of the initial fire or injuries to personnel. Therefore, it is critical to accurately capture the pressure and temperature of the process system in a fire blowdown event.

A Blowdown analysis is typically carried out using simple thermodynamic software. While these software provide cost-effective approaches to studying process systems, they are unable to accurately compute the local effects of heat transfer and fluid behavior. Moreover, stress concentrations on the pressure vessel due to nozzle effects are not captured correctly using classical approaches and advanced analysis is required. Therefore, to accurately capture the rupture time of pressure vessels, scrubbers, etc., during a fire scenario, advanced approaches such as Computational Fluid Dynamics (CFD) are required.

In this paper, we will demonstrate a new approach in analyzing onshore and offshore process systems in an extreme fire scenario. Heat transfer through the vessel shell as well as the internal fluid will be accounted for by computing the rates of conduction, convection, and radiation. The time history of temperature and pressure as well as the change of phase of the fluid inside the pressure vessel will be captured in the analysis. A degradation study of the structural steel will be performed to predict the time of rupture of the process system to internal pressure using advanced analysis. Moreover, the effect of jet fire momentum on the vessels and their supports will be computed according to industry guidelines.

The methodology is validated by comparison with experimental data. The rupture time and rupture location will be computed and design recommendations will be provided to increase the strength of the process system at rupture locations. The current approach could provide accurate results to compute the response of process systems such as high-pressure vessels, scrubbers, etc., during offshore or onshore fire scenarios.

12TH GLOBAL CONGRESS ON PROCESS SAFETY

Assessing Safety Culture via the Site Safety Index™

Julia V. Bukowski, Ph.D.
Dept. of Electrical & Computer Engineering
Villanova University, Villanova, PA 19085
julia.bukowski@villanova.edu

Denise Chastain-Knight, P.E., CFSE
exida
64 N. Main St., Sellersville, Pa 18960
dchastainknight@exida.com

Prepared for Presentation at
American Institute of Chemical Engineers
2016 Spring Meeting
12th Global Congress on Process Safety
Houston, Texas
April 11-23, 2016

Assessing Safety Culture via the Site Safety Index^{TM}

Julia V. Bukowski, Ph.D.
Dept. of Electrical & Computer Engineering
Villanova University, Villanova, PA 19085
Julia.Bukowski@villanova.edu

Denise Chastain-Knight, P.E., CFSE
exida

Keywords: Safety Lifecycle, safety culture, systematic failure rate, site-specific safety performance, PFDavg

Abstract

How can a company establish a baseline measurement of its safety culture against which to gauge improvement? The Site Safety Index^{TM} (SSI) quantifies in part (on a scale of 0 – 4) the degree to which a company's end-user practices support the attainment and retention of an appropriate safety culture for operations and maintenance. Further, the SSI can be used to appropriately adjust parameters that directly impact measures of safety such as probability of failure on demand (PFDavg). Thus, the impacts of safety culture can be further quantified and the effects of changes to safety culture can be assessed.

1. Introduction

Consider the following hypothetical scenario. A company assesses the risk intrinsic to one of its plant processes (given its current level of mitigation) to be greater than the tolerable risk dictated by company policy or regulatory standards. Further mitigation to reduce the process risk to an overall risk at or below tolerable risk level will require a safety instrumented function (SIF) which will result in a risk reduction factor (RRF) of at least 110, i.e., a SIF of safety integrity level (SIL) 2. The company may choose to design and implement the required SIF itself or to deploy an existing SIF design and implementation.

In either case, the company will need to ascertain that the SIF meets its risk reduction requirements. There are several ways to do this. For example, it may rely on proven-in-use data to claim the company has met its risk reduction goal. This proven-in-use data may come from another company or the database of a consortium of companies. Alternately, the company may predict the risk reduction and SIL level by using manufacturers' published data or other available databases perhaps in conjunction with a commercial tool. If by either or both means of risk reduction/SIL verification the company determines the risk reduction goal is met by the SIF, is this alone sufficient for the company to state that, when implemented in its plant, its process risk will be at or below tolerable levels? **No!**

When the company implements the SIF in its process, the actual risk reduction achieved by the SIF will be determined not only by the SIF attributes taken into account by the verification methods but also by the company's safety culture, i.e., by the overall impact of end-user practices that take place within the company in general and especially within the plant/site itself. Further, these practices may differ from site to site even in the same company.

For example, if the risk reduction/SIL was estimated from proven-in-use data, unless "systematic" failures were eliminated, these data inherently contain the impacts of the end-user actions/omissions which took place at the plant or plants during the time period in which the proven-in-use data were collected. When the company implements the SIF, it may actually experience the same, higher or lower risk reduction than that estimated from proven-in-use data depending on whether the company's end-user current practices are the same, better or worse than those practices in the plants *when* and where the proven-in-use data were collected. Especially for consortium-based data, it is not generally known exactly from where the proven-in-use data were gathered much less what the specific end-user practices were at those plants. Thus, estimated risk reduction/SIL based on proven-in-use data is difficult to adjust to more accurately represent the actual risk reduction/SIL the company will experience.

If the risk reduction/SIL was predicted based on the manufacturers' published failure rates or other failure rate data, these rates are likely predicated on the assumptions that the end-user faithfully performs all required actions to sustain the safety integrity of the equipment. Such actions include, but are not limited to, insuring that

- the equipment implementing the SIF is appropriate to the task and environment,
- the equipment is properly installed and calibrated,
- the equipment is correctly functioning when installed,
- all in-service maintenance is correctly and completely performed on schedule,
- the equipment is maintained so that no ageing occurs prior to proof testing,
- proof testing is completely and correctly performed,
- the equipment is replaced when its useful life can no longer be extended by maintenance and refurbishment.

Note that the list extends beyond the usual concepts of maintenance alone. Any required action that is omitted, delayed, or performed incompletely or imperfectly will reduce the risk reduction (and perhaps SIL) achieved in the SIF's actual use in the plant. However, in the case of predicted risk reduction/SIL, which is based on manufacturers' data or Failure Modes, Effects and Diagnostics Analysis (FMEDA) [1], it is possible to adjust the initially calculated risk reduction/SIL to account for a company's end-user practices. This adjustment occurs through the use of a SSI.

This paper explains the way in which the SSI is measured and illustrates how the SSI can impact typical safety parameters such as PFDavg. A second paper [2] explains how the SSI is used to adjust the initially calculated risk reduction/SIL.

2. Notation

FMEDA	failure modes, effects and diagnostic analysis
IEC	International Electrotechnical Commission

PFDavg	average probability of dangerous failure on demand
RRF	risk reduction factor
SIF	safety instrumented function
SIL	safety integrity level
SSI	Site Safety Index^{TM}

3. Measuring a Site's Safety Culture

3.1 Why a "Site" Safety Index?

The reader may wonder why the terminology "Site" Safety Index is used as opposed, perhaps, to "Company" Safety Index. It is well documented that a safety product may exhibit a safety performance less than that specified by the manufacturer or by a FMEDA [3, 4] and that even within the same company, failure rate estimates derived from field failure data can vary significantly from site to site [5]. Thus, while company-wide policies and procedures will impact the safety achieved, the possibly different ways company-wide directives are implemented on a site-by-site basis requires that a safety index be measured at the site level.

3.2 How is the Site Safety Index^{TM} Evaluated?

There are currently two ways that SSI can be evaluated: via an external audit, which provides the most accurate measure of SSI, and via a self-administered questionnaire, which relies on less information than the audit and provides an approximate measure of SSI. These two evaluation methods are detailed next.

3.2.1 Via an External Audit

The international standard IEC 61511 [6] lists the requirements that must be met for the entire Safety Lifecycle process. In an on-site audit, it can be verified that the site meets these requirements in combination with additional detailed requisites for the operations and maintenance activities. Additionally, all safety and lifecycle procedures are audited to be sure: a) that the procedures are adequate to support the requirements and b) that they are being correctly followed at the site. The outcome of the audit is a report which includes the SSI score obtained as well as recommendations for changes and improvements. The interpretation of the measured SSI level itself is discussed below.

3.2.2 Via a Self-Administered Questionnaire

A site that is interested in acquiring an approximate measure of its current state of safety culture may choose to complete a questionnaire of about 40 items which represents a subset of the information gathered in a complete audit. These questions cover activities and practices spanning a portion of the lifecycle of the SIF and include installation and commissioning, safety validation, operations and maintenance, SIF modification and decommissioning. Table 1 lists the questions for the operations and maintenance phase of the lifecycle. The complete set of questions is available at [7]. The questions, which are answered Always, Often, Never, or Not

Applicable, are subject to weighted scoring. The total score received determines an approximate SSI level for the site.

Table 1 Questions Regarding Site Activities and Practices during the Operations and Maintenance Phase of the SIF Lifecycle

Do your operation and maintenance procedures include competency requirements for those performing tasks to maintain the "as designed" functional safety of the SIF?
Do your operation and maintenance procedures include the information which needs to be maintained on device failures, proof test results, false trips, demand rates and restore times on the SIF?
Do your operation and maintenance procedures include procedures for fault diagnostics and repair after a failure in the SIF?
Do your operation and maintenance procedures require instruments to be rebuilt or replaced at the end of their useful life?
Do your operation and maintenance procedures require the use of a computer tool like SILStat™ to track demands, false trips, failures and maintenance performance?
Do your operation and maintenance work tasks proceed only in accordance with the relevant procedures?
Are operators that work with the SIF in their area trained to ensure they understand how the SIF functions and what hazards are being protected against?
Do you perform periodic proof-tests using the written procedure?
Is your frequency of proof tests decided using the PFD_{avg} calculation?
Are scheduled proof test performed on schedule?
Do you repair any deficiencies found during the proof testing in a safe and timely manner?
Do you maintain records of proof tests including descriptions of tests and inspections performed, equipment tested, dates, and names of personnel performing work?
Do you maintain records of proof tests including "as found" results of the tests and inspections?

The basis of the validity of using a questionnaire for determining SSI is the accepted practice per IEC61508 of using a questionnaire to determine the β factor that describes the extent of common cause failure to be modelled when evaluating the safety performance of redundant SIF architectures [8].

3.3 Interpreting the SSI Level

The SSI is a number from 0 – 4 which is an indication of the level of site activities and practices that contribute to the safety performance of SIF's on the site. Table 2 details the interpretation of each SSI level. Note that the levels mirror the levels of SIL assignment and that SSI 4 implies that all requirements of IEC 61508 and IEC 61511 are met at the site and therefore there is no degradation in safety performance due to any end-user activities or practices, i.e., that the product inherent safety performance is achieved.

Table 2 Details of the Interpretation of Each SSI Level

SSI Level	Description
SSI 4	Perfect - Repairs are always correctly performed, Testing is always done correctly and on schedule, equipment is always replaced before end of useful life, etc.
SSI 3	Superior - Repairs are usually correctly performed, Testing is usually done correctly and on schedule, equipment is usually replaced before end of useful life, etc.
SSI 2	Typical - Repairs are mostly correctly performed, Testing is mostly done correctly and on schedule, most equipment is replaced before end of useful life, etc.
SSI 1	Good - Repairs are often correctly performed, Testing is often done and often on schedule, some equipment is replaced before end of useful life, etc.
SSI 0	None - Repairs are not performed, Testing is not done, equipment is not replaced until failure, etc.

4. An Example of the Impacts of SSI

The specific ways in which SSI modifies key variables (e.g., failure rate, proof test effectiveness, etc.) which impact PFDavg, SIL and RRF, as well as other safety metrics, is explained in detail in [2] where additional examples are presented. For the purposes of this paper, Figures 1 and 2 provide a simple example of how SSI can impact safety performance. In Figure 1, computations were performed under the assumption that SSI level is 4 and thus end-user activities and practices do not degrade the safety performance of the SIF. Note that in this case the SIF SIL level is computed as SIL 2 based on PFDavg, and the RRF is computed as 113 – slightly greater than the requirement of the opening hypothetical example. In Figure 2, the analysis is repeated but this time the SSI level is assumed to be 3. Now, based on PFDavg, the SIL performance is a SIL level 1 and the RRF is computed as 98 – insufficient for the requirement of the hypothetical example. This shows how the impacts of site activities and practices can be quantified and how effects of changes to these activities and practices can be measured.

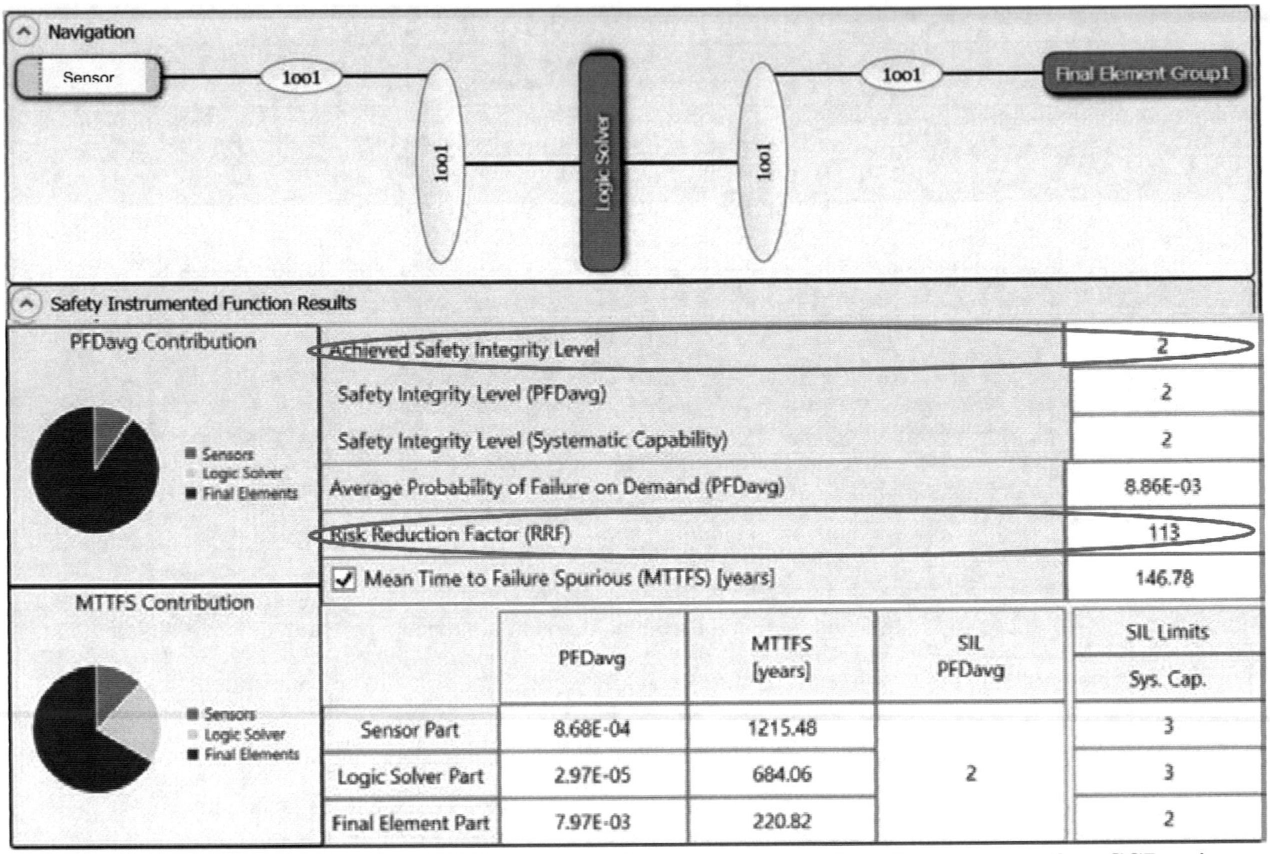

Figure 1. SIL verification/RRF computations performed under assumption SSI = 4.

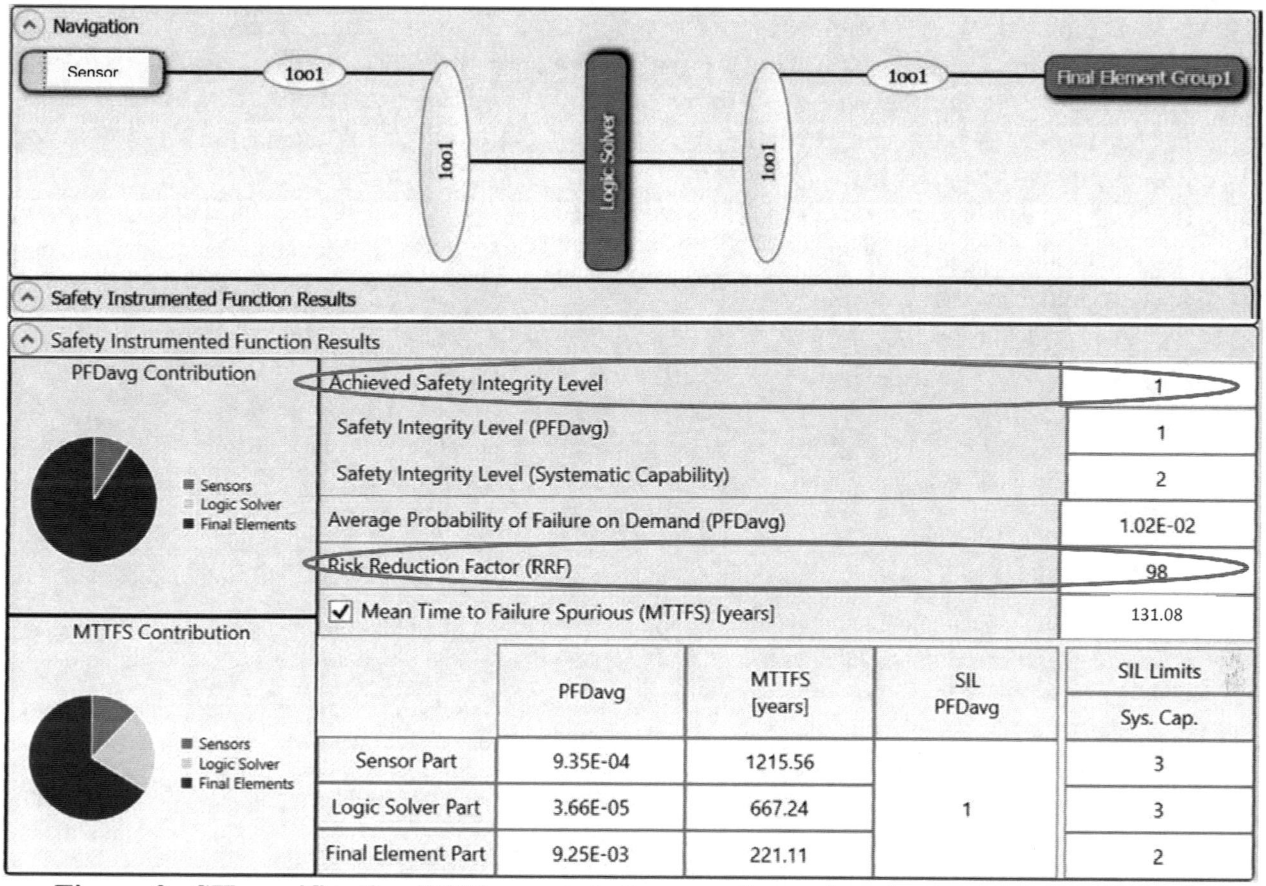

Figure 2. SIL verification/RRF computations performed under assumption SSI = 3.

5. Conclusions

The SSI provides a simple means to measure how well site-specific end-user activities and practices support the achievement and retention of SIF maximal safety performance. Further they can be used to adjust metrics of SIF safety performance from the ideal achievable to a more realistic measure of what is actually attained at a specific site given the site's activities and practices. This last concept is explored more fully in a separate paper [2].

6. References

[1] W. M. Goble and A. C. Brombacher, "Using a Failure Modes, Effects and Diagnostic Analysis (FMEDA) to Measure Diagnostic Coverage in Programmable Electronic Systems," *Reliability Engineering and System Safety*, Vol. 66, No. 2, Nov 1999, pp. 145-148.

[2] J. V. Bukowski and L. Stewart, "Quantifying the Impacts of Human Factors on Functional Safety Failure Rates," *Proceeding AIChE 12th Global Congress on Process Safety*, Houston, TX, April 2016.

[3] J. V. Bukowski, W. M. Goble, and I. van Beurden, "Product Failure Rates vs Total Failure Rates at Specific Sites: Implications for Safety," *Proceeding AIChE 11th Global Congress on Process Safety*, Austin, TX, April 2015.

[4] J. V. Bukowski, W. M. Goble, R. E. Gross and S. P. Harris, "The Effects of Maintenance Actions on the PFDavg of Spring Operated Pressure Relief Valves," *Proceedings of the ASME 2014 Pressure Vessels And Piping Division Conference*, Anaheim, CA, July 2014.

[5] W. M. Goble, "Maintenance capability: Does it matter in the long term?" Hydrocarbon Processing, June 1, 2013.

[6] IEC 61511, *Application of Safety Instrumented Systems for the Process Industries*, Geneva, Switzerland, 2nd edition, 2004.

[7] www.exida.com/SSI

[8] IEC 61508, *Functional Safety of electrical / electronic / programmable electronic safety-related systems*, Geneva, Switzerland, 2010.

441381 Fire Hazard Associated with Polymers and Application of Polymer Nanocomposites for Flame Retardancy

Abstract:

Due to advances in polymer science, their widespread application is found in daily products including vehicles, electronics, clothing, furniture *etc.* However, most of the items that first ignite in any fire scenario includes mostly polymeric materials according to NFPA reports which clearly elucidates the importance into looking for flame retardancy approach for polymers. Polymer nanocomposites is a relatively new approach undertaken for flame retardancy and has the potential to bring multi-functionality including improved thermal behavior, physical barrier effect, catalytic charring effect and better material performance. The objective of this paper is to initially provide the current state-of-the-art review on polymer nanocomposites in terms of their fire retardancy. The paper describes the hazards associated with burning of polymers, flame retardant mechanism, synthesis and characterization of polymer nanocomposites with focus on thermal analysis. Different kinetic model has been studied and applied for analysis of thermogravimetric analysis data. The kinetic parameters of the polymer degradation have been identified. Primarily polystyrene-silica nanocomposite has been prepared by in-situ method and were characterized for their morphology and thermal behavior using characterization methods, such as Scanning Electron Microscopy (SEM), X-Ray Diffraction (XRD), Thermogravimetric Analysis (TGA), and Differential Scanning Calorimetry (DSC). Initial results show nanocomposites with promising fire-retardancy properties.

441530 Applying QRA Study to Select and Design Effective Release Prevention Safeguards

Abstract:

In the past two decades, Quantitative Risk Analysis (QRA) has become a staple safety analysis in offshore oil and gas industry. A comprehensive QRA study could be a powerful tool if the results are understood and changes are implemented to improve safety in the engineering design. Conventional QRA study relies on generic leak frequency and parts count method. One major limitation of conventional QRA study is that it is focused primarily on the loss of containment of hazardous fluids and what happens when they are released. To reduce the risk to acceptable level, usually post-release mitigation measures are addressed. It is a challenge to tackle the effectiveness of release prevention safeguards. This paper presents an approach to address the dominant risk contributors identified in conventional QRA study. The aim is to increase confidence and facilitate decision making in the design selection of loss of containment prevention safeguards.

Sensitivity analysis is usually conducted as part of a QRA study to evaluate the effect of input parameters. However, in most cases, sensitivity analysis focuses on the mitigation measures, such as safety system locations or ratings, personnel distribution or modification to the QRA model itself. Means of release scenario prevention is rarely investigated to the point that it is beneficial to the operation or to the design project. This paper proposes that in order to maximize the use of QRA results, decision makers should investigate means of scenario prevention by performing reliability analysis (e.g., fault tree analysis, human reliability analysis) on the top contributing scenarios. This paper presents some examples of reliability analysis performed on top risk contributors. This paper also presents suggestions on how the results can potentially be applied for the benefit of the operation or design project. The outcome from the QRA study in combination with reliability analysis will increase decision makers' understanding of the top risk contributing scenarios. This approach can be utilized to identify opportunities for improvement in the prevention safeguards.

441536 Managing the Uncertainty in Blast Load Calculations with Enhanced Technical Safety Analysis

Abstract:

The safety of personnel working within onshore and offshore oil and gas processing facilities with respect to vapor cloud explosion (VCE) scenarios revolves around the risk associated with the event. Conducting a risk assessment is quite typical within the offshore and onshore loss prevention communities. The consequence of a VCE event is a very complex phenomenon however. It varies based on several blast load parameters. With improved understanding of blast loading parameters such as blast overpressure, impulse, time lag, net blast loading, etc. structural designers have gained increased confidence in their design and have reduced conservatism in their blast designs. However, there is a significant gap in the risk assessment of onshore and offshore facilities between the process safety and the structural design groups. Using structural insight in combination with rigorous process safety risk evaluation techniques we can reduce the gap between the risk and consequence of a blast design, providing the stakeholders with a more accurate understanding of the risk and consequence of a blast event.

This paper will attempt to address the impact of uncertainty in pressure exceedance curves and;

- how the confidence level impacts the design blast loads;
- how the slope of the exceedance curves impacts the blast design loads;
- two different blast events can result in similar design blast loads for a given risk;
- how a small change is frequency can lead to a significant change in design blast loads;
- how the negative phase of a blast scenario could lead to failure of some specific type of structures;
- how the response of long wall or piping structure is impacted by time-lag of the design blast load;
- how the design blast pressure and design blast impulse impacts a structure;
- how non-structural members, typically not considered in blast design, impact the blast design;
- how simultaneous application of blast load on different parts of structure, taking into account time-lag, affects the blast design.

Currently for a blast event, both blast pressure and blast impulse are defined as independent variables whereas in reality they are coupled. The correlation between the impulse and pressure will also be discussed in the paper. Defining appropriate pressure time-history is critical for structural design. Rise time, positive-phase and negative-phase duration are unknowns in a blast design. How an appropriate time-history is established will also be addressed.

Prior knowledge of implications of a blast event for an associated risk is helpful during structural blast design. Early involvement of structural engineers within the process safety procedure is a recommended solution to such questions. Enhanced loss prevention of offshore and onshore structures can only be established by always keeping the consequence for a blast event for a given risk in mind.

441537 Occupant Vulnerability for Process Plant Building Design Due to Secondary or Domino Events

Abstract:

Statistics indicate that it is in fact not uncommon to have an accidental explosion on an offshore platform, refinery or petrochemical installation. The focus for several years has been on occupied buildings and their resistance to blast loads from accidental releases of hydrocarbons. Common practice dictates that domino effects may not need to be considered given that a large catastrophic release is equivalent in consequence. However, the catastrophic events are filtered out in a risk assessment due to their unlikelihood and as a result, domino effects are not captured. For example, a small release of highly reactive fuel, can lead to a small explosion which can easily damage nearby piping and vessels which can in turn become catastrophic events. The design and assessment of piping systems and vessels against blast loads are of paramount importance and require advanced analysis capabilities. In order to have a safe, i.e. blast-resistant process piping system design, an advanced nonlinear analysis in time domain is required so that we may properly capture the dynamic response of the piping and vessels subjected to blast loading. This type of analysis, in general, should take into account (1) Blast drag load and blast overpressure; (2) Nonlinear material properties, i.e. thermal and strain-rate dependence; (3) Effects of non-structural masses and adjacent piping systems; (4) Pipe supports; (5) Failure of flanges and piping detail; (6) Effects of Operating Temperature on Material Properties; (7) Blast direction/Ignition location/Attenuation sensitivity; (8) Shielding effect, e.g. pipes behind large objects such as vessels; and (9) Effects of pipe insulation, i.e. increased pipe diameters. This paper first discusses the methodology/techniques to account for such effects in finite element analysis for a safe design. The paper then uses case studies to demonstrate the methodology we develop to show how advanced simulation techniques are applied as tools for piping blast response to minimize the risk of secondary failure due to initiating events. These case studies involve a piping system that is simulated from global to detail modeling. The goal of the paper is to discuss how to achieve a safer design for process piping systems by using advanced analysis.

441629 Assessment of the Mechanical Integrity of a Pipeline with Metal Loss, Using Finite Element Method

Abstract:

Following the guidelines proposed by the ASME B31G and API579 standards, for the assessment of systems that compromise its mechanical integrity, given the existence of defects associated with the presence of corrosion damage, the probability of failure of a pipeline, which transports hazardous materials associated with the Oil & Gas industry, and is exposed to pitting corrosión damage was calculated when it overcomes its yield stress. For this, a methodology that combines the corresponding structural analysis of the system and subsequent reliability analysis was proposed. Through the first, the calculation of the von Mises stress present in areas where the pitting defects was made; for this, the corrosion growth model to be used in the present work was defined by implementing the correlation proposed by the Southwest Research Institute (SwRI) to find out the corrosion rate growth in annual terms. Subsequently the pipeline section where the study would be carried out was defined, based on the results obtained in a previous study, which analyzed information related to corrosion damage present on the pipeline, collected through the In Line Inspection (ILI) technique. This allowed to restrict the test area to a total length of 80 cm and 1/8 section of pipe, lastly simulating the system and obtaining the loads associated with the von Mises stress affecting corrosion defined areas. For reliability analysis it was necessary to find the probability distribution function associated with the von Mises stress on the system studied; defining an experimental design which sought to evaluate different values of the random variables defined on the model proposed by SwRI, thus simulating the system a sufficient number of times in order to find the distribution of von Mises stress for this case study. With the distribution function defined, system reliability analysis was carried out through Monte Carlo simulations; thus, it was established that the limit state function would have a conventional structure comparing resistance with loads. Thus, the resistance was set as the yield stress of the material, while the load would be the von Mises stress, and thus the random variable from which random numbers governed by the probability distribution found previously were generated. As mentioned previously, it was defined that the present system fails when the von Mises stress exceeds the yield stress of the material; finding that for the time range evaluated (year one, five, ten and fifteen), the probability of failure is behaved as expected, increasing year by year if no maintenance is not performed to the pipeline section evaluated.

441539 An Improved Risk-Based Approach for Accidental Loadings

Abstract:

Accidental loading such as fire, blast, ship impact, and dropped objects constitute the great majority of potential and actual fatalities in both onshore and offshore operations. In order to help prevent human loss and for increased safe design or assessment of an onshore/offshore asset, the risks of these accidental loadings should be quantified, in terms of both probability/frequency and consequence aspects. In this paper we propose a unified risk-based approach for the assessment of all accidental loadings mentioned above.

This approach will address both human and asset risk assessment against accidental loadings. Technical safety and process safety inputs such as hazard curves, frequency exceedance curves, and return periods are where the process all begins. Based on these inputs, an event frequency analysis is carried out to determine how often one should expect the event to happen, i.e., the frequency aspect of the assessment. Depending on the type of the accidental loadings and the subject of the risk assessment (human or asset), the output can include a fragility curve, damage frequency matrix, and individual risk matrix. Additionally, a consequence analysis is carried out to determine the severity levels of the event. In a risk-based consequence analysis, the severity levels are fully determined based on the risk associated with the event. Based on both event frequency and consequence, the risk ranking is identified using a risk matrix. If the risk is not acceptable according to the risk matrix, mitigation/repair/strengthening options are then developed.

As an example, a risk-based structural fire analysis is compared with the conventional approach which is widely used in the industry. In a conventional Passive Fire Protection (PFP) analysis using Ductility Limit Analysis (DLA) according to API RP 2FB, fire loads are deterministically applied to a structure whose response is then analyzed. The initial PFP scheme is developed based on the analysis and then optimized based on input from all disciplines (e.g. technical safety, operations, structures, etc.). This approach is sometimes misinterpreted as a "risk-based" approach; however, it does not take into account the frequency aspect of the risk assessment. In a risk-based PFP analysis using DLA, fire scenarios are developed in a particular target zone based on exceedance curves (i.e., input from technical safety discipline). The ductility level analysis is performed to determine the structural consequence. If personnel safety is of interest, the consequence of the structure is then linked to individual risk to determine fatalities. The amount of PFP to be applied on the structure is fully based on the risk that is produced by the fire scenarios in the target zones or areas.

Followed by the approach, a new perspective on safe design of onshore/offshore structures for accidental loadings is outlined to estimate the associated risk to potential targets such as human personnel as well as asset components. The proposed assessment methodology will contribute towards identifying the mitigation measures and safety-critical procedures and equipment, and toward a safer design.

441540 Occupant Vulnerability for Temporary Blast Resistant Modular Buildings

Abstract:

Human fatality and injury in the process facility due to overpressure effects can be categorized as (a) direct injury to body due to pressure change, (b) secondary injury resulting from fragment or debris impact, and (c) tertiary injury as a result of body being thrown by explosion drag on stationary objects. Secondary and tertiary effects are typically the dominant causes of fatalities in the process facilities. Direct injuries may result from change in internal pressure as a result of wall/roof response or penetration of blast overpressure into the building. Existing studies allow for an estimation of occupant vulnerability to blast on typical default building construction types. There exists no such building occupant vulnerability calculation method for modular buildings. This paper seeks to develop an improved understanding of the impact of blast loads on modular building and a more comprehensive understanding of the vulnerability to personnel who are within this type of building if they are subjected to a blast load. The study will be carried out in analytical basis which will include basic and advanced numerical methods along with engineering judgment and accident investigation for estimation of vulnerability. A numerical model can include deformable crash test dummies as well as many types of non-structural components in order to estimate occupant vulnerability and maximum individual risk. The outcome of this study will include new response criterion limits related occupant vulnerability in modular buildings as well as design and retrofit guidelines for critical non-structural components.

441541 Occupied Building Blast Retrofit and Upgrade for Personnel Safety

Abstract:

Accidental blast events at petrochemical facilities such as BP Texas City have shown that the occupied buildings near the process areas can be vulnerable to blast and collapse even under low blast overpressures. Additionally, existing permanent and portable buildings may exhibit insufficient capacity to resist blast loading during facility siting studies per API RP 752 and 753. Although it is not possible to completely eliminate risk from blast loading, it is possible to minimize the individual risk, which is a probability an occupant will experience fatality due to an explosion event, by applying retrofit and upgrade strategies to increase the blast-resisting capacity of an occupied building and meet the safety performance criteria.

There are several ways to decrease individual risk due to explosion with respect to occupants in a building:

1. Moving the building from a high blast region near the process area to a lower blast region, and design a new building;
2. Retrofitting existing occupied building;
3. Replacing existing occupied building with a blast resistant modular building; and
4. Replace the existing occupied building with a new blast resistant building.

The first and forth options can be inefficient from construction cost and time perspective. Additionally, the first option may require purchasing an additional land and obtaining permit. The second and third options are significantly cost and time efficient, and causes less interruptions for the operations. However, these options require analysis and engineering judgment from a highly skilled and experienced engineer. In this paper, case study will be presented where a variety of retrofit and upgrade options will be discussed from perspectives such as design, constructability, cost and individual risk.

Implementing Engineered Controls to Reduce the Likelihood of High Severity Scenarios

Alek Hamparian
Kazarians & Associates, Inc.
100 West Broadway, Suite 970, Glendale, California
Alek.Hamparian@Kazarians.com

Mardy Kazarians
Kazarians & Associates, Inc.
100 West Broadway, Suite 970, Glendale, California
MKazarians@Kazarians.com

Prepared for Presentation at
American Institute of Chemical Engineers
2016 Spring Meeting
12th Global Congress on Process Safety
Houston, Texas
April 11-13, 2016

Implementing Engineered Controls to Reduce the Likelihood of High Severity Scenarios

Alek Hamparian
Kazarians & Associates, Inc.
100 West Broadway, Suite 970, Glendale, California
Alek.Hamparian@Kazarians.com

Mardy Kazarians
Kazarians & Associates, Inc.

Keywords: Hazard Identification and Risk Analysis, Process Safety

Abstract

The Center for Chemical Process Safety (CCPS) *Guidelines on Inherently Safer Chemical Process A Life Cycle Approach* (2009) defines a hierarchy of controls that can be used to decrease the risk of a system. These guidelines include a hierarchy of layers of protection that can be used in Process Hazard Analyses (PHAs) to determine adequacy of the existing safeguards. This approach can be particularly useful for companies that have potentially high severity scenarios for which only a procedural layer of protection is available.

Releases of hazardous materials have occurred that could have been prevented by implementation of engineered safeguards. From a review of these incidents, it had become evident that the owners and operators of the affected facilities relied on procedural safeguards to prevent occurrence of those events. Examples of such cases have involved gasoline, chlorine, and anhydrous ammonia, which are three commonly used hazardous materials at facilities in the United States.

Process Hazard Analysis (PHA) provides a thorough and efficient method for systemically reviewing complex systems for safety concerns. A commonly applied PHA method is the Guide-Word Style Hazard and Operability (HAZOP) approach. HAZOPs are a team effort and risk determination varies based on the experiences and safety culture of team members. Some teams may determine that procedural safeguards are adequate to prevent against a scenario that could lead to a severe event (e.g., liquid release of a volatile toxic material such as chlorine). However, accidental releases have been experienced demonstrating inadequacy of procedural safeguards to minimize the risk of a severe event.

The above noted CCPS guidelines consider procedures as the least effective protection layer. An active engineered safeguard is considered a more reliable protection layer. The purpose of this paper is to point out the potential for a pitfall in risk acceptance where a high severity consequence scenario is concluded to be of acceptable risk based solely on operator action. It is recommended for owner and operators of such facilities to require mitigation feature other than procedural controls for high severity scenario regardless of the risk ranking.

1 Introduction

Releases of hazardous materials have occurred in a wide range of systems with safeguards in place. A number of these events have occurred in systems that relied on procedures and operator intervention as the primary safeguard. The Center for Chemical Process Safety (CCPS) [1] and the American Industrial Hygiene Association (AIHA) [2] have each published guidelines on the topic of hierarchy of controls which can be used as a qualitative measure to establish the effectiveness of risk reduction measures. Both CCPS and AIHA have ranked engineered controls as a higher level safeguard than human intervention. The October 23, 2009 explosion at the Caribbean Petroleum Corporation's facility in Puerto Rico shows the importance of implementing engineered controls in addition to human intervention. The Chemical Safety Board (CSB) [3] investigated that event and recommended for similar facilities to add engineered controls such as "automatic overfill protection systems" to prevent future accidents.

Facilities that contain certain hazardous materials above a certain quantity must comply with the Process Safety Management (PSM) [4] and Risk Management Plan (RMP) [5] process safety regulations that require a Process Hazard Analysis (PHA). The purpose of a PHA is to identify potential hazards posed by a system, the safeguards that are in place to minimize the likelihood or severity of the event and risk rank the scenarios that can lead to adverse consequences. In some instances, the probability qualitatively assigned to an adverse scenario with only procedural safeguards may be underestimated. This underestimation of probability can lead to a lack of urgency to implement engineered safeguards to prevent an adverse consequence.

In addition to PHAs, some companies voluntarily perform Layer of Protection Analysis (LOPA), which is a more detailed analysis than PHA. LOPA is used to semi-quantitatively assess whether sufficient safeguards exist to meet a target risk level. In some instances, teams may use LOPA in such a way as to show that adverse consequences are unlikely by crediting conditional modifiers that may not be applicable to the scenario. This can lead to a determination that additional safeguards are not necessary as risk tolerance has been met.

It may be prudent to establish rules to implement engineered safeguards in potentially severe scenarios that rely only on procedural safeguards. The purpose of this paper is to describe the establishment of a criterion in which potentially high severity scenarios are identified that rely primarily on procedural safeguards and how to add an engineered safeguards to reduce the likelihood of the scenarios occurrence.

2 Industry Incidents

Many incidents have occurred in the chemical process industry where a procedural safeguard was the main layer of protection.

The CSB has performed many incident investigations and provides recommendations to different organizations with the main purpose of learning from past events to enhance safety. A recent investigation focused on the CAPECO incident of October 23, 2009 when a gasoline storage tank was overfilled leading to the formation of a vapor cloud and explosion [3]. In Appendix B of that investigation report [3], CSB lists twenty-two major accidents that have occurred in the process industry since 1962 that involved hydrocarbon storage tank overfill. A majority of these accidents involved processes that did not have an automatic overfill protection system and only relied on human intervention to prevent an overfill.

Similar events have occurred at facilities that handle anhydrous ammonia, a highly hazardous material. Ammonia is a toxic material with a low boiling point that vaporizes upon release into the atmosphere. Inhalation of ammonia vapors can result in irritation of the mucous membranes of the nose, throat, and lungs. If a plume of ammonia vapors with a concentration above 5,000 ppm is inhaled, sudden death may occur [6]. Many refrigeration systems use anhydrous ammonia as the refrigerant in the United States. In ammonia refrigeration systems, compressors provide the motive force to circulate the ammonia. Compressors require oil for lubrication, which by design leaks into ammonia side of the system. Entrained oil is removed from the system from the oil pots where oil separates from ammonia and settles in the pot. In order to ensure proper ammonia refrigeration system cooling efficiency, these oil pots must be drained periodically to remove the oil on the ammonia side of the system. This is achieved by opening a drain valve into an open pale. System pressure moves the oil. The operators are required to be vigilant and close the valve as soon as oil composition starts to change indicating that the pot is drained of oil and ammonia may follow.

Releases of ammonia have occurred from ammonia refrigeration systems due to oil drain operations. An engineered safeguard that could prevent this type of release is a self-closing emergency stop valve. It is designed such that an operator must hold the valve open by holding on to it and when operator's hand is removed the spring loaded self-closing emergency stop valve immediately closes. If an emergency stop valve is installed on an oil drain valve, it can limit the amount of ammonia released to atmosphere in the event that the drain operation is performed too long and ammonia starts to release once the oil has been drained from the pot.

The accident at CAI, Inc. and Arnel Inc., [7] is another example of an accident involving a system without an engineered safeguard. The CSB performed an incident investigation on this accident and issued recommendations [7]. According to the CSB, an initiating cause of the explosion was determined to be that a steam valve was left open to a tank containing flammable liquid. This steam valve was periodically used to heat the contents of a tank that stored ink and then manually shutoff by an operator once the desired temperature of the material was reached. On November 22, 2006 the steam valve was forgotten open and the ink in the storage tank overheated and vaporized. The vapor cloud found an ignition source and a large explosion occurred. The CSB recommended that CAI, Inc. install safety controls that would prevent overheating of flammable or combustible liquids.

3 Hierarchy of Controls and its Application

3.1 Hierarchy of Controls

CCPS has published "Inherently Safer Chemical Processes, A Life Cycle Approach", which describes the importance of making chemical processes inherently safer [1]. The CCPS states the following, "A chemical manufacturing process is inherently safer if it reduces or eliminates the hazards associated with materials and operations used in the process and this reduction or elimination is permanent and inseparable" [1]. The CCPS has described methods of making chemical processes safer and an important concept of the guidelines is levels of inherent safety. The CCPS [1] describes the levels of inherent safety as follows from most effective to least effective:

First Order Inherent Safety Measures –
- Avoid Hazards

Second Order Inherent Safety Measures –
- Reduce Severity;
- Reduce Likelihood

Layers of Protection –
- Apply Passive Safeguards;
- Apply Active (Engineered) Safeguards;
- Apply Procedural Safeguards

It is important to note that the least robust layer of protection listed by the CCPS is the application of procedural safeguards. This means that the application of active safeguards is a more robust layer of protection than procedural safeguards. The highest or first order of inherent safety listed is avoiding hazards. This is defined as elimination of a hazard. The two second orders of inherent safety listed are reducing severity and reducing likelihood. The severity of a hazard can be reduced by a number of methods including reducing the quantity of the material in a system, reducing the pressure or temperature within a process, substituting the material for a less hazardous material, etc. The likelihood of a hazard's occurrence can be reduced by a number of methods including using more corrosion resistant material, redesigning a process from four vessels to requiring two vessels, etc. The next order of inherent safety after the first and second orders are layers of protection. Layers of protection are ranked in order as passive safeguards, active safeguards, and procedural safeguards with procedural safeguards being the least robust and passive safeguards being the most robust. An example of a passive safeguard is designing the vessel within a system with a maximum allowable working pressure that is higher than the highest pressure that can potentially be reached within the system. An example of an active safeguard is an automatic overfill protection system that shuts off flow to a tank when high level is reached in the tank. An example of a procedural safeguard is relying on operators to shut off the flow to a tank when a certain level is reached.

The AIHA also addresses hierarchy of control from the standpoint of occupational safety. The "Z10-2012: Occupational Health and Safety Management Systems" Standard, defines hierarchy of controls which can be summarized as follows:

A. Elimination;
B. Substitution of less hazardous materials, processes, operations, or equipment;
C. Engineering controls;
D. Warnings;
E. Administrative controls; and
F. Personal protective equipment

There are many similarities between the CCPS's inherent levels of safety and the AIHA hierarchy of controls. Similar to the CCPS, the AIHA hierarchy of controls considers engineering controls as more robust than human intervention, warnings and administrative controls. The AIHA hierarchy of controls is written for occupational safety whereas CCPS's hierarchy of controls is written specifically for process safety applications. In both hierarchy of controls, the first item listed is eliminating a hazard. The second order of inherent safety listed by the CCPS is "reduce severity, or reduce likelihood", which is very similar to the second hierarchy of control listed by AIHA. Substituting a less hazardous material translates to reduced severity. Substituting less hazardous operations can be considered as a measure that reduces likelihood. The third hierarchy of control listed by AIHA is "engineering controls", which can be considered the same as the first two layers of protection listed by the CCPS, "Apply passive safeguards, Apply active safeguards". The third layer of protection listed by the CCPS, "Apply procedural safeguards" can be considered a combination of the fourth and fifth hierarchy of controls listed by AIHA, "Warnings; Administrative controls". The one major difference is that the CCPS does not call out alarms specifically as the AIHA does with the *warnings* hierarchy of control. The CCPS's inherent safety measures refer to alarms in the procedural controls as all alarms require personnel intervention.

An example of an engineered control referred to by AIHA in regards to occupational safety is a shutoff system for an automated laser welding machine that shuts off immediately upon detection of limbs within the welding area. This is an example of an engineered control designed to prevent personnel exposure to laser burns. An automatic overfill protection system that is considered an engineered safeguard by the CCPS for process safety applications is also considered an engineering controls according to the AIHA hierarchy of controls.

Regulatory bodies involved with PSM and RMP have taken notice of the importance of hierarchy of controls. Current drafts of California's Process Safety Management [8] and Risk Management Plan [9] regulations for refineries will specifically require hierarchy of hazard controls to be formally addressed. The draft regulations state that an analysis must be performed on all existing processes at a refinery that pose a process safety hazard. The analysis team must document the possible inherent safety measures for each hazard that can be implemented to reduce risk. The analysis team must review the feasibility of each level of inherent safety and must document why an inherent safety measure cannot be implemented for the process if that decision is made.

3.2 *Application of Hierarchy of Controls*

The most preferable time period to apply the hierarchy of controls to a system is during its design phase when changes can be made without severe economic impact. In addition to the design phase of a system, the hierarchy of controls can be applied throughout a system's life cycle when changes are considered. However, the upcoming regulations in California will be requiring application of the concept to built and operated systems.

To be able to apply hierarchy of controls to a system, it is only rational to establish what hazardous scenarios are possible and what safeguards are in place as protection. Process Hazard Analysis (PHA) provides a systematic method to identify and rank potential hazard scenarios and to identify the layers of protection in place to minimize the risk. By conducting the PHA, one can claim that all potential hazard scenarios that a team of experts could identify have been addressed systematically.

The hierarchy of controls can be used during a Process Hazard Analysis (PHA) to determine the adequacy of available safeguards to prevent each release scenario identified. This can be completed by identifying the available safeguards for scenarios that are postulated to have severe consequences and questioning whether or not inherently safer measures can be implemented. In addition, the hierarchy of controls can be applied when a recommendation is made during a PHA. The hierarchy of controls can act as guidance to choosing a higher level control measure to reduce the risk of a scenario. The hierarchy of controls can be reviewed while making any major changes to a system to assess if there are inherently safer options than the change being proposed. The hierarchy of controls can also be reviewed during an incident investigation to determine if a higher order of control should be implemented to prevent the recurrence of an incident.

The thought process of applying hierarchy of controls for a gasoline producer may be the following:

1. Currently there is a need for gasoline as vehicle fuel since a majority of automobiles sold in the United States require gasoline as the energy source. Therefore, this is a viable business venture that society at large needs. Eliminating gasoline from the line of products produced and sold by the company is not a prudent business decision because its safety risk is relatively well understood and has historically been accepted by society. Therefore, eliminating the hazard (i.e., moving out of gasoline business) as the very first item of the hierarchy of control is not a prudent option.
2. The next hierarchy of controls addresses reducing severity and likelihood. In order to reduce the severity of a gasoline storage tank related mishap (e.g., overfill and vapor cloud explosion), the hazard radius of a vapor cloud would need to be reduced. This can be done by reducing the pool surface of released material, as an example. Pool area is a function of the berm walls around the storage tanks and height of the walls. As another example, the owner may decrease the flow rate into a tank allowing the operators more time to discover and shutoff the inflow. These changes may not be feasible or the risk reduction gained from them may be smaller than other options (e.g., automatic shut-off on high level).

3. In order to reduce the likelihood of a gasoline storage tank overfill, the number of times the storage tank filling operation is performed can be reduced. This will require larger storage tanks which may increase the quantity of gasoline release in case of tank failure. Thus, the risk reduction gained from decreasing frequency of fills may be offset by the increase in risk due to larger tank size.
4. Engineered controls are the next available hierarchy of control. The first type of layer of protection based on CCPS's hierarchy of controls is a passive safeguard. One commonly used passive safeguard for large hydrocarbon storage tanks is secondary containment (a berm). The purpose of a berm is to contain a spill. However, in the case of gasoline, the light ends within the mixture would vaporize resulting in a vapor cloud formation with the potential for an explosion. As discussed above, reducing the pool size may reduce the evaporation rate but better risk reduction can be obtained by other means.
5. The next hierarchy of control is an active safeguard such as an automatic overfill protection system that will shut off the pump filling the storage tank when high level is detected. For a built and operated system, this option may provide the best risk reduction.
6. Procedural safeguards are the next hierarchy of control after active safeguards. The procedural safeguard for a gasoline storage tank filling operation would be operator monitoring of the level on the storage tank, installation of a high level alarm, and shutting off the feed pump manually at a certain level.

4 Process Hazard Analysis

4.1 Use of Process Hazard Analysis

A Process Hazard Analysis (PHA) is a hazard analysis tool that must be performed for systems covered under the PSM and RMP regulations [4 and 5]. PHA is a systemic tool for identifying the hazards of a system. PHA also provides a means to establish the potential severity of each hazard scenario, the available safeguards to prevent the hazard and likelihood of the scenarios occurrence. By applying risk ranking protocols, the risk gap of each scenario is identified in the PHA process.

The most common PHA method is HAZOP, which is typically performed in a brainstorming session with the participation of team members with experience in different fields necessary to properly analyze a system.

An important part of a PHA is establishing the risk of a hazard scenario. Risk is defined as "the combination of the expected frequency (events/year) and severity (effects/event) of a single incident or a group of incidents" [10]. Before a PHA study commences, the management responsible for the system must decide what risks they are willing to accept and what risks they will consider unacceptable. The acceptable and unacceptable risks are then translated into a risk tolerance matrix. The risk tolerance matrix contains severity and likelihood pairs that are acceptable and severity and likelihood pairs that are considered unacceptable and require risk reduction. This risk matrix is then used by PHA teams in all the PHA studies owned and operated by the management. The risk matrix provides a basis to identify risk gaps and whether or not system upgrades should be considered to improve safety.

4.2 Augmenting PHAs with Hierarchy of Controls

PHA results can vary from team to team. People working in the chemical process industry have different career experiences, which can result in different risk estimations during a PHA. It is possible and the authors' experience in leading PHA studies corroborates this observation that the likelihood of a severe consequence may be underestimated by PHA Teams. For example, operators who work with a system on a daily basis and have not experienced any incidents may be under the impression that the probability of a specific high severity scenario occurring is very low. Operators may feel that a system has been performing well for many years and if an issue did arise, the operators are knowledgeable enough to stop the progression to the bad event. Also, some PHA team members may believe that certain initiating causes of hazardous consequences are very unlikely based on their experience. Many of these issues can lead to underestimation of the risk of a severe consequence. Therefore, it is possible that the risk of a high severity scenario is estimated within the acceptable range allowing reliance on procedures based safeguards with no other safeguard. Due to the fact that risk ranking issues such as the one described above exist, it may be prudent for companies to use hierarchy of controls as the bases to require that all high severity scenarios to include at least one independent layer of engineered safeguard protection.

5 Layer of Protection Analysis

Layer of Protection Analysis (LOPA) is a tool that can be used to semi-quantitatively estimate the likelihood of a high severity scenario. Typically, LOPA is applied to the high severity scenarios identified during a PHA in order to enhance the accuracy of scenario likelihood. As part of the LOPA process, typically the LOPA Team estimates probabilities on an order of magnitude basis. The values that are estimated are the initiating cause likelihood, failure probabilities of the safeguards, occupancy factor, ignition probability, time at risk, and other conditional modifiers [11]. The initiating event likelihood is an estimate of how likely it is that the initiating cause occurs that can initiate the postulated chain of events. For example, according to the CCPS [12] the probability of a basic process control system (BPCS) failure is 0.1 per year. Also, if the safeguard has special features (e.g., it is designed to be of high reliability) the team may assign 0.01 probability of failure.

Occupancy factor refers to the amount of time personnel are within the hazard zone. For example, the CCPS [13] states that the probability of one person being present in a hazard zone is 0.1 for an area where operators spend 17 hours on average per week. Ignition probability refers to the probability that a released flammable material finds an ignition source resulting in a fire or explosion. Ignition probabilities will vary based on many factors such as the properties of the released material in terms of the released temperature, hot equipment near the release point, etc. Time at risk factor applies to non-normally operating modes of a system. Other conditional modifiers may also be used depending on the scenario. For example, some companies may take a conditional modifier stating that a catastrophic compressor seal failure due to liquid being sent to the suction side may occur once every ten times.

LOPA teams may credit additional conditional modifiers and enabling conditions resulting in a low risk level within the acceptable range. One example is crediting a time at risk factor for scenarios that can only occur during startup. Another likelihood reduction can be achieved through the application of occupancy factor for scenarios where generally personnel are not present within the hazard zone. Application of these factors raises an important question about the possibility of reliance on procedural safeguard as the only safeguard against a high severity consequence. Similar to the PHA, management may mandate that for all high severity scenarios, the LOPA team should consider an engineered safeguard regardless of the likelihood and risk ranking.

Startup is an interesting case as a safety risk contributor. The CCPS states that time at risk is not advised to be used during startup, "Use of a time-at-risk enabling condition in such a situation may lead to a conclusion that the average risk meets a facility's risk criteria; however, the peak risk imposed on persons exposed to the hazard may be inappropriately high" [13].

6 Limitations and Benefits

6.1 Increased Complacency Potentially Leading to Adverse Consequences

A potential adverse outcome of adding engineered safeguards to a system is an increase in operator complacency, which may lead to an adverse consequence. Operators involved in tasks such as storage tank filling may become complacent if an engineered safeguard is available as a backup in case they do not shut off the filling operation in a timely manner. Many engineered safeguards are meant to serve as a final barrier to prevent the occurrence of a release. The knowledge that a final barrier such as an engineered safeguard is available may lead to personnel relying on that safeguard and not being as attentive to system operation as before the addition of the engineered safeguard. Increased complacency may unintentionally lead to an adverse consequence as typical engineered safeguards have a reliability of approximately 90%. This means that approximately 10% of the time that they are required to prevent an adverse consequence, they may not work. If operators become complacent and rely on engineered safeguards to prevent incidents, an incident may occur if the engineered safeguard does not function as required. For this reason, it must be stressed to personnel prior to installation of engineered safeguards that they are not meant to be a replacement of a safeguard (i.e., engineered safeguard replacing a procedural safeguard). Adding engineered safeguards to a system must be understood to be a measure to reduce risk significantly, which it cannot accomplish if it only replaces a procedural safeguard.

6.2 Engineered Safeguards Are Not Perfect

Engineered safeguards should not be considered as means of eliminating the potential for an adverse consequence. Engineered systems are not guaranteed to work every time they are needed. Incidents have occurred in the chemical industry in which an engineered safeguard did not work. One such incident occurred on August 14, 2002 in which a chlorine hose ruptured during a chlorine transfer and resulted in a release of over 48,000 pounds of chlorine [14]. The chlorine transfer system did have automatic shutoff valves that were supposed to activate if chlorine was detected at nearby chlorine detectors. The chlorine detectors did sense chlorine and sent a signal to the automatic shutoff valves to close. However, all automatic shutoff valves did not close as they had experienced some internal corrosion. The automatic shutoff system was not tested properly as part of the Mechanical Integrity Program. Management at the site was not aware of the corrosion issue within the system and did not take proactive measures to ensure the automatic shutoff system activates properly in the event that it was needed.

Another example of an engineered safeguard failing to work is the explosion at the Buncefield Oil Storage Depot on December 10, 2005 [15]. A gasoline storage tank was overfilled and the automatic overfill protection system did not activate as it was not placed into service properly after its previous test. Maintenance errors such as this can happen and it is important for personnel working with engineered safeguards to understand its functionality, and test and maintenance procedures in order for it to activate when needed.

The above two examples of an engineered safeguard failing to activate should not lead to a conclusion that adding engineered safeguards to a system does not reduce risk. Engineered safeguards do indeed reduce risk if the Maintenance Program of the system to which it is added is updated to include the proper procedures and testing frequency recommended by the manufacturer of each piece of equipment associated with the engineered safeguard. It should also be noted that personnel must be aware of the functional details of the engineered safeguard system to ensure that it is operable so that it can prevent an accident when needed as it is designed to.

6.3 Resources

Adding engineered safeguards to a system requires resources from many departments. Engineered safeguards added to an existing system may need to be designed according to the Instrumentation, Systems, and Automation Society (ISA) Standard 84.00.01-2004 [16]. The ISA 84 Standard has many detailed requirements for installing Safety Instrumented Systems (SIS). In order for these requirements to be properly met, input is needed from many different groups within a company.

6.4 Financial Aspect

Adding engineered safeguards may prevent financial losses incurred due to major accidents. Some may argue against adding engineered safeguards to a system that has been found to be of low risk but high consequence severity due to the immediate financial strain it may have on a company. It is true that adding an engineered safeguard requires some financial investment and the system must be included in the periodic test and maintenance program. However, a major accident can cost millions of dollars in monetary losses and other non-tangible impacts (e.g., personnel injuries and reputation impact). For example, in 1999 an explosion occurred due to a gasoline storage tank overfill at the Thai Oil Company that cost $22.3 million and resulted in seven fatalities [3]. The storage tank only had a high level alarm, which was not heard by onsite personnel. The automatic overfill prevention system would have cost a small fraction of that monetary loss, not to mention the loss of life that was experienced.

An opinion that is increasingly being accepted by industry management and safety professionals is that safety is an investment, not a cost. According to OSHA's Business Case for Safety Webpage [17], "Employers that implement effective safety and health management systems may expect to significantly reduce injuries and illnesses and reduce the costs associated with these injuries and illnesses, including workers' compensation payments, medical expenses, and lost productivity." We can extrapolate this statement to process safety. "Z10-2012: Occupational Health and Safety Management Systems" Standard [2] is generally accepted as being a very good guidance document to implementing an effective safety and health management system. A very important section within that standard is the one on hierarchy of controls. Therefore, implementing a system in which engineered safeguards are added for high severity scenarios where only procedural safeguards exist would be a step towards improving process safety. This can result in a decrease in incidents at a chemical processing facility, which can have a direct positive impact on facility profitability.

Adding engineered safeguards to a system does not necessarily mean that future accidents will be prevented. However, it will reduce the likelihood of a major accident occurring at a facility. The investment required to add and maintain an engineered safeguard can be significantly less than the costs associated with an accident.

7 Conclusion

In the chemical process industry, many potentially high severity tasks exist for which procedural safeguards are the main layer of protection. The hierarchy of controls guidance developed by the process industry ranks engineered safeguards as a more robust layer of protection than procedural safeguards. Multiple chemical process industry accidents have occurred in which the procedural safeguard layer of protection failed and an engineered safeguard did not exist. These accidents could have been prevented if the associated system had an engineered safeguard as a layer of protection in addition to the procedural safeguard.

The PHA and LOPA methodologies are tools that can be used to identify high severity scenarios and the existing safeguards. It is possible, to apply these methods properly and arrive at a low risk conclusion for high severity scenarios that solely rely on a procedural safeguard. It is recommended in this article that these methods be augmented by owners and operators responsible for highly hazardous operations with a mandate that all high severity scenarios to include at least one engineered safeguard. This should ensure that regardless of the estimated likelihood of the scenarios, the operators are not the only safeguard preventing the high severity release.

References

1. *Inherently Safer Chemical Process: A Life Cycle Approach.* Center for Chemical Process Safety, American Institute of Chemical Engineers, New York, NY, 2nd Edition, December 2008.

2. *Z10-2012: Occupational Health and Safety Management Systems*. American Industrial Hygiene Association, Falls Church, Virginia, June 2012.

3. *Final Investigation Report – Caribbean Petroleum Tank Terminal Explosion and Multiple Tank Fires, Bayamon, Puerto Rico, October 23, 2009*. U.S. Chemical Safety and Hazard Investigation Board, Report 2010.02.I.PR, October 2015.

4. *Process Safety Management of Highly Hazardous Materials*. Occupational Safety and Health Administration, Federal Code of Regulations, 29 CFR, §119.119.

5. *Risk Management Programs for Chemical Accidental Release Prevention.* U.S. Environmental Protection Administration, 40 CFR Part 68.

6. *Safety Data Sheet (SDS) for Anhydrous Ammonia*. Hill Brothers Chemical Company, HBCC SDS No. CA10000, Revision Issued: 5/14/2015.

7. *Investigation Report – Confined Vapor Cloud Explosion – CAI, Inc. and Arnel Company, Inc. – Danvers, Massachusetts*. U.S. Chemical Safety and Hazard Investigation Board, Report 2007-03-I-MA, May 2008.

8. *Process Safety Management for Petroleum Refineries*. California Occupational Safety and Health Administration, California Code of Regulations, Title 8, §5189.1, Draft Version 5.0, September 14, 2015.

9. *California Accidental Release Prevention (CalARP) Program*. California Governor's Office of Emergency Services. Title 19 of CCR Division 2, Chapter 4.5, Draft, September 24, 2015.

10. *Guidelines for Hazard Evaluation Procedures*. Center for Chemical Process Safety, American Institute of Chemical Engineers, New York, NY, 3rd Edition, April 2008.

11. *Layer of Protection Analysis – Simplified Process Risk Assessment* Center for Chemical Process Safety, American Institute of Chemical Engineers, New York, NY, October 2011.

12. *Guidelines for Initiating Events and Independent Protection Layers in Layer of Protection Analysis*. Center for Chemical Process Safety, American Institute of Chemical Engineers, New York, NY, February 2015.

13. *Guidelines for Enabling Conditions and Conditional Modifiers in Layer of Protection Analysis*. Center for Chemical Process Safety, American Institute of Chemical Engineers, New York, NY, November 2013.

14. *Investigation Report – Chlorine Release – DPC Enterprises, L.P.* U.S. Chemical Safety and Hazard Investigation Board, Report No. 2002-04-I-MO, May 2003.

15. *Buncefield: Why Did It Happen?*. United Kingdom Health & Safety Executive, February 2011.

16. *ANSI/ISA-84.00.01-2004: Functional Safety: Safety Instrumented Systems for the Process Industry Sector*. Instrumentation, Systems, and Automation Society (ISA), September 2004.

17. https://www.osha.gov/dcsp/products/topics/businesscase/. Occupational Safety and Health Administration, Business Case for Safety, Accessed on February 8, 2016.

Maintaining Consistency in the Determination of Consequence Severity Rankings for Process Hazard Analyses

Kirk Busby
Kazarians & Associates, Inc.
100 West Broadway
Glendale, CA 91210
Kirk.Busby@kazarians.com

Prepared for Presentation at
American Institute of Chemical Engineers
2016 Spring Meeting
12th Global Congress on Process Safety
Houston, Texas
April 11-13, 2016

Maintaining Consistency in the Determination of Consequence Severity Rankings for Process Hazard Analyses

Kirk Busby
Kazarians & Associates, Inc.
100 West Broadway
Glendale, CA 91210
Kirk.Busby@kazarians.com

Keywords: Hazard Identification and Risk Analysis, Process Safety, Consequence Modeling

Abstract

A Process Hazard Analysis (PHA) is an important tool for identifying and managing the hazards associated with a chemical process. One of the most critical aspects of the analysis is the risk ranking of scenarios, which is used to determine which events pose a high risk and may require further mitigation. The risk ranking of a scenario is built on two factors – severity and likelihood. The severity of a scenario is a measure of the potential worst-case impact on facility operation, facility personnel, the public, or the environment, while the likelihood is an estimate of the probability with which an event may occur, given the presence of the safeguards acting to prevent or limit the severity of the event.

Severity of a scenario is generally expressed in qualitative terms. While a typical approach to assigning consequence severity rankings involves categorizing events by impact (e.g., recordable injury, permanent disability, or fatality), this approach relies on the subjective judgment of the analysis team in determining the exact impact of a specific event. Experience has shown that, because of the subjective nature of this step, different PHA study teams have had varying opinions on the potential severity level of a given scenario. It is the aim of this study to provide guidelines to improve consistency of severity rankings both within a single analysis, and across multiple analyses with separate PHA study teams. It is possible to increase the consistency in consequence severity rankings through application of standardized guidelines. A method has been developed to aid PHA facilitators based on dispersion modelling of various conditions typically encountered in a PHA. The underlying data is based on groups of chemicals with similar properties and other conditions (e.g., break size, break location, etc.).

1. Introduction and Background

Process Hazard Analysis (PHA) is an important tool for identifying and managing the hazards associated with a chemical process. The goal of a PHA is to use a systematic methodology to identify each potential hazard source within the process. A typical PHA methodology is the guideword-style Hazard and Operability (HAZOP) study. To ensure completeness, the HAZOP methodology breaks down a complex system into individual nodes, making it easier to thoroughly analyze each section of the process. Further, within each node, hazards are postulated according to a standard combination of guidewords and process deviations. For example, when the guideword 'No' is paired with the deviation "Flow," all scenarios within the node which lead to a loss of flow are considered.

Once a scenario is identified, the risk posed to the facility must then be determined. This is normally achieved through the risk ranking process. To determine the risk ranking, each scenario is assigned a severity and a likelihood level. This pair is then used in a risk ranking matrix, which defines the risk targets developed by company management prior to the start of the PHA. Assigning a risk ranking is one of the most important facets of the PHA process, as it is used to determine which risks are acceptable given the safeguards currently in place, and which require further actions to ensure that the risk is maintained at a tolerable level.

Currently, there is a large potential for variance in risk ranking determination, specifically with regard to consequence severity. While severity determination is meant to be a qualitative ranking, there are a number of benefits to improving the consistency of severity rankings among similar release scenarios. Kazarians & Associates, Inc. has developed a method that has proven effective in reducing variance among consequence severity rankings through the use of pre-determined dispersion modeling results.

2. Current Methods for Consequence Severity Determination

It is common practice to estimate the worst case consequence severity for the identified potential release scenarios. For this, it is assumed that all existing active safeguards are unavailable. Typically, the severity ranking is a qualitative determination made by the PHA study team based on their understanding of the potential impact a given scenario may have on the safety of facility personnel and the public, as well as the environment or business operations. The text provided in defining each severity level is generally used to determine the consequence severity. Table 1 provides an example of severity level definitions available to the PHA study team.

Definitions such as those in Table 1 can provide a good starting point for consequence severity selection. However, use of these limited definitions can result in inconsistent severity rankings, both within an individual PHA, as well as among PHAs. This variation stems from the fact that the PHA study team generally carries the chain of events to the release condition (i.e., flange leak, pipe rupture, etc.) whereas the definitions of consequence severity level is often in terms of level of harm (i.e., first aid injury, fatality, etc.). This then requires the study team to envision how the release will ultimately result in harm, which means that they have to take into consideration factors such as source conditions, dispersion, and ignition, which are often complex phenomena by themselves. Therefore, this leads different PHA study teams to hold varying opinions of what the impact of a certain event may be. Typically, these opinions are based on the experience of the personnel present during the PHA.

Table 1. Typical Consequence Severity Guidance

Consequence Severity	Impact on Personnel	Impact on the Environment	Impact on business
1	Minor first aid injury	Slight spill onto soil, release to flare	< $10,000
2	Recordable injury	Small spill onto water, moderate spill onto soil	< $100,000
3	Permanent disability	Moderate spill onto water, large spill onto soil	< $500,000
4	Fatality	Moderate spill into an environmentally sensitive area	< $1,000,000
5	Multiple fatalities	Large spill into an environmentally sensitive area	< $5,000,000

The following example is taken from PHAs conducted for chlorination systems at two facilities operated by the same organization. The example demonstrates the difference operational experience can make in the severity ranking of a given scenario. During PHA sessions for the two systems, the PHA team encountered the same scenario in which a storage tank is exposed to an external fire, resulting in a pressure increase within the tank, ultimately resulting in tank rupture. In this scenario, the two teams have arrived at an identical endpoint (i.e., release scenario). While the release conditions (i.e., temperature and pressure) will be the same for both cases, the first team concludes that a single fatality may result, the second believes that multiple fatalities may result. The two teams may record the scenarios according to Figure 1.

In each case, the PHA study team has arrived at the same chain of events leading up to an integrity failure of the storage tank, and release of chlorine into the atmosphere. However, the final consequence statement and therefore risk rankings vary based on the estimated severity of the release.

Inconsistency in severity rankings can cause a number of problems following a PHA. Most importantly, they can lead to an over- or under-estimation of the risk a scenario poses. A consequence with the potential to cause severe injury that has only been identified as a first aid issue could result in insufficient safeguarding to protect against the event. This could ultimately lead to major safety incidents during the operating lifetime of the process. Safeguards are typically added when the risk ranking of a scenario does not meet the pre-determined corporate risk target. A lower severity ranking may indicate that the appropriate safeguards are in place, when in reality an additional protection may be necessary.

Team 1:

Causes	Consequences	Safeguards	S	L	RR	HAZOP Recommendations
Greater than 140°F in the Chlorination Room caused by a fire	▪ Pressure buildup in the Chlorine Storage Tanks. ▪ Potential for loss of containment of a Chlorine Storage Tank if the relief valve fails to lift. Release of chlorine into the Chlorination Room. Potential for personnel exposure.	▪ Relief Valves will lift to prevent overpressure. ▪ Detectors alarm locally and on SCADA. ▪ Scrubber activated by detectors at 3 ppm. ▪ Plant housekeeping practices prohibits storage of any combustible materials in the Chlorination Room.	4	D	M	

Team 2:

Causes	Consequences	Safeguards	S	L	RR	HAZOP Recommendations
Fire inside the Chlorine Building.	Potential for increased pressure in the storage tanks. Potential to rupture the storage tank. Potential for release of chlorine inside the building. Potential for personnel and public exposure.	1. PSVs on the storage tanks set at 225 psig. 2. Chlorine detectors alarm at 1 ppm. Scrubber system will start at 3ppm on any one detector. 3. Good housekeeping practices.	5	E	M	No recommendations necessary. Safeguards are considered adequate.

Figure 1. Identical PHA Scenarios with Differing Consequene Statements and Risk Ranking

A second and more subtle consequence of inconsistent severity ranking is the potential discrepancy between processes or facilities. If two PHA study teams assign different severity rankings to similar events on two similar systems, there is a potential for one system to require an extra safeguard to be installed which is not in place on the second system. This discrepancy between units could potentially lead to confusion later in the life of the plant. If a group of operators is responsible for both systems, there could be confusion as to which setup is correct, as well as the incorrect assumption that both systems are configured in the same manner. Additionally if an incident occurs in the system with fewer safeguards, there is a potential for increased regulatory scrutiny and fines due to the fact that the second system is not as well protected as the first.

3. Improving Consistency through Dispersion Modeling

To reduce variation among consequence severity rankings, it is possible to provide PHA teams with improved guidance on the potential impacts of a release. Dispersion modelling is a tool which is commonly used to determine the hazardous area resulting from a specific chemical release scenario. There are a number of sophisticated dispersion models and software packages currently on the market which can provide a picture of the hazard distance presented by various chemical release events. Many of the commercially available packages allow the user to input specific data such as the temperature, pressure, quantity, and release orifice size, and calculate the released quantity, pool size, vaporization rate, and resulting hazard distance. Depending on the model chosen, the user may be able to customize the scenarios to a greater degree, including such factors as terrain type, degree of confinement, explosive characteristics, etc.

Currently, some PHA teams already use dispersion modeling to assist in performing a more in-depth analysis for select scenarios with complex release characteristics. These studies are useful, but require a large amount of input data to be determined by the PHA team in order for them to generate accurate results. Therefore, it is much too time consuming to use such an analysis for every release scenario postulated during a PHA.

To combine the time and cost efficiency of PHA with the increased accuracy afforded by the use of dispersion modeling, the dispersion calculations can be performed generically prior to beginning the PHA discussions. To utilize this method, a database is developed for each hazardous chemical which may be released, and dispersion calculations are performed for the range of temperatures, pressures, phases (liquid or gas), and break sizes which may be encountered during the PHA. Every possible combination of parameters does not need to be considered, as the PHA team may quickly interpolate between results when needed. With a pre-developed matrix in front of them during the PHA sessions, participants can quickly relate a set of known release conditions to an approximate hazard zone, which will allow them to choose a more refined consequence severity. Some large chemical processing and manufacturing corporations have already begun to adopt such methods to improve the consistency of their PHAs.

The method developed by Kazarians utilizes PHAST modeling software by Det Norske Veritas [Reference 1] to create databases for both flammable and toxic chemicals. Specifically, the common alkanes (i.e., methane through hexane), hydrogen, hydrogen sulfide, anhydrous ammonia, and chlorine have been analyzed, in both liquid and gaseous form. In each case, a generic outdoor release event is postulated for varying temperatures, pressures, spill quantities, and release sizes. From this basic input data, PHAST is capable of generating information on the maximum dispersion distance, the flammable range, the range of explosive overpressure, and toxic effects. For each release event, the maximum distance of each hazard type is compared, and the greatest distance is selected. For example, for a specific set of release conditions, the maximum distance at which an injury may occur due to thermal radiation is 200 feet, while the maximum injury distance due to explosive overpressure is 250 feet. For these release conditions, 250 feet will be selected and recorded as the worst-case injury hazard distance. The final result is a set of eight tables for each chemical, with four liquid phase tables and four vapor phase tables. Tables 2 and 3 below show the four hazard distance tables for methane in the liquid phase.

Table 2A. Methane Liquid Injury Hazard Distance (ft)

	Source Temperature (F)			
Release (lbs)	**-250**	**-200**	**-150**	**-117**
10	110	100	140	160
100	350	230	320	350
1,000	610	510	710	780
10,000	1,520	1,130	1,570	750
100,000	4,060	2,480	3,350	3,350
500,000	7,690	4,260	5,360	5,360

Table 2B. Methane Liquid Injury Hazard Distance (ft)

	Source Pressure (psi)				
Break Diameter (in)	**50**	**150**	**300**	**400**	**450**
0.25	20	30	30	30	30
0.5	110	110	110	110	110
0.75	180	210	210	210	210
1.0	210	280	310	310	310
1.5	280	410	480	510	510
2.0	380	540	640	670	700
3.0	510	770	970	1,070	1,100
4.0	610	1,000	1,330	1,430	1,460
6.0	840	1,430	2,020	2,180	2,250

Table 3A. Methane Liquid Fatality Hazard Distance (ft)

	Temperature (F)			
Release (lbs)	**-250**	**-200**	**-150**	**-117**
10	40	40	60	70
100	240	90	130	150
1,000	530	210	310	340
10,000	1,390	470	690	1,690
100,000	2,990	1,070	1,500	1,500
500,000	5,290	1,860	2,400	2,400

Table 3B. Methane Liquid Fatality Hazard Distance (ft)

	Pressure (psi)				
Break Diameter (in)	**50**	**150**	**300**	**400**	**450**
0.25	20	20	20	20	20
0.5	40	50	50	50	50
0.75	110	140	140	140	140
1.0	140	210	240	240	240
1.5	210	340	400	440	440
2.0	300	470	570	600	630
3.0	440	700	890	990	1,030
4.0	530	930	1,260	1,350	1,390
6.0	760	1,350	1,940	2,110	2,170

Tables 2 and 3 present two sets of hazard distances. One represents the distance inside which an injury is expected to occur, and one inside which it is possible for fatalities to occur. Additionally, the sets have been subdivided into Tables A and B, which provide dispersion information for Release Quantity vs Temperature, or Break Diameter vs Pressure. The PHA team may elect to use either of the tables based on what information is available for a given scenario. For example, Break Diameter vs Pressure tables may be more useful for a pipe break scenario, while Release Quantity vs Temperature may be easier to use in the event of a seal failure or liquid overfill scenario.

There are a few notable trends which can be identified from Tables 2 and 3. Specifically, it can be seen that in the tables relating Break Diameter and Pressure, the maximum hazard distances reach a constant value as pressure increases. This is due to the release reaching what is known as "sonic velocity." [Reference 2] When a material under pressure is released through an orifice, the velocity of the release is initially dependent on the pressure within the vessel. Higher release pressure will typically result in a higher release velocity. However, once the release velocity has reached sonic velocity, further increases in pressure will no longer cause the velocity to increase. Put simply, no more material can fit through the release orifice, no matter how much force is behind it.

4. Defining Exposure Thresholds

The final dispersion distances reported in Tables 2 and 3 can vary greatly depending on what thresholds values are chosen for injuries and fatalities. This analysis used three potential sources of impact to the population: overpressure due to explosion, thermal radiation due to fire, and exposure to toxic vapors. Each threat has a unique hazard zone that is determined by its effects on both the human body and surrounding structures.

There is a great deal of research performed on the various effects on the human body caused by exposure to overpressure, thermal radiation, or toxic chemicals [References 3 through 10]. From this research, threshold values can be estimated to determine the exposure limits for injuries and fatalities. Table 4 presents the threshold values used in this article to differentiate between effects.

Table 4. Exposure Threshold Definitions

Explosion Overpressure	
Effect	**Threshold Value**
Injury	0.5 psi
Fatality	2 psi
Thermal Radiation	
Effect	**Threshold Value**
Injury	5 (kW/m^2)
Fatality	10 (kW/m^2)
Toxic Exposure	
Effect	**Threshold Value**
Injury	ERPG-2
Fatality	ERPG-3

It is important to note that these values represent conservative estimates of the potential impact an event may have on a person. For example, the threshold values for thermal radiation exposure are time dependent. Exposure to 10 kW/m^2 will not result in an immediate fatality, but prolonged exposure of approximately 60 seconds could result in death. This value is chosen as it represents the possibility of a fatality since it requires rapid evacuation from the hazard zone.

For each scenario modeled in PHAST, the distances to each threshold value described in Table 4 are compared, and that which results in the largest hazard distance is selected as the worst-case hazard zone. Toxic effects are only calculated for those chemicals which pose a significant risk when inhaled or contacted.

5. Using Hazard Distance Tables During PHA Sessions

Once developed, hazard distance tables can be used during PHA discussions to quickly determine an appropriate severity ranking based on the potential impact to facility personnel and the public. To convert the hazard distance tables into a form which may be more easily used during the PHA, the PHA facilitator may review the company risk matrix, as well as the facility layout prior to starting the PHA sessions. During this time, the facilitator should convert the dispersion distances in Tables 2 and 3 into charts providing guidance on the potential worst-case severity of different release events.

For example, a hypothetical company has developed a risk matrix with the consequence severity rankings described in Table 1. The system under analysis is an LNG storage tank and associated piping, which are located in an isolated part of the facility, but approximately 700 feet from the closest residences. Using this information, the PHA facilitator may translate the worst-case hazard zones in Tables 2 and 3 into a set of tables with suggested consequence severity rankings which are tailored to the system under review. Table 5 provides an example of these consequence severity tables for the specific LNG system described above.

Table 5A. Methane Liquid Consequence Severity Rankings

	Source Temperature (F)			
Release (lbs)	**-250**	**-200**	**-150**	**-117**
10	1	1	2	2
100	3	2	2	3
1,000	4	3	3	4
10,000	4	4	4	5
100,000	5	5	5	5
500,000	5	5	5	5

Table 5B. Methane Liquid Consequence Severity Rankings

	Source Pressure (psi)				
Break Diameter (in)	**50**	**150**	**300**	**400**	**450**
0.25	1	1	1	1	1
0.5	1	1	1	1	1
0.75	2	2	2	2	2
1.0	2	2	3	3	3
1.5	2	3	4	4	4
2.0	3	4	4	4	4
3.0	4	4	4	5	5
4.0	4	5	5	5	5
6.0	5	5	5	5	5

When translating the calculated hazard zone to consequence severity, it is important to note that the dispersion modeling calculations are performed generically, and their results may need to be adjusted based on the specifics of the system under review. The relationship between hazard distance and similar consequence categories (i.e., first aid injury, reportable injury, and permanent disability) should be developed with the primary decision maker for the facility (i.e., company management). To accomplish this, the facilitator should record the decision making process used to relate hazard distances to consequence severity levels, and present these decisions to company management prior to beginning the PHA sessions, and adjust the severity tables based on comments received. .

Additionally, the model calculations make a number of assumptions which may or may not apply to the system being analyzed. Most importantly, the calculations assume that releases occur outdoors, and outside of any secondary containment areas. Indoor releases can have a significantly reduced hazard zone outside of the building. This may depend on the ventilation capabilities of the building, or doors normally kept open. Toxic material may build to higher concentrations within the building than those reported in the hazard distance tables. Similarly, the presence of secondary containment limits the surface area of a liquid pool, which may result in a smaller vapor cloud than an uncontained liquid spill. To adequately assess the severity of a chain of events, these deviations from the assumed model must be accounted for. In cases such as these, the hazard distance tables should be used as a starting point, but adjust their consequence severity ranking according to the physical systems present in the field.

During the PHA discussions, once a scenario has been identified and the release conditions are established, the consequence severity tables can be consulted to assist in selecting the most appropriate consequence severity ranking. To use the tables, a process engineer or other person familiar with the design of the system under consideration should determine the physical parameters of the release. It must first be determined whether the released material is in the liquid or the vapor state. Subsequently, the release conditions must be specified in terms of temperature, pressure, quantity spilled, or breach diameter, depending on what information is most readily available to the team. This determination is not intended to be a rigorous calculation, but rather a quick estimation of the conditions of the release.

Once the release parameters have been agreed upon, the team may then use the consequence severity tables to determine the worst-case severity ranking. For example, in the LNG system described above, it is postulated that a full break occurs in a one inch diameter pipe containing LNG at 170 psig. Given the information at hand, the team consults Table 5B, and identifies 150 psig as the closest pressure rating found on the tables. From the Table, the team selects a severity ranking of 2, indicating that a recordable injury is the anticipated worst-case consequence.

6. Conclusion

Kazarians & Associates, Inc. has developed a method which uses pre-developed consequence severity tables to refine the judgement of PHA teams with regard to the impact that a scenario may have on personnel. This method has been applied, resulting in a more accurate determination of consequence severity. When applied across multiple teams and studies, this increase in accuracy will lead to consistency in severity ranking among similar PHA scenarios. In turn, this will lead to consistent application of safeguards across multiple systems, resulting improved plant operability and avoiding the potential for confusion among operating staff.

7. References

[1] Phast Modeling Software, version 7.11.33.0. DNV GL AS.

[2] *Guidelines for Hazard Evaluation Procedures*, published by the Center for Chemical Process Safety of the American Institute of Chemical Engineers, 3rd Edition, April 2008.

[3] *OGP Risk Assessment Data Directory: Vulnerability of Humans*. The International Association of Oil and Gas Producers. Report No. 434-14.1. March 2010.

[4] Toups, Harry J., *Fires and Explosions - Fundamentals and Design Considerations* [PowerPoint Slides]. Retrieved February 26, 2015 from www.sache.org / links / Pike21Jul2004 / Fires%20and%20Explosions.ppt.

[5] Clancey V J, 1972. Diagnostic features of explosion damage, 6th Intl. Meeting on Forensic Sciences, Edinburgh, Scotland.

[6] ALOHA 5.4.3 Modelling Software, Office of Emergency Management – U.S. EPA and Emergency Response Division – National Oceanic and Atmospheric Administration.

[7] *Handbook of Chemical Hazard Analysis Procedures.* Federal Emergency Management Agency, U.S. Department of Transportation, U.S. Environmental Protection Agency.

[8] Hockey, S M, and P J Rew. *Review of Human Response to Thermal Radiation.* HSE Contract Research Report No. 97/1996. Health and Safety Executive. July 2, 1996.

[9] Raj, Phani K. *A Review of the Criteria for People Exposure to Radiant Heat Flux from Fires*. Technology & Management Systems, Inc. October 25, 2006.

[10] National Fire Protection Association. (2013). *NFPA 59A: Standard for the Production, Storage, and Handling of Liquefied Natural Gas (LNG).*

441564 Application of Open Source Software Tools in the Process Safety Education

Abstract:

Process safety professionals have used, developed, and contributed to open source software tools. They have fascinated more and more notice from the society. The free software will enhance the collaboration and motivate the scientific and technical development. The software users can use and even modify the software modules. A lot of applications can be found in the real industrial practice so the engineering learning process should follow this trend. Before starting the career, the students should get to know how to take the full advantage of these tools in their future jobs because the future engineers should know how to find the free software packages in order to solve process safety problems.

This presentation will tell the audience how to integrate the open source software into process safety education or learning practice. The examples of modeling projects which can be practiced by the students in the classroom will also be introduced. The modeling will give the students more opportunities to recall their knowledge of mass and energy balance, thermodynamics, heat transfer, reaction kinetics, transport phenomena, process safety, statistics and numerical methods. These projects will also help them to learn how to conduct the risk assessment and process safety management through different kinds of model development. These chemical engineering education activities will allow them to explore computer aided tools such as computing and process simulators throughout the curriculum.

441688 Beyond Quantitative Risk Analysis Results - Part I: Explosions and Blast Phenomena Characterization

Abstract:

The present paper illustrates the results of a Quantitative Risk Analysis (QRA) case study of a refinery. The main purpose of a QRA is to evaluate the risk levels (both individual and societal risk) of a process due to Loss Of Containment scenarios (LOCs). A QRA can be used to optimize the risk reduction measures to be implemented to comply with a given tolerable risk level. Moreover, the development of a detailed QRA is the best opportunity to perform a specific analysis addressing structural response due to overpressure impacts caused by explosions and also estimating the Time To Failure (TTF) of structures and equipment due to thermal radiation of fires.

Part I of this paper is focused on illustrating how a detailed analysis of QRA results can provide valuable information for characterizing the blast phenomena parameters for a given location. The analysis gathers three key sources of information: (1) peak overpressure, maximum and minimum positive phase durations and impulses as a function of the cumulative frequency of occurrence; (2) total number of LOC scenarios which contribute to the impact of explosions (including vapor cloud explosions due to both immediate and delayed sources of ignition); and (3) the identification of each LOC scenario and associated individual frequency of occurrence. This information is valuable for ensuring the blast loading function for a subsequent structural response analysis (e.g., equivalent elasto-plastic Single Degree Of Freedom analysis), and also provides specific information regarding which LOCs contribute most on overpressure impact.

441690 Beyond Quantitative Risk Analysis Results - Part II: Fires and Domino Effect Characterization

Abstract:

The present paper illustrates the results of a Quantitative Risk Analysis (QRA) case study of a refinery. The main purpose of a QRA is to evaluate the risk levels (both individual and societal risk) of a process due to Loss Of Containment scenarios (LOCs). A QRA can be used to optimize the risk reduction measures to be implemented to comply with a given tolerable risk level. Moreover, the development of a detailed QRA is the best opportunity to perform a specific analysis addressing structural response due to overpressure impacts caused by explosions and also estimating the Time To Failure (TTF) of structures and equipment due to thermal radiation of fires.

Part II of this paper is focused on illustrating how a detailed analysis of QRA results can provide valuable information for characterizing potential domino-effects due to thermal radiation of fires at a given location. The analysis gathers three key sources of information: (1) thermal radiation as a function of the cumulative frequency of occurrence; (2) total number of LOCs which contribute to the impact of fires; and (3) the identification and characterization of each LOC scenario and associated individual frequency of occurrence. This information not only is valuable for ensuring the heat radiation for a subsequent analysis intended to estimate the TTF and associated conditions remaining in the system (e.g., Wall Dynamics Analysis), but also provides specific information of which LOCs contribute most to the heat radiation impacts. This information ensures which LOC scenarios need to be addressed for optimizing the risk reduction decision-making process. Additionally, the accurate estimation of potential equipment failure due to thermal radiation provides added value for predicting when and how impacts such as BLEVEs (Boiling Liquid Expanding Vapor Explosions) are expected to occur.

441700 A Leadership Vision for Accelerating Learning from Past Lessons to Create a Future Free of Incidents

Abstract:

The practice of process safety management (PSM), involves the analysis and understanding of what can go wrong, what the effects could be, and what should be done to prevent or mitigate an incident. That is: (1) what are the causes (human errors and equipment malfunctions)? (2) What are the potential consequences and impacts (to people, property, process, reputation)? and (3) what are the risk controls needed to prevent, detect, control, mitigate, or respond to such events (safeguards, layers of protection)?

Lessons learned from industrial disasters, as well as successes by industry leaders in preventing such incidents, form the basis for a practical framework for a PSM system as a holistic management tool for managing hazards and their risks. Quality design/development and effective implementation of a robust risk management system driven by a proactive culture of leadership to reduce operational risks and improve performance, enables organizations to avoid future incidents.

However, a conventional approach of longer-term continual improvement of operational risk performance could easily be outpaced by the rapid changes in technology advances, organizational and human factors, and the dynamics of the market. Operating companies, service providers and regulators alike are challenged by such changes and their potential impacts. Lessons of disaster investigations from the process industry have been full of learning opportunities. In fact, many of the lessons and their root causes are common across most process safety incidents and near misses ... yet, 15-20 years has proven to be an unacceptable period for gradual learning.

This presentation will present some common lessons as a priority for establishing improvement strategies as organizations strive for operational excellence. Available proven risk management methodologies and best practices for operational risk reduction will be summarized. Given that the global industry has made remarkable progress toward developing proven solutions aimed at prevention, organizations must now commit to accelerating the adoption of these solutions before at-risk operations experience near future incidents.

The focus of the paper will be the challenge to accept a leadership position to resolve existing but perhaps unidentified risks, then to commit to action to proactively identify hazards and their associated risks, and implement risk reduction measures before losses occur. A framework of a robust risk-based management system, incorporating all of the necessary elements based on lessons learned and including supporting leadership and management processes to achieve risk reduction performance objectives, will be presented.

Attendees will benefit from practical takeaways of how each management system element can be developed with "leadership designed in" to control the process safety risks at three opportune times: before the incident (prevention), during the incident (emergency management) and after the incident (incident investigation, root cause analysis and lessons learned).

The presentation will pose a challenging question: "Knowing what we already know, why would we want to take years to gradually reduce our risks, when we could accelerate the learning from past lessons and prevent incidents using available best industry practices"?

This paper will draw upon the authors' extensive experience in designing, developing and implementing PSM systems.

441698 Analytics to Reveal Meaurable Factors to Process Safety Performance

Abstract:

Analytics on a Large Industry Data Set Reveal the Leadership and Organizational Factors that Drive Down Risk and Reduce Incidents

With access to a multi-year data set of over a million incidents from 1000's of operating facilities worldwide in the Energy and Chemicals industry, the research team mined the data set to explore the differences in risk management practices and behaviors which most effectively explain the variation in process safety outcome performance. By modeling and examining predictive relationships, the study identifies four attributes - worker engagement, process discipline, risk sensitivity, and learning-mindedness – as critical for reducing incidents.

Findings from this study support a progressive and integrated approach and suggest that companies that are proficient in driving process safety performance are companies that, to a greater degree, have successfully engaged their workforce, strengthened their operating discipline, sharpened their risk sensitivity, and become learning-minded. A learning-minded organization manages each incident and responds to each failure deliberately to identify and fully leverage improvement and learning opportunities.

While this study provides statistical evidence to support the strong correlation between leadership, organizational behavior attributes and safety performance, it also emphasizes the organizational factors that are necessary to enable and sustain effective risk mitigation efforts. By deriving these insights from a data set of routine field-level activities, operational measurements may be created to measure progress and drive focus on engagement, discipline, risks, and learning.

This paper will conclude by sharing examples of how companies have applied these findings to establish quarterly and monthly measurements to drive leadership and organizational focus in these critical areas and ultimately to improve outcome performance.

441960 Coupling Dynamic Pressure Relief System Design with Experimental Vent Flow Regime of Dicumyl Peroxide / Toluene System

Abstract:

Abstract

A safe process design requires knowledge of the chemical reactivity of the desired reactions as well as the undesired reactions due to process upset conditions. Techniques used for sizing pressure relief systems for reactive systems under thermal runaway conditions include direct empirical scaling of experimental data obtained from a vessel with very low thermal inertia, analytical methods, and computer simulations using reaction kinetic models developed from adiabatic calorimetry data.

Pressure relief size estimation is based on the required flow capacity through the relief device at a specific set pressure, in order to keep the pressure in the vessel below the maximum allowable accumulated pressure. The flow capacity of the relieving device is highly dependent on the quality of the vent flow regime, whether it is liquid, all-vapor/gas, or two-phase. This requires knowledge of the physical properties of the contents inside the vessel at the time of venting. For a reactive system, these properties can be difficult to determine.

In this paper, the pressure relief size of a reactive system containing 50% dicumyl peroxide in toluene will be estimated using the dynamic simulation method. The vent flow regime of the reactive system will be determined from measured data, according to the methods described in the DIERS (Design Institute of Emergency Relief Systems) project manual. An automatic pressure tracking adiabatic calorimeter (APTAC^{TM}), which is a low thermal inertia calorimeter, will be used to conduct vented tests. The pressure relief size estimates of the above dynamic simulation method will be presented and compared with the DIERS simplified method.

441701 We Did a Quantitative Risk Assessment Study. Now What Do We Do with It?

Abstract:

Quantitative Risk Assessment studies have been extensively used in risk based regulatory regimes and they are becoming more common as a replacement for facility siting requirements in Process Safety Management (PSM) covered regulatory regimes and as a corporate risk management tool. Besides assessing risks to society and personnel, it can also assist in practical risk reduction implementation in design projects and operating assets. This paper intends to discuss these practical risk reduction implementation to show value to projects and operating assets, and how implementation can be achieved to using specific examples. The following examples will be discussed in the paper and will outline a broader set of QRA implementation improvements:

- Operating modes input to isolatable inventories and failure criteria
- Failure mechanisms from Hazards and Operability studies, and corrosion or mechanical integrity surveys
- Survivability criteria for manned and unmanned buildings for upgrades or siting challenges
- Specialized consequence analysis such as uncommon release material, indoor releases, and elevated receptors.
- Spill containment application
- Thermal protection and passive fire protection application
- Personnel and structural cryogenic protection application
- Risk based fire and gas detection mapping
- Escape, evacuation and rescue impacts
- Operational risk reduction using performance standards

The objective of this paper would be to better understand how to utilize QRA studies by improving their content and by practical implementation (as necessary) of mitigating risk reduction measures in design projects and to existing assets.

442063 Effective Implementation of PSM to Reduce Operational Risks

Abstract:

Ultimate success of process safety management (PSM) is measured through effective operational risk reduction. Usually, there is a lot of emphasis on putting the various elements of PSM in place through adoption of established international standards and best practices. Many of these standards and best practices are very well documented and element leaders/champions are appointed. Unfortunately, many of these elements and element champions are in silos and thereby not contributing to the overall reduction in operational risk. The risk is not reduced until the documentation is converted into reality on the ground.

Also, there is a lot of emphasis on doing high quality technical studies and quantitative process hazards analysis but very little is done to follow up on implementing the recommendations and results of these studies. The operational risks are not reduced until the recommendations are actually implemented. So, despite significant amount of effort, the improvement in the effectiveness of PSM implementation is limited.

Learning from this experience, SCG Chemicals has adopted a practical, bi-focal approach that is driven with a specific purpose in mind — operational risk reduction. The short term focus was to address deviations from design intent and established procedures in the field. If these items are not addressed today, they can contribute to a potential catastrophic incident tomorrow. This enhanced operational discipline and prevented 'normalization of deviations'. The longer term approach focuses on redoing structured PHAs and using the risk matrix to establish the levels of acceptable/unacceptable risks. Management decisions were made to address priority of the unacceptable risks with a rigorous follow up on implementing recommendations. This process has contributed to better outcomes in increased effectiveness of PSM while optimizing the money, time, and effort required. This paper will share the experience of SCG, along with DuPont as a partner, and the journey in implementing PSM effectively to reduce operational risks.

442151 Probit Models for the Determination of Damage Probability of Storage Tanks Due to Flooding Impact

Abstract:

Global climate changes may significantly influence temperatures and precipitation patterns, increasing the frequency of severe meteorological events which may lead to flooding. Flooding may trigger severe accidents on population among urban as it may do on chemical process industries too, causing the damage of equipment and the loss of containment of hazardous materials which might transform into toxic dispersions, fires, explosions or severe environmental contamination. This type of events are known as Natech events (Natural-Technological). This type of events may lead to extremely severe consequences bringing up the need to develop approaches for the risk assessment of Natech scenarios. Recently important efforts have been dedicated to develop tools that include Natech events in the quantitative risk assessment framework. In order to allow a quantitative risk assessment of a Natech event, a key point is the definition of tools for the estimation of equipment damage probability on the basis of intensity parameters of flooding. Past accident data analysis evidenced that horizontal and vertical storage tanks were often damaged in Natech events triggered by floods. Because of this the present study presents specific probit models for the damage probability of horizontal and vertical storage tanks due to flooding intensity parameters (flooding water height and flooding water).

Probit models in this study were derived on the basis of a mechanical model used for the assessment of the vulnerability of storage tanks due to the impact of flooding. For the determination of damage probability Montecarlo simulations were applied over the mechanical model in order to obtain a more accurate value and a probabilistic distribution of this parameter in dependence of the flood impact vector. Once the damage probability distribution is determined the probit method was applied using a logarithm approach with the impact flood vector (water height and water velocity) as the dose variables and the probability of damage as the response variable. A logarithm regression using the minimum square method was utilized to determine the probit constants for the model. Since the mechanical model used as basis has a huge dependence on the geometry of the storage tank being studied, it was find the damage probability will be affected and its value will change. To make sure this affections will be taken into account in the probit modeling a relationship between the geometry of the tanks and the probit constants was established using a polynomic regression. Once again the minimum square method was used and these two variables were related using a second order polynomic approach. Probit constants were determined for two substances (gasoline and LPG) and two types of storages tanks (vertical and horizontal).

The results of this study are a useful tool for the development of the quantitative risk assessment of Natech events due to flood impacts in storage parks. Also contributes to the reduction of the uncertainty in the framework for the QRA of Natech events triggered by floods.

442178 Analysis of Fire Extinguishing Methods in Engine Control Rooms

Abstract:

The study of electrical fires is a major concern for fire safety in the industry and more particularly for oil industry. In petrochemical facilities, the fire hazard may cause severe consequences on safety functions inside the fire compartment and furthermore may lead to the loss of one or several compartments, as just as structural damage on the facilities. From the OECD Fire Database a majority of the fire ignition is due to electrical sources (close to 50%, with a significant contribution attributed to electrical or electronic cabinets, with an occurrence of more than 10% of all fire events). Because of the above mentioned, it is necessary to implement tools to study these phenomena without using an experimental method. For example, the simulation of the dynamics of combustion and fire spread, allows an estimation of the effects on the structure of the case study preemptively. This project addresses the comparison of different fire protection philosophies in engine control rooms, using Fire Dynamics Simulator as support tool. The project studies three different cases, one of them without a protection system, another a clean agent as extinguishing tool and a final one, which implements an inherently safe design. For the first case study, the obtained data for the room temperature and the heat release ratio indicate a potential risk for the personnel and the equipment of the facility. It is expected that the study of the remaining cases show the significant advantages of each fire extinguishing philosophy.

12TH GLOBAL CONGRESS ON PROCESS SAFETY

The Study of HAZOP Revalidation Approach

Jing Xu
SINOPEC Research Institute of Safety Engineering
No.218, Yan'an san Road,Qingdao,China
xuj.qday@sinopec.com

Chuntao Ge
SINOPEC Research Institute of Safety Engineering
No.218, Yan'an san Road,Qingdao,China

Yi Zhang
SINOPEC Research Institute of Safety Engineering
No.218, Yan'an san Road,Qingdao,China
Zhangyi.qday@sinopec.com

The Study of HAZOP Revalidation Approach

Jing Xu
SINOPEC Research Institute of Safety Engineering
No.218, Yan'an san Road,Qingdao,China
xuj.qday@sinopec.com

Chuntao Ge
SINOPEC Research Institute of Safety Engineering

Yi Zhang
SINOPEC Research Institute of Safety Engineering

Keywords: HAZOP, HAZOP revalidation, update HAZOP,PHA, MOC

Abstract

Abstract: HAZOP evaluation method is used to analyze the risk of the process by the worldwide chemical enterprises and design units. However, Standard Specification for many countries stipulate the enterprise needs to carry on the process hazard analysis once again (revalidating) in several years later. The goal of periodic revalidation is to identify the cumulative changes of risk over time and to improve understanding of risk. The ultimate goal is to avoid accidents. But there is no systematic approach to guide the HAZOP revalidation at present. If we redo HAZOP blindly (performing a new HAZOP similar to the initial one) will only be time-consuming. This paper introduces the concept of HAZOP revalidation which is based on the analysis of the initial HAZOP and discusses many of the factors that need to be considered for the revalidation such as the change management, accident analysis and HAZOP report quality. At the same time, the specific procedure and approach of the HAZOP revalidation are proposed. The panel of experts scoring method is used to determine the revalidating approach and then decides whether to redo or update the initial HAZOP. If the scoring results suggest redo the HAZOP, and then conduct the Hazard and Operability Analysis in accordance with the complete HAZOP procedures; if the scoring results suggest update the initial HAZOP and then modify and supplement previous HAZOP analysis results to meet enterprise current status. This approach is not only time-saving and more targeted which can guide enterprises to conduct better and faster HAZOP revalidation.

1 Introduction

Hazard and Operability Study (HAZOP) is a kind of process safety analysis methods which is invented by Britain's imperial chemical industries (ICI) in the 1960s[1]. It is carried out by a suitably experienced multi-disciplinary team (HAZOP team) during a set of meetings to identify potential risk and operational problems in the system.

However, the safety status of enterprise is not invariable after the HAZOP.Many changes may occur in the later production process inevitably. The safety status of enterprise will change if the technological process, equipment, safety equipment, operating procedures, personnel, organization structure, and environment changes in the entire life cycle. So getting through the HAZOP analysis is not once for all. HAZOP Revalidation is to conduct a comprehensive evaluating for process hazards situation of the system during the next few years after the first time HAZOP analysis. To eva`luate the validity of the analysis results and determine whether meet the requirements of laws and regulations,and improve the process safety management level of enterprises at the same time. The goal of regular revalidation is to identify the change of the risk as time goes on and to improve the understanding of the risk. The ultimate goal is to avoid the accident. However, HAZOP analysis of a typical chemical process takes a few weeks to a few months. When in the HAZOP revalidation, whether can take advantage of the existing HAZOP analysis report to make the HAZOP revalidation more simplified and efficient and guarantee the results conform to the enterprise current situation. More and more companies try to simplify the method of revalidation, rather than redo HAZOP blindly[2].

2 Status analysis of HAZOP revalidation

HAZOP technology has developed more than 50 years getting a lot of project application, and become one of the general evaluation method on process safety domain at present. Petrochemical companies and design companies worldwide use HAZOP method to analyze the risk of petrochemical process. Such as Dow chemical, Rhodia company, BP, Bayer, Royal DSM, SINOPEC, CNPC, CNOOC and so on[3].

OSHA(the Occupational Safety & Health Administration) put forward this revalidation concept at Process Safety Management in 1992. Claim that “After the first process hazard analysis, revalidation should be done at least every five years, to ensure consistency of process hazard analysis and the current process” [4]. In the risk management procedures of EPA(Environmental Protection Agency） and aEuropean SEVESO Ⅱ have corresponding requirements in the instruction. CCPS(Center of Chemical Process Safety) published a book Revalidating process hazard analyses Defined that the revalidation process attempts to protect this investment by identifying and building upon the still pertinent portions of the prior PHA .Corrections are made and new content is added to the PHA as required,and the result are documented to serve as the new PHA[5]. In the United States, the large-scale petrochemical enterprises have done 2 ~ 3 times HAZOP, identifying the potential risk in the system, found a lot of potential safety problems. The study of HAZOP technology started late in China. Now a lot of enterprises done HAZOP before will conduct HAZOP revalidation. Guidelines for process safety management of petrochemical corporations (AQ/T3034-2010) standard requires: Guidelines for process safety management of petrochemical corporations (AQ/T3034-2010) standard requires: Companies should conduct a process hazard analysis in the process during construction and should to finish the process hazard analysis to confirm again and updated every three year[6]. The difference between Revalidation and first HAZOP is the former is conduct on the basis of the practical operating experience. The key for the first time HAZOP is to perfect the design and the retrieval permissions[7]. HAZOP Revalidation is to compare the results of the first HAZOP whether still applicable to the current environment, more dependent on site operation actual situation and record filing, etc. There is no systematic method to guide how to HAZOP review again. However, blindly redo HAZOP is not only time-consuming, also need a lot of resources. More and more companies expect to revalidate on the basis of the first HAZOP analysis.It can be more simple and efficient, and at the same time guarantee the quality of process hazards analysis.

3 The Study of HAZOP Revalidation Approach

3.1 The key factor of HAZOP revalidation

HAZOP revalidation using redo or update way is decided by many factors, such as whether there is a big defect of the previous process hazard analysis, whether change management system is perfect, whether there is a major accident happened, whether the information is complete, new standards, etc.

The way for HAZOP revalidation depends on the following three aspects：

a. The design change number, the size of the problem, process complexity, change process hazard analysis quality and management of change since the last HAZOP analysis；

b. The quality of Previous HAZOP analysis, including process safety information comprehensiveness, the comprehensiveness of HAZOP analysis, HAZOP record content integrity, HAZOP results application effect, etc；

c. Previous accidents and device status: the type of process safety accidents, statistical accident frequency, Consequences severity and the reason for the accident , etc.

3.2 The process of HAZOP revalidation

a. Establish revalidation team, the composition of team members are as follows[8]:

- the group leader: trained, rich experience in organization HAZOP analysis; Responsible for communication between HAZOP team and project management personnel; Review plan
- Process equipment technician: familiar with the past project, past and present design/operational problems and change the content
- Site operator: familiar with the situation, the operation experience
- owner (user) : familiar with the operating environment, the consequences of deviation and the degree of danger of deviation
- Experts: providing related professional knowledge. Technology, automatic control experts shall be invited to assist analysis group for part analysis
- Recorder: review record, record the related changes, and update or supplement HAZOP records, to assist the analysis team leader for planning
- Maintenance personnel (if required)

b.Collect information, including the HAZOP report, change management (MOC) data, the previous accident report, HAZOP recommendations, process engineering and safety information, operation procedures, etc

c.Expert assessment team scoring method. Details see Table 1 - the marking table for determining HAZOP revalidation methods and Table 2 - HAZOP revalidation method
d.According to the examination way to implement. Redo HAZOP or update HAZOP.

The flow to determine the specific revalidation way is shown in figure 1

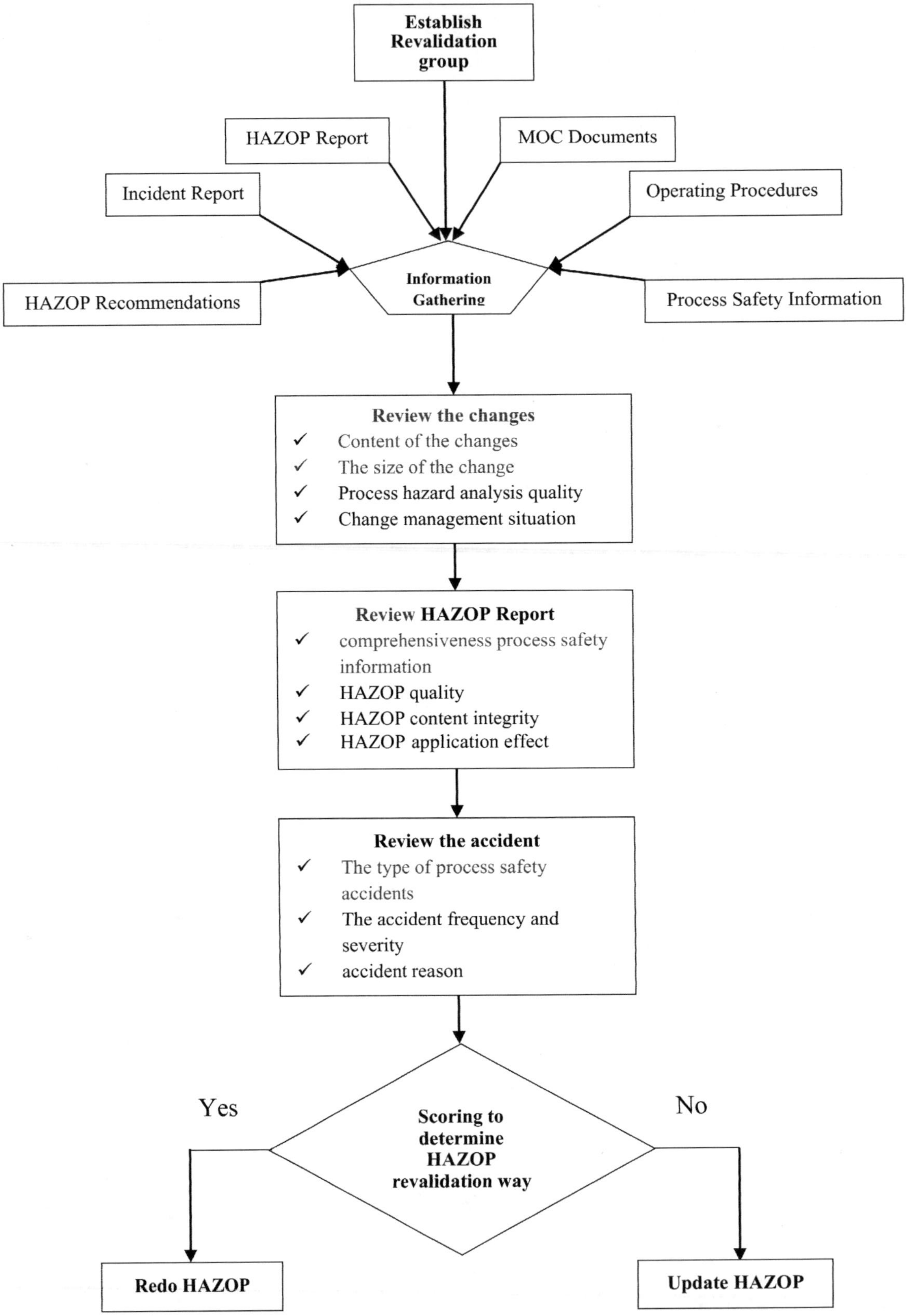

Fig.1 HAZOP revalidation approach decision diagram

Redo HAZOP is not introduced in the paper, the most effective way to update the HAZOP is to collect all data and accident information related to the change, to review the preliminary HAZOP Suggestions, to view the process pipeline flow chart meters (PID), understanding the changes and the defects of the system[9]. Such as process flow, equipment, safety equipment, operating procedures, operating conditions, location and environmental changes, and keep record by the recorder. According to the situation of change size, change management and change of process hazards analysis quality lead revalidation group to carry out different levels HAZOP for change unit by the team leader. Update the HAZOP results that conform to the enterprise the status quo. Update the HAZOP is time saving and maximum use of existing resources compared to redo HAZOP.

Table 1. HAZOP revalidation reference score table

Review items		**Index description**	**Reference scores**
Change Review	Change content (max)	Change of Process/tank type/operating procedure	15*
		Change of safety facilities	10
		Change of pipeline, pump, valve	5
		No change	0*
	Scale of change	Change of working hours more than one month/ 5 types of job post	10*
		Change of working hours about 2 weeks/3 types of job post	5
		Change of working hours about one week/2 types of job post	2
		No change	0*
	Quality of hazard analysis of change	No hazard analysis of change	10*
		Bad hazard analysis of change(lack of pertinence, record, etc)	5
		Good hazard analysis of change(More targeted, detailed record)	2
		Eexcellent hazard analysis of change(effective method, comprehensive analysis)	0*
	Management of change	Records and files are deficient and not combined with operating procedure or job training partly after change	5*
		Records and files are complete and combined with operating procedure or job training partly after change	2
		Complete and electronic records and files,which combined with operating procedure or job training after change	0*
HAZOP quality review	Comprehensiveness of process safety information	Deficient data or big change of site layout	5*
		Partly complete data or small change of site layout	2
		Complete data and less or no change of site layout	0*

	Integrity of HAZOP record	Deficient node description, design intent, drawing name, date, participant and the content of the standard form, bad reference	5*
		Partly complete node description, design intent, drawing name, date, participant and the content of the standard form, good reference	2
		Complete node description, design intent, drawing name, date, participant and the content of the standard form, wonderful reference	0*
	Rationality of HAZOP	Deficient deviation and guide word, unreasonable nodes, bad suggestions and measures	15*
		Partly complete deviations and guide words, reasonable nodes, common suggestions and measures	10
		Complete deviations and guide words, more reasonable nodes, good suggestions and measures	5
		Perfect deviations and guide words, most reasonable nodes, appropriate suggestions and measures	0*
	Application effect	Weak effect, less use of risk identification	5*
		Good effect, general use of risk identification	2
		Perfect effect, effectively use of risk identification	0*
Review of accidents		Vast or serious process safety accidents	30*
		Some specific process safety accidents	15
		Some accidents which has nothing to do with the process	5
		No accidents	0*
* reference point， according to the actual situation within the range of its rated arbitrary integer (interval)			

Table 2. HAZOP revalidation approach determination table

Scores	Review way
⩾60	Redo HAZOP
<60	Update HAZOP

4 Application examples

A spherical tank area of PetroChina has four 3000 m3 spherical tank, two 5000 m3 spherical tank, storage for butene 1, butadiene. The company has completed the HAZOP analysis in 2010. Device in service conduct HAZOP analysis every four years according to the company policy. HAZOP revalidation use the proposed method, completed HAZOP analysis efficiently.

Table 3. Application of HAZOP revalidation approach determination table

Review items		practical situation	Scores
Change Review	Change content (max)	No change	0
	Scale of change	No change	0
	Quality of hazard analysis of change	Good hazard analysis of change(More targeted, detailed record)	2
	Management of change	No change	0
HAZOP quality review	Comprehensiveness of process safety information	Partly complete data or small change of site layout	2
	Integrity of HAZOP record	Perfect deviations and guide words, most reasonable nodes, appropriate suggestions and measures	0
	Rationality of HAZOP	Complete deviations and guide words, more reasonable nodes, good suggestions and measures	5
	Application effect	Good effect, general use of risk identification	2
Review of accidents		No accidents	0
Scores			11

5 Conclusion

This paper introduces the concept of HAZOP revalidation which is based on the analysis of the initial HAZOP and discusses many of the factors that need to be considered for the revalidation such as the change management, accident analysis and HAZOP report quality. At the same time, the specific procedure and approach of the HAZOP revalidation are proposed. The panel of experts scoring method is used to determine the revalidating approach and then decides whether to redo or update the initial HAZOP. If the scoring results suggest redo the HAZOP, and then conduct the Hazard and Operability Analysis in accordance with the complete HAZOP procedures; if the scoring results suggest update the initial HAZOP and then modify and supplement previous HAZOP analysis results to meet enterprise current status. The method has been applied for a spherical tank area of PetroChina. Conclusion is spherical tank area further implement the last countermeasures of HAZOP in view of the actual operation situation, do not need the modified or added. To satisfy the needs of the enterprise for better and faster process hazards analysis.

6 References

[1]Dunjó J, Fthenakis V, Vílchez J A, et al. Hazard and operability (HAZOP) analysis. A literature review[J]. Journal of Hazardous Materials, 2010,173(1):19-32.

[2]Smith K E, Whittle D K. Six steps to effectively update and revalidate PHAs[J]. Chemical engineering progress, 2001,97(1):70-88.

[3]WAN Gujun, LI Qi, ZHANG Guangwen, et al. Implementing HAZOP to Control Major Accidents (Ⅰ)[J]. Safety Health & Environment, 2009,9(011):2-4.

[4]ZHAO Jianmin. Discussion About HAZOP Revalidation [J]. Modern Occupational Safety, 2012(134):83-85.

[5]L. F W, K. W D. Revalidating process hazard analyses[M]. Center for Chemical Process Safety, 2001.

[6]State Administration of Work Safety. AQ/T 3034-2010 Guidelines for process safety management of Petrochemical corporations[S]. 2010.

[7]FU Jianmin, ZHAO Dongfeng, CHEN Guoming, et al. Quantitative study over the probability of HAZOP analyzing technique in petro-chemical plants[J].Journal of Safety and Environment,2008,8(6):130-134.

[8]Herbert I. Learning the lessons-retrospective HAZOPs[J]. Offshore Europe, 2011.

[9]Wagner T, Champion J. A work process for revalidating LOPAs and other risk analyses[J]. Process Safety Progress, 2012,31(2):122-129.

The Importance of Process Safety:

A Young Engineer's Perspective

Waheed Wakil
Smith & Burgess
Waheed.Wakil@smithburgess.com

Travis Bruner
Smith & Burgess
Travis.Bruner@smithburgess.com

Prepared for Presentation at
American Institute of Chemical Engineers
2016 Spring Meeting
12th Global Congress on Process Safety
Houston, Texas
April 11-23, 2016

The Importance of Process Safety:

A Young Engineer's Perspective

Waheed Wakil
Smith & Burgess
Waheed.Wakil@smithburgess.com

Travis Bruner
Smith & Burgess
Travis.Bruner@smithburgess.com

Keywords: Process Safety Management, Young Engineers, Overpressure, OSHA, LOPA

Abstract

Most young engineers experience a defining moment or moments in their career that shape their attitude towards importance of Process Safety Management (PSM). Experiencing a plant upset first-hand knowing that there were actual capital or human life impacts or both from an incident at the site, one soon realizes early in their engineering career why process safety is so important. It becomes evident that the most successful corporations are those who are willing to invest large capital costs into PSM documentation projects or projects that have come as a result of findings identified during a study, allowing them to prevent incidents from occurring. The findings generated during the course of a study may result in simple documentation update action items, or they may require a multi-million dollar investment. For young engineers, this exposure puts into perspective the value of their work, and emphasizes the importance of a thorough and adequate study.

1 Introduction

Many young engineers have a point early in his or her career when process safety hits home. Different companies develop their own unique strategies towards safety that they are comfortable with, which results in considerably different safety cultures within industry. Unfortunately, it is all too common for engineers to treat process safety as an afterthought, with priority given to efficiency and profitability. However, as we will see, treating safety with the utmost importance can prevent costly and life-threatening incidents.

2 First Hand Experiences

2.1 Organic Peroxides Plant Background

Two major upsets at a chemical plant within a year may very well serve as a defining "moment" in a young engineer's career that forces him to reflect on and realize the importance of dedication to process safety. A fellow Process Safety Consultant at Smith & Burgess who was a process engineer at a plant that produces organic peroxides was fortunate enough to live through two such accidents unscathed. His story is a lesson and reminder that even the most disastrous accidents can be prevented with due diligence, a safety-focused company culture, and proper attention to basic Process Safety Management.

Based on the nature of the chemicals produced, a Process Hazard Analysis for this particular plant would have revealed scenarios with severe consequences. However, the safety culture at this plant was not adequately focused in preventing and mitigating upsets. Not only were poor preventative maintenance practices used, but lack of written emergency procedures and poor training prevented adequate mitigating protection layers after the initiating events. A few relevant OSHA requirements are worth noting here as failure to meet these requirements was likely the reason for loss of containment:

- Per OSHA 29 CFR 1910.119(f)(1), "The employer shall develop and implement written operating procedures that provide clear instructions for safely conducting activities involved in each covered process consistent with the process safety information...".
- Furthermore, per OSHA 29 CFR 1910.119(j)(2), "The employer shall establish and implement written procedures to maintain the on-going integrity of process equipment."
- Lastly, per OSHA 29 CFR 1910.119(n), "The employer shall establish and implement an emergency action plan for the entire plant in accordance with the provisions of 29 CFR 1910.38. In addition, the emergency action plan shall include procedures for handling small releases."

If these basic elements of PSM would have been implemented at this plant, both accidents may have been prevented or mitigated.

Organic peroxides are highly unstable – their O-O bond breaks very easily, and that is why their decomposition properties are so useful in initiating polymerization reactions. This useful property can also be very dangerous. When an organic peroxide is in its pure form, the rate at which its exothermic decomposition generates heat is likely to be greater than the rate at which the heat can be dissipated to its surrounding environment if proper heat transfer equipment are disabled, resulting in a situation called self-accelerating decomposition in which the reaction continues to accelerate due to its own generated heat.

The organic peroxides produced in this particular plant are characterized by very low Self Accelerating Decomposition Temperatures (SADT), defined as the minimum temperature at which an organic peroxide would experience self-accelerating decomposition after one week of storage. As such, glycol cooling jackets on vessels and piping in this plant maintained operating temperatures in the ballpark of -30 °C (-22 °F) to prevent the possibility of self-accelerating decomposition. For this heat transfer to work properly, the organic peroxides must be continuously flowing in the process piping and equipment and cannot become stagnant. At the same time, the glycol that cools the organic peroxide must also continuously flow. Stoppage of flow for either component may be disastrous. As we will see in the following incident descriptions, loss of this cooling results in rapid increase of temperature and decomposition reaction rate, resulting in a deflagration that cannot be mitigated by a typical relief device such as a safety valve or a rupture disk. To prevent loss of containment and subsequent damage to surrounding structures, this scenario must instead be prevented or mitigated using written operating procedures, regular inspection and maintenance of equipment, and employee training.

2.2 Organic Peroxides Incident #1

What went wrong?

Organic peroxides are so thermally unstable that precautions must be taken when packaging and storing the material. Packaging must occur continuously to prevent the material from becoming stagnant in piping and the temperature must be monitored and controlled at or below the recommended storage temperature. If material remains stagnant without sufficient cooling, organic peroxides can reach their SADT temperature and undergo a violent reaction, which usually results in rupturing the packaging, dispersing peroxide, as well as discharging liquid and gaseous decomposition products considerable distances. The heat generated during this reaction has the potential to auto ignite flammable vapors.

An operator was transferring organic peroxides to 55 gallon drums, which are immediately transported by forklift to an icehouse for storage. An incident occurred which caused the operator to abandon the task in the packaging building to provide assistance elsewhere. While the operator was away, the product remained stagnant in the filter where glycol cooling was inefficient. This caused the material to exceed its SADT. The material began to decompose as temperature continued to increase. When the operator returned the material suddenly exploded propelling the head of the drum through the air. The remains of the drum were embedded in the ground and lights in the facility were shattered. The operator was behind a blast shield, but still suffered severely burnt hands and damage around the ear. Although no life threatening injuries were inflicted, the plant suffered considerable downtime and costly repairs to the filter, packaging equipment, and process piping.

How could it have been prevented?

Proper implementation of PSM standards likely could have prevented this incident from occurring - in particular, OSHA 29 CFR 1910.119(f)(1), mentioned previously. Documented operating procedures that clearly state not to leave the packaging unit unattended during operation to avoid stagnant organic peroxides, inefficient cooling, and the potential for explosion could have prevented (or, at the very least, limited) the disastrous effects of this incident.

2.3 Organic Peroxides Incident #2

What went wrong?

This plant utilizes large plug flow reactors that are jacket-cooled. On one particular day, the site was hit by a thunderstorm with wind speeds high enough to break a tree branch that would damage a power line, resulting in total loss of electricity to the plant. Electric power was vital to this plant as it kept peroxide flowing through the reactors and piping and it kept glycol flowing through the jacketing to keep the peroxide temperatures down to safe levels. The plant had a backup generator, but only a single operator was aware of the location of its switch. It just so happened that it was this operator's day off.

In the event of an emergency such as this, interlocks were in place to purge out residing organic peroxides in the equipment using control valves with actuators regulated by instrument air. Once glycol and peroxide flow was lost in the event of a power failure, the peroxides would rapidly exceed their SADT. The purging action would dump the peroxides into waste water drains that are gravity fed to open tanks, where it is much safer for the peroxides to reach their SADT. However, the instrument air that controlled the valves was supplied from air storage, and during the purging, the instrument air ran out. Nitrogen was used as backup for air during emergency situations, but due to poor inspection and maintenance for cost savings, there was damage to the backup nitrogen header that had gone unnoticed. This damage caused a leak in the header when it was exposed to the nitrogen pressure, rendering the header useless as there was not enough pressure to actuate the purge control valves properly. At this point, purging was halted. The catastrophic failure was now inevitable. With stagnant peroxides in piping and equipment, and no jacket cooling, fluid temperatures rapidly sored past the SADTs and eventually resulted in a deflagration. The reactor top heads failed first, and the building in which the reactors were located was designed to fail at the ceiling and the outer wall. The building and its contents were destroyed, and the plant suffered 6 months of downtime. Fortunately, operators were protected on the other side of a blast wall, and no one suffered major injuries.

How could it have been prevented?

Looking at the number of Independent Protection Layers (IPLs) that were in place at the time of failure, namely the operator's response to activate the backup generator and interlocks that functioned to purge out residing peroxides, there were not enough layers of protection in place to prevent a deflagration. Table 1 summarizes the highest possible credits that could be taken for each IPL based on CCPS's LOPA book [1].

Table 1: IPL Credit Summary

Initiating Event	Initiating Event Frequency	Number of IPL Credits Required for Category IV Consequence	Maximum Number of IPL Credits Available
Loss of Cooling	Frequency ≥ $1X10^{-2}$	**2.0**	Operator Response for Backup Generator 0.5
			SIL 1 SIF 1.0
			Total **1.5**

Like many such accidents, it is not uncommon for multiple independent failures to cascade into a catastrophic event, especially when a key to a successful emergency response from operators is proper training and a key to proper function of mechanical equipment is proper maintenance and inspection. To pinpoint PSM measures that would have likely prevented this accident, we can start by calling on OSHA 29 CFR 1910.119(n) referenced previously in this paper. The first broken link in the series of events leading up to the accident was the lack of training amongst operators with respect to activating the backup generator in the event of loss of power. This should have been a basic emergency response action that every operator at the plant should have been trained on and been aware of, but unfortunately, no one at the plant at the time of the failure was trained on how to activate the generator. The failures cascaded into the second phase of emergency response when the instrument air was exhausted at exactly the wrong time. To address the issue of the backup nitrogen header leak, we call upon OSHA 29 CFR 1910.119(j)(2), "The employer shall establish and implement written procedures to maintain the on-going integrity of process equipment." To cut costs, plant management had reduced regular inspection and maintenance of the backup nitrogen header, and it proved disastrous on this day when they needed it the most.

3 Conclusion

As illustrated in the incident descriptions, a poor safety culture is likely to result in poor preventative inspection and maintenance cycles and lack of clear emergency operating procedures and training. In most process plants, inadequacies in PSM elements can result in disastrous consequences. For a young engineer, it sometimes takes a real-life experience to change their perspective on PSM from one of passive compliance to one of active responsibility. Each element of PSM is important and critical, from the most obvious and simple, like written procedures on activating a backup generator, to the most expensive and extensive, like redesigning a flare system. It makes every detail in every process safety analysis that much more important and that much more rewarding when the benefits are apparent.

References

[1] CCPS. *Layer of Protection Analysis: Simplified Process Risk Assessment.* Center for Chemical Process Safety, American Institute of Chemical Engineers, New York, NY, 2001.

Scaling Up Safely: Making Smarter Decisions about Rate Dependency

Zubin Kumana, P.E.
Smith & Burgess LLC
7600 W. Tidwell Rd, Ste 600
Houston TX 77040
Zubin.Kumana@smithburgess.com

Venkata Badam, P.E.
Smith & Burgess LLC
Venkata.Badam@smithburgess.com

Achilles Arnaez
Smith & Burgess LLC
Achilles.Arnaez@smithburgess.com

Prepared for Presentation at
American Institute of Chemical Engineers
2016 Spring Meeting
12th Global Congress on Process Safety
Houston, Texas
April 11-23, 2016

Scaling Up Safely: Making Smarter Decisions about Rate Dependency

Zubin Kumana, P.E.
Smith & Burgess LLC
7600 W. Tidwell Rd, Ste 600
Houston TX 77040
Zubin.Kumana@smithburgess.com

Venkata Badam, P.E.
Smith & Burgess LLC
Venkata.Badam@smithburgess.com

Achilles Arnaez
Smith & Burgess LLC
Achilles.Arnaez@smithburgess.com

Keywords: Process Safety, Safe Design, and Venting and Emergency Relief

Abstract

Over the lifetime of a facility, it is almost a certainty that the throughput of a process will be increased, often repeatedly. Rate-dependency is frequently used as a way to identify pressure relief system sizing that would be affected by the scaling up of process throughput. This paper will define rate-dependency as it relates to pressure relief systems and describe the various ways in which rate-dependent classifications may be affected by the mechanical and thermodynamic constraints of the process and equipment. In addition, examples of when the "rate-dependent" designation may no longer apply due to common assumptions and methodological choices will be provided.

1 Introduction

In the refining and chemical processing industries, increasing throughput through a processing unit is often considered as a means of maximizing value with existing capital or minimal additional investment. Often, the throughput will be increased many times over the lifetime of the facility. The pressure relief system for a process is designed in part based on the throughput of the unit; therefore, the change in the throughput may affect the adequacy of the pressure relief system design.

Verifying the design of the process unit's pressure relief system for these increased rates is typically done by identifying overpressure scenarios as either rate-dependent or rate-independent. A rate-dependent overpressure scenario is one whose required relief rate varies with changes in process throughput. A common example is a blocked outlet scenario on a separator. With higher the process throughput, the quantity of mass that must be relieved to limit the pressure to the allowable accumulation is expected to increase.

Conversely, rate-independence refers to overpressure scenarios whose required relief rate is independent of process throughput. A common example is exterior fire, where the required relief rate depends on the heat of vaporization of the liquid mixture inside the vessel, the density of the vapor and liquid phases, and the wetted area of the vessel the fluid is residing in. Changing the feed rate to the vessel is not expected to result in a change to the required relief area for the exterior fire scenario. In some circles, these scenarios may also be termed "equipment-dependent". While the equipment and piping remains constant, the required relief rate remains constant. However, as rate scale-up increases, the equipment and piping may be changed to accommodate the additional capacity.

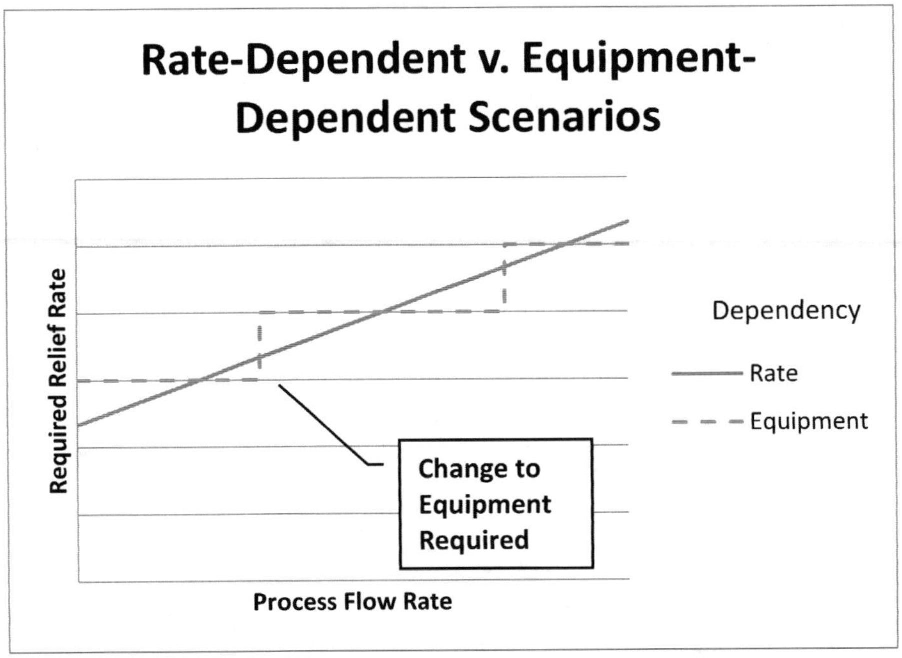

Figure 1: Rate-dependent scenarios increase in required relief rate (and therefore area) with process flow rate increase, while equipment-dependent scenarios are effectively constant until a change to equipment is required.

1.1 How Rate Dependency is used in Pressure Relief System Design

The common practice in pressure relief system design is to select a unit process basis that the user believes represents both the actual operation of the process as well as the worst case for the relief system design – specifically, the case that will result in the largest required areas and piping diameters. Since process units can have a range of operating modes and rates, selection of these cases tends to favor the following:

- Modes of operation with lighter feed, as this can result in more vaporization and therefore larger vapor and two-phase relief loads
- Higher process throughput rates; often the peak rate achieved, or some percentage above the peak rate

The conscientious user might separately specify a worst-case heavy liquid relief load basis in order to properly design the liquid handling capabilities of the disposal system.

As the pressure relief system design is completed, the engineer performing the design will typically identify overpressure scenarios as rate-dependent as the analysis proceeds. This designation is then used when the throughput rate is modified in the future. This allows engineers to minimize the amount of recalculation that needs to be performed, by only performing new calculations for the overpressure scenarios that are dependent on rates; the scenarios that had been identified as rate-independent are believed to be static and are left unchanged.

A generalized expression for pressure relief device sizing is given in API Std 520 Part I (9th ed.) Annex B, equation B.5, rearranged to solve for required area [1]:

$$A = \frac{W}{G\prod[K]} \qquad \text{(Eq. 1)}$$

Where

G	is the theoretical mass flux through the nozzle
W	is the mass flow through the PRV
A	is the effective discharge area
Π[K]	is the product of all applicable correction factors

The scale-up of the pressure relief system design is typically done using the assumption that the only change is to the flow rates through the unit; variables such as fluid composition, operating pressures and temperatures are assumed to remain constant. This effectively limits the relationship between the throughput rate and the relief system required areas to a linear one; as the fluid mass flux is considered constant, and the applicable correction factors are generally constant (exceptions are always possible based on Reynolds' number and actual backpressure resulting from flow), the values of A and W are directly proportional.

For the evaluation of the pressure relief system design, this has led to the following widespread practice:

1. The user takes the heat and material balance at the previous throughput rate
2. The user obtains a list of calculations that had been identified as rate-dependent
3. The user takes the percentage increase in flow and factors this value into the required relief rates
4. The user re-evaluates the pressure relief device requirements vs. capacity (along with other associated concerns where relevant, e.g. pressure losses if calculated at required relief rates)

This process rarely involves revalidating the process unit simulation to confirm the heat and material balance. Typically this practice has been followed for small changes to throughput rates. However, once the procedure has been implemented, the potential exists for repeated applications to increase the degree of deviation from a validated baseline even further. This poses a danger, as excessive extrapolation and repeated application of the scale-up process without carefully considered heat and material balance revalidation can quickly lead to larger and larger deviations, and the potential for undersizing the pressure relief system.

This paper focuses on rate scale-ups that do not involve any change to the as-built system. The rate scale-ups considered are therefore limited to the maximum throughput that the existing equipment can support. Debottlenecking the equipment in the process unit (rotating equipment, exchanger heat transfer area, and pipe sizing) may be required to accommodate further rate scale-up. As a capital project, this effort typically includes development of a new project baseline heat and material balance along with a review of the pressure relief system design.

2 Proper Identification of Rate-Dependency

2.1 How to Identify Rate-Dependency

Smarter decision-making with regard to rate dependency begins with understanding the ways that the traditional definition of rate-dependency may be invalidated by subsequent modeling choices in the determination of the required relief rate.

A general rule of thumb often passed around among novices in pressure relief system design is that overpressure scenario required relief rates that result from equations are rate-independent, and required relief rates that come from a heat and material balance (HMB) are rate-dependent. However, while it may be generally true for many facilities, there are aspects that are commonly overlooked in this approach, and some engineers may take subsequent steps in the required relief rate calculation that can change the classification. These additional steps are often taken as a means to "sharpen the pencil" when the initial determination of the required relief rate results in an undersized pressure relief device. These may include such actions as:

- Taking credit for an outlet that would not be blocked as a result of the scenario (including secondary causes like controller action and shut off systems)
- Reducing heat transfer into the system due to elevated bubble point temperatures
- Limiting flow rates based on the hydraulic limits of the piping, including valves and fittings
- Limiting flow rates to the maximum capacity of rotating equipment that can deliver the stated pressure at the point of relief

For example, if determining the required relief rate in the event of a blocked liquid outlet on an exchanger by taking the normal feed rate as the required relief rate is found to result in a required relief area greater than the currently installed relief area, the engineer may review the inputs and determine that when discharging against the relief pressure, the capacity of the feed pump is reduced below the normal feed rate, potentially making the existing installed relief area adequate. In this example, the overpressure scenario has changed from rate-dependent to equipment-dependent.

Table 1 shows common sources for determining the required relief rate for an overpressure scenario, and potential additional modeling choices that can change the classification.

Table 1. Examples of Dependency and Factors That May Change Classification†

Overpressure Scenario	Rate Commonly Derived From	Dependency for Common Derivation	Factors that may change Classification
Closed outlets	HMB/Simulation, Exchanger duty	Rate-Dependent	Limited by pump/compressor, valve, or pipe capacity
Reflux failure	HMB/Simulation, Exchanger duty	Rate-Dependent	
Cooling failure	HMB/Simulation, Exchanger duty	Rate-Dependent	
Overfilling	HMB	Rate-Dependent	Limited by pump or valve capacity
Inadvertent opening of manual valve	Line capacity	Equipment-Dependent	Credit taken for volume outflow
Failure open of inlet control valve	Valve capacity	Equipment-Dependent	Credit taken for normal minimum flow or outflow
Check valve failure	Valve or line capacity	Equipment-Dependent	
Abnormal heat input	Valve or exchanger capacity, HMB/Simulation	Rate-Dependent	
Exterior fire	Fluid properties, Vessel area	Equipment-Dependent	
Hydraulic expansion (pipe)	Fluid properties, Pipe area	Equipment-Dependent	
Hydraulic expansion (exchanger)	Fluid properties, Exchanger duty	Equipment-Dependent	Exchanger duty driven by hot side flow rate
Tube rupture	Exchanger properties, fluid properties	Equipment-Dependent	
Tube leak	Exchanger properties, fluid properties	Equipment-Dependent	
Plate weld failure	Exchanger properties, fluid properties	Equipment-Dependent	
Column power failure (Complex)	HMB/Simulation, Exchanger duty	Rate-Dependent	
Column power failure (Simple*)	Fluid properties, reboiler duty	Rate-Dependent	
Loss of Absorbent	HMB/Simulation	Rate-Dependent	

*Simple column failure includes overpressure scenarios where the required relief rate can be determined as a function of the reboiler duty and heat of vaporization of the residing fluid.
†Not a comprehensive list

3 Rate-Dependent Factors Influencing Pressure Relief System Design

Although the common practice is to size the pressure relief system as a linear function of process flow rate, the reality is more complex, as other factors that are themselves influenced by process flow rate can also impact the pressure relief system design, primarily through their impact on pressure relief device sizing. These factors can impact the pressure relief device sizing by causing additional deviations in relief rate calculations for rate-dependent overpressure scenarios (i.e., nonlinear relations), as well as creating changes in required relief rates for overpressure scenarios previously categorized as equipment-dependent.

A quick review of three factors influencing pressure relief device sizing will aid in understanding the limits of scaling up the pressure relief system design basis: composition, operating conditions, and heat exchanger duties. Note that this list is intended for illustrative purposes of the effects and is not an all-inclusive list; there are other factors that can influence the design of the pressure relief system that may change during a process rate scale-up.

3.1 Composition

3.1.1 Composition and Pressure Relief System Design

Fluid composition impacts the pressure relief device sizing in two significant ways.

1. Firstly, it can directly impact the determination of the required relief rate; for example, fire calculations use the latent heat of vaporization for a mixture, and depending on the particular modeling choices selected (e.g. inclusion of sensible heat, vaporization interval selected) may include further dependencies on heat capacity, bubble point temperature, vapor to liquid volume ratio, and other variables. Orifice flow calculations (such as are found in tube rupture calculations) depend on the density or molecular weight, compressibility, and specific heat ratio for the fluid.

2. Secondly, the composition of the fluid mixture directly impacts the relationship between the pressure of the fluid and the specific volume of the fluid, which governs the determination of the critical mass flux across the relief device nozzle (G in equation 1).

3.1.2 How Scaling Up the Process Can Affect Composition

It has been commonly observed that the compositional data in a heat and material balance is rarely adjusted when a scale-up of the unit throughput is performed. This assumption can be invalidated by any of the following effects:

1. Changes in the operating pressure and temperature – Changes in operating temperature and pressure can have an impact on the composition by affecting the separation of the vapor and liquid phases within the process equipment. Increasing the operating pressure can raise the bubble point temperature of the fluid mixture, requiring additional heat to achieve the same separation point. Decreasing the operating pressure can reduce the bubble point temperature of the fluid mixture, possibly resulting in more vapor flashing off, with a higher molecular weight.

2. Limitations on feed sources – In order to maximize the throughput for the specific process unit, feeds from multiple sources may be required. If the process unit has excess capacity but insufficient feed is available from on-site units, feed may be supplemented from offsite or from tankage. These feeds may have variations in compositions, and as the new feed is added, the blend ratio can change. For example, a gas oil hydrotreating unit (GOHT) may have additional capacity, but due to changes in product demand the crude column system may be producing a higher proportion of lighter products. To take advantage of the slack

in the GOHT, additional gas oil may be imported, resulting in a change in composition of the feed.

3.2 Operating Conditions

3.2.1 Operating Conditions and Pressure Relief System Design

The operating conditions within equipment in the process unit can directly impact the pressure relief device sizing primarily through the temperature's effect on the fluid conditions (for overpressure scenarios that model the required relief rate using its operating temperature). Temperature also influences the maximum mass flux across the pressure relief device. Since the relief device is sized at the relief pressure, there is no similar direct effect of operating pressure.

More significantly, changes in the normal operating temperature and pressure can affect the required relief rate calculations for overpressure scenarios that require these values as inputs, such as modeling flow across an orifice, flow through piping or valves, or where separation processes are occurring. For example, inadvertent opening of a valve into a separator may involve flow of the fluid mixture across a valve and series of pipe fittings, followed by a separation where the vapor is expected to overpressure the downstream vessel, while the liquid will slowly fill the vessel. The change in overall vapor quality can both affect the required rate to be relieved while the vessel is filling as well as affect the available time for operator intervention – in some cases making the required relief rate better, and in other cases making it worse. Another example is the potential for changes in composition, temperature, or pressure to change the net positive suction head available (NPSHA) for pumps, requiring adjustments to liquid levels in order to maintain proper pump operation, thereby affecting equipment-dependent overpressure scenarios.

It is even conceivable that such a change may alter the applicability of an overpressure scenario, where the existing operating conditions are already near the threshold of applicability – from either direction. Changing the operating pressure may affect the suction pressure of a pump or compressor, thereby altering the deliverable pressure to a downstream system, whose determination of applicability for a control valve failure or inadvertent opening of a manual valve overpressure scenario may depend on the comparison of this pressure to the downstream system maximum allowable working pressure.

Changes to operating conditions can also affect the specification of the pressure relief device. Operating pressures that were previously at the limit of the acceptable range that may now exceed the operating limits of the pressure relief device may require alternative relief device specification (e.g., if the operating pressure exceeds 90% of the PRV set pressure, this may require replacing a conventional PRV with a pilot operated PRV). Significant changes to operating temperature may also affect the set pressure of the pressure relief device. For a spring loaded valve, a new cold differential set pressure may be required.

3.2.2 How Scaling Up the Process Can Affect Operating Conditions

The operating conditions can be impacted in the following ways during a scale-up:

1. The mechanical characteristics of rotary equipment – For example, for a given installed impeller diameter, as the flow rate is increased, the operating point of the rotating equipment must move out on the pump performance curve, resulting in a lower delivered pressure. (This can be countered by changing out the pump impeller, but that is outside the scope of this paper).
2. Piping hydraulic limitations – As flowrate is increased, the pressure drop is expected to increase. With changes in the pressure drop between equipment, the operating pressure profile may be affected. For flow in the turbulent region, for the same diameter of piping, the pressure drop normally follows a trend of

$$\frac{F_1}{F_2}\frac{v_1}{v_2} = \frac{Q_1}{Q_2} = \sqrt{\frac{\Delta P_1}{\Delta P_2}} \quad \text{(Eq. 2)}$$

   Where

F	is the mass flow rate
v	is the specific volume of the fluid
Q	is the volumetric flow rate of the fluid
ΔP	is the pressure drop through the pipe

3. Heat transfer – As flowrates change, when running up against the limits of the heat transfer equipment, changes in temperature and mass quality may result. For example, applying the same duty to a fluid entering at a higher feed rate will soak up more sensible heat and result in a smaller vapor fraction exiting the equipment.

3.3 Heat Exchanger Duties

3.3.1 Heat Exchanger Duties and Pressure Relief System Design

Heat exchanger duties can potentially cause deviations in the relief device sizing basis if the process duties are used as inputs in determining the required relief rates. Note that some users choose to use the design duties of the exchangers in lieu of the process duties. This may mitigate the impact of scale-up (up to the point where new heat transfer equipment is required). Common examples of overpressure scenarios impacted by this factor are blocked outlets on heat exchangers, thermal expansion, and simplified column calculations (i.e. boil-up).

The other route through which heat exchanger duties can impact the pressure relief system sizing is by impacting the operating conditions of the process (see section 3.2.1).

3.3.2 How Scaling Up the Process Can Affect Heat Exchanger Duties

Scaling up the process flow rate through an exchanger can significantly impact the exchanger heat duty if it is limited by flow instead of limited by heat transfer area and overall heat transfer coefficient ("UA-limited"). For the same heat exchanger, the flow pattern correction factor F_T is expected to be constant. In these cases an increase in process fluid flow can result in increased duty, up to the point where the exchanger UA becomes limiting. Note that for forced convection, since the Nusselt number is a function of the Reynolds number, the value for the overall heat transfer coefficient U is not constant; rather, it is itself a weak function of flow rate (as well as composition and operating conditions). If the exchanger duty is limited by UA, then the exchanger heat transfer duty can be still changed by differences in log mean temperature difference (ΔT_{lm}). The temperatures changes may come about as described in section 3.2.2.

3.4 Extent of Deviations and Overall Impact on Pressure Relief System Design

The above discussions have focused on how rate scale-ups can impact the pressure relief system design in ways beyond the typically assumed linear relationship between mass flow rate and required relief area, as well as impacting scenarios that might otherwise not be classified as rate-dependent. While it can be expected that minor changes in flowrate would have a correspondingly minor impact on composition, temperature and pressure, and heat exchanger duties, and therefore a minor impact on the pressure relief system design, it is important to recognize that the extent of deviation from linear behavior can increase with the extent of the process throughput scale-up. Additionally, repeatedly performing scale-ups on a previously scaled-up process, without revalidating the heat and material balance, can lead to the same result in a less obvious manner.

4 Examples

4.1.1 Examples of Compositional Impact on Pressure Relief System Design

For equipment and systems where relief rate is determined by calculating a rate of vaporization, the relief rate typically increases as the proportion of lighter components within the feed increases. Even where the relief rate is not determined by calculating a rate of vaporization, a lighter feed may not change the required relief rate mass flowrate but might instead decrease the relief device capacity, on a mass basis. For example, crude units can be operated with significant variations in the crude slate. These changes in feed composition could cause different effects for relief device design within the crude unit, even with no change in overall unit throughput.

The sizing scenarios for crude column systems commonly include partial power failures that result in continuing crude feed and heat input from the fired heater, with loss of all overhead and pumparound cooling, among other effects (such as backflow and desalter water carryover). In these scenarios, the relief stream is based on the lighter portion of the crude feed, which typically consists of the offgas, liquefied petroleum gas (LPG), naphthas, kerosene, and/or diesel, etc., depending on the heating capability of the feed heater relative to the feed. Therefore, a change in crude slate that increases the proportion of naphtha cuts in the crude while decreasing the proportion of atmospheric gas oil (AGO) cuts can result in a direct increase on the relief rate of the crude column, whereas a change that only changes the heavier portions of the crude feed may not have a significant effect on the crude column. Note that composition changes that do not affect the crude column may affect another system, such as the vacuum column.

4.1.2 Examples of Changes in Operating Conditions on Pressure Relief System Design

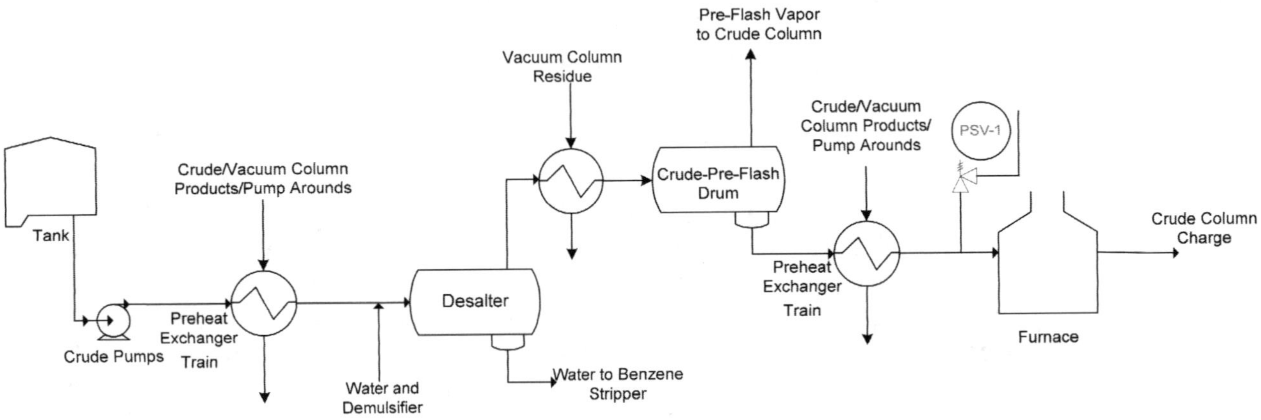

Figure 2: Crude Feed Pre-heating System Process Flow Diagram

The process flow diagram in Figure 2 represents the feed pre-heating section of a Crude Column system. Feed from tankage is pre-heated in a series of exchangers against hot fluid from Crude/Vacuum Column products and pump arounds; the crude feed is then treated for salt content in the Desalter before sending to the Crude Pre-Flash Drum. The Crude Pre-Flash Drum operates significantly lower than the Desalter, allowing the crude to flash. The flashed vapor is directly sent to Crude Column, while the liquid from the drum is sent through another train of exchangers, the furnace, and then to the Crude Column.

a. Effect of Operating Pressure Change on Composition

PSV-1 on the Flashed Crude pre-exchangers is sized for Blocked Outlet, and this case is usually selected as rate-dependent. However, a change in operating pressure of the Crude Pre-Flash Drum, with or without the throughput change, affects the composition of flashed liquid and vapor at the Crude Pre-Flash Drum, because the bubble point of the crude stream components is directly related to the operating pressure. In this case, as the fluid composition is changed, an assumption of a linear relationship between changes in throughput and required area may not be accurate.

b. Effect of Operating Temperature Change on Composition

Similarly, a change in the throughput may result in change in product rates, Crude Column product/pumparound rates and Vacuum Column product/pumparound rates, which could result in change in the heat input at the Crude Preheat exchangers. Similar to pressure, a change in Pre-Flash Drum inlet temperature, may affect the composition of Crude Pre-Flash Drum liquid and vapors, and the rate dependent assumption may no longer apply for Blocked Outlet case for PSV-1.

5 Conclusion

The economic driver for maximizing the production from process units is ever-present. However, it is vitally important to ensure that the techniques used to evaluate the pressure relief system design are appropriate and suitable to the task. Awareness of the assumptions inherent in rate-dependent classifications, and the potential to invalidate this classification with further "pencil-sharpening", is a starting point. Understanding the ways in which process throughput scale-ups can impact other factors in determining the design of the pressure relief system, such as affecting the fluid composition, operating conditions, and heat exchanger duties, and their impact on required relief rate determination or even scenario applicability, helps to illuminate potential problem areas when evaluating a process throughput scale-up, and informs the decision on whether a revalidation of the heat and material balance is appropriate. Altogether, these considerations should enable engineers to make smarter decisions about rate-dependency and ensure safer operation during process throughput scale-ups.

6 References

[1] American Petroleum Institute, *API Standard 520: Sizing, Selection, and Installation of Pressure-relieving Devices Part I – Sizing and Selection*, 9th edition, July 2014.

442298 Learning Lessons from a Big Chemical Disaster in Korea

Abstract:

Learning lessons from a big chemical disaster in Korea

Kwnaghee Lee and Il moon*

Department of Chemical and Biomolecular Engineering, Yonsei University, 50 Yonsei-ro, Seodaemun-gu, Seoul 120-749, Republic of Korea

On September 27, 2012 a massive chemical dispersion disaster occurred at the Gumi National Industrial complex in the southern city of Korea. About 8 tons of hydrogen fluoride gas leaked as the valve of tank lorry was opened due to a worker' mistake. The operators at the factory had never experienced such a big chemical disaster, and the fire station in the area did not have enough equipment at the time of chemical accident, so that an initial counter measurement could not be conducted smoothly. Due to this accident, 5 workers were killed and about 3,600 local residents had to receive a medical treatment at hospitals due to the accident. And the damage proceeded to the second and third stage, causing 17.7 billion KRW of damage to enterprises and spoiling the crops on 323.8ha of land. After the Gu-mi disaster, law and regulations, government organization, and industrial safety system were completely revised and culture have leaded positive effects towards managing and responding to chemical disasters. The governmental action has been led to make more practical plan and manual to control and respond to chemical accident. The governmental reorganization against chemical disaster such as Joint inter-agency Chemical Emergency Preparedness Center which is a specialized agency focused on responding to chemical disasters has also made to increase technicalities of chemical accidents response. In addition, the relevant laws and regulations has clarified the different acts and programs on control, management and assessment of chemicals along as well as its accident. Moreover, private and public enterprises have been also making efforts to systematically manage safety issues and build high-level safety culture. In this paper, we discuss the changes attitudes against chemical accidents from various viewpoints and the successful case in which the relevant agencies responded to major chemical accidents more efficiently after those changes. In order to improve the initial response system against chemical accidents continuously, it is required to strengthen the role of on-site coordinator, establish the Chemical Accident Investigation Committee on a national level. This study covers the big systematic changes in laws, regulations, business culture for the chemical safety management after a big chemical release accidents.

442301 Study on Thermal Radiation of Enclosed Ground Flare

Abstract:

Flare system is an important part of the safety systems in refineries or petrochemical facilities, and on oil and gas platforms. There are generally two types of flares: the elevated flare and the enclosed ground flare. Enclosed ground flares get more and more applications because of its high efficiency, flame invisibility and low nose. During an emergency flaring situation, a portion of the heat produced by the flame will radiate to its immediate surroundings.The amount of radiation heat emitted by flares is important in the design and location of flare systems. But there are no theoretical or empirical methods for predicting the radiation of enclosed ground flares in current literatures. Predicting the thermal radiation becomes one of the greatest challenges in designing a groud flare system. To investigate the radiation of the enclosed ground flare, the flaring process of a series of flare gases were simulated using CFD method.The temperature and radiation field was calculated.Results shows that the radiation flux at ground level were below 1.58kW/m2; at the height above 9m, the 1.58kW/m2 radiation isopleth can reach 60-70 meters radially from the enclosed ground flare axis,so the design of towers or other elevated structures containing operating platforms should consider radiat.

442313 Simple Objects As Indicator of the Severity of Vapour Cloud Explosions

Abstract:

Simple objects such as instrument boxes and oil drums are often found on petro-chemical plant sites. In the event of an accidental vapour cloud explosion (VCE), they serve as reliable indicators of the explosion overpressure at their positions provided there is a good understanding of their behaviour in a VCE.

Such objects were used to great effect in estimating the overpressure magnitude in the Buncefield incident in 2005. In large scale tests conducted since the Buncefield incident (both as part of a joint industry project and subsequently), such objects were located at varying distances from the cloud, as well as some within the cloud. Overpressure measurements were made at or adjacent to the location of these objects and many have been analysed using advanced finite element analysis. The objectives were to (i) gain a better insight into the response of these objects when subjected to different types and magnitudes of pressure loading, (ii) validate finite element models; (iii) assess the suitability of different finite element modelling techniques for predicting the response of these structures and (iv) develop pressure impulse (PI) diagrams which can be used to provide a rapid assessment of explosion magnitude through an examination of the damage level.

This paper presents an overview of the tests performed and the analysis results with comparisons with the actual behaviour of the objects.

Best Practices for Writing Operating Procedures and Trouble-Shooting Guides

Lauren Madden, Process Safety Manager
REC Silicon
Moses Lake, WA 98837
e-mail: Lauren.Madden@recsilicon.com

William G. Bridges, President
PROCESS IMPROVEMENT INSTITUTE, INC. (PII)
1321 Waterside Lane, Knoxville, TN 37922
Phone: (865) 675-3458
Fax: (865) 622-6800
e-mail: wbridges@piii.com

Prepared for Presentation at
12th Global Congress on Process Safety
Houston, TX
April 10-13, 2016

UNPUBLISHED

Best Practices for Writing Operating Procedures and Trouble-Shooting Guides

Lauren Madden, PSM Manager
REC Silicon

RECSiLICON

William G. Bridges, President
PROCESS IMPROVEMENT INSTITUTE, INC. (PII)

Keywords: Procedures, Human Factors, Risk control, process safety management, PSM

Abstract

There is no complete, best practice guideline or textbook for writing operating procedures and trouble-shooting guides. This paper presents the proven, best approach for developing accurate operating procedures and for ensure the format of the page and steps are optimized to reduce human error rates. This approach and the 35 rules established in this paper are based on the foundations set by Swain and others in the 1970 for control of human error rates, but uses experiences from more than 60 sites where this approach has been successfully followed. The approach and rules for developing procedures for responding to process deviations (such as those needed for Human IPL) are again the best approaches found and have been applied successfully since the early 1990s; these are many times called operational trouble-shooting guides. The paper also discusses how and when to integrate trouble-shooting guides into distributed control systems. Several case studies are provided that show the gains from following this approach. The guidelines in this paper build upon ones presented in 1999 at CCPS.

Background

Operating procedures have always been crucial to the safety, quality, and productivity of process systems. And with the advent of new safety and quality standards such as OSHA's process safety management (PSM) regulation, OSHA's personal protective equipment (PPE) regulation, EPA's risk management program (RMP) regulations, and the ISO 9000 quality standard, many companies are facing the daunting task of developing or upgrading their procedures to satisfy varied and sometimes complex or conflicting requirements.

Although there are overlapping characteristics that the various regulations and standards share, each approaches the procedure-writing process from a somewhat different perspective (e.g., quality or safety, protection of workers or protection of the public/environment), and the required level of detail for documenting the procedures differs greatly between these regulations and standards. For instance, the ISO 9000 quality standard advocates documenting procedures and work instructions that impact quality and suggests keeping the procedures as simple as possible, while the OSHA PSM regulation requires detailed procedures that address all operating modes (startup, shutdown, etc.) and that address operating limits, consequences of deviations, means to avoid hazards, safety and health considerations, and safety systems and their functions. How can companies develop and maintain procedures that ensure productivity and simultaneously satisfy the different regulations and standards? The key is to remember that the ultimate goal of the regulations and standards is consistent: reduce human errors that can impact quality, productivity, and/or safety.

Procedure-related errors are errors that occur because some characteristic of the procedure caused task performance to fail. ***This is currently the most critical human factor at most sites since 90% of accidents have at least one root cause related to mistakes within procedures.*** Reducing these procedure deficiencies can reduce human error rates by a factor of 2 to 10, normally.[1, 2]

The concepts and rules presented in this paper are applicable to all industrial settings. Most importantly, this approach **is effective in satisfying both safety and quality concerns.** With this approach, procedures can and have been developed that provide a sound basis for training and that provide the users and supervisors with an understanding of what can go wrong and how it can go wrong. This approach also provides management with an understanding of the inherent risk (to quality, productivity, safety, etc.) of an activity (i.e., the activity addressed by the procedure) and an appreciation of the safeguards that are being relied upon to mitigate this risk.

An extensive case study is presented by REC Silicon, which illustrates this approach, along with supporting the benefits of the accurate and clear procedures. But the same is true at more than 60 sites where the same approach has been followed.

Fundamentals of Writing Effective Procedures

When developing procedures, or any written material, always remember that the burden of written communication is on the writer, not on the reader. It should never be up to the reader to decipher the meaning of a message. Engineers who write procedure manuals may "tend to include information that is more relevant to them than it is to the operators actually running the unit" [Sutton, 1992].[3] When writing procedures, your guiding principle should always be: if the written procedures are difficult to read, difficult to locate (e.g., if titles are do not have a clear meaning), or inconvenient to use, they will seldom be used [Swain, 1983].[1]

Deficient Procedures are the most prevalent problem in process industries since procedures have not traditionally been developed from the perspective of optimizing human factors; instead, procedures have been traditionally developed to meet a compliance requirement to have written procedures. Examples of procedure deficiencies (inaccuracies) include:

- Incorrect/incomplete/nonexistent (most procedures we have audited have been only 70-75% accurate – the inaccuracies include missing critical steps, steps as written are not what needs to be done, or the steps are out of sequence)
- No/misplaced warnings (for a example, a warning should Never **contain** the action to take; it should instead **emphasize** the action to take)
- Poor format and presentation rules

Figure courtesy of Process Improvement Institute, Inc. (PII), copyright 2003-2016[4]

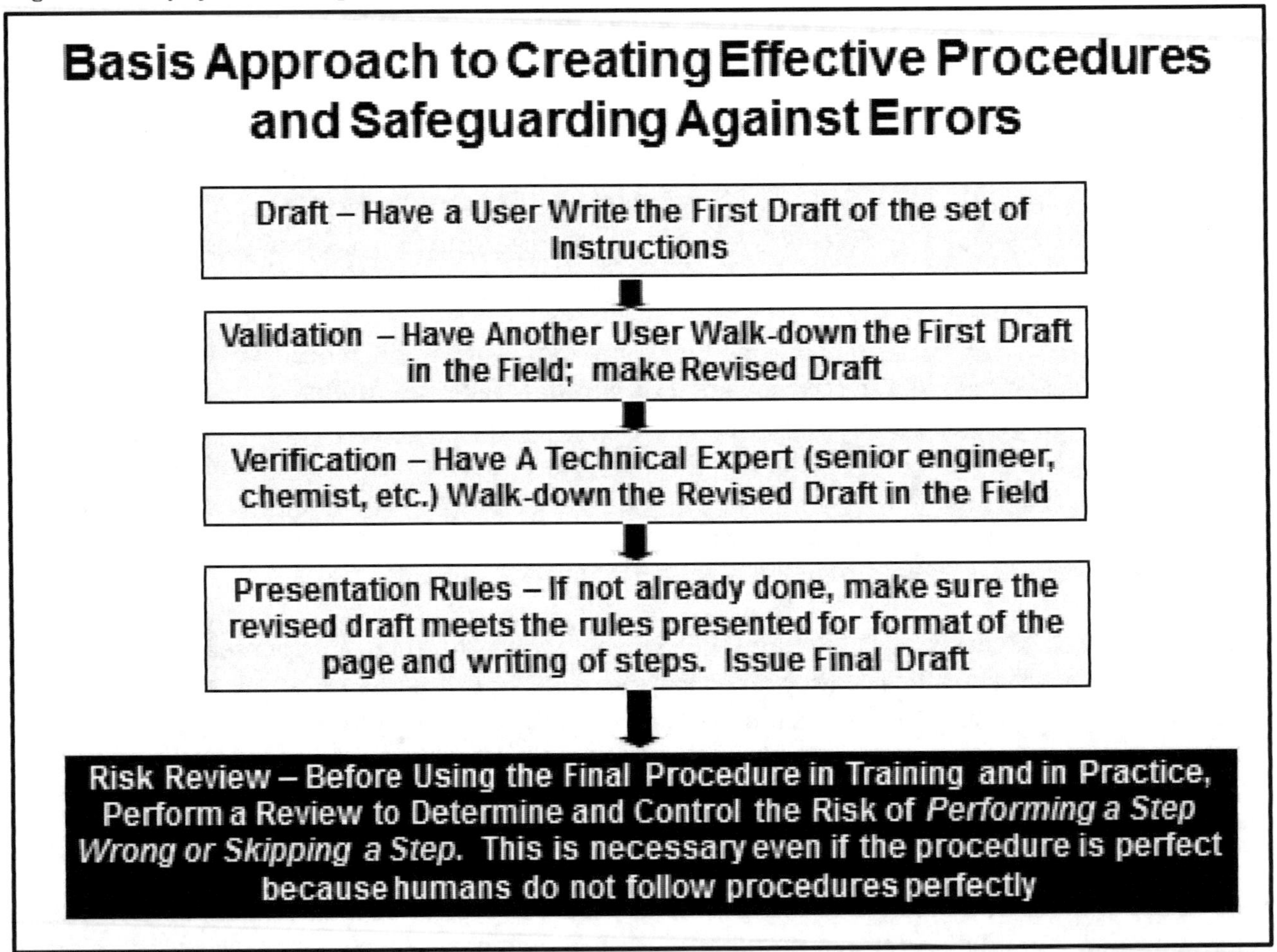

The flowchart above describes the basic sequence of steps for writing effective procedures. Note that the some basic rules are contained in the flowchart, such as "Operators writing operating procedures" and "Maintenance craft-persons write maintenance procedures" and "Lab technicians write lab procedures." Procedures are Not written by engineers or department superintendents. This is a critical rule because it ensures the procedures are written in the common language of the users and is written at the right level of detail.

Universal Purposes of a Procedure[4]

1. A written guide to accomplish initial training of employees
2. A quick reference manual for experienced (trained) employees

The Right Information to Write – Content is KEY[4]

The Content of the procedure controls the human error related to "procedure-based" error – if and content is wrong, the trainee is very likely to learn the step wrong (if the trainer is using the procedure as a basis or when the trainee refers the to the procedure step that is wrong). So:

If the content of the step is wrong it does not matter if the presentation is clear; but if the content is perfect, you can reduce error rates a further 50 to 70% by presenting the information properly.

Ask experienced mechanics or operators to describe a simple task that they perform. You will quickly see that even simple tasks can involve many steps.

You should consider the following questions when writing each procedure step:

- How is the step performed? Does the action involve interacting with a computer terminal, an automatic controller, or simple devices (gauges, valves, micrometers)?
- Can the actions be performed as written and in the sequence written? Can the equipment be operated as specified? Can the steps be physically performed? Does the user have the training or experience to understand and carry out the action using the information available, or is additional information needed?
- Does the user need to be alerted of any potential hazards (Cautions or Warnings) or need any supporting information (Notes) before performing the action?
- Does the user need to know specific operating ranges or limits to:
    - Perform this action
    - Recognize the successful completion of the action
    - Recognize an actual or potential problem
    - Make a decision

For example, does the user have to verify temperature or pressure visually? Is the user waiting for a computer command or verification? How should the process react? Will exceeding a limit result in an accidental chemical release, personal injury, or quality concern? What is the chance of not operating within safe operating limits, and what would be the consequences? If operation of the component is critical and a deviation will probably cause a failure, state the corrective actions. If failure is unlikely, keep going.

- Is needed information found on an instrument, panel, or monitor or is it in the procedure or another source such as a graph, table, drawing, or specification sheet? Should this information be included in the procedure or be referenced?
- What is the next logical step? How is the next step affected by what is performed in the current step?
- What are the results of improper task performance? If an action is critical, spell it out in detail. For example, when starting a special pump that may bind or cavitate if not vented and primed properly, the procedure may require more information than simply stating "start the pump."
- Is the action frequently performed? Is it an action that is easily overlooked? Is this a complex piece of critical equipment that has a low frequency of use? These questions can help determine the level of detail in your procedure.
- Is the action performed so infrequently or is it so complicated that the user is not sure how to do it? Is the action so complicated that nobody is ever certain it's done right the first time?
- Is the decision point clearly defined if a decision is required? Unclear decision points can cause arguments and delays in performing actions.

To address the concerns listed above, a few simple rules are needed for developing content:

- The procedures should be written at the level of someone who has just completed the basic training for that task. Do not write the procedure for someone just hired or for the 10+ veteran.
- Clearly identify ahead of time what activities need procedures to help ensure error rates or controlled low enough, and identify which activities are considered common "skills" of all of the staff.

  Example: Starting a pump is typically considered a "skill." This means a procedure step can simply be "Start the benzene recycle pump (P-119)." The procedure does not need to explain how to start the pump. But, starting a pump is a skill that nonetheless must be learned; it requires doing several sub-steps related to positioning of intake and discharge valves, how the discharge pump is throttled at startup, checking local pressure gage, etc. So, a training module is needed for starting pumps of various types, and the operator needs to learn this necessary skill. Once the skill is learned, it will be applied to so many various cases and so often, that a procedure is no longer needed as a "refresher" on how to do "start the pump."

- The first draft of the procedure must be walked-down in the field by another user. (Simply reading through the procedure does not catch the enough of the mistakes.)
- The revised draft should be reviewed in the field by a technical staff person, such as a process engineer (for operating procedures).

Next:

- The procedure needs to follow the page format and step writing rules described in the next section. This ensures the accurate steps are clearly presented.

The target is to reach an accuracy of 95% or better (so, no more than one wrong or missing step out of 20 steps). The following have been observed in the field by PII staff by direct data collection at more than 110 sites/plants around the world, based on walk-down review of several to dozens of procedures at each unit/plant:

- At 95% accuracy or better and when the same procedures follow 80% of the rules for procedure clarity (presented next), then most users will follow the written procedures and will try to keep the procedures up-to-date.
- At about 85% accuracy or less, about half of the users stop using the procedures
- At about 70% accuracy or less, less than 10% of the users will refer to the procedure or will try to keep it up-to-date. ***So, the written procedures are useless.***
- Unfortunately, the typical Operating procedure walked down by PII staff (accompanied by senior operators, a process engineer, and a shift supervisor) is about 75% accurate (so one step in four is missing or wrong). Usually these inaccurate procedures also follow less than half of the best practices for procedure format, presented next.

PHA/HAZOP of procedures

Just as critical as getting the content accurate and presenting the steps clearly, is performing a PHA/HAZOP of the written steps to determine if you have enough independent protection layers (IPLs) against errors that the "imperfect" operators will eventually make when using the "near perfect" written procedures. This analysis is described in detail in Chapter 9 of the *Guidelines for Hazard Evaluation Procedures*, 3rd edition, 2008[5]. **For an update on the methods for such analysis and for a discussion of the extra emphasis on PHA of procedures by US OSHA, refer to the more recent paper (Bridges, Marshall; 2015).**[6 and 7, 8]

Procedure Clarity

The importance of the accuracy of the procedure steps were discussed above. The clarity of the procedure steps (how they are written and how the page is formatted) is also important. Following best practices for step and page format reduce human errors by a multiplying factor of 3 to 5.[1, 9]

The best practice rules for writing and validating procedures have been published for many years *(see Bridges & Williams, 1997).*[11] These rules include best practices for formatting of the pages and steps; these have been gradually improved over the past decades and now are incorporated into PII's training materials.[4]

The Table on the next is a checklist (in auditing format) based on the current set of best practice rules for developing accurate and clear operating, maintenance, and other work-instructions (procedures).

PROCEDURE QUALITY CHECKLIST *(courtesy PII, copyright 2003-2016)*[4]

#	Issue	Response
	Procedure Content Checklist	
1	Is the procedure drafted by a future user of the written procedure? (Engineers should not author procedures to be used by operators or maintenance staff.)	
2	Is the procedure validated by a walk-down in the field by another future user of the procedures?	
3	Is the procedure reviewed and commented on by technical staff (engineers or vendors)?	
4	Is the procedure checked versus the Page and Step format rules below?	
5	Is a hazard review of step-by-step procedures performed to make sure there are sufficient safeguards (IPLs) against the errors that will occur eventually (when a step is skipped or performed wrong)?	
6	Is the content measured using "newly trained operators" to judge the % of steps that are missing, steps that are confusing or wrong, and steps that are out-of-sequence? (A score of 95% accuracy of content is good; 98% should be the targeted average.)	
	Page Format Checklist	
1	Is the title of the procedure the largest item on the page?	
2	Is the procedure title clear and consistent with other titles, and does it uniquely describes the topic?	
3	Are the document control features the smallest items on the page?	
4	Are temporary procedures clearly identified?	
5	Is white space used effectively? • Is there one blank line between each step? • Does indentation help the user keep their place? • Are the margins large enough to reduce page congestion?	
6	Is type size is 12 pt font or larger?	
7	Is mixed case used for words of steps, with ALL CAPS used only for special cases (such as IF, THEN, AUTO, and WARNING)?	
8	Is the step number very simple (single level of number used)? Only an integer?	
9	Have sections or information not necessary to performing the steps been moved to the back or to another part of the manual or training guide?	
10	Are section titles bold or larger than the text font? Do sections have clear endings?	
11	Is the decision on electronic presentation versus hard copy correct? Are electronic linkages to procedures clear and accurate and easy to use? If paper is chosen, is the color of the paper appropriate?	
12	Is the overall page format (such as Outline format or Two-Column [T-Bar] format) appropriate to the use of the procedure?	
13	Are play script features added for tasks that must be coordinated between two or more users? • Play script is normally used when there are two or more hand-offs of responsibility for steps.	
14	Are rules followed for formatting of Warnings, Cautions, and Notes? (See annotated rules, such as Warnings are for worker safety and Warnings must clearly standout from rest of page.)	
	Step Writing Checklist	
1	Is each step written as a command?	
2	Is the proper level of detail used throughout? This is judged based on: • *Who* will use the procedures • Same level of detail used in similar procedure steps	
3	On average, is there only one implied action per instruction? Best practice is to average 1.2.	
4	Does the procedure indicate when sequence is important? • If sequence matters, each step should be numbered (with an integer or letter) • If sequence does not matter, bullet lists should be used	
5	Are only common words used? Apply "education" level test (5 grade reading level is best)	
6	Do all acronyms, abbreviations, and jargon aid understanding? • Develop a list of such terms for use in procedures *and* communication. • Use terms that users use (within reason)	

7	Is each step specific enough? No room left to guess/interpret: • The meaning of a word or phrase (Check vs. Make sure) • The intent of a step or series of steps • A desired quantity or value • To what equipment the step applies	
8	Is the procedure free of steps that require in-your-head calculations? • Values expressed as ranges rather than targets with error bands • Conversion tables, worksheets, or graphs provided where needed	
9	Are graphics to the user's advantage? • No explanatory paragraphs or lengthy instructions that could be replaced by a picture • No impressive graphics that provide no real advantage	
10	Are references to the user's advantage? • No lengthy explanations or instructions that could be replaced by branching to a reference • No references to a procedure that references still another • No gaps or overlaps between this procedure and a referenced document • If branching, must branch to a procedure, not to a specific step in a procedure	
11	Are rules followed for writing warnings, cautions, and conditional steps? • No more than 2 per page • No actions within a warning or caution (actions must always be numbered steps) • Warnings and Cautions contain descriptions of potential consequences	

Each of these rules is explained in detail in the course *Writing Effective Operating and Maintenance Procedures*, by PII.[4] (Visit www.piii.com for more details.) Each rule is backed by experimental or observed data that indicates there is a clear advantage in error reduction by using the rule. Below are a couple of examples from others on procedure writing rules.

> ***Proper Page Format:*** *In one study, Haney took a procedure presented in narrative format and converted it to column format. He then asked experienced technicians who had been completing the procedure using the narrative format to complete it as presented in the column format. He found that when the technicians used the column format, their errors were reduced by a factor of 3 compared with the narrative format [Haney, 1969].*[10]

> ***Including Supplemental Checklists:*** *Checklists can be effective aids to performance and often should be used to supplement procedures presented in any format. Checklists can be particularly effective in terms of preventing errors of omission. Some of the literature supports using checklists only for simple procedures that do not need to be performed under time or other constraints [Sutton, 1992].*[3] *However, there is also support for using checklists for emergency procedures in which a core set of steps is used to respond to emergency situations, particularly when the actions follow a simple sequence of steps for shutting down a process [Connelly, 1992].*[12] *Overall, experimental data have shown that the probability of errors of omission and commission can be reduced by a factor of 3 to 10 when checklists are used as reminders of key steps in a work instruction [CMA's A Managers Guide to Reducing Human Error, 1990].*[9]

For more data on the history and development of these rules, see Bridges & Williams (1997)[11] and especially refer to originating data and discussion for most of the rules (Swain, 1983).[1] **The procedure on the next page follows the rules**; this is a T-Bar format. Several other page formats are also valid.

Unloading Monomer from Tanker to Storage

STEPS	DETAILS
1. Wear standard PPE, plus rubber gloves and full-face organic respirator.	Standard PPE includes hard hat, safety glasses with side shields, and steel-toe shoes.
2. Weigh in the tanker.	Record weight on GROSS line of form.
3. Check bill of lading.	... to verify correct type of material is in tanker.
4. Sign in tanker driver.	Driver must sign in as a visitor and be escorted at all times.
5. Take Certificate of Analysis to QC.	***CAUTION:** To avoid contamination, **DO NOT** unload until you receive approval from **QC**.*
6. Spot tanker.	A) Direct driver to location. B) Verify brake is set. C) Chock the wheels on at least one side between the two rear axles. D) Ground the tanker (attach grounding strap to an unpainted metal surface).
7. Have the storage area operator make sure the storage tank can hold contents of tanker.	***WARNING:** Failure to perform this step can result in an overflow and spill of hazardous Monomer.* Perform **Tank Gauging Procedure (SOP-01-804)** or check the storage tank load cell readout. Storage tank operating limits are stenciled on the control panel.
8. Place 5-psig nitrogen pad on tanker.	A) Connect from local 5-psig nitrogen drop to aft bottom nitrogen connection on tanker. B) Open nitrogen supply valve. C) Then open valve at tanker.
9. Connect Monomer unloading hoses.	A) Verify hoses and gaskets are in good condition. (Hoses are normally stored on hose rack at unloading spot.) B) Connect from tanker to air pump. (Use aft bottom connection on tanker. Remove cap.) C) Connect from air pump to storage tank. (Connect to storage tank line labeled "from TW.")

Trouble-Shooting Guides (Procedures for Responses to Critical Deviation Alarms) [13]

Trouble-shooting guides are a special form of operating procedure; they are written for the actions that we want the operators to take to recover from a process deviation, *before an emergency situation occurs.* They are called guides since rarely can we predict the process conditions at the time the action is required. Trouble-shooting guides (and training and drills of the action) are required for any action that is considered a Human IPL. The Action Limit is what we show as the Min or Max in a Trouble-Shooting Guide. The action limit triggers the demand to use the trouble-shooting guide.

- The response is typically still possible, but it is time dependent. The time available is called the process safety time (PST). The operator must complete the diagnosis, make the necessary change(s), and make sure they are out of harm's way by the end of the Maximum Allowable Response Time (MART). [13]
- We usually set an alarm or a pre-alarm to trigger this action.
- This is usually before the shutdown triggers (ESD occur automatically) or release points (PSV set points) are reached
- The Min and Max shown in the Trouble-shooting Guide table is not the absolute safety limit for a system, but is instead some value that leaves us some time to take action to prevent from reaching the absolute limits.
- There is still time to prevent or avoid the final consequence that could occur if we reach the ultimate limits of the process. Usually, we want the MART to be ½ or less of the PST, and we want MART > 10 minutes for trouble shooting in the field/plant and MART > 5 min for trouble-shooting only from the control room. [13]

If the unit has a good PHA/HAZOP, then it is best to extract information from the HAZOP (or What-If) analysis tables to start the development of each guide. The guide is then finished with input from the process experts.[4]

Examples of Creating a Trouble-Shooting Guide from a HAZOP Table

HAZOP Table Entry	**Trouble-Shooting Guide Entry**
Cause: *Bypass valve is open or passing*	Make sure the bypass is tightly closed
Safeguard: *Isolation valves for the vessel*	Isolate the vessel, if necessary
Safeguard: *Relief valve*	Make sure the relief valve block valves/relief path are open

The key categories of information needed in a trouble-shooting guide are:[4]

- IMMEDIATE ACTION (by system or by operator)
- DECIDE IF ALARM is REAL
- FINDING and FIXING the CAUSE
- FIX or BYPASS PROBLEM

The example below is of a trouble shooting guide for one critical alarm/action.

Trouble-Shooting Guide

Alarm or Indicator:	**PAL 4446 – Low Pressure Alarm for Suction of Organic Feed Pump 40-PM-18.445**

Action Limit:	5 kPa		
Consequence:	Possible pump seal failure, releasing or spraying organic waste into the berm.		
Process Area:	FB&D Incinerator; Liquid Organic Liquid Feed	**Oper. Mode:**	Normal
Drawing #s:	D-400-PI-013		

IMMEDIATE ACTION (by system or by operator)

- DCS should shut down the organic feed pump (40-PM-18.445).
- From the DCS display, MAKE SURE the organic feed pump is shutdown.
- HAVE the field operator check for leaks near the organic feed pump.
- IF there is a large leak/release, THEN use the ESD switch to shutdown the unit and then follow/complete the shutdown and isolation procedure, OPS-ESD-117.
- IF there is a minor leak or no leak, THEN:
    - COMPLETE the rest of the trouble-shooting,
    - and DECIDE how to contain the leak for now,

DECIDE IF ALARM is REAL

- From the DCS, CHECK the pressure and feed tank level trends. IF the trends indicate the alarm if valid, THEN continue with finding the cause or fixing or bypassing the problem.

FINDING and FIXING the CAUSE

- CHECK valves upstream of the organic feed pump to see if any are closed too far, including checking ESD valves.
- CHECK, by feel with hand, if the heat tracing is on; IF Not, then TURN ON or open heat trace valves
- MAKE SURE nitrogen to the pump seal is at the normal operating pressure.
- CHECK if the line if plugged or frozen (skill)

ABC Chemical Company Prepared by: *Printed copy of this procedure is good for one job task duration.*	OPS-76-TSG-233	Incinerator Unit Revised 8/24/2015 Printed 2/21/16 Page 1 of 2

Trouble-Shooting Guide

Alarm or Indicator:	**PAL 4446 – Low Pressure Alarm for Suction of Organic Feed Pump 40-PM-18.445**

- CHECK if the line if plugged or frozen (skill)
- CHECK if the level is actually low, use the Organic Feed Tank.
- IF the cause is low level in the feed tank, THEN resolve the problem if necessary based on cause that is found (skill)

FIX or BYPASS PROBLEM

- IF necessary, SHUT DOWN the Unit to allow fixing of the problem.
- FOLLOW the proper procedure to resolve problems (repair procedure, line clearing procedure, etc.)
- IF the decision is made to continuing operation without all equipment in normal condition, THEN:
    - FOLLOW Temporary Operating mode, if there is a temporary procedure already written for this possible problem/condition
    - FOLLOW MOC procedures to obtain approval for any non-standard temporary operating procedure or mode

END

Optimal presentation of trouble-shooting information.

Ideally (or ultimately) such trouble-shooting information should be imbedded in the basic process control system (DCS) so that the operators can access the information on demand, with the click of mouse or key.

Comments on Current Industry Consensus Guides for Writing Procedures

None of the consensus textbooks available for procedure writing cover all of the major points of this paper. The best "guidelines" for writing procedures currently exist in the course notebooks used by PII (and likely in the course notebook used by ABS Consulting, as they had similar origins).

Notably, the AIChE/CCPS textbook, *Guidelines for Writing Effective Operating and Maintenance Procedures*, 1996[14]:

- has NO description at all on the development and use and format of Trouble-Shooting Guides. This should have been a major chapter in the textbook.
- mentions only about half of the rules listed in this paper (and the related PII course notebook) for formatting of pages and steps
- emphasizes technical staff or professional procedure writers writing procedures, rather than the Users themselves writing the procedures (it violates the first rule of procedure writing)
- Fails to cover the competency needed by procedure writers.
- recommends and emphasizes having subsections in each work instruction/procedure (between the Title of the task and Step 1) rather than relegating such information either to other sections of the operating manual or rather than turning this information into steps.
- fails to mention the need for a PHA of procedures that focuses on hazard review of deviations from the steps; this is required by best practices and by US OSHA for PHAs under PSM.

This guideline textbook is not worth the investment.

As a better example, but one that nonetheless still falls short of what is needed, the ASM Consortium *Guidelines for Effective Procedural Practices,* 2010[15]:

- has good discussion on the need and thought processes related to handling abnormal situations (trouble-shooting or response to critical deviations), but there is little description on the format and detailed content of the trouble-shooting guides; in fact, the guideline fails to differentiate between routine procedures for startup, shutdown, and online maintenance, and those used for response to critical alarms (trouble-shooting)
- mentions only about half of the rules listed in this paper (and the related PII course notebook) for formatting of pages and steps
- mentions that experience users themselves should write their own procedures, but it does not emphasize this rule and leaves the authorship open to engineers or a committee. And it specifically mentions use of retired senior operators for writing operating procedures. While this is one good way to write the procedures (while also capturing the knowledge from this experienced group is prior staff), we have found that the current users do a better job overall, once they have achieved competency in how to write procedures.
- mentions the competency needed for procedure writers in general.
- mentions the need for a PHA of procedures, but does not reference the method for this in Chapter 9 of *Guidelines for Hazard Evaluation Procedures*, CCPS/AIChE, 3rd edition, 2008,[5] or other related papers on the topic. Also, the description of the PHA is misleading because it focuses on how PHAs should be used in writing procedure content (which is fine for trouble-shooting guides) but fails to emphasize that a step-by-step analysis of deviations from the steps are needed to ensure that the process has enough IPLs against mistakes that can occur when imperfect humans use the procedure.

Though this guideline has deficiencies, it is worth the investment and provides useful insight not found elsewhere. Just be sure to address the deficiencies noted above.

Implementation of this approach

PII staff has helped more than 60 sites implement the approach above. The typical approach for successful implementation is:

1. Perform a gap analysis to determine the average accuracy of procedure content, average clarity (as judged versus the procedure format rules, and identify any examples of excellent procedure writing.
2. Convince management of the problem and get their buy-in on the solution.
3. Train (1.5 to 2 days for operators; 1 day for lab or maintenance staff) and coach the users of procedures (normally for 3 to 8 days, to have them achieve full competency) so they can be effective authors. Normally, 20 to 25% of the users (hands-on operators, hands-on maintenance and inspection staff, lab staff) need to be trained as authors. Not all of those trained will make good writers; the organization needs to judge who does the best jobs. About 15 to 20% of the staff should end up being authors.
4. Get organized, like a project, for tracking and scheduling of the effort. Use master spreadsheets such as those used at REC Silicon (shown later).
5. Use a Traveler form (described later) to make sure each new procedure gets a proper amount of review by other users and by technical staff (such as engineers)
6. Reward procedure writers for their efforts. Especially ensure there are no disincentives for taking the extra effort to write procedures.
7. Track the progress on the effort; fix problems as they arise; enforce schedules (to show management cares).
8. Perhaps use to specialized software make it easy to output the content of procedures into different user-friendly formats such as training tools, training competency records, and field checklists. WordPerfect (formerly from Corel) made this relatively easy with the use of Primary and Secondary merge files; MS Word makes this more difficult. As a result, some specialized software have been developed to help keep formats consistent and to help reformat the content, on demand.
9. Issue procedures through a document control system that does not add extra workload.
10. Of course, ensure there is JSA performed for procedures that have "personal injury" potential and perhaps a PHA of the steps in the procedures whenever there is "process safety" potential.
11. Train all affected workers in the new procedures.
12. Identify the best writers to become the trainers of new authors, over the long run.

Overview of REC Silicon

REC Silicon is one of the world's largest manufacturer and suppliers of

- Silane Gas and specialty silicon gases.
- Granular Polysilicon from Fluid Bed Reactor technology for solar photovoltaic applications
- Ultra-pure float zone Polysilicon for the electronic industry

REC Silicon's Product and Service Offerings

Solar Grade
Polysilicon with various applications for the solar industry

Silane Gas
Product used in production of semiconductors, flat panel displays & thin film technologies

Electronic Grade
Polysilicon delivered to the semiconductor industry as well as other electronic market segments, such as optics & MEMS

Float Zone
High-purity polysilicon delivered to the power electronics industry

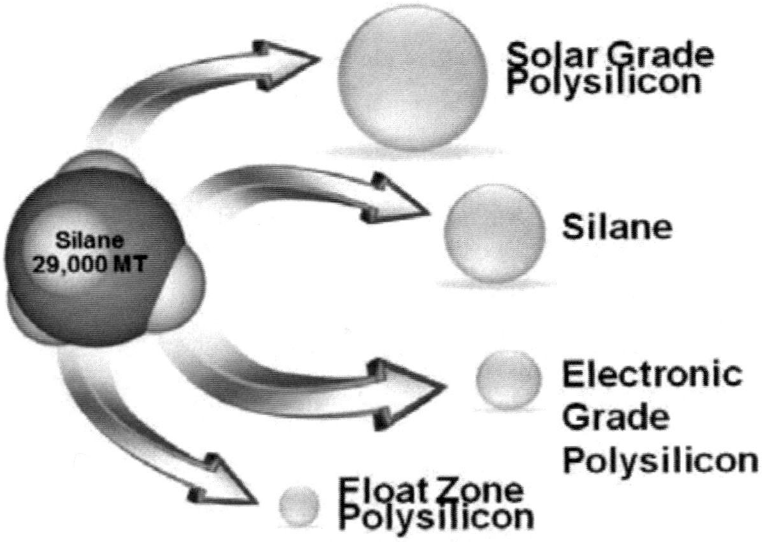

Moses Lake, Washington, USA 396 Employees	Silicon III, IV FBR	Silane gas: 22,000 MT Granular silicon: 16,300 MT
Butte, Montana, USA 255 Employees	Silicon II	Silane gas: 7,000 MT Siemens grade: 3,600 MT

Implementation at REC Silicon

REC production facilities in the USA have been actively implementing process safety for many years, with a renewed interest in 2013 in meeting industry best practices (which goes well beyond US OSHA PSM requirements). To that end, REC commissioned gap analyses versus best practices at both the Butte and Moses Lake production sites. One common finding was that the operating procedures and maintenance procedure for both sites needed improvement with respect to accuracy of the steps as well as clarity (page format and step format). The following were some of the issues with the procedures and related management systems, per the audit results:

- Procedure Inaccuracy (procedure content was less than 80% accurate)
- Technical Staff (such as process engineers) were viewed as owners, not shop floor
- Annual review process ineffective
- Procedure formats were poor; did not follow best practices for reducing human error
- Inconsistent use of Warning, Caution, and Note
- Procedure Naming – not consistent
- Tried to take on all modes of operation in one procedure title
- Procedure accessibility was difficult in the legacy database; multiple searches needed to find tool for job
- Procedure Safety Reviews – no formal methods were applied for procedure safety reviews; i.e., there were no PHA of procedures with process hazards or JSA of tasks with personal hazards

Procedure Author Training – *the first step in any procedure writing effort*

In early 2014, REC Silicon adopted Process Improvement Institute's (PII's) procedure author training module (*Writing Effective Operating and Maintenance Procedures*[4]) with minor customizations. The training consists of 2 days of classroom lecture and about 6 exercises, followed by practical procedure development and field validation steps in the coaching phase. The classroom course covered all aspects of procedure writing with the focus on optimizing the aspects of the procedures that will lower the likelihood of human error when using the procedures. Topics included:

- human factors overview
- getting the content right
- page format rules and selection
- step format rules
- addressing operating limits (including creating trouble-shooting guides
- nuances of emergency shutdown and emergency operating procedures

- implementation guides and high level procedures for the task of getting all of the work instructions written
- examples/exercises on each topic

The coaching consisted of the instructors from PII, who are expert in organizing procedure writing efforts, helping the procedure writers organize and write operating and maintenance procedures, and trouble-shooting guides. Coaching was for 2+ days. About 20 staff from the Moses Lake and Butte sites attended the initial author training and coaching session.

REC Silicon has revised its management systems to require this competency for all personnel developing or changing procedures and to reach full operator certification. Operators must demonstrate an active role in the procedure MOC process to qualify for recertification every 3 years, otherwise they must repeat the classroom and field training. The unit trainers have taken over leading the procedure author training.

The benefits we have seen from these changes include transfer of ownership of procedures from the technical day staff to the operator since these are the people performing the work in the field. This is necessary to have accurate procedures. **The emphasis on intentional competency development for what is required to write effective operating procedures has greatly improved the quality of our procedures and our operator training and certification program.** The quality of the annual reviews and audits has improved tremendously with the rewritten procedure format and content changes.

Safe Operating Limits and Consequence of Deviation

REC Silicon has developed an engineering standard that governs the content and format for this type of special safety troubleshooting procedure. *CCPS Guidelines for Engineering Design for Process Safety*[16] is the basis for determining the safe operating limits. Production engineers are the owners of these procedures. Engineering competency is required to develop the safe operating limits table based on the available process safety information. A consequence of deviation is then developed by the Process Hazard Analysis team. Finally involvement of certified operators is required to develop steps to correct or avoid exceeding the safe limit. This combined effort produces the safety troubleshooting guide, which complies with the approach listed earlier in this paper.

REC silicon has risk ranked its process areas and is focusing on developing safety troubleshooting guides for high risk process first. To date, for the Moses Lake Facility, thirty six (36) guides have been developed. These have been developed in conjunction with PHA updates to the safe operating limits. We have not yet implemented the formal training and competency testing of these safety procedures. This is planned for the next phase of the initiative.

How REC Silicon approached the re-write effort:

To begin this initiative, each operating department listed all of the existing operating procedures into the procedure development tracking form. Each department also developed a list of "skills" that are used across multiple procedures and that operators or maintenance staff are required to attain before using the written procedures that were developed for more complex tasks (such as starting a pump, or hookup of a chemical transfer hose).

The workers (users) then screened each procedure for safety, equipment damage, or quality risk. This assisted in determining whether 2, 3 or 4 peer walk-downs were required on the validation traveler workflow. Since most of the REC Silicon procedures covered more than one mode of operation, the term "parent" was developed for the original procedure and each procedure was rewritten and broken out into a separate "child" procedures. The procedure re-write assignments were divided up between all four rotating shifts and the superintendent assigned certified authors (e.g., operators trained/coached as procedure writers) from their pool. Each department is managing their pool of qualified authors. For example, our fluidized bed reactor (FBR) unit has 11 qualified authors. Silane, which is a much larger department with 3 production units, has 56 qualified authors. Cross shift validation was utilized to ensure all operating approaches were considered. The assigned author was responsible for consolidating red line comments.

The traveler process introduced by PII was customized to consider information that is marked for removal from the parent procedure. **A Traveler form shown to the right accompanies each new or revised procedure through the writing and editing/review process.** When complete, the traveler form is then kept in the QA file for the procedure to validate long-term how the initial procedure was developed.

Procedure Development Traveler - Moses Lake			
New Document Number:		Area:	
Procedure Title:			
Procedure Type:			
Author:			
Initial Draft Completion Date:			
Author Walkdown Date:			
Author Revision 1 Completion Date:			
Peer 1:		Shift:	
Peer 1 Walkdown Date:			
Optional Peer 2:		Shift:	
Peer 2 Walkdown Date:			
Optional Peer 3:		Shift:	
Peer 3 Walkdown Date:			
Optional Peer 4:		Shift:	
Peer 4 Walkdown Date:			
Author Revision 2 Date:			
Engineer/Verifier:			
Verification Completion Date:			
Author Revision 3 Completion Date:			
QA Person:			
Author Revision 4 Completion Date:			

Images and layouts copyrighted by PII, 2008-2016[4]

A color coding process has been developed to indicate if the content can be permanently deleted, or must be moved into the unit training manual. The unit trainers have been responsible for tracking and reporting weekly progress on the initiative. **See an example of the overall tracking form below.** A total of 184 "parent document" procedures for Moses Lake were tracked through the development, review, and revision process (along with about 200 child documents).

Images and layouts copyrighted by PII, 2008-2016[4]

Procedure Development Tracking Form - Moses Lake

							Initial Development			Validation				
Item	New Doc #	Old Doc #	Area	Procedure Title	Proc Type	Author(s)	Initial Draft	Author Walkdown	Author Rev 1	Peer 1 Walkdown	Opt. Peer 2 Walkdown	Opt. Peer 3 Walkdown	Opt. Peer 4 Walkdown	Author Rev 2
1	FBP.3.xxx	FBP.3.160	FBR	Hydrogen Booster Compressor Cleanup	Cleanup	LT/FG	3/19/14	3/19/14	3/20/14	3/21/14	3/21/14			
2	FBP.3.xxx	FBP.3.160	FBR	Hydrogen Booster Compressor Operation and Op Restrictions	Operation	LT/FG	3/20/14	3/20/14	3/20/14	3/21/14	3/21/14			
3	FBP.3.xxx	FBP.3.160	FBR	Hydrogen Booster Compressor LOTO	LOTO	BCA/OL	3/19/14	3/19/14	3/21/14					
4	FBP.3.xxx	FBP.3.160	FBR	Hydrogen Booster Compressor Standby	Standby	BCA/OL	3/20/14	3/20/14	3/21/14					
5			4.0	STC/TCS from Receiving Tank to STC/TCS Storage Transfer Through Metering Line	Transfer	ME/RR	3/19/14	3/19/14	3/20/14	3/20/14	3/21/14	3/21/14	3/21/14	
6			4.0	STC/TCS from Receiving Tank to STC/TCS Storage Transfer Through Bulk Line	Transfer	ME/RR	3/19/14	3/19/14	3/20/14	3/20/14	3/21/14	3/21/14	3/21/14	
7			4.0	H2 Recycle Compressor Switching to Standby	Standby	ME/RR	3/20/14	3/20/14	3/20/14	3/20/14	3/21/14	3/21/14	3/21/14	
8			4.0	H2 Recycle Compressor Cleanup	Cleanup	ME/RR	3/20/14	3/20/14	3/20/14	3/20/14	3/21/14	3/21/14	3/21/14	

From PII's procedure author training,[4] the Quality and Content checklist was modified and used as quick reference guide to make sure that all the content development and formatting rules are being followed during the re-write process. We used Microsoft Word and made templates for the Tbar, Flat and Consequence of Deviations.

Procedure Management of Change

Through this project REC Silicon has adopted formal safety review methods for procedures. These methods standardize the approach to determine if a procedure is safe to use as written or if revisions are required. These details are outlined in a completely new REC Silicon policy - Procedure Management of Change PSM.30.004.01. The REC Silicon IT department developed a custom tool to execute these requirements in an electronic workflow. The benefits from these changes are formally transferring competency to lead and document job safety analysis to the shop floor supervisors. Prior to this effort, JSA was traditionally a review lead by our safety specialists.

Overall Summary of the procedure writing initiative:

For the Moses Lake facility, 398 procedures (of which 184 were unique and 214 were similar to the 184 unique ones) for normal and routine operations, various maintenance tasks, and trouble-shooting guides have been developed through this project (new and rewrites combined.) Similarly, for the Butte facility, 116 procedures (of which 24 were unique and 92 were similar to the 24 unique ones) for normal and routine operations, various maintenance tasks, and trouble-shooting guides have been developed through this project (new and rewrites combined.) In addition, about 56 existing procedures (that were in "okay" state) were updated and the format improved.

We do not have any data yet for reduction in error rates or increase in production, quality or safety; however we have seen great improvements with the new formats with the ability to conduct job observations in the field. Our job observations assess accuracy of the work being performed, accuracy of the written procedures, and an opportunity for on the job operator training.

This type of project can seem overwhelming for a manufacturing site. It is essential to form a project team with key decision makers from all the correct disciplines and have a project sponsor that can remove barriers. The way REC Silicon was able to accomplish this type of effort was dedicating time to meet each week, setting short term goals, and tracking KPIs to meet those goals. The procedure writing initiative at REC Silicon is still very much "a work in progress". Leaders of the project have to make time to support the staff to keep the momentum going on this initiative and also on the related culture change, until it becomes "the way the work always gets done."

Quotes from workers involved in initiative:

- Operators are receiving better work instructions
- Sequence of steps is well understood
- Much easier to audit operator compliance to procedure
- Standardizes how we do work – protects our equipment from damage, protects our people, protects the environment
- No longer need 20 year operator tribal knowledge and interpretation to complete procedure
- Procedures are not cluttered anymore, they just have the necessary information
- Procedure element now requires involvement of many shop floor workers
- Procedures are really safe to use
- JSA technique is alive on the shop floor
- Procedure can be trusted
- Technical staff no longer doing all procedure work. Freed up to optimize processes and realize other cost saving benefits.

Conclusions

Having clear and accurate procedures is a must. Poor procedures show up as one of the root causes in 90% of the accidents. The approach described in this paper has worked for more than 35 years for one of the authors, and continues to work today. Controlling risk is mostly about controlling human error rates, and having accurate and clear procedures are one key to this control. One other gain from the approach described in this paper that the users become the owners, and therefore they have a vested interested to keep the procedures up to date and clear.

No doubt, writing all of the procedures for a facility is a large task. But, it can be accomplish by site staff, with support from process safety or quality control leadership, after enough users become competent authors; attaining this competency takes less than 2 weeks of work, on average.

Acronyms Used

AIChE– American Institute of Chemical Engineers
ASM – Abnormal Situation Management consortium
CCPS – Center for Chemical Process Safety (a division of AIChE)
CMA – Chemical Manufacturer 's Association, now American Chemical Council (ACC)
FBR – Fluidized Bed Reactor
HAZOP – Hazard and Operability Analysis
IPL - Independent Protection Layer
JSA – Job Safety Analysis
LOPA – Layer of Protection Analysis
MART – Maximum Allowable Response Time
MOC – Management of Change
NUREG – Nuclear Regulation, US Nuclear Regulatory Commission
OSHA – Occupational Safety and Health Administration, US Department of Labor
PHA – Process Hazard Analysis
PII – Process Improvement Institute, Inc.
P&ID – Piping & Instrumentation Diagram
PSI – Process Safety Information
PSM – Process Safety Management
PST – Process Safety Time
SOP – Standard Operating Procedure

References

1. Swain, A. D., Guttmann, H. E., *The Human Reliability Handbook - with Emphasis on Nuclear Power Plant Applications*; NUREG/CR-1278, U.S. Nuclear Regulatory Commission, Division of Facility Operations, 1983.
2. Gertman, D., Blackman, H., Marble, J., Byers, J. and Smith, C., *The SPAR-H Human Reliability Analysis Method,* NUREG/CR-6883, U.S. Nuclear Regulatory Commission, Office of Nuclear Regulatory Research, 2005.
3. Sutton, Ian S., "Writing Operating Procedures," *Proceedings of the International Conference on Hazard Identification and Risk Analysis, Human Factors and Human Reliability in Process Safety*, AIChE, Center for Chemical Process Safety, New York, NY, 1992.
4. *Writing Effective Operating and Maintenance Procedures*, Course Notebook, Process Improvement Institute, Inc., 2003-2016.
5. *Guidelines for Hazard Evaluation Procedures*, 3rd Edition, 2008, CCPS/AIChE.
6. "Necessity of Performing Hazard Evaluations (PHAs) of Non-normal Modes of Operation (Startup, Shutdown, & Online Maintenance)", W. Bridges and Mike Marshall (US OSHA), *18th Annual International Symposium, Mary Kay-O-Connor Process Safety Center, College Station, TX, October 2015.*

7. "How to Efficiently Perform the Hazard Evaluation (PHA) Required for Non-Routine Modes of Operation (Startup, Shutdown, Online Maintenance)," W. Bridges and T. Clark, *7th Global Congress on Process Safety*, Chicago, AIChE, March 2011.

8. Bridges, W.G., et. al., "Addressing Human Error During Process Hazard Analyses," *Chemical Engineering Progress*, May 1994

9. *A Manager's Guide to Reducing Human Error*, 1990, Chemical Manufacturers Association (CMA)

10. Haney, R. W., "The Effects of Instructional Format on Functional Testing Performance," *Human Factors*, 11, 1969, pp. 181-187.

11. "Effective Procedure Writing – A Two-Part Series (Part 1: Reducing Risk of Worker Errors; Part 2: Addressing Process Safety Management Requirements for Operating Procedures)," *Chemical Engineering Progress*, 1997, W. Bridges with T. R. Williams (both of JBFA).

12. Connelly, C. S., "Emergency Operating Procedure Writing for Oil Refinery Process Units," *Proceedings of the International Conference on Hazard Identification and Risk Analysis, Human Factors and Human Reliability in Process Safety*, AIChE, Center for Chemical Process Safety, New York, NY, 1992.

13. *Guidelines for Initiating Events and Independent Protection Layers*, 2015, CCPS/AIChE.

14. *Guidelines for Writing Effective Operating and Maintenance Procedures*, 1996, CCPS/AIChE.

15. *ASM Consortium Guidelines – Effective Procedural Practices*, 2010, ASM Joint R&D Consortium.

16. *Guidelines for Engineering Design and Process Safety*, 2nd Edition, 2012, CCPS/AIChE.

12TH GLOBAL CONGRESS ON PROCESS SAFETY

How Process Safety Information Affects the Quality of a Process Hazard Analysis

Michael Mayo
Smith & Burgess LLC
7600 W. Tidwell Rd, Ste 600
Houston TX 77040
mike.mayo@smithburgess.com

Dick Baum
dick.baum@smithburgess.com

Prepared for Presentation at
American Institute of Chemical Engineers
2016 Spring Meeting
12th Global Congress on Process Safety
Houston, Texas
April 11-23, 2016

How Process Safety Information Affects the Quality of a Process Hazard Analysis

Michael Mayo
Smith & Burgess LLC
7600 W. Tidwell Rd, Ste 600
Houston TX 77040
mike.mayo@smithburgess.com

Dick Baum
dick.baum@smithburgess.com

Keywords: PHA, PSI, MI

Abstract

This paper is a tutorial that shows the importance of how Process Safety Information affects the quality of a Process Hazard Analysis. The Process Hazard Analysis (PHA) is the one requirement of the PSM standard that has the facility stop and review a process in its entirety to ensure safe operation. However, a high quality PHA is dependent on accurate process safety information (PSI) that is consistent with the current equipment state (Inspection History, Changes from original design, operating conditions etc.). A PSI sub-element, relief system design basis documentation, requires significant amounts of other PSI and if organized with forethought, will greatly improve the quality, effectiveness and, potentially, speed of a PHA. If the relief systems documentation were performed in a system that stored all of the other relevant PSI and was cross referenced with current inspection records (MI systems), then this subset of PSI would contain much of the critical information needed for a PHA. The system would have to be designed such that it was accessible to the PHA team and organized to match the PHA workflow. This paper will present a proposed organization of PSI and then show how with a trained PHA facilitator and organized and accessible PSI/MI data, specific items found in PSM audits of PHAs would have been avoided.

1 Introduction

From defining the Causes, developing the Consequence scenarios and ending with Safeguards, accurate PSI information is required to provide the most credible assessment of risk. The best practice of minimizing the assumptions used during a PHA study has two significant benefits;

- Management can be assured (feel good) that the PHA process utilized all relevant and up-to-date PSI information thereby reducing potential inaccurate assumptions and,
- It will reduce time spent for discussions between participants to come to common conclusions and speed up the PHA process while maintaining the integrity of the study.

As learned from many PSM audits, PSI information is typically spread throughout many locations and maintained by numerous groups;

PSV files - Engineering

Mechanical Equipment Data/Inspection reports - Engineering, Maintenance, Inspection Departments

Critical Instrument Logs - Engineering or Controls

In a perfect world, maintaining a common database to store all up-to-date PSI information would be ideal, but the delegation of maintaining and access would be cumbersome. There are many people involved with PSI information and each has their own responsibilities and actions for validating the data. One way to reduce the roadmap for locating PSI data is to maintain relief systems documentation in such a way that it provides most all relevant data in a simplified format and links to other sources of PSI information.

This poster will walk through the PHA process and how readily available PSI information can be located with robust relief systems documentation.

2 Causes

Typically, the detailed cause of a PHA would not reference back to any PSI data. The failure or misalignment of valves are always credible occurrences that are documented. Where PSI information becomes relevant is in the cause category. The example shown in figure 1 indicates the following;

- Inlet PCV-001 on methane supply to V-001 reactor fails open

Cross-referencing to the P&ID indicated a figure 2, it is noted there is a PSV-002 upstream of PCV-001, a vessel, V-001 with its own relief device, PSV-001. A quick check during the PHA study of the relief systems documentation for PSV-002 will show that the set pressure of 400 psig is indeed above the MAWP of V-001, 285psig.

This validates the cause category of high pressure. Had the set pressure of PSV-002 been documented at 250 psig, this cause probably belongs in another category, i.e. more flow.

Processes: 1. V-001 Reactor

Nodes: 1. Feed streams

Deviation	Causes	Consequence	Risk			Safeguard
			S	L	RR	
1. High Pressure	1. Inlet PCV-001 on methane supply to V-001 reactor fails open	1 Potential for high pressure in V-001 reactor, potential **flange leak**, loss of containment, explosion, jet fire, employee injury, damage to equipment and loss of production				1. PSV-001 on V-001 reactor
		2 Potential for high pressure in V-001 reactor, potential **rupture**, loss of containment, explosion, jet fire, employee injury, damage to equipment and loss of production				

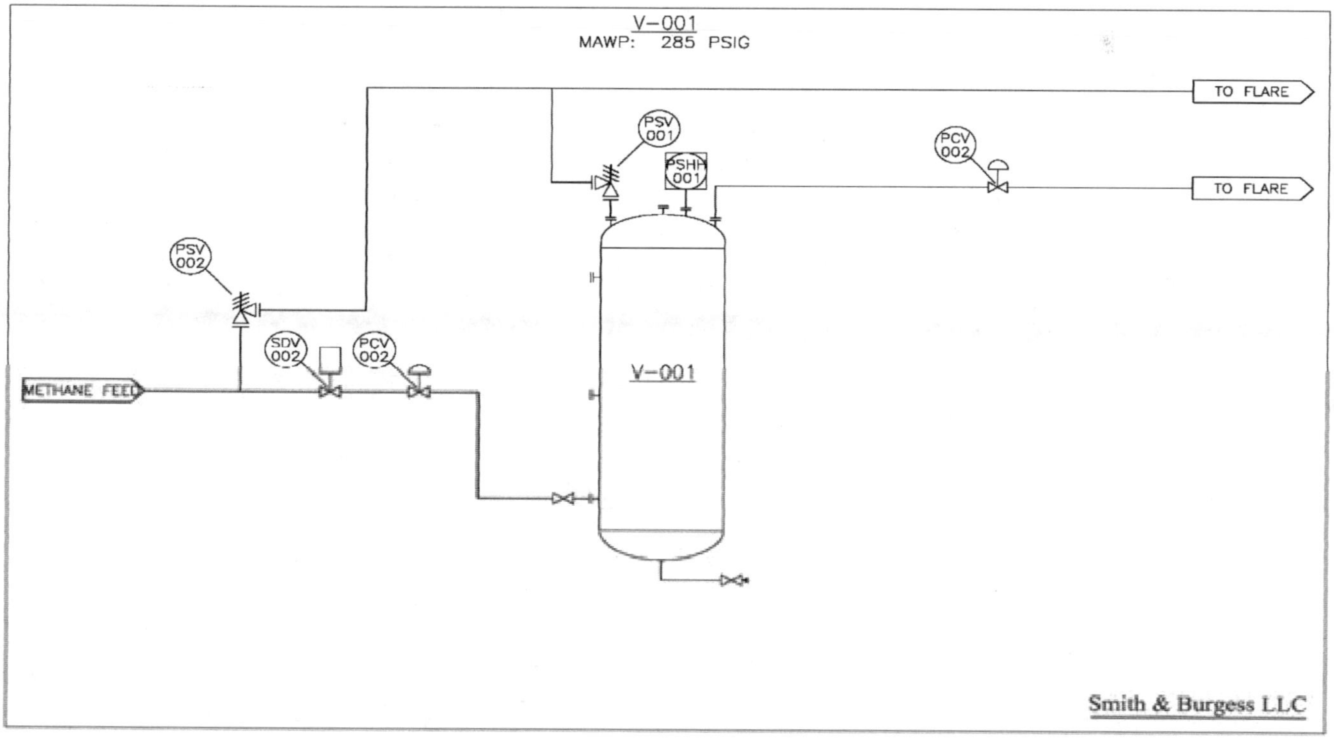

3 Consequences

The consequences developed during the PHA can benefit greatly from accurate and readily accessible PSI information. In the high pressure deviation we can use the PSI information in establishing the overpressure ratio to assist in pinpointing the consequence severity levels. Using the PSI information for PSV-002, a 400 psig supply into vessel V-001 with a MAWP of 285 psig, results in an overpressure ratio of 1.404. Based on your client's guidelines for overpressure and probabilities, this could result in either a flange leak or vessel rupture.

The consequence severity of these two scenarios is quite different but with the PSI data well documented and easily obtained, accurate severities can be obtained during the PHA.

Processes: 1. V-001 Reactor

Nodes: 1. Feed streams

Deviation	Causes	Consequence	Risk			Safeguard
			S	L	RR	
1. High Pressure	1. Inlet PCV-001 on methane supply to V-001 reactor fails open	1 Potential for high pressure in V-001 reactor, potential **flange leak**, loss of containment, explosion, jet fire, employee injury, damage to equipment and loss of production				1. PSV-001 on V-001 reactor
		2 Potential for high pressure in V-001 reactor, potential **rupture**, loss of containment, explosion, jet fire, employee injury, damage to equipment and loss of production				

PSV-002 set pressure can help determine consequence severity

PSI data for V-001 can help determine the consequence severity

4 Safeguards

The safeguards listed during a PHA have the most critical requirement for validation and accuracy during a PHA. They are the protection layers that mitigate or prevent major accidents from occurring and thus leave no room for assumptions or quantitative assessments. PSI information in relief system documentation should be current and accurate for the current operating parameters. A well-documented and arranged system will allow for easy access during the PHA. Information from other PSI data such as P&ID's does not always reflect the purpose of the safeguard. For example, many P&ID's show the governing case of a pressure relief device, which may or may not satisfy the scenario release and could lead to inaccurate assumptions being made during the PHA. (Fire case vs overpressure)

Processes: 1. V-001 Reactor

Nodes: 1. Feed streams

Deviation	Causes	Consequence	Risk			Safeguard
			S	L	RR	
1. High Pressure	1. Inlet PCV-001 on methane supply to V-001 reactor fails open	1 Potential for high pressure in V-001 reactor, potential **flange leak**, loss of containment, explosion, jet fire, employee injury, damage to equipment and loss of production				1 PSV-001 on V-001 reactor set at XXX
		2 Potential for high pressure in V-001 reactor, potential **rupture**, loss of containment, explosion, jet fire, employee injury, damage to equipment and loss of production				PSV-001 set pressure can determine the likelihood

5 Conclusion

Well documented PSI information is critical to the PHA process and provides the validation needed to ensure that the process is being operated and maintained in the best possible safe state. The examples shown indicate how comprehensive pressure relief files can provide much of the needed PSI data or arranged to provide the roadmap for data locations through links to other files. The additional benefit for having accessible PSI information is a smooth transition during the PHA and reduced costs and time for the PHA process.

442025 Experimental Measurements of Flammability at Elevated Conditions Using Thermal Criterion

Abstract:

Accurate measurements of flammability limits for combustible gases and vapors are necessary for the prevention of fire and explosion. The flammability limit, which is the maximum or minimum fuel concentration at which a gas mixture is flammable in a given atmosphere, has been studied for decades. However, due to the measurement conditions and apparatus set difference, the flammability values provided by different experimental methods are inconsistent, especially at elevated conditions. In this study, the flammability limits of methane-air were measured in an 8-L closed cylindrical vessel with upward flame propagation at maximum initial temperature of 300°C and maximum initial pressure of 2 bar. During the experimental measurement, a new thermal criterion was used to define if the gas is flammable and the experimentally determined flammability limits were compared with existing data reported in the literature. Finally, the temperature and pressure effect on flammability was investigated and compared with the modified Burgess-Wheeler Law.

Use of Process Incident Data for Accident Sequence Precursor Analysis

Erin P. Collins
JENSEN HUGHES
333 E. 46th Street, #11L
New York, NY 10017
ecollins@jensenhughes.com

Bijan Najafi
JENSEN HUGHES
2105 S Bascom Ave, Suite 350
Campbell, CA 95008-3278
bnajafi@jensenhughes.com

Prepared for Presentation at
American Institute of Chemical Engineers
2016 Spring Meeting
12th Global Congress on Process Safety
Houston, Texas
April 11-23, 2016

Use of Process Incident Data for Accident Sequence Precursor Analysis

Erin P. Collins
JENSEN HUGHES
333 E. 46th Street, #11L
New York, NY 10017
ecollins@jensenhughes.com

Bijan Najafi
JENSEN HUGHES

Keywords: accident sequence precursor, data, risk analysis

Abstract

Investigations of recent process incidents have identified multiple process safety management deficiencies that led to the accident. Furthermore, audits of the Safety and Environmental Management System (SEMS) program under the U.S. Department of the Interior's Bureau of Safety and Environmental Enforcement (BSEE) have shown that the "management of risk" elements such as Hazards Analysis and Management of Change (MOC) are not consistently implemented as tools to manage risk. In addition, the requirement to perform hazards analysis is frequently documented and often referenced as a corrective action, rather than as a "system driver". There are several pieces to risk assessment and risk management that allow facility operators to evaluate and measure what is actually occurring in their facilities, including the review of historical event data to identify what went wrong and for near misses, what almost happened. This paper describes techniques that have been used in the commercial nuclear power and other industries to collect and apply historical experience data to improve safety and operability. An approach using accident sequence precursor analysis of incident data and risk analysis techniques is recommended to identify potential accidents before they occur as part of a comprehensive treatment of historical data.

1 Introduction

There are several pieces to risk assessment and risk management that allow facility operators to evaluate and measure what is actually occurring in their facilities, including the review of historical event data to identify what went wrong and for near misses, what almost happened.

For example, the U.S. Chemical Safety Board (CSB) investigation of the February 18, 2015, explosion at the ExxonMobil Refinery in Torrance, California, identified multiple process safety management deficiencies that led to the accident and a serious near miss. In the case of the near miss, "...large pieces of debris from the explosion were thrown into other units of the refinery…One of these pieces of debris hit scaffolding in the refinery's alkylation unit, narrowly missing a tank containing tens of thousands of pounds of modified hydrofluoric acid, or HF. The CSB determined that had the debris struck the tank, a rupture could have been possible, resulting in a potentially catastrophic release of extremely toxic modified HF into the neighboring community." [Ref. 1]

Such near misses are not unique to the chemical process industry, but have been documented in the offshore oil industry as well. As described by EnergyWire [Ref. 2], "Partial failures happen regularly on oil rigs, with a worker catching the errors and correcting them before a major blowout. These are not always reported to regulators, but they are logged carefully and stored as raw data on rigs and onshore in Houston."

The Discussion Paper on Precursor Analysis by the research organization Resources for the Future [Ref. 3] identifies the specific challenge faced by the oil and gas industries of "how to reduce the risk of low-frequency oil spill events, and especially high-consequence events like the Deepwater Horizon accident, when historical experience contains few oil spills of material scale and none approaching the significance of the Deepwater Horizon." Based on a review of successful programs in the nuclear power, aerospace and other industries, the paper recommends the consideration of accident precursor analysis as a means for providing "a risk-informed oversight capability focused on preventing catastrophic oil spills and based on data from ongoing operating experience on the Outer Continental Shelf (OCS)."

The ASP approach is similar to Layer of Protection Analysis (LOPA) [Ref. 4] or Bowtie Analysis in that it postulates accident scenarios and the paths that could lead to a potential accident, but differs in that it uses insights from actual historical near-miss event data to identify the branch points which, if different, could have led to a severe accident.

This paper discusses the collection of historical event data, various programs that undertake such data collection, and proposes steps for use of this data toward risk assessment and management using ASP methods.

2 Collection, Evaluation and Application of Historical Events Data

While HAZOP and FMEA are useful tools for postulating incidents and their consequences, historical event data provides important input to the risk assessment process beyond the development of accident frequencies. Incident data collection and analysis can provide multiple qualitative insights, such as information for developing and refining risk models and a better understanding of system operation and failure (including the human-machine interface and efficacy of procedures and practices).

The CSB investigates incidents at chemical process facilities, produces investigation reports such as the one for the West Fertilizer Company Fire and Explosion in April 2013 [Ref. 5] and conducts public meetings to present and discuss their findings, as with the February 2015 ExxonMobil Refinery explosion [Ref. 6].

Following the Deepwater Horizon accident, the Deepwater Presidential commission report [Ref. 7] found at the time that "Although the raw incident data are online, MMS last produced an analysis of offshore incidents—critical data for promoting the safety of offshore operations—for calendar year 2000."

Probably for this reason, incident analysis was highlighted in the discussion of the Workplace Safety Rule in 2009 [Ref. 8] and one of the key areas where it was expected that the U.S. Department of the Interior's Bureau of Safety and Environmental Enforcement (BSEE) would be focusing their attention during the second Safety and Environmental Management System (SEMS) audit cycle in 2015 was on "Developing representative leading indicators for SEMS performance, while continuing to track industry accepted lagging indicators for evidence that SEMS programs are driving an improvement in safety performance." [Ref. 9] The development of these representative indicators by its nature must come from the review of historical evidence.

The narratives and timelines of previous accidents and near-misses can indicate what triggered the event, the automatic system response and failure, subsequent system failures and their causes, and worker involvement either to cause, exacerbate or mitigate the event. These insights in turn feed back into the data collection process by indicating what trends are of interest to monitor and, therefore, what categories of data should be collected.

The Workplace Safety Rule announcement by MMS in 2009 [Ref. 8] indicated that "As a result of MMS research conducted on accident panel investigations and reports, incident analysis, and [Incidents of Non-Compliance] INCs, it appears that equipment failure is rarely the primary cause of the incident or accident. This is due to technological advances which have provided industry with very efficient and reliable equipment for finding, producing, and transporting offshore oil and gas. However, in most cases, accidents and oil spills can be traced to human error and/or organizational failures." This is consistent with the Deepwater Presidential commission report [Ref. 7], which identified several human error-related issues, including the following:

- Numerous changes to the temporary abandonment procedures for securing the well were made in the two weeks leading up to the operations plan for the worker activities, but "[t]here is no evidence that these changes went through any sort of formal risk assessment or management of change process"

- "The drilling crew and other individuals on the rig also missed critical signs that a kick [unplanned influx of gas or fluids] was occurring. The crew could have prevented the blowout—or at least significantly reduced its impact—if they had reacted in a timely and appropriate manner. What is not now clear is precisely why the crew missed these signals."

- "The crew should have diverted the flow overboard when mud started spewing from the rig floor. While that ultimately may not have prevented an explosion, diverting overboard would have reduced the risk of ignition of the rising gas. Considering the circumstances, the crew also should have activated the blind shear ram to close in the well. Diverting the flow overboard and/or activating the blind shear ram may not have prevented the explosion, but likely could have given the crew more time and perhaps limited the impact of the explosion."

While the human involvement in incidents cannot be denied, the ATHEANA method of human reliability analysis (HRA) developed by the US Nuclear Regulatory Commission [Ref. 10] discusses the situation of "error-forcing contexts" or EFCs in which the plant conditions and so-called "performance shaping factors" (context-specific influences on reliable human performance) conspire to make human error nearly unavoidable. The ATHEANA document provides examples of these EFCs that have occurred in past US nuclear plant accidents, such as mismatches between the actual plant conditions and the operator job aids (e.g., training, experience) during the Three Mile Island event. EFC elements were also identified by ATHEANA for each of the four stages of information processing (i.e., detection, situation assessment, response planning, and response implementation) that describe the range of human activities required to respond to abnormal or emergency conditions. The review of past events and near-misses considering these EFC elements may indicate that organizational factors and gaps in procedures or training may in fact be the root cause.

An independent investigation of the Deepwater Horizon by the UC Berkeley Center for Catastrophic Risk Management stated that "[T]he Macondo system's good record of lost time accidents could not provide effective precursors for the major impending system accident. Efforts to prevent system accidents must address a broader set of organizational issues and hazards associated with all of the elements that comprise a given system. The good record of lost time accidents plausibly led to complacency that contributed to the disaster." [Ref. 11]

A program such as ASP looks at historical events from the perspective of commonality of potential outcomes and identifies the organizational issues that can cause near misses to become severe accidents.

3 Chemical Process and Nuclear Power Industry Experience

Examples of the projects that exemplify the type of incident reviews that factor industry experience into lessons learned are the Accident Sequence Precursor (ASP) program [Refs. 12, 13 and 14], the AIChE CCPS Process Safety Incident Database (PSID) [Ref. 15] and Process Equipment Reliability Database (PERD) [Ref.16], the EPRI Fire Events Database (FEDB) project [Refs. 17 and 18], the EPRI Guidelines for PRA Data Analysis [Ref. 19], the AIChE CCPS Guidelines for Process Equipment Reliability Data [Ref. 20], and the EPRI/NRC-RES Fire HRA Guidelines [Ref. 21]. Jensen-Hughes has been an active participant (lead in some cases) in the development of all of these projects except PERD and PSID.

3.1 Accident Sequence Precursor (ASP) Program

The ASP Program began in the early 1980s under the auspices of the U.S. Nuclear Regulatory Commission in response to the Three Mile Island accident. ASP is concerned with the identification and documentation of operational events that have involved portions of accident sequences and with the estimation of associated frequencies and probabilities. Identification of precursors requires the review of operational events for instances in which plant functions that provide protection against severe accidents (core damage, in the case of nuclear plants) have been challenged or compromised. The intention is to understand how these 'near misses' could have propagated into a severe accident had the circumstances been slightly different and thereby inform the risk modeling, data collection and operational insights.

[NOTE: The following explanation of the US NRC's ASP program is extracted from Reference 11.]

Reports of plant incidents are reviewed to identify those potentially involving failures in systems that provide protective functions against severe accident-related initiating events or the initiating events themselves. Those events selected undergo one- or two-engineer review(s) to determine if the reported event should be examined in greater detail. Events determined to be potentially significant as a result of this initial review are then subjected to a thorough, detailed analysis. This extensive analysis is intended to identify those events considered to be precursors to potential severe accidents, either because of an initiating event or because of failures that could have affected the course of postulated off-normal events or accidents. A variety of other data sources about the same event are also reviewed and compared to thoroughly understand the contributors to each event.

Quantification of accident sequence precursor significance involves the determination of a conditional probability of a subsequent severe accident given the failures observed during an operational event. This probability is estimated by mapping failures observed during the event onto the ASP accident sequence models (event trees and linked fault trees modified to reflect the event), which depict potential paths to the severe accident, and by calculating a conditional probability of the severe accident.

The information gained from the ASP process not only provides quantitative data, but provides insights into the types of events that should be included in risk models. In addition, events that provide insights into unusual failure modes with the potential to compromise continued safety functions but are not considered to be precursors are documented by ASP as "interesting" events.

Building upon this effort, the National Academy of Engineering Program Office undertook the Accident Precursors Project in 2003 to examine the complex issue of accident precursor analysis and management. This seven-month project was designed to document and promote industrial and academic approaches to detecting, analyzing, and benefiting from accident precursors, as well as to understand public-sector and private-sector roles in using precursor information. [Ref. 10]

3.2 AIChE CCPS Process Safety Incident Database (PSID)

The purpose of the CCPS PSID is to pool process safety incident experience among participating companies so they can learn from the experiences of others without suffering the consequences of failures, while minimizing corporate liability. PSID includes process safety incidents with a potentially important lesson to be learned from incidents that did or could have resulted in a fire, explosion, fatality, multiple injuries, significant release of hazardous materials, or any other unique process safety incident (including near-misses).

3.3 AIChE CCPS Process Equipment Reliability Database (PERD)

The equipment reliability database is intended to provide high quality, valid, and useful data to the Hydrocarbon Processing Industries (HPI) and Chemical Process Industries (CPI). This type of data provides input to analyses of availability, reliability and equipment design improvements, maintenance strategies and life cycle cost determination. The database includes Lessons Learned and Contributing Factors, which are data input requirements.

3.4 Fire Events Database (FEDB) Project

The FEDB is a collection of fire events that occurred in nuclear power plants since 1967. These events were reported to the U.S. Nuclear Regulatory Commission or the industry groups such as the Institute for Nuclear Power Operations (INPO) and the Nuclear Energy Institute (NEI). These events were reviewed to extract information on the equipment involved in the fire initiation, response of fire prevention and mitigation systems, response of the fire brigade and the consequences of the fire (degraded incipient or catastrophic). Guidance was also provided for the interpretation and categorization of fire event data. Since its publications the event history has been used, at time in combination with tests and a process called Phenomena Identification and Ranking Table or PIRT, to help define and characterize accident initiation, progression, mitigation and consequences, including:

- Types and characteristics, e.g., timing, of accident initiators,
- Sequence of events, including prevention/mitigation and human interface that helps define accident scenarios (or sequences) with risk implications,
- Severity of the event, e.g., heat release rate of an electrical panel fire or zone of influence from a high energy arc fault in a high voltage switchgear,
- Mitigating systems response, e.g., reliability and effectiveness of the fire brigade response to various types of fires,
- Cable damage and quantification of effects from fire, such as spurious indication of instrumentation and control systems leading to failure of accident prevention/mitigation systems or inadequate/erroneous operator response, and
- Qualitative insights on the types of fires that are most likely to occur, their duration, severity and how the plants address them.

3.5 EPRI Guidelines for Data Analysis for PRA

The EPRI Data Guidelines discusses the nuclear power industry approach to collecting, analyzing and providing summary statistics for several major categories of data types: initiating event data, component reliability data, unavailability (downtime) data and common cause failures. Initiating events are quantified using a combination of historical evidence of event occurrence, both from the specific plant and across the nuclear industry.

3.6 AIChE CCPS Guidelines for Process Equipment Reliability Data

Many of the aspects of data analysis discussed in the EPRI Data Guidelines were identified thirteen years earlier for the chemical process industry in the AIChE CCPS Process Equipment Reliability Data Guidelines. This data book built on the Chemical Process Quantitative Risk Assessment (CPQRA) Guidelines to show the process engineer how to make quantitative estimates of the risk of hazards identified, which can in turn identify the major contributors to risk. This project helped demonstrate the need for process industry data collection efforts like PERD and PSID.

3.7 EPRI/NRC-RES Fire Human Reliability Analysis (HRA) Guidelines

Although its focus is on human reliability, the Fire HRA Guidelines discusses the need to review past incidents experienced at a facility to understand whether the event was human-caused or how the workers' responded to the event as it evolved. These insights not only assist the HRA, but provide an understanding of how workers perceive their role during an accident and how they can influence accident progression, for good or for ill. These findings can then be fed back into the improvement of organizational and safety practices for accident prevention.

4 Considerations for Historical Data Analysis Application in the Chemical Process Industry

The Discussion Paper on Precursor Analysis for Offshore Oil and Gas Drilling [Ref. 3] identifies incident reports required under the Incident Reporting Rule (30 CFR 250.188) for oil, gas and sulphur operations in the outer continental shelf as an example of industry precursor data that would allow a system failure of concern [such as "significant uncontrolled escape of hydrocarbons" (SUEH)] to be evaluated through event trees based on generic commonality of events (such as those used in the US NRC ASP program).

Further, the Discussion Paper recommends an approach for the design of an ASP-enabled incident reporting system, which has been made more general here for application to a wider variety of chemical processes:

a) Define the class of events (such as SUEH) related to the system failure of concern.

b) On the basis of PRA models, design the set of generic event trees that have these events as possible endpoints.

c) Ensure that all nodes in these generic event trees are captured in an incident reporting system, such as PSID or under 30 CFR 250.188.

d) Design a control loop, possibly based on the Potential Incident of Non-Compliance (PINC) checklist inspection system, to ensure coverage of the incident reporting system, with appropriate sanctions for noncompliance.

e) Set up an analysis team to map the key even-related incidents onto generic event trees.

f) Set up a review board comprising industry and government agencies.

g) Design feedback mechanisms for involving industry in efforts to optimize industry-wide performance as measured by the ASP indicators.

h) Design an automation plan to streamline the ASP analysis.

The following additional tasks should be considered for translating these steps into industry guidance:

a) Documentation of insights and development of a draft guidelines for a selected process facility (to be defined) reflecting items a) through h) above.

b) Testing, validation and review of the guidelines. This may be done using one or more well documented industry events.

c) Final guideline for the selected process facility.

d) Adaptation of the guideline to other processes.

5 Conclusion

A recent article in CEP Magazine [Ref. 22] profiles five chemical process industry accidents that, while not as severe as Bhopal or Deepwater Horizon, still had serious results. The article states that "Accidents such as those discussed in this article are more likely to occur. These incidents demonstrate that even small mistakes can have disastrous consequences. Therefore it is imperative that engineers learn from previous incidents to reduce their likelihood of recurring." Near misses, like smaller scale events, are also instances that can provide valuable insights on lapses in process safety management.

As stated in EnergyWire [Ref. 2]. "ASP would use the data from these partial failures, including the failure of rig workers to respond appropriately, to compute the likelihood of a catastrophe. It would identify problems when they occur so regulators can stop them from happening again."

6 References

[1] "U.S. Chemical Safety Board Finds Multiple Safety Deficiencies Led to February 2015 Explosion and Serious Near Miss at the Exxon Mobil Refinery in Torrance, California," News Release, U.S. Chemical Safety and Hazard Investigation Board, Washington, DC, 16 January 2016.

[2] Vaidyanathan, G., "GULF SPILL: Could risk analysis prevent future deepwater disasters?" EnergyWire, Environment & Energy Publishing (E&E), http://www.eenews.net/stories/1059963200, 20 April 2012.

[3] Cooke, R.M., Ross, H.L. and Stern, A., "Discussion Paper - Precursor Analysis for Offshore Oil and Gas Drilling, From Prescriptive to Risk-Informed Regulation", RFF DP 10-61, Resources for the Future, Washington, DC, January 2011.

[4] Guidelines for Initiating Events and Independent Protection Layers in Layer of Protection Analysis, American Institute of Chemical Engineers (AIChE), Center for Chemical Process Safety (CCPS), New York, NY, 2015.

[5] "West Fertilizer Company Fire and Explosion", Final Investigation Report 2013-02-I-TX, U.S. Chemical Safety and Hazard Investigation Board, Washington, DC, January 2016.

[6] Presentation slides from CSB Public Meeting on ExxonMobil Refinery Explosion, U.S. Chemical Safety and Hazard Investigation Board, Washington, DC, 13 January 2016.

[7] Deep Water: The Gulf Oil Disaster and the Future of Offshore Drilling, Report to the President, National Commission on the BP Deepwater Horizon Oil Spill and Offshore Drilling, January 2011.

[8] The Bureau of Ocean Energy Management, Regulation and Enforcement, Office of Public Affairs, Fact Sheet, The Workplace Safety Rule on Safety and Environmental Management Systems (SEMS).

[9] Memorandum from Doug Morris, Chief, Office of Offshore Regulatory Programs to Brian Salerno, Director BSEE, SEMS Program Summary—First Audit Cycle (2011-2013), July 23, 2014.
http://www.bsee.gov/uploadedFiles/BSEE/Regulations_and_Guidance/Safety_and_Environmental_Management_Systems_-_SEMS/SEMS%20Program%20Summary%208132014.pdf

[10] Technical Basis and Implementation Guidelines for A Technique for Human Event Analysis (ATHEANA), NUREG-1624, Rev. 1, U.S. Nuclear Regulatory Commission, Washington, DC, May 2000.

[11] Final Report on the Investigation of the Macondo Well Blowout, Deepwater Horizon Study Group (DHSG), Center for Catastrophic Risk Management (CCRM), University of California – Berkeley, Berkeley, CA, March 2011.

[12] Minarick, J. W. and Kukielka, C. A., Precursors to Potential Severe Core Damage Accidents: 1969-1979, A Status Report, NUREG/CR-2497, U.S. Nuclear Regulatory Commission, Washington, DC, June 1982.

[13] Belles, R.J. et al., Precursors to Potential Severe Core Damage Accidents: 1998, A Status Report, NUREG/CR-4674, US Nuclear Regulatory Commission, Washington, DC, July 2000.

[14] Phimister, J.R., Bier, V.M., and Kunreuther, H.C. (Eds.), Accident Precursor Analysis and Management: Reducing Technological Risk Through Diligence, The National Academies Press, Washington, DC, 2004.

[15] Presentation slides, Overview: Process Safety Incident Database (PSID), American Institute of Chemical Engineers (AIChE), Center for Chemical Process Safety (CCPS)

http://www.aiche.org/ccps/resources/psid-process-safety-incident-database

[16] American Institute of Chemical Engineers (AIChE), website "About PERD"

http://www.aiche.org/process-equipment-reliability-database-perd/about-perd

[17] EPRI 1000894, Fire Events Database for U.S. Nuclear Power Plants: Update Through 1999, Electric Power Research Institute (EPRI), Palo Alto, CA, October 2000.

[18] EPRI 1025284, The Updated Fire Events Database: Description of Content and Fire Event Classification Guidance, Electric Power Research Institute (EPRI), Palo Alto, CA, July 2013.

[19] EPRI 3002000774, EPRI Guidelines for PRA Data Analysis, Electric Power Research Institute (EPRI), Palo Alto, CA, December 2013.

[20] Guidelines for Process Equipment Reliability Data with Data Tables, American Institute of Chemical Engineers (AIChE), Center for Chemical Process Safety (CCPS), New York, NY, 1989.

[21] EPRI 1023001/ NUREG-1921, EPRI/NRC-RES Fire HRA Guidelines, Electric Power Research Institute, Palo Alto, CA and U.S. Nuclear Regulatory Commission, Washington, DC, July 2012.

[22] A. Ness, "Lessons Learned from Recent Process Safety Incidents," CEP Magazine, American Institute of Chemical Engineers (AIChE), New York, NY, March 2015.

442384 Quantitative Risk Analysis of the Expansion of the Chlorine Room of the Industrial Water Plant of the Petrochemical Complex Jose Antonio Anzoategui of Petroquimica Venezuela Pequiven S.a

Abstract:

Quantitative risk analysis of the expansion of the chlorine room of the industrial water plant of the petrochemical complex Jose Antonio Anzoategui of Petroquimica venezuela PEQUIVEN S.A.

Venezuela Petrochemical PEQUIVEN SA has a management responsible for risk control, which makes this quantitative risk analysis to estimate the risk to employees and third parties for process safety events that may occur as a result of the loss of containment in the facilities and thus indicate recommendations aimed at reducing the level of risk associated with such events. Quantitative risk analyzes are made from the standard rules of Petroleos de Venezuela , PDVSA as IR-S-02: criteria for quantitative risk analysis and other international standards such as Center of Chemical Process Safety Guidelines for Chemical Process Quantitative Risk Analysis and the netherlands organization Yellow book: Methods for calculation of physical effects, through the computational tool PHAST RISK 6.7 . Developed by DNV Company.

PEQUIVEN has three complexes nationwide, in which fertilizers for the food industry, olefins and plastics for residential construction, among other products are produced. The Jose Antonio Anzoategui complex will make an expansion of the chlorine room of the industrial water plant and because of the dangers associated with chlorination process, a quantitative risk analysis has requested to engineering management risks.

The process of a quantitative risk analysis (QRA) started setting goals, objectives and depth of the study, once these are established, and in conjunction with materials, inventory, environmental conditions and the description of the process, one proceeds to identify scenarios. In this case, the HAZOP tool was applied, which gives the place a list of possible scenarios, of which those that could cause the worst consequences. With the previously selected scenarios, the model building on which the estimated frequency and consequence will be held is started. The frequency is estimated based on a review of probability theory study according to Chemical Process Safety Fundamentals with Applications Standard, through a theoretical data extracted from the norm Guidelines for Process Equipment Reliability Data, a probability of occurrence of the scenarios is established.

The estimation of Consequence answer a series of complex mathematical models that simulate the behavior of these events based on the data of the others that occurred already trough PHAST RISK software. Once the simulation is made, the software generates a series of results of contours risks in function of exposure to chlorine gas to the personnel working in this complex and surrounding facilities, allowing emit a series of observations than are aimed to decrease level of Risk associated with the expansion of the chlorine room.

12TH GLOBAL CONGRESS ON PROCESS SAFETY

Assessing Cognitive Human Errors for the Process Industries

Erin P. Collins
JENSEN HUGHES
333 E. 46th Street, #11L
New York, NY 10017
ecollins@jensenhughes.com

Bijan Najafi
JENSEN HUGHES
2105 S Bascom Ave, Suite 350
Campbell, CA 95008-3278
bnajafi@jensenhughes.com

Prepared for Presentation at
American Institute of Chemical Engineers
2016 Spring Meeting
12th Global Congress on Process Safety
Houston, Texas
April 11-23, 2016

Assessing Cognitive Human Errors for the Process Industries

Erin P. Collins
JENSEN HUGHES
333 E. 46th Street, #11L
New York, NY 10017
ecollins@jensenhughes.com

Bijan Najafi
JENSEN HUGHES

Keywords: cognitive, human error, human reliability analysis

Abstract

Through the analysis of actual incidents and the understanding of the underlying human errors that have contributed to them, it is apparent that 'first generation' HRA methods are appropriate for some instances but have their limitations. For example, failure to recognize the significance of high fluid returns during the negative pressure test and failure to perceive the importance of drill pipe pressure during well monitoring activities were key contributors to the Deepwater Horizon incident. These diagnosis and decision-making errors are not well addressed by first generation HRA methods and require particular timing, error mode and performance shaping factor considerations. This paper provides an introduction to HRA methods and then discusses those available for evaluating cognitive errors and their applicability to the chemical process industry.

1 Introduction

Layer of Protection Analysis (LOPA) provides a simplified risk analysis technique for a select set of postulated incident scenarios, evaluating an initiating event (IE) and subsequent independent protection layers (IPLs) as a series of probabilities of failure on demand (PFDs). Human error is assessed in LOPA as either an IE or IPL. Techniques that have been recommended for human reliability assessment (HRA) by the CCPS Guidelines for IEs and IPLs in LOPA [Ref. 1] are generally considered in other industries as 'first generation' HRA methods, meaning that they are appropriate for some instances but have their limitations. These limitations have become more apparent through the analysis of actual incidents and the understanding of the underlying human errors that have contributed to them. For example, failure to recognize the significance of high fluid returns during the negative pressure test and failure to perceive the importance of drill pipe pressure during well monitoring activities were key contributors to the Deepwater Horizon incident. These diagnosis and decision-making errors are not well addressed by first generation HRA methods and require particular timing, error mode and performance shaping factor (PSF) considerations. Several HRA quantification methods are available to assess the human error probability associated with these so-called "cognitive" aspects of a task, as well as the equipment manipulation.

While many HRA approaches had their genesis in the nuclear power industry, they have been successfully applied to other industries, such as chemical weapons destruction plants and nuclear fuel transport and storage, and their results have been accepted during peer and regulatory agency reviews. In each instance where HRA is applied to a new industry, however, an evaluation must be made regarding the degree of applicability of the standard error types, PSFs and quantification methods. Since the error types and PSFs have an underpinning in human factors, psychology and cognitive science, they generally apply regardless of the industry. However, it is important to evaluate the applicability of HRA tools and techniques to ensure that industry-specific issues are being appropriately addressed.

This paper provides an overview of HRA principles and methods, then discusses those available for evaluating cognitive errors and their applicability to the chemical process industry.

2 HRA Principles and Methods

HRA provides a framework to evaluate, prioritize, quantify and improve the human interface with complex systems. HRA tools and techniques were pioneered by the commercial nuclear power industry, but are based on fundamental precepts of human reliability and data collection and analysis that are not technology specific. They have been applied to an increasing number of facilities across various different industries, such as aerospace [Ref. 2], chemical weapons destruction [Ref. 3], HVAC system optimization, nuclear fuel disposal facility design risk evaluation [Ref. 4] and detection and suppression activities in response to fire events in the military [Ref. 5]. This experience and repetition have allowed lessons learned to be identified and fed back into safe practices, the development of a safety culture, and the ability to meet standards [Ref. 6] and regulatory goals (which are quite stringent in the nuclear power industry).

This history and associated lessons learned indicates that wider application of HRA in the chemical process industry could provide greater understanding of risks (both accident and economic), how to control them (by improving procedures, focusing maintenance and surveillance efforts on the actual errors), and how to develop a safety culture that focuses on human error prevention.

The steps of the HRA process discussed below are adapted from NUREG-1921, the EPRI/NRC-RES Fire Human Reliability Analysis Guidelines [Ref. 7]. The process begins with an identification and definition task to identify actions performed by various workers and to define the actions in terms of human failure events or HFEs. These actions may include both failure to perform a required task or undesired actions with deleterious impact on system performance and safety. For the chemical process industry, this identification and definition step could be expected to be very closely tied with the subsequent step of qualitative analysis because the same information gathering that helps define the events also feeds into the development of an "operational story" or HFE narrative that assists in further understanding the events.

A sound qualitative analysis allows the HRA to provide feedback to the facility operator on the factors contributing to both the success and failure of a worker action. This involves the evaluation of performance shaping factors (PSFs), which are situation-specific influences on human error probabilities (HEPs) and include:

- Cues and indications that indicate a system problem
- Timing of the task
- Procedures and training
- Complexity of the task
- Workload, pressure and stress
- Human-Machine Interface (HMI)
- Environmental factors, such as heat or confined spaces
- Special equipment, such as tools and protective gear

These PSFs are used to flesh out the HFE narrative and identify the key elements of the HFEs that require further understanding through the performance of subject matter expert interviews.

Finally, quantification or quantitative analysis is performed by applying the information gained from identification, definition and qualitative analysis of HFEs toward the estimation of HEPs. Quantification can be done at varying degrees of detail and rigor. The three levels of analysis performed in HRA are: Screening, Scoping and Detailed.

Screening analysis involves the assignment of a preliminary and conservative expert judgment value as a means for "screening" out the risk-significant HFEs from the non-contributing ones.

The Scoping approach, discussed in NUREG-1921, was developed to provide additional rigor to a semi-screening process, but still results in conservative values for initial quantification.

Detailed analysis is performed by evaluating the action steps, timing and PSFs in detail and assessing the probability of errors in the cognitive aspects, or diagnosis and decision-making process for the correct action to take, as well as the execution aspects, such as the equipment manipulation steps.

Several HRA quantification methods are available and it has been found to be most effective if the analyst evaluates the strengths and weaknesses of each method against the nature of the human error being evaluated to make the best match. Having a large HRA method toolkit to draw from allows the analyst the flexibility to assess the driving failure mechanism behind the error rather than simply filling out a checklist.

The following HRA quantification methods are most widely used in the nuclear power industry, and are contained in Electric Power Research Institute's (EPRI's) HRA Calculator software [Ref. 8]:

- Human Cognitive Reliability/Operator Reliability Experiment (HCR/ORE) [Ref. 9], which focuses on the cognitive or decision making errors that are constrained by time;

- Cause-Based Decision Tree Method (CBDTM) [Ref. 9] that addresses cognitive errors as well but from the standpoint of the error-causing mechanisms, such as unclear instrumentation or lack of procedural direction;

- The Technique for Human Error Rate Prediction (THERP) [Ref. 10], which has been used for many years for the evaluation of task performance failures (also called execution errors) that can occur either before or after an accident;

- The Accident Sequence Evaluation Program (ASEP) [Ref. 11] that provides a simplified version of THERP for human errors that contribute to but occur prior to the accident; and

- The SPAR-H method [Ref. 12], which only permits highly simplified assessments of cognitive and action errors (and has not been used in the latest generations of nuclear plant risk assessments since it does not allow the depth of analysis to support safety improvements).

For HRA applications other than nuclear power, such as chemical weapons destruction plants and nuclear fuel transport and storage, other quantification methods have been successfully applied and their results accepted during peer and regulatory agency reviews. These include the Cognitive Reliability Error Analysis Method (CREAM) [Ref. 13], and the Human Error Assessment and Reduction Technique (HEART) [Ref. 14]. As the name implies, CREAM focuses on the decision-making errors by providing HEPs for a standard set of task types, but also includes adjustment factors to evaluate facility and scenario-specific influences. HEART contains a list of generic task types and their HEP estimates that are modified using factors associated with 'error producing conditions'.

While the application of a method and generation of HEP values is important, the primary output of the HRA is the identification and prioritization of the human error issues of greatest risk significance and the development of risk insights. The benefit of conducting more detailed HRA is in gaining a better perspective on what influences human failure and therefore what can be done to improve human performance.

The decomposition of human error into cognitive vs. execution tasks and a detailed evaluation of these tasks has proven to be particularly important since the approaches to their improvement are different. For example, more instrumented processes attempt to provide information to the operator that allow him/her to diagnose the situation and take the appropriate action. In other cases, however, the process may not include such instrumentation or it may be not be available due to failure, so diagnosis becomes a critical issue. Even when there is no instrumentation or early warning system provided for an action, the HRA can provide an understanding of the cues required for diagnosis and allow these to be included in training and procedures/manuals. These cues may even be sensory in nature, such as the sound, smell or sight of a process that is going awry. This does not mean that the HRA will recommend instrumenting everything. If correctly done, the HRA instead allows the most critical information used by the worker to be identified and suggests how it should be presented or emphasized to optimize performance.

The reliability of execution tasks is generally improved by providing a formal set of procedures and training that are continually reviewed and refined. However there may be instances where procedures and training are not yet formalized, such as when the facility is a new design or is in the pre-operational testing phase, or because of historical precedent or preference. In these instances, the worker action is evaluated in HRA as a so-called 'recovery' action, where the worker utilizes previous experience with tasks they consider to be similar and applies them to a new situation. These insights are important to evaluating the reliability of task performance and can even be subsequently captured in procedure modifications and training enhancements. The risk benefit assessment might even recommend conducting a design change to the human-machine interface to improve the execution task reliability.

In summary, HRA is

- Widely used
- Applied across a variety of industries, which demonstrates that it provides value
- Allows the user to quantify and prioritize the human influence on safety
- Provides the means for improving process safety and reliability in the short term and a safety culture to reduce future errors over the longer term.

3 Cognitive Errors in the Process Industries

The Deepwater Presidential commission report [Ref. 15] identified several human error-related issues that led to the severity of the incident, including the following:

- Numerous changes to the temporary abandonment procedures for securing the well were made in the two weeks leading up to the operations plan for the worker activities, but "[t]here is no evidence that these changes went through any sort of formal risk assessment or management of change process"
- "The drilling crew and other individuals on the rig also missed critical signs that a kick [unplanned influx of gas or fluids] was occurring. The crew could have prevented the blowout—or at least significantly reduced its impact—if they had reacted in a timely and appropriate manner. What is not now clear is precisely why the crew missed these signals."
- "The crew should have diverted the flow overboard when mud started spewing from the rig floor. While that ultimately may not have prevented an explosion, diverting overboard would have reduced the risk of ignition of the rising gas. Considering the circumstances, the crew also should have activated the blind shear ram to close in the well. Diverting

the flow overboard and/or activating the blind shear ram may not have prevented the explosion, but likely could have given the crew more time and perhaps limited the impact of the explosion."

These issues can be considered as relating to cognitive human errors since they concern the lack of procedural guidance that assists with diagnosis and the decision-making on the part of the crew that appears to have been ill advised or ill informed. Hubbard and Embrey [Ref. 16] assessed these issues from a human factors standpoint and they raised the following concerns related to the cognitive errors during the incident:

- Why were procedures not used – is this typical?
- What was the effect on decision making and on practice of the pressure to get the well tested and capped?
- Why did the crew use the mud gas separator rather than pipework that would have discharged the mud more quickly?
- What emergency training did the team have? Did they have training or job support to help them identify and respond to escalating situations?

Questions such as these are key to the evaluation and assessment of cognitive errors.

A recent article by Ness in the CEP Magazine [Ref. 17] outlined lessons learned from several process accidents, many of which involved cognitive human errors:

- ARCO Channelview wastewater tank explosion, 1990 – "The wastewater tank was not considered part of the operating plant. Hence, the management and workers did not understand that a chemical reaction was taking place in the tank, generating oxygen. The lack of understanding enabled a series of poor decisions, such as discontinuing the nitrogen purge…no management of change (MOC) review of these decisions, and no pre-startup safety review."
- Terra Industries ammonium nitrate explosion, 1994 – "…the lack of procedures for shutdown and monitoring the equipment during shutdown led operators to perform actions that sensitized the AN solution and provided energy to the decomposition reaction."
- Formosa Plastics vinyl chloride release, 2004 - "The operator overrode an interlock, which led to a release of hot, pressurized VCM. Formosa Plastics did not have comprehensive written standards…for managing interlocks on the vessels. Employees were unprepared for a major accident at the facility."

These incidents demonstrate faulty understanding of the process and lack of procedural direction, which in turn led to improper decision-making and catastrophic consequences.

4 Treatment of Cognitive Human Error in Process Guidelines and Studies

The CCPS *Guidelines for Preventing Human Error in Process Safety* [Ref. 18] mentions cognitive failure assessment methods such as CBDTM, CREAM and HEART but does not provide much guidance to the user for implementing them. Furthermore, in the ten years since that guidance was issued, the Deepwater Horizon incident has occurred and much has been learned in various industries on the analysis and mitigation of human error using HRA.

The recently issued CCPS *Guidelines for Initiating Events and Independent Protection Layers in Layer of Protection Analysis* [Ref. 1] states that "developing PFD values for human IPLs is not straightforward." Note that PFDs as defined in these guidelines include not only the HEP, but the PFD of the cue for response, namely the "instrumentation or equipment used to detect the process deviation and/or take action to return the process to a safe state."

The IPLs for human actions that are cited in these guidelines are:

- Human response to an abnormal condition
- Human response to an abnormal condition with multiple sensors and indications and the operator has >24 hours to accomplish the required response action
- Adjustable movement-limiting device – this is actually equipment that is intended to prevent operator misoperation of equipment beyond a limit that would initiate a scenario of concern
- Personal Protective Equipment (PPE) – safety gear that acts as a barrier between a person and a hazard.

In an HRA context, the first two IPLs would be considered Human Failure Events (HFEs), but the specification of sensors/indications (cues) and time available for response would be evaluated in detail for both cases, in addition to other scenario-specific PSFs, such as procedures and training. Most importantly, the time available would be balanced against the time *required* for the action to assess the time margin and the HEP.

Note that the *nature* of the human response is not specified for these IPLs, so there is no differentiation between cognitive and execution errors, which have different aspects and methods for evaluation.

The adjustable movement-limiting device IPL would not itself considered to be a human failure, but would be considered as a mechanical form of interlock that would need to be defeated (either through another human action to leave it out of position after maintenance or through equipment failure) to allow the human error to occur. A Quantitative Risk Assessment (QRA) model allows such equipment and human failures to be functionally linked.

PPE is addressed in HRA not as an HFE but as a performance shaping factor on an existing human response action since it is considered to cause sight, sound and manipulation limitations on the human action, and can even lead to increased stress.

Papers presented in recent years discuss application of HRA to the petroleum industry through the use of the SPAR-H quantification method [Refs. 19 & 20]. Gould, Ringstad and Van De Merwe state that "The main advantages of SPAR-H is that it provides clear guidance on assigning numerical weights to the performance-shaping factors, is not very resource-demanding, and includes worksheets that promote analyst consistency." While SPAR-H provides a useful screening analysis of factors related to cognitive error, such as Available Time, Procedures and Ergonomics, it would seem to be best considered as the first step toward identifying risk-significant cognitive human errors, which would then need to be evaluated in greater detail. The benefit of detailed HRA lies in identifying the driving mechanisms behind the error that can then be applied to steps for error likelihood reduction.

Another current study [Ref. 21] has successfully applied the CREAM cognitive error evaluation technique to the cargo loading process of LPG tankers "to systematically predict human error potentials for designated tasks and to determine the required safety control levels on-board LPG ships." The human cognition and action context are determined in accordance with Common Performance Conditions (CPCs) which provide a well-structured and comprehensive basis for characterizing the circumstances and to define sets of possible error modes and error causes.

5 Information Processing Model for Cognitive Activities

Assessment of HFEs can be guided by a model of higher-level cognitive activities, such as an information processing model. The model that has been used by the authors is based on the discussion in Chapter 4 of the ATHEANA method [Ref. 22] and consists of the following elements:

- Monitoring and detection - Both of these activities are involved with extracting information from the environment. Also, both are influenced by the characteristics of the environment and the person's knowledge and expectations. Monitoring that is driven by the characteristics of the environment is called data-driven monitoring. Monitoring initiated by a person's knowledge or expectations is called knowledge-driven monitoring. Detection can be defined as the onset of realization by operators that an abnormal event is happening.

- Situation awareness - This term is defined as the process by which operators construct an explanation to account for their observations. The result of this process is a mental model, called a situation model that represents the operator's understanding of the present situation and their expectations for future conditions and consequences.

- Response planning - This term is defined as the process by which operators decide on a course of action, given their awareness of a particular situation. Often (but not always) these actions are specified in procedures.

- Response implementation - This term is defined as the activities involved with physically carrying out the actions identified in response planning.

The final bullet concerns the execution of the response, but is predicated upon the successful cognitive assessment of the course of action.

By assessing these features of a human action, the analyst can then select from the available HRA models to evaluate the HEP and moreover, identify the driving issues leading to error that can be factored into error prevention.

6 Conclusions

Recent studies are showing that existing HRA methods can be applied to assess cognitive human errors in the process industries, but these insights have yet to be incorporated into industry guidelines. While initial assessments can be made with simplified tools, the key to addressing and preventing major incidents is the identification of underlying human error mechanisms so that risk-mitigating solutions that truly address the diagnosis and decision-making process can be developed.

7 References

[1] *Guidelines for Initiating Events and Independent Protection Layers in Layer of Protection Analysis*, Center for Chemical Process Safety, American Institute of Chemical Engineers, New York, NY, 2015.

[2] Collins, E.P. and Fragola, J.R., "Advances in Human Error Evaluation", 1995 Reliability, Availability, Maintainability Symposium (RAMS), Washington, DC, January 1995.

[3] Umatilla Chemical Agent Disposal Facility Quantitative Risk Assessment, Report No. SAIC-00/2641. Volume I, Science Applications International Corporation (SAIC), Abingdon, MD, 2002.

[4] Yucca Mountain Project Canister Receipt and Closure Facility Reliability and Event Sequence Categorization Analysis, 060-PSA-CR00-00200-000-00B, Bechtel SAIC Company, LLC (BSC), November 2008.

[5] Garvey, M. (Dept. of the Navy), Joglar, F. and Collins, E.P. (JENSEN HUGHES, Inc.), "HRA for Detection and Suppression Activities in Response to Fire Events", 2014 Annual Reliability and Maintainability Symposium (RAMS), Colorado Springs, CO, January 2014.

[6] ASME/ANS RA-Sa-2009, Addenda to ASME/ANS RA-S-2008, "Standard for Level 1/Large Early Release Frequency Probabilistic Risk Assessment for Nuclear Power Plant Applications", The American Society of Mechanical Engineers (ASME), New York, New York, 2009.

[7] Electric Power Research Institute and U.S. Nuclear Regulatory Commission, EPRI/NRC-RES Fire HRA Guidelines, EPRI 1023001/ NUREG-1921, July 2012.

[8] Electric Power Research Institute, The EPRI HRA Calculator® Software Manual, EPRI 3002004030. Palo Alto, CA, May 2014.

[9] EPRI TR-100259, An Approach to the Analysis of Operator Actions in Probabilistic Risk Assessment, Electric Power Research Institute (EPRI), Palo Alto, CA, June 1992.

[10] Swain, A. D. and Guttman, H., Handbook of Human Reliability Analysis with Emphasis on Nuclear Power Plant Applications, NUREG/CR-1278, Washington, DC: U.S. Nuclear Regulatory Commission, August 1983.

[11] Swain, A., Accident Sequence Evaluation Program Human Reliability Analysis Procedure, NUREG/CR-4772, U.S. Nuclear Regulatory Commission, Washington, DC, February 1987.

[12] The SPAR-H Human Reliability Analysis Method, NUREG/CR-6883, Idaho National Engineering Laboratory for U.S. Nuclear Regulatory Commission, Washington, DC, August 2005.

[13] Hollnagel, E., *Cognitive Reliability and Error Analysis Method – CREAM*, Elsevier, New York, NY, 1998.

[14] Williams, J., HEART - A Proposed Method for Assessing and Reducing Human Error, 9th Advances in Reliability Technology Symposium, National Centre of Systems Reliability, University of Bradford, Bradford, UK, 1986.

[15] Deep Water: The Gulf Oil Disaster and the Future of Offshore Drilling, Report to the President, National Commission on the BP Deepwater Horizon Oil Spill and Offshore Drilling, January 2011.

[16] Hubbard, A. and Embrey, D., "Deepwater Horizon – Summary of Critical Events, Human Factors Issues and Implications," Human Reliability Associates, Ltd., 23 September 2010.

[17] Ness, A., "Lessons Learned from Recent Process Safety Incidents," CEP Magazine, American Institute of Chemical Engineers (AIChE), New York, NY, March 2015.

[18] *Guidelines for Preventing Human Error in Process Safety*, American Institute of Chemical Engineers (AIChE), Center for Chemical Process Safety (CCPS), August 2004.

[19] Gould, K.S. and Ringstad, A.J (StatOil) and Van De Merwe, K. (Det Norske Veritas), "Human Reliability Analysis in Major Accident Risk Analyses in the Norwegian Petroleum Industry", Proceedings of the Human Factors and Ergonomics Society 56th Annual Meeting, San Diego, CA, October 2012.

[20] Boring, R.L (Idaho National Laboratory) and Øien, K. (SINTEF), "Defining Human Failure Events for Petroleum Risk Analysis", Probabilistic Safety Assessment and Management PSAM 12, Honolulu, HI, June 2014.

[21] Akyuz, E. and Celik, M., Application of CREAM human reliability model to cargo loading process of LPG tankers, Journal of Loss Prevention in the Process Industries, Vol. 34, pgs. 39-48, Elsevier Ltd., 2015.

[22] Technical Basis and Implementation Guidelines for A Technique for Human Event Analysis (ATHEANA), NUREG-1624, Rev. 1, U.S. Nuclear Regulatory Commission, Washington, DC, May 2000.

442486 Critical Considerations in the Vacuum Protection of Butane Spheres

Abstract:

Vacuum protection of pressure vessels, including spheres, is not well defined in the ASME Boiler and Pressure Vessel Code (BPVC) Section VIII, Division I or API Std. 520. Vacuum protection of pressure vessels usually follows API Std. 2000 on low pressure tanks for guidance.

Vacuum can occur on butane storage spheres at temperatures of 25 to 30 F due to butane condensing. This paper addresses the challenges in vacuum protection of butane storage spheres, including both the positive pressure rating of vacuum devices and sizing of vacuum vents for a butane condensing scenairo.

442493 Assessing Risk in the Supply Chain Using the Bowtie Method

Abstract:

After a number of recent incidents involving the transportation and storage of hazardous materials, many industry end-users wish to understand their level of risk exposure during the supply chain. This exposure can either be via actual ownership or when the product can be associated with the concerned party. Ultimately, this may mean that an end-user's procurement group will want to see that their suppliers of products and services are performing as needed in the control of these risks.

A qualitative risk assessment is a technique by which these risks can be identified, understood, and evaluated. This paper will describe the application of one such method, where the bowtie barrier analysis is central to the means by which this assessment is completed. The barriers identified are those necessary to prevent and mitigate undesired events that could result in a fatality or a catastrophic incident to the environment, public, or a company's reputation.

The overall method employed in this process includes the following steps:

1. Hazard identification
2. Bowtie diagram development and barrier identification
3. Barrier screening to identify safety critical elements - those that provide a credible reduction in the risk of fatalities or other catastrophe
4. Performance standard development for the safety critical elements

A quantitative risk assessment can also be completed to provide additional risk knowledge to aid in prioritization and decision-making. The performance standards are then enforced appropriately throughout the supply chain. This will ensure that the transportation and handling of hazardous materials meet the purchasing company's expectations with regards to proper risk management.

442386 Reference Framework for the Application of Quantitative Risk Analysis for Hydrocarbon Pipelines, Coupled with Uncertainty Treatment Methods: Techniques for Analysis and Modelling Methods

Abstract:

Risk assessment is imperative to reduce consequences of accidents generated by leaks on pipelines. After a failure in an urban pipeline, the contents can be released rapidly forming a spreading pool in the surface or a vapor cloud that can lead to different events such as fires and explosions, affecting the population in the area. Generally, deterministic assessments are carried out where certain assumptions are made in the models used to calculate failure and event probabilities, event consequences and risk values. This causes the existence of uncertainty that is not considered and therefore it generates results only for "mean" or "worst-case" scenarios. Not taking uncertainty into account can lead to decisions with an unknown degree of conservatism: overestimate results could increase cost of decisions and underestimate results can lead to poor risk management. There is a lack of a clear directive to use methods of uncertainty assessment in risk analysis and for that reason, a framework is needed for carrying out quantitative risk assessment coupled with uncertainty treatment methods in order to handle, reduce, and properly inform uncertainty to the decision makers.

Uncertainty appears when we try to represent a system using different analysis techniques and therefore, depends on how accurately this process can be done. This in turn relies on how complex is the system under analysis and which techniques can be used. As part of the framework mentioned above, a classification is suggested to adequately select methods in probabilistic risk analysis to take into account different types of uncertainty in different kinds of representation and system analysis techniques. The classification is first based on the complexity of the system to be analyzed and then in the complexity of the modeling technique that is used to deal with system complexity. System complexity depends mainly on three variables: 1) understanding about the system (variables and its interactions) 2) the number of elements in the system 3) complexity of system interactions. Depending on system complexity, different techniques to represent it and analyze these systems in risk analysis are proposed and then, methods to propagate uncertainty are mentioned. A comparison of methodologies is presented to provide the engineer with tools to use adequately these methods in different situations and provide guidelines to deal with uncertainty in risk assessment.

Dynamics Operational Risk Management in Organizational Design, the Challenge for Sustainability

Salvador Ávila Filho
Universidade Federal da Bahia, UFBA
Rua Aristides Novis, Federação 02, Salvador Bahia
avilasalva@gmail.com

José Filipe Michel Gagliano Ferreira
Universidade Federal da Bahia, UFBA
Rua Aristides Novis, Federação 02, Salvador Bahia
gagliano.fm@gmail.com

Carlos Roberto de Oliveira Sousa
Universidade Federal da Bahia, UFBA
Rua Aristides Novis, Federação 02, Salvador Bahia
carlososousa.cs@gmail.com

Ricardo Araújo Kalid
Universidade Federal do Sul da Bahia, Itabuna
Ricardo.kalid@gmail.com

Prepared for Presentation at
American Institute of Chemical Engineers
2016 Spring Meeting
12th Global Congress on Process Safety
Houston, Texas
April 11-23, 2016

Dynamics Operational Risk Management in Organizational Design, the Challenge for Sustainability

Salvador Ávila Filho
Universidade Federal da Bahia, UFBA
Rua Aristides Novis, Federação 02, Salvador Bahia
avilasalva@gmail.com

José Filipe Michel Gagliano Ferreira
Universidade Federal da Bahia, UFBA
Rua Aristides Novis, Federação 02, Salvador Bahia
gagliano.fm@gmail.com

Carlos Roberto de Oliveira Sousa
Universidade Federal da Bahia, UFBA
Rua Aristides Novis, Federação 02, Salvador Bahia
carlososousa.cs@gmail.com

Ricardo Araujo Kalid
Universidade Federal do Sul da Bahia
Itabuna
ricardo.kalid@gmail.com

Keywords: business risk and corporative, human factors and organizational, tools, sustainability

Abstract

The maintenance of the organizational sustainability requires acknowledgment that business risks are dynamic and, therefore, depend on the balanced analysis of issues relating to product and process with people and culture. The analysis of the dynamic risk occurs in three steps as described as follows. The first analyzes the balance between the human and social factors with process and product factors, generating preliminary recommendations from probable scenarios through techniques such as LOPA and PRA with recommendations to avoid accidents and technical failures. Risks related to the environmental and human aspects are analyzed with PRA-SH to anticipate future events that may cause technical failure or evoke a stressor environment in the workplace. The second step requires knowledge of the organization in its aspects of management, technical and non-technical, inputs and productivity benchmark to define the assumptions and criteria used to project performance of operations, including the skills required by the business. The result of this study is the definition of technical, human, organizational and natural factors for the sustainable operation of production system. The techniques applied in this second stage are: HAZOP and HAZOP-Social; requirements for the task (GMTA, PADOP); events cause and consequence analysis; and analysis of barriers for events (BOWTIE). In the third stage of dynamic risk analysis rescues the criteria and assumptions established in the previous step by defining the management model and detailing the technology used. In this phase, it will be handled operational routines, asset control, control of failures and possible crisis. Thus, it will designed the systems of technical area (equipment, processes, products and control modes), interface (task, management, and human-machine interface), and social area (culture and biases, cognitive decision, stress analysis and behavior). The factors that make up the control of operations are analyzed in the standard procedure way and in the failure analysis using the following techniques: equipment (FMEA, FMEAH); product and process (monitoring and statistical control); task in complex mode (GMTA, CARMAN, PADOP); man-machine interface and control room (PRA-SH and Regional Criteria for control room design); culture (human reliability committee for organizational change and archetypes analysis); cognitive decision (study of cognitive failure); stress and decision (LOPA); and, stress and occupational health (LESHA). These techniques for analysis of operations in a dynamic way have metrics based on the operator's speech from the results of routine by reflecting the occupational consequences (accident), environmental (pollution and waste), climate (CO2 emissions) and economic (competitiveness losses based in the rates of consumption and productivity). The result of these measurements allows the routine adjustments in the design of operations, the organization's project settings and business project. A deployment exercise will be proposed for the chemical industry. Diagram will be presented reflecting the dynamic in the management of operational risks.

Glossary

ANP - National Petroleum Agency
FMEA – Failure Mode and Effect Analysis
FMEAH – Human Failure Mode and Effect Analysis
FPSO – Floating Production Storage and Offloading
FTA – Fault Tree Analysis
GMTA – Goals-Means Task Analysis
HAZOP – Hazard and Operability Study
HAZOP-Social – Social Hazard and Operability Study
HSE – Health, Safety and Environment
LESHA – MultiLayer Stress Effect on Health & behavior Assessment
LODA – MultiLayer Operation Decision Assessment on Progressive Stress
LOPA – MultiLayer Operation Protective Assessment
MDA – Methyl Diphenyl Amine
PADOP – Patterns and Procedures in the Operation
PHR – Preliminary Hazard Analysis
PRA – Preliminary Risk Analysis
PRA-SH – Socio-Human Preliminary Hazard Analysis
THERP – Technique for Human Error Rate Prediction

1 A discussion on reliability and risk – an introduction

1.1 Industrial risk analysis

The initial thought on the issue of risk and subsequently his management was primarily driven by the financial sector, specifically post-World War II [1][2]. However, in parallel with accelerated development of industry and processes increasingly complex - which implies exposure to workers, society and the environment to hazards ever growing - this issue gained notoriety in the industrial environment.

The occurrence of major industrial accidents in Brazil, and around the world, became the subject Risk Analysis and Management more popular to meet the demands of society and governments about the safety of industrial processes. It is in this context [3] that is needed better management and mitigates the impacts of accidents. One reason for the importance of developing risk management was the change in constraints for operating license after the Seveso accident in Italy (1976).

Schneider and Martins [4] define risk as a result of the frequency of occurrence and the consequences generated by an unwanted event, so that risk reduction is by reducing the frequency of undesirable events and the mitigation of the consequences. The risks that can be quantified are risks of injury to the person, damage to property and environment, and risk of loss of reputation. Risk management depends on the level that is being analyzed: equipment, process unit, facilities, and other levels.

According to Bharatiya [5], an industrial hazard example is the handling of flammable products that can be released causing fire and explosion and resulting impacts on workers and third parties, as well as economic losses such as: profit loss caused by shutdown plant; damage to equipment; penalties for release of toxic material, etc. In this regard, it is necessary application of qualitative and quantitative techniques of risk analysis and its review to ensure a safe working environment for employees and neighborhood.

Qualitative techniques of risk analysis such as PHA/PRA (Preliminary Hazard Analysis / Risk) and HAZOP (Hazard and Operability Studies) allow identifying and assessing risks and hazards associated with industrial processes, and identify operational problems through scenarios identification. According to Fleming and Garcia [6] PHA/PRA technique performs a risk assessment for each identified scenario, develops mitigation measures and specific studies to calculate the probability and analyzes the effects through the vulnerability of these scenarios. The HAZOP allows analyzing risk scenarios using a procedure that generates questions and answers in a structured way by guiding words that enable the identification of possible deviations of the operation. The review team examines the links between the causes and consequences of these deviations.

Regarding the semi-quantitative techniques for risk assessment, there is a tool for layers of protective analysis (LOPA) that is used in the safety integrity level setting against unwanted events in safety instrumented system [7]. His concept was first treated by Center for Chemical Process Safety [8].

In quantitative analysis, the various methods are distributed from the analysis of consequences, which is measured the extent of damage caused by adverse events, to quantify the frequencies and probabilities of each event causing accidents [4]. Among the most widely used techniques are Vulnerability, Fault Tree Analysis (FTA) and LOPA. The techniques can be combined in a study of quantitative risk analysis.

The development of quantitative criteria can follow internationally recognized parameters. In general, the aim should be to develop criteria also based on the existing company culture and the most effective process in this direction would be a hybrid of internal and external internationally accepted criteria [9].

1.2 Risk analysis services

Brazil has been placing on a world stage major events that generate therefore great responsibilities. Mega-events such as road races, international shows, World Cup soccer (2014) and the Olympics (2016) demand new ways to manage the risks. Experiences with carnivals indicate the need for organization in public management to meet the security needs of the population, quality in the health area, mobility, and social life.

For each event has a positive impact on society and economy, it is important to have a control of risks in the event of scenarios. According to Correia [10], the factors that pose major difficulties in quality control and risks in the event derive from the fact of being subjective experiences and difficult to measure.

Each event is unique and presents its own security risks. Major events bring large crowds with factors to be analyzed for risk management [11]: demand complex planning with different stages in the design and operation; have different participating organizations that serving people of different customs; It works with many volunteers and inexperienced staff; It has national and international media projection; demands less than the time required for the complexity of the services; It has regional, state and national economic impact; and requires careful consideration for risk management.

For large events are ephemeral, not admit failures, delays, and its facilities and spaces are in most provisional, at the same time these experiences are characterized by subjectivity, It is more efficient to analyze the risk qualitatively [10]. In this way the analysis generates a more tangible and simple approach, easy to study and implementation by the organizing companies.

1.3 Integrating the reliabilities: equipment, operation and human

The first steps in the study of reliability were been conducted in the early twentieth century (the end of the World War I), but the concept of reliability as it is treated today only began to take shape during the World War II in an attempt to improve the reliability of missiles [12]. For this classic model of reliability, Kardec and Nascif [13] define reliability as "the probability of the system or device to perform the function successfully required, under specified conditions of operation and a certain time".

The development process of reliability engineering was important in product development and process improvements. In the latter case your application has a strong relationship with the evolution of maintenance management acting not only in the transition between corrective, preventive and predictive models, but above all, the definition of maintenance strategies where each model has a degree of importance that depends on the results of investigations into the system failure patterns. That is, even if the corrective, preventive and predictive models represent, respectively, the 1st, 2nd and 3rd phase of maintenance management, any one of them may be applicable depending on the operational context of the organization and its failure patterns. The concept of reliability then goes to lead the efforts for efficient maintenance.

To meet the demands of analyzing the nature of the failure and its consequences several tools were developed to improve the reliability of equipment. The most used are: Failure Mode and Effect Analysis (FMEA), and Root Cause Failure Analysis, Method of Analysis and Troubleshooting and Reliability-Centered Maintenance.

However, while many companies design the best systems based on this reliability model, taking into account the great difficulty in maintaining the availability and reliability of equipment and machinery at satisfactory levels and the cases of accidents with environmental impacts and fatalities, it is clear that such practices are not sufficient to maximize system reliability and/or guarantee the effectiveness of crash prevention barriers. Therefore, in order to fill this gap, many studies have been conducted to include the role of human in this process, what we call human reliability analysis. These studies began to gain strength from the sixth annual meeting of the Human Factors Society which had as its theme the quantification of human error. So was born the first formal method for human reliability analysis – the THERP technique [14].

According to Pallerosi *et al.* [15], human reliability can be defined as "the probability that a person does not fail to comply with a task in a given period of time, in appropriate environmental conditions and available resources to do it". The author also classified two types of procedures to quantify the human failures. One is called the prediction methods and includes tools as TESEO, HEART, SPAR-H, THERP, ASEP, MPPS and others. The other method is classified as evaluation methods and use theories and probability distributions as Gamma, Weibull, Exponential, Binomial and other distributions applied to each type of task analyzed.

The elements or factors that contribute to human error happen are related to the cognitive and operational processes. These elements of the cognitive process of decision-making are: perception, memory, abstraction, emotions, stress, fatigue, knowledge, skills, communication, tasks, procedures, organizational culture etc. Ávila [16] discusses these factors in the operator's performance in the control of socio-technical processes in task execution and learning stage. These factors can be measured by individual results in the task accomplishing, from the quantitative approach presented in verification methodology of learning parameterized (RGA - Reliability Growth Analyses).

The human factors analysis allows the projection and the calculation of risk and also the calculation of human reliability. Among the analytical techniques are: HAZOP SOCIAL, PRA-SH, FMEAH, LODA, etc. In the organizational dimension, the API 770 standard [17] aims to assess the company culture and the impact on human and operational reliability diagnosing aspects of policy, procedures, man-machine interface, and issues relating to the roles and task settings. The latter issues are present in David Embrey (2000), Frank Lees (1996), Rasmussen (1997) and Avila studies (2009).

The understanding of cognitive processes is essential to increase the human and operational reliability in organizations. Study the social, economic and emotional utility [16], cognitive function [18][19] and models for decision-making [20] are required to analyze the commitment and quality of the decision.

Therefore, approaches to socio-technical areas of reliability should not be an isolated study for each functional part of the industry [21]. The reliability of each part is interconnected with each other and their level of integration depends on the level of automation, the type of process, the manager's management style and the environment promoted by the organization. By understanding about the relationship between the functions of these systems makes it possible to study the integrated reliability. This knowledge enables the formatting of factors that increase the performance of staff and equipment leading to greater organizational efficiency.

1.4 Routine management and crisis management

The failure analysis and reliability project should be carried out at various stages of the project lifecycle - from conceptual design (detailing and assembly), through operation (control of operations and routine) to crisis management. The study of fault begins with the perception of signals from the process, people, equipment and tasks and advances in the analysis of incidents and / or accidents that cause financial loss (or image). These events, if not treated properly, can lead to loss of control in a chain reaction and It requires contingency actions not always known by the leaders, characterizing a crisis situation.

According Constantinides [22], crisis management is a systematic attempt to identify and detect potential crisis scenarios and take action to prevent them, contain its effects or interruption, and finally the recovery of the system normality. Muller [23] highlights the need for early warning systems projects of undesirable events and resolve them before they turn into a real crisis. He added that in case of failure of preventive measures is essential to take an immediate action.

2 Performance of GRODin and publications

The Group of Dynamics and Operational Risks (Grodin) located at the Federal University of Bahia intends to conduct diagnostic systems from applied research and works to develop solutions to complex problems in industrial sectors and services over the integration of academy, industry, and Public sector. Its constitution allows multidisciplinary making the investigative process dynamic starting in the subjective transformations of human and social behavior until the objective accomplishing of the task in industrial facilities with the search of metrics for decision making in organizations.

Currently (2015) the group consists of 25 masters, 5 doctoral and 10 graduate students with graduate completion forecast for masters and doctors around 10 per year. Large areas of concentration of Grodin's publications are: (a) Process Safety; (b) Human Factors; (c) Reliability and Risk analysis in socio-technical systems; (d) Energy Efficiency and Sustainability.

(a) Process safety

Publications performed in process safety seek always discuss the inclusion of human and organizational factors among the aspects considered in the accident process. Thus, several articles were discussed and presented at conferences as described: (a1) risk analysis of the factors that define human performance for the tasks in equipment and processes classified as critical [24]; (a2) the analysis of the performing task environment, the relationship between man and technology that result in failures and the discussion about leadership and the moment of decision [25]; (a3) analysis of underlying causes (human error) which are involved in the accident process (economic bias, risk management, conflicts between policy and practice, cultural traits, task planning, contracted management) [26]; (a4) analysis of decisions of the operation under progressive stress on the sinister situation [27]; (a5) analysis of high frequency and lower impact accidents considering the inclusion of human factors and complexity of the task [28]; (a6) analysis of social relations in oil platform and the implications for an enabling environment to an accident [29]; (a7) analysis of organizational change, drafting checklist, impact analysis of possible process safety events and the need for a more detailed discussion of the human factors [30]; and (a8) risk analysis of public services in cultural mega events (carnival) [31] and sporting events (World Cup) [32].

(b) Human factors

The publications performed on Human Factors always tries to discuss the analysis of the intrinsic variability of the man in an organizational and cultural environment and that brings consequences as gains or losses for the society and industry. Thus, several articles were discussed and presented at conferences as described: (b1) usability of procedures based in operators opinion of the fertilizer industry [33]; (b2) complex task evaluation taking into account the requirements of criticity [34]; (b3) cognitive deficit analysis during the performance of executive function in performing tasks [35]; (b4) social assumptions that change cognitive decision for the review of ergonomics standards in the control room [36]; (b5) analysis of archetypes and their derivatives to avoid improper behavior on safety of industrial plants [37]; (b6) Operator behavior analysis and leadership in emergency situations, high stress caused by chain reactions [27]; (b7) trace analysis of the Brazilian and global culture, biases, and bad habits that hinder the establishment of operational discipline [38]; (b8) stress analysis in industry, health care and diving and its consequences on the task, behavior and health [39]; (b9) changing operator behavior based on social dynamic culture, difficult to detect changes before impacting events happening like accidents with economic, human and environmental losses [40].

(c) Reliability and risk in socio-technical systems – complex processes

Publications performed in Reliability and Risk Analysis in complex processes discusses how to detect and calculate the risks and failures always considering the presence of human performance factors. Thus, several articles were discussed and presented at conferences as described: (c1) relationship between loss of industrial processes and connection to the cognitive processes with impacts measurement [24]; (c2) the operational failure exergy relating the energy loss in cognitive failure with the loss of energy in the industrial process heat systems [41]; (c3) evaluation of failure in complex processes, complex tasks and complex social relations in the chemical industry [34]; (c4) dynamic risk analysis for industrial sustainability [42]; (c5) human factors and equipment failures analysis using tools such as SPAR-H and Weibull ++ in the nuclear and chicken industries [43][44]; (c6) Analysis of socio-technical reliability in the refining industry [21]; (c7) risk analysis of the pre-salt considering: phenomenology, reliability model and risk factors [45][46].

(d) Energy efficiency and sustainability

Publications performed in energy efficiency and sustainability seek to discuss what are the opportunities to reduce consumption of energy and water and ensuring the sustainability of business. Thus, several articles were discussed and presented at conferences as described: (d1) mass balance and energy in cooling towers [24]; (d2) audit in organic effluent and rainwater to prevent process losses; (d3) audit of environmental deviations in industrial routine [47]; (d4) contamination control program at source by mapping effluent [48]; (d5) application of mass balance techniques and energy to identify opportunities to reduce energy and water losses in the industry; (d6) a discussion of systems and cooling towers, challenges to avoid energy losses and water; (d7) identification and calculation of thermal performance and energy efficiency in equipment and industrial processes [49]; (d8) Analysis of loss of condensate and steam heating systems and energy recovery [50].

3 A vision of sustainable organization based on risk

Maintaining organizational sustainability requires an understanding that the business risks are dynamic and the operational risks depend on a balanced analysis of issues related to product and process with the people and culture issues. So, to start the discussion on sustainability in high-risk industry a management model will be presented.

An efficient risk management of business develops premises to organizational project considering social-technical aspects. This risk management draws an industrial operation format including multidisciplinary analyzes of different levels of normality (default to failure and the crisis). It is intended with this model to measure the effects and impacts on society and then review the criteria and procedures adopted by project operations.

3.1 Inputs in the Risk Management and Organizational Design

Important inputs for the analysis of dynamic risk are: entrepreneur demand (vision and mission), technological conceptual design (product and process), analysis of the availability of resources and production factors (energy, materials and people), and external demands represented by the law and behavior of the stakeholders in the existence of this business. The risk analysis stage to address the dynamic risk intent to insert a discussion about the process and product taking into account the type of task (or the procedure), the cognitive models for decision (local or multicultural) and the social behavior of internal and external community of the organization.

3.2 Step one: Risk Management - socio-technical risk analysis

The first step of socio-technical risk analysis balances human-social factors with process and product factors generating preliminary recommendations. In the case of technical risks are evaluated the product, process and the likely scenarios through LOPA and PRA and getting as result recommendations to prevent accidents and technical failures. For non-technical risks such as environmental and human aspects is performed an PRA-SH to anticipate future events related to human errors that can cause technical failure or generate an stressor environment at the workplace. At this time are considered the risks of events that involving cognitive faults and failure on the task resulting from social processes.

3.3 Step two: organizational project - detailing of technical and non-technical project

The second step requires knowledge of the detailed organization design in the technical and non-technical aspects to define the assumptions and criteria used for the design of operations. Thus, the organizational project aims to open the product technology, process, service, and control to discuss which the necessary demands on leadership, behavior, attitude, and skills.

The intrinsic variability policies, humanities, social and variability of nature (climate, biota, etc.) are scored as important for maintaining the environment in which the industry is installed and intends to stay in a sustainable manner. The organizational design analysis requires detailed information of technology, knowledge in integrated management system, consumption analysis and energy benchmark, water, materials and people (productivity) to establish performance indicators. The result of these studies is the definition of technical, human, organizational, natural factors for the sustainable operation of the production system. The techniques applied in this second phase are: HAZOP, HAZOP-Social; requirements for Task (GMTA, PADOP); cause and effect analysis of the events; barriers analysis of events (BOWTIE).

PADOP is a method that starts with the analysis of the requirements for feeding the following phases: environment, planning and execution of the task. Others steps such as task control and conformity analysis under review are performed in operational routine (see item 3.4).

3.4 *Step three: design and control of operations*

In the third stage of dynamic risk assessment is intended to rescue the criteria and assumptions set out in the previous step, organizational design, where technology is detailed and an integrated management mode is defined. In the project of operations, it will be handled operational routines, asset control, and control of failures and possible crisis. So, it will be designed systems for operation in the technical area (equipment, processes, products and control modes), the interface (task management, and human-machine interface) up to non-technical area (culture and biases, cognitive decision and stress and behavior analysis). The results of these measurements allow adjustments on the routine of the design of operations and business and organization projects.

These factors and elements that make up the control of the operations are analyzed on standard procedure mode and failure analysis in the following areas and using the following techniques: equipment (FMEA, FMEAH); product and process (monitoring and statistical control); task in a complex way (GMTA, CARMAN, PADOP); man-machine interface and control room (PRA-SH and regional criteria for control room projects); culture (human reliability committee for organizational change and archetypes analysis); cognitive decision (study of cognitive failure); stress and decision (LOPA); and stress and worker health (LESHA). These techniques for analysis of dynamically operations present measurements and indicators that are based on operator's speech. This speech is announced from the routine results and reflects the occupational consequences (accident), environmental (pollution and waste), climate (CO2 emissions) and economic (loss of competitiveness based on consumption rates and productivity).

In this case, the PAPOD technique intent to compare the standard designed for task control rates with measurement of the process after the task performed. It is worth noting that these standards include variables of integrated management (safety, quality, environment, occupational health, etc) as well as socio-humans aspects (quality of communications). In case of deviations it will be realized preventive corrections with necessary revision to the task or written procedure.

4 Application of exercise

An implantation exercise is proposed based on the application of different technical and management systems in research in various chemical process industries. Activities and analyzed processes involve energy products (oil and LPG), chemical (sulfuric acid, methyl diphenyl amine – MDA, and explosives), petrochemicals (basic petrochemicals and fertilizers), polymers (polycarbonates) and plastic (plastic bottles).

These various research opportunities involve aspects of routine at the work place and characteristics of management that promote the company's results close to normal or fault situations. The experiences are gathered in an integrated way toward sustainability considering that risks are dynamic and that the systems are socio-technical.

Some of these techniques are applied with quantitative results indicating the intensity of the problem and the volume of investment to be taken and others indicate the existence of the failures and suggest the construction of safeguards, but in a qualitative way.

This exercise will only treat of techniques relating to the socio-environmental-human risk and it will quote only within the management system proposed and which evaluations (of process and product) should be interconnected.

4.1 Inputs in the risk management and organizational design

For this model that considers the dynamic risk, the entrepreneur discusses the social variability in the possibility of failure and practices a vision and mission more sustainable, which admits the existence of a multidisciplinary approach to the definition of the operating pattern. This discussion is part of the control of the operational routine. So, inputs to this management system from experienced cases it is suggested.

Entrepreneur demand

Case 1: the entrepreneur demand to identify the routine deviation is a requirement to draw up barriers that prevent systemic failure. To discuss the causes of failures that lead the operation to a mode of low productivity, an entrepreneur of the LPG industry recognizes the need to analyze the routine. The operator's speech feeds the construction of the fault tree for routine events and together with the task analysis identify the GAPs of competence that lead to failure [51].

Case 2: the perception of the manager must be attentive to behavioral deviations that can lead to accidents. In this case, the corporate HSE manager of the explosive industry believes that the study of human typology is necessary to change the criteria in risk management. This belief is based on the possible change of behavior, which caused two consecutive accidents not expected in different countries, after a history of 20 years without accident. This entrepreneur corroborates the following statement: "cultural changes alter the behavior of the workers and there is a need to study new criteria for dynamic risk management".

Analysis of the availability of resources and production factors (water, energy and people)

Production factor: energy. The water crisis in the state of São Paulo [52] and the reduction of energy quotas for petrochemicals due to its unavailability caused by climate change indicates that the analysis of availability is critical input in the organizational design to address the variability of nature.

Production factor: water. CEOs of Brazilian refineries are being instructed by National Agency of Water (for now voluntarily) for an actual 20% reduction in water consumption. This highlights the urgent need to change behavior and decisions in this organizational design.

Production factor: people. The opening speech of the Brazilian Association of Risk and Reliability (ABRISCO) indicates that the Brazilian engineer does not know and does not quantify industrial risks indicating the unavailability of appropriate skills to quantify investments in safeguards to prevent accidents.

External demands legislation for new behaviors of stakeholders

Stakeholder – National Petroleum Agency (ANP). After the FPSO production ship accident in the city of São Mateus, an investigation realized by the ANP [53] indicates the need to discuss the issues of contract management, omissions, commitment, crisis management and decision in stress. These issues are part of social- human variabilities.

Stakeholder – University which reflects demand for workers. In Risk Management is important to define the requirements necessary from technology and local culture for planning the task (PADOP). From the master's degree studies on the task analysis it was found that most of the 450 operators of a fertilizer industry accepted the importance of establishing social and human premises to carry out the critical task.

Stakeholder – Multinational oil companies. From international meetings with reliability consulting groups that serve the oil industry, it was found out the need to discuss regionalities to change the ergonomic criteria in control room. This indicates that the CEOs of major companies (Chevron, Statoil, BG) are reflecting on the question of risk review to include in their analysis the dynamic and social aspects (cultures and people).

4.2 Step one: risk and opportunity management - socio-technical risk analysis

After the hazard analysis and technical risks by conventional application of the techniques APP and APR (to business, processes and products) it is natural to use the vulnerability analysis in the most critical scenarios to indicate the level of impact on human and nature. Similarly to analyze the social and human risk seeks to detail the various performance shaping factors [54] that indicate the level of consequence in case of human errors.

Fault analysis (in 20 cases) in the chemical process industry indicates that elements such as organizational climate, leadership, skills, knowledge, and decision-making processes are related to human error in performing the task.

Several cases involving deviations, failures, incidents and accidents were analyzed and allowed the construction of the anatomy of failure in complex mode relating human and organizational factors [26]. Among the cases are: director's behavior in the plant starting management (polycarbonates); deviations in mitigation of waste generation reaction (MDA); operational control to avoid stoichiometric reaction stop (MDA); skill in the perception of vacuum loss in sulfuric acid unit; failures caused by improper turn change (MDA); investigation of effluent contamination source (MDA, sulfuric acid, lab, fertilizers); diagnosis to map the gaps in the process (MDA, polycarbonates, LPG); negotiation of new operating parameters without causing downtime (MDA); improper use of individual protection equipment and omission by fear of guilt (MDA); oring diameter change at startup leading to failures with solvent losses (polycarbonate); and others.

Based on the investigation of cases the initiators factors to human error are [26]: design criteria, risk management, economic bias, cultural traits, task planning, and conflict between policy and practice. It is also important to identify the social dangers to avoid accidents before assembling the installation and make the hires to operate the industrial site.

A specific case was applied at the stage of design of an oil exploration platform in northeastern Brazil during the graduate student research in the area of security and human factors analysis in 2013 [55]. After seven risk analysis sessions for human and social factors, the group of six experts (socio-technical) listed the following factors that can cause human errors in this installation that come into operation in 2015: difficulties in routine leadership; non-commitment to the safety result; low perception to failure; and improper internal communication.

It is worth listing the technique steps: identification of social and human dangers, social and human risk analysis, risk classification, interaction between risks (interaction matrix to correct risks), the final risk map, and treatment of social and human risk.

An item that is repeated in the workplace discussion refers to the effect of stress on the effectiveness of thc task [56] and occupational health [57]. Thus, it interconnects the worker performance factors to the result of task with economic results for the industry. The causal relationship between work environment, cognitive fault and failure in the task are motivating ongoing research that result in techniques and applications.

Technique 1: analysis of social relations in the workplace. Application: oil platform where the rituals of loading and unloading, the distance and isolation are analyzed to list critical topics that can cause stress and human error [58] [29]. Technique 2: protocol for the analysis of stressor environment. Application: industry [56], nursing services [39] and diving services in deep waters. Technique 3: task analysis [33]. Application: chemical process industry. Technique 4: analysis of cognitive impairment [35]. Application: chemical industry.

4.3 Step two: organizational project - detailing of technical and non-technical project

Step 1 of this management model for sustainability is related to the fit project of product, process, and business, in a discussion of technology and economic viability. The analysis of social and human risk in the task is performed along with the APR technique of this future operation.

Step 2 is the organizational design that includes the adaptation of the technology for the future operation that includes product and process aspects, and non-technical issues related to teamwork in performing the task (culture and people). At this stage (detailing phase) are carried out studies of the conceptual design (made in step 1) and details the risks of failure to cause economic loss and / or the company's image.

Application of HAZOP-Social

The process deviation analysis during operation indicates the possibility of danger containment loss related to the industrial activities. The Social HAZOP tool works similarly to the conventional technique [16]. In its application the cognitive decision-making is tested through the deviations and recommendations are announced to prevent failure of attention, memory, competence, managerial decisions and others. Social HAZOP was applied to a draining operation of a decanter vessel in the refining industry as having maximum result the accident with fatality and with the critical lack of attention, centralized management profile and commitment to the task.

Diagnostic of skills gap

Identification of fault root cause in the chemical industry that has characteristics of processes and complex tasks [59] [34] requires the study of process variables and production beyond the qualitative and quantitative analysis of operator's speech [60][61]. This discourse analysis allows the organization to verify the main failure modes in socio-technical systems by locating the critical variables in the tasks, equipment, skills and operator behavior. When comparing the type of fault carried out (based on competence offered) with the demand for knowledge and skills in technology (competence demanded) have been related what are the gaps in competence.

This technique was applied in PhD research [51] of the gas compression unit (LPG) in northeastern Brazil. In this case, the database was collected from one year of statistical analysis of process and production variables, and four months of shift book analysis in low performance periods. Diagnosis of gaps indicates the lack of skills in the following areas [25][62]: Task control (abstraction and logical); concepts (thermal and material exchange); failure (rotary, motorized valve and burner); skills (thermal cycles, pressurized system and logistics); process control and safety (supervisory and risk in the routine).

The gap of skills already identified in the LPG plant may result from lack of commitment of the people, the team cooperation and inappropriate profile of leadership. A classification model of commitment, the leader's behavior and aggregation of the team is suggested by Avila [51][25][62].

Research on behavioral intrinsic variability

Demand for leaders in chemical industry and oil organizations during the GCPSs events indicates the importance of analyzing the changing worker behavior as a result of Dynamic Culture. A methodology is proposed to find the biases of global and regional culture that promote behavioral changes towards the accident [40]. This method was structured to identify potential changes in teams of explosives chemical company with unusual situations accident in facilities overseas. The suggested steps with the respective models are: (1) analysis of social culture; (2) analysis of the social and human types to work; (3) analysis of the dynamic factors of culture or affective socio-economic utility; and (4) evaluation of the result in the worker's behavior performing the critical task.

Archetypes and regionalities analysis

Entrepreneurs from different areas (chemical and oil) have discussed the ergonomic criteria for control rooms and raise the possibility that tacit aspects influenced by behaviors are not being considered. This hypothesis is due to continuity failures in critical operations such as plant startup and shutdown. These questions led the researchers to analyze the control room construction criteria based on the ergonomics requirements in the EEMUA and ISO standards [36]. The result of the research is the mapping of critical situations that can result in accidents.

In the case of a LNG industry control room were related aspects of local and global culture that impact on the cognitive processing leading to failure in decision making. This tool depends on the study of product risks and social risks (APR and APRSH), the analysis of the major failures and relations with human error in similar facilities, and brings results in a cross-matrix, which relates the social influences, cognitive functions with interfaces and control room devices.

The regional or global behavior that can turn into social risk (eg, default and inadequate leadership) may be due to cultural biases from primary or secondary archetypes. A case of industrial gases plant is analyzed and actions are suggested in the form of ritual to neutralize or deflect improper behavior [37].

Variability of climate change

The urgency of actions to reduce GHG emissions, fuel consumption and water motivate researchers in techniques seek to adjust the demand for water and energy in relation to confirmation of the offer. The demand adjustment actions involving routine management, knowledge of techniques for thermal performance analysis and new technologies such as process intensification or modularization of projects. The research conducted (LAB, Olefins, Urea) indicate that the equipment involved in research and routine adjustment are: cooling towers, boilers, furnaces, condensate systems and unit operations with their reflux [49][50].

4.4 Step three: design and control of operations

The Control of Operational Routine depends on the knowledge of appropriate standards for the maintenance of a stable process with the appropriate cost. It is increasingly announced by managers of high-risk companies that the economic result is guaranteed if the image is positive - that is, the non-occurrence of events related to accidents.

The definition of operating standard to routine management is part of a policy initiated by the management of technical and non-technical risks. So, the foreknowledge keeps the assets in normal mode. The occurrence of deviations, failures and incidents in the industrial operation can cause a crisis situation that requires a new leadership profile (crisis and contingency management).

Cases of task analysis

In LPG industry there are high-risk activities that can cause fire or explosion. Purification of LPG avoids failure through moisture removal. The moisture-removing columns have a saturation point and are removed from operation to be regenerated. This, in turn, requires the moisture expulsion using heated LPG itself resulting from passage through a furnace.

During the investigation concerning the study of root cause analysis in LPG process [51] was made the analysis of the operation in the task, startup and breakdown in regeneration furnaces using techniques such as GMTA[20], CARMAN[63] and PADOP[26][33]. These techniques used for fault analysis in the task involves the logic and the nature of the failure, the chronology of events, connectivity between factors beside the point where failure goes out of the social mode (eg. communication) to the physical mode in GLP [51] and Polycarbonate [60].

The task control actions include the procedure review (requirements, targets, or steps) with training. When the risk, complexity of processes, the complexity of social relationships and stress levels are high (in part or in full), the training actions has no effectiveness demanding action on culture [56][33] or to the redesign of equipment and processes.

Cases of failure in equipment and process

The benzene hydrogenation is an intermediate step in the production of caprolactam. The recycle gas is maintained at constant pressure to control the reaction through a centrifugal turbo-compressor of the brand Demag. Repeated interruptions on Monday (after the weekend) raised the possibility of having human error in the operation and/or maintenance which did not allow the effectiveness of technical actions.

A FMEAH study was performed to identify the failure mode which problems of perception, attention and expertise were occurring in routine production. The method resembles the traditional FMEA, but includes analysis of human error in the equipment failure mode [64].

The degradation systems, changes in technology, organizational changes [30] and climate change may cause variations in the process which results in the modification of the possible event scenario, which in turn increases the risk of accidents. In this regard, through the signals emitted in routine, seeks to review the product risk, process and operations through the application of new techniques such as APR, APRSH, HAZOP, Social HAZOP.

Human performance case in state of stress

A startup case of sulfuric acid purification discussed the influence of stressful environment. The elements that make up the work environment to achieve a state of stress that affects human performance are: task risk level; health impacts, the cost of business and social environment; function of the task in the process; and analysis of the psychological and organizational types. The protocol indicated that the level of stress was 4.8 exceeding the stress level manageable (under 4).

An accident analysis in progressive stress

An accident involving LPG furnace, as discussed in this paper, was used to build a fault tree of technical and non-technical events in different levels of stress.

The application of LODA technique [27] from gradually more dangerous events indicates that the decision of the shift varies according to the level of stress. The progressive stress is promoted by the need for decision-making by building possible increasingly serious consequences scenarios at the following levels: deviation from the routine, lack of process, lack of equipment and task, accident and sinister.

Unfortunately, the psychological unpreparedness of staff in dealing with contingencies, especially the leaderships, makes the human error probability goes from 10%, in steps of process deviations and lack of control of process, to 50% in sinister steps.

Cognitive failure cases and vices in the industry operators behavior

In a startup of a chemical industry reactor [35] are analyzed psychological functions in the executive work of the task. From social relationship models and the executive function analysis it was identified the following types of failure on the team with priority level 1: lack of knowledge and difficulties in memory recovery.

In the sulfuric acid industry [40] were analyzed behavioral addictions through fault trees that indicated the need for vaccines for cognitive diseases, among which: "frozen" perception, low motivation to cooperate, not enabled memory, commitment corrupted, communication and leadership inadequate, and very simple mental map. This finding allows the construction of behavior adjustment programs.

In industrial gases and chemical industries were found two types of behaviors in the control room which allowed the analysis of two different groups archetypes. Some archetypes favor a robust safety culture and others hinder their establishment [37].

Root-cause analysis in complex systems

The operator speech analysis is a tool that links events occurred in the shift such as: problems with equipment, product quality, double-shifts, missing and other facts. The form of communication, the statistics of events, the possibility of omission and team cooperation are some of the aspects included in this speech.

On several occasions in the chemical process industries [60] were applied root cause analysis techniques in complex systems. In the production of alkyl amines it was noted that the delays were the result of errors in the setup of the campaign. In the production of methyl diphenylamine it was observed that the interruptions were strongly influenced by the formation of groups that are not communicated with each other and by the lack of knowledge of the importance of recycle line for the quality in the process. In polycarbonates production sought to reduce loss of methylene chloride. In this case the operator's speech has enabled to review: procedures (from the task analysis) and manager behavior in the plant startup in control room.

In LPG industry in which the concern is the leakage events, fire or explosion, the operator's speech helped to identifying critical areas, gap of competence for decision-making, and team behavior in emergency situations [51].

5 Conclusion

The model of operational risk management requires technical and social aspects are correlated at different levels of normality / abnormality of the processes with the consequences and impacts that produce in their own business, on environment and society. By multidisciplinary involved in the analysis of these risks, the definition of management strategies should take into account the dynamics of the risks in business.

In some cases analyzed were used the operator' speech as a tool for building the event fault tree in the routine that jointly with task analysis identified the gaps of competence that lead to failure where the criteria and procedures adopted can be reviewed in the design of operations.

In another case (explosives factory) was found to be necessary to investigate the human type to change the criteria in risk management. In this case, the entrepreneur recorded accidents with serious injuries to the worker after a history of 20 years without accident (plants located in different countries) and admitted that cultural changes alter behavior which highlights the need to consider new criteria for dynamic risk management.

From the perspective of production factors such as water and energy, is also recorded vulnerability in the sustainability of the business in major change situations in the availability of these resources requiring that teams know how to prioritize investments to overcome such difficulties.

The applications of various techniques mentioned in this work are structured in a sustainable model of risk management. This model requires generation of multidisciplinary expertise (socio-technical aspects) to properly handle the dynamics of risk.

6 Annex

Figure 1 – Risk Management Model for Sustainable Organization

7 References

[1] G. Dionne Risk Management: history, definition, and critique. Risk Management and Insurance Review, 16 (2013), 147-166.
[2] J. L. Vesper. *Risk Assessment and Risk Management in the Pharmaceutical Industry: Clear and Simple.* Parenteral Drug Association, New Vernon, 2006.
[3] M. R. Martins *et al.* Undesirable events Consequences analysis during the transport, loading and unloading of Liquefied Natural Gas (LNG) on ships. Proceedings of Brazilian Association of Risk - ABRISCO. Rio de Janeino, 2013.
[4] A. M. Schleder and M. R Martins. Methodology for quantitative analysis of the consequence of flammable leaks and / or toxic environments with barriers or confined. Proceedings of Brazilian Association of Risk - ABRISCO. Rio de Janeino, 2013.
[5] B. K. Bharatiya. Projet Hazard Analysis In Hydrocarbon Processing Industry. Proceedings of 4th Latin American Conference on Process Safety - CCPS. Rio de Janeino, 2012.
[6] P. V. Fleming and C. B. Garcia. Industrial and Environmental Risk Assessment With Preliminary Hazard Analysis (PHA) and Fuzzy Logic. Proceedings of Brazilian Association of Production Engineering - ENEGEP. 1999.
[7] R. Freeman. Quantifying Lopa Uncertainty. Process Saf Prog 31(2012), 240-247.
[8] CCPS. *Guidelines for Safe Automation of Chemical Processes.* Center for Chemical Process Safety, American Institute of Chemical Engineers, New York, NY, 1993.
[9] J. N. Shan. Understanding and Developing Quantitative Risk Criteria. Proceedings of 8th Global Congress on Process Safety, Houston, 2012.
[10] A. H. L. Correia. Strategic Marketing of Sporting Event. International Seminar of Sports Event Management. Ministério da Juventude e do Desporto. Centro de estudos e formação desportiva. Lisboa, 2001.
[11] Worksafe Victoria. Advice for Managing Major Events Safely. 1st Ed. Austrália. 2006. Available at: https://www.worksafe.vic.gov.au/__data/assets/pdf_file/0010/12502/major_events_organisers_advice.pdf. Accessed on January 3, 2015.
[12] F. S. Flogiatto and J. L. D. Ribeiro. *Reliability and Industrial Maintenance.* Elsevier: ABEPRO, Rio de Janeiro, 2009.
[13] A. Kardec and J. Nascif. *Maintenance – Strategic Function.* 4th Ed. Qualitymark Editora, Rio de Janeiro, 2013.
[14] R. L. Boring. Fifty Years of THERP and Human Reliability Analysis. U.S. Department of Energy National Laboratory. U.S.A., 2012.
[15] C. A. Pallerosi; B. P. M. Mazzolini and L. R. Mazzolini. *Human reliability - Concepts, Analysis, Evaluation and Challenges.* AllPrint Editora, São Paulo, 2011.
[16] ÁVILA FILHO S.; PESSOA F.L.P. and ANDRADE J.C.S. Social HAZOP at an Oil Refinery. Process Saf Prog - 32 (2013), 17-21.
[17] D. K. Lorenzo. API770 – A manager's guide to reducing human errors, improving human in the process industries. API Publishing Services, Washington, 2001.
[18] J. Reason. *Human Error.* Cambridge University Press, New York, 2003.
[19] R. Stenberg Junior. *Cognitive Psychology.* 3rd Ed. Artmed, Porto Alegre, 2008.
[20] E. Hollnagel. *Human Reliability Analysis Context And Control - Computers And People Series.* Academic Press Inc, San Diego, 1993.
[21] S. Ávila Filho. Reliability analysis for socio-technical system, case propene pumping. Engineering Failure Analysis, 56 (2015), 177-184.

[22] P. Constatinides. The Failure of Foresight In Crisis Management: a Secondary Analysis of the Mari Disaster. Technological Forecasting & Social Change, 80 (2013), 1657–1673
[23] R. Muller. Corporate Crisis Management. Long Range Planning, 18 (1985), 38-48.
[24] S. Ávila Filho *et al.* Risk Management and Prevention Program to Avoid Loss of Process (LP3) based on Human Reliability, Practical Application and Measurement. Proceedings of 7th Global Congress on Process Safety – GCPS. Chicago, 2011.
[25] S. Ávila Filho and A. C. Carvalho. Human Factors Assessment to Avoid Accident at LPG Industry. Proceedings of 6th Latin American Conference on Process Safety – CCPS. Rio de Janeiro, 2012.
[26] S. Ávila Filho. Assessment of Human Elements to Avoid Accidents and Failures in Task Perform, Cognitive And Intuitive Schemes. Proceedings of 7th Global Congress on Process Safety – GCPS. Chicago, 2011.
[27] S. Ávila Filho. Dependent Layer of Operation decision analyzes (LODA) to calculate Human Factor, a simulated case with PLG event. Proceedings of 7th Global Congress on Process Safety – GCPS. Chicago, 2011.
[28] S. Ávila Filho. Review of Risk Analysis and Accident on the Routine Operations in the Oil Industry. Proceedings of 5th Latin American Conference on Process Safety – CCPS. Cartagena de Indias, 2013a.
[29] S. Ávila Filho. Risks in the Platform: Social and Human Guidelines Avoid Accidents and Alter Standards in Oil Production. Revista Proteção, 2013b.
[30] S. Ávila Filho; E. T. Dantas and J. Duarte. The Requirements & Tools to Treat Process Safety Risk as Result of Organizational Change. Proceedings of 9th Global Congress on Process Safety – GCPS. Austin, 2015a.
[31] S. Ávila Filho, *et al.* Analysis of risk scenarios and mapping of communication processes to avoid incidents in Salvador Carnival in 2015/16. Proceedings of Brazilian Association of Risk – ABRISCO/PSAM. Rio de Janeiro, 2015a.
[32] S. Ávila Filho, *et al.* Risk Analysis in Sports Mega-events, UFBA Project in the World Cup. Proceedings of 34th Brazilian Association of Production Engineering - ENEGEP. Curitiba, 2014.
[33] S. Ávila Filho; C. R. Sousa and A. C. Carvalho. Assessment of Complexity in the Task to Define Safeguards Against Dynamic Risks. Proceedings of 5th International Conference on Applied Human Factors and Ergonomics – AHFE. Las Vegas, 2015.
[34] S. Ávila Filho. Failure Analysis in Complex Processes. Proceedings of 19th Brazilian Chemical Engineering Congress – COBEQ. Búzios, 2012.
[35] S. Ávila Filho and C. Costa. Analysis of Cognitive Deficit in Routine Task: as a Strategy to Reduce Accidents and Increase Industrial Production. Proceedings of European Safety and Reliability Conference – ESREL. Zurich, 2015.
[36] S. Ávila Filho and F, L. P. Pessoa. Proposition of review in EEMUA 201 & ISO Standard 11064 Based on Cultural Aspects in Labor Team, LNG Case. JofPM, 3 (2015), 6101 – 6108.
[37] S. Ávila Filho, M.L.A. Menezes. Influence of Local Archetypes on the Operability & Usability of Instruments in Control Rooms. Industrial Production. Proceedings of European Safety and Reliability Conference – ESREL. Zurich, 2015.
[38] S. Ávila Filho, *et al.* Influence of Brazilian Culture in the Workers' Behavior and Operational Discipline. Proceedings of Brazilian Association of Risk – ABRISCO/PSAM. Rio de Janeiro, 2015b.

[39] K. R. V. Nunes, *et al.* Working Hours Excessive and its Influence on Accidents, Productivity and Decision Making of Nursing Technicians in an ICU. Proceedings of Brazilian Association of Risk – ABRISCO/PSAM. Rio de Janeiro, 2015b.

[40] S. Ávila Filho; C. C. Silva and C. R. O. Sousa. Worker Behavior Changing by Dynamic Culture. Proceedings of 9th Global Congress on Process Safety – GCPS. San Antonio, 2013.

[41] S. Ávila Filho. The Exergy of Operational Failure, a Philosophical Discussion with Basis Practices. Proceedings of 19th Brazilian Chemical Engineering Congress – COBEQ. Búzios, 2012.

[42] S. Ávila Filho, J. R. N. Lopes and C. R. O. Sousa. Strategies to Industry Sustainability. Annuals of 20th World Renewable Energy Forum – WREF. Denver, 2012.

[43] A. C. Alcoforado and S. Ávila Filho. SPAR-H Application in the Crushing Operation of Uranium Ore. Proceedings of Brazilian Association of Risk – ABRISCO/PSAM. Rio de Janeiro, 2015.

[44] S. S. Dórea and S. Ávila Filho. Failure Investigation in Operations of Chicken Industry with Based Human and Technical Factors. Proceedings of Brazilian Association of Risk – ABRISCO/PSAM. Rio de Janeino, 2015.

[45] M. N. G Abreu and S. Ávila Filho. Assessment of Risk in the Exploration and Production of Oil at Pre-Salt Layer – Brazil, a Discussion about Social-Technical Factors. Proceedings of Rio Pipeline Conference & exposition. Rio de Janeiro, 2015.

[46] M. N. G Abreu and S. Ávila Filho. Assessment of Reliability and Risk at Dynamic Systems: a Discussion of Social-Technical Factors in the Exploration and Production of Oil at Pre-Salt Layer – Brazil. Proceedings of Global Congress on Process Safety – GCPS. Austin, 2015.

[47] S. Ávila Filho, *et al.* Environmental Deviation Assessment, Record, Failure, Probability and Investment. Chemical Eng Trans, 26 (2012a), 441-446.

[48] S. Ávila Filho, *et al.* Program to Avoid Product Leakage at Source, Fertilizer Case. Chemical Eng Trans, 26 (2012b), 447-452.

[49] S. Ávila Filho, *et al.* Iniciatives of Energy Efficiency Toward Sustainability, Opportunities & Challenges In: Rio Oil Gas Conference & exposition. Rio de Janeiro, 2014.

[50] S. Ávila Filho, *et al.* Avoiding Loss of Energy in a Petrochemical Industry, Operation and Design. Proceedings of World Renewable Energy Congress – WREC. Linköping, 2011.

[51] S. Ávila Filho. Etiology of Operational Abnormalities in Industry: a Model for Learning. Doctoral thesis - Chemistry School. Federal University of Rio de Janeiro. Rio de Janeiro, 2010.

[52] ANP. Investigation Report of an Explosion Incident in FPSO Ship in São Mateus Town on November 02, 2015. National Agency of Petroleum, Natural Gas and Biofuels. Available at: http://anp.gov.br/?pg=78836&m=&t1=&t2=&t3=&t4=&ar=&ps=&1452081674560. Accessed on January 15, 2016.

[53] R. Bonalume Neto. 30% Loss of Water During Distribution Contributes to Crisis in São Paulo. Sustainability Forum. Folha de São Paulo. São Paulo, 2014.

[54] D. K Lorenzo. API770 – A manager's guide to reducing human errors, improving human in the process industries. API Publishing Services, Washington, 2001.

[55] M. P. Barroso. The influence of human factors in APP and APR risk analysis techniques: a study case of an oil drilling platform in the northeast. Master's Dissertation – Polytechnic School. Federal University of Bahia. Salvador, 2013.

[56] G. G. Gonçalves, *et al.* Protocol to evaluate the influence of stress on the task involving human types, behavior and health: An exercise proposal. ESREL 2015. Zurick. 2015.

[57] S. Ávila Filho. LESHA – Multi-Layer progressive stress & impact assessment on health & behavior. Proceedings of 9th Global Congress on Process Safety. San Antonio, 2013.
[58] S. Ávila Filho. Human Errors Assessment, Social Environment at Oil Platforms. Proceedings of 4th Latin America Conference on Process Safety. Rio de Janeiro. 2012.
[59] C. Perrow. *Normal accidents: living with high-risk technologies.* NY: Basic Books, 1984. 453 p.
[60] S. Ávila Filho. Methodology to Minimize Generation of Waste at Source from the Investigation of Operational Abnormalities. Master's Dissertation – Polytechnic School. Federal University of Bahia. Salvador, 2004.
[61] E.S. Drigo, S. Ávila Filho and C. A. O Sousa. Operator discourse analysis as a tool for risk management. Proceedings of European Safety and Reliability Conference – ESREL. Zurich, 2015.
[62] S. Ávila Filho. Diagnosis in Socio-Technical Systems Aiming to Increase Human Reliability, a LPG Case. Proceedings of International Applied Reliability Symposium – ARS. Fortaleza, 2011.
[63] D. Embrey. Preventing human error: developing a consensus led safety culture based on best practice. London: Human Reliability Associates Ltd. 2000.
[64] S. Ávila Filho, *et al.* Human Factors Analysis in Turbocharger Equipment Failure FMEAH in Chemical Plant. Proceedings of 27th Brazilian Maintenance Congress. Rio de Janeiro, 2012.

12TH GLOBAL CONGRESS ON PROCESS SAFETY

Water and Energy availability analysis, a proposal for management for chemical industry

Salvador Ávila Filho
Federal University of Bahia – UFBA
Salvador – Brazil
avilasalva@gmail.com

José Rafael Nascimento Lopes
National Service of Industrial Learning – SENAI
Salvador – Brazil
jrafael.lopes@gmail.com

Jean marcel Prazeres Silva
Federal University of Bahia – UFBA
Salvador – Brazil
jmarcel_7@yahoo.com.br

Prepared for Presentation at
American Institute of Chemical Engineers
2016 Spring Meeting
12th Global Congress on Process Safety
Houston, Texas
April 11-23, 2016

Water and Energy availability analysis, a proposal for management for chemical industry

Salvador Ávila Filho
Federal University of Bahia – UFBA
Salvador – Brazil
avilasalva@gmail.com

José Rafael Nascimento Lopes
National Service of Industrial Learning – SENAI

Jean Marcel Prazeres Silva
Federal University of Bahia – UFBA

Keywords: water availability, energy analysis, sustainability

Abstract

The evidences that the chemical industry is not ready to deal with interruption of operations resulting from climate change and the complexity of decisions is increasingly apparent. In the state of São Paulo / Brazil, petrochemical industries are using served water from sewage treatment plants to supply the cooling tower makeup water. However, industries are not striving to analyze the availability of water and energy to suit the demand to supply.

The setting of premises to build scenarios of water availability, energy and skills is required. This will allow to make strategic plans, control programs, techniques and methods of investigation that aspects of products and processes are considered as well as the characteristics of the culture and behavior of managers and employees.

The water supply scenario and its use in the agriculture, livestock and cities indicate that there will be reduced availability of water for industry. As these consumer sectors are disorganized and there are losses in the distribution of water in the order of 30%, it is up to industry to worry about meeting their demand while reducing your water consumption.

The aim of this paper is to present an analysis model of availability of water and energy in the chemical industry, along with a behavioral adjustment model to enhance a large reduction in water consumption and energy supply to meet the expectations of these. The reduction in water consumption and energy depends on: management actions; switching procedures; behavior changes in the short term; application reuse techniques and reprocessing wastewater and energy in the medium term; and development and installation of technologies to increase the supply of water and energy and to reduce demand for water and energy in the long run.

Initially normal trend of events without intervention in the provision of management or demand management is considered. If occur climate change altering the supply of water and energy, on the other is the population and industrial growth by increasing demand for water and energy. In this case the industry will become hostage to a situation of lack of water and energy when looking at demand in relation to supply. In our case this will occur in 2022.

The analysis of availability compensates for the reduced supply with even greater reduction in demand. The analysis method of availability is divided into the following steps considering the comparison between the supply of water and energy with climate change with the demand for water and energy with preventive actions. Should perform prediction of what should happen in the future to adjust the present with the routine programs. This methodology is divided into the following steps: (1) Definition of current and future supply scenarios based on IPCC; (2) Identification of current demand scenario and future to compare Offer; (3) Analysis of the availability of water, energy and competence; (4) Culture Analysis and People for behavioral change; (5) Analysis of Routine and Integrated Management; and (6) Compliance Analysis of consumption of water and energy. A case of the chemical fertilizer industry are tested.

1 Introduction: Context and Background

Water is essential for life and it is the base of almost all human activities. In this context, water is used in power generation processes, domestic use, agriculture, cattle raising, and industrial sector among others. Provide access to clear water in quantity and quality represent one of the major contemporary challenges [1].

According to the UN World Report, the demand for water is expected to grow by 55% worldwide by 2050 due to the fast urbanization mainly in developing countries. The report also states that the world's water supply will suffer a deficit by 40% until 2030 if the water management do not improve. The agricultural sector will produce about 60% more food, and the industry sector will increase water demand up to 400% until 2050. Developing countries such as Brazil, Russia, India, China and South Africa will drive this [2].

The concern about water scarcity increased in Brazil after the water crisis in the state of São Paulo the most populated area located in the southeast of the country. This region from Brazil instead of the Northwest was an area that there is not this kind of problems. The water crisis didn't harm only the life of the local population, but also the industrial sector which account for 32% of the Brazilian GDP [3].

The industry needs to be alert and prepared to deal with this kind of event, because this issue affect the running and growth of production [4]. Moreover, scarcity and water pollution have amplified some planetary boundaries known by scientists as safe operating space for humanity. Be in the planetary boundaries is a precondition for sustainable development. According to Johan Rockstrom *et al.* freshwater problems is influencing and affecting some planetary boundaries as biodiversity, food security, population health, and the ecological functioning, such as climate regulation and carbon sequestration [5].

According to ANEEL (National Electric Energy Agency), approximately 75% of Brazilian power generation come from the use of renewable sources mainly hydropower with 62% [6]. Water is also a driving force in the generation of electricity in thermal power generation (steam generation), and necessary to chemical and petrochemical industries for heating (thermal energy) or cooling, among others services [7]. Thus, the availability of water is associated to the availability of energy in socio-economic activities in Brazil and the world.

The lack of structure industrial sector to deal with the interruption of operations resulting from non-availability of water and energy (effects of climate change) is increasingly evident. The structure is related to technological aspects (e.g., automation) and of complex decisions. Many decisions involve multiple factors such as climatic and behavioral aspects.

Although industries are trying to develop projects to reuse of water from sewage treatment plants to supply cooling tower makeup water as was done in the state of São Paulo during the water crises, it is not a tendency. The reuse of wastewater in the industrial sector has been considered synonymous with economic gains, in addition to the positive image transmitted. Nevertheless, the concern is the concentration of pollutants and the possibility of increased losses by corrosion. Although reuse of water from sewage treatment demonstrate that management procedures are related to the basic principles of preventing water shortages, energy and environmental impacts, it isn’t clear a commitment to analyze the availability of water and energy in order to adequate demand over supply considering scenarios for 2035 [4].

2 Method

The method intend to compare the supply of water and energy to their demand in the present time and a projection for the future - forecasting. This future scenario permit to take place the necessary actions to maintain the water and energy availability over time – backcasting [8]. Furthermore, prediction scenarios of greenhouse will help adjust the current emissions through routine management programs and new technologies.

Thus, short and long-term actions can be preventive as reducing the emission of greenhouse gases (replacement of energy by renewable or decrease their use or reduction of fuel burning) or corrective as the reduction of water consumption.

The analysis of availability method (shown in Figure 1) is divided into the following steps. (1) Definition of current and future supply scenarios based on the IPCC. (2) Identification of the current demand scenario and future to compare with the supply. (3) Analysis of the availability of water, energy and competence. (4) Culture Analysis and People for behavioral change. (5) Analysis of Routine and Integrated Management. (6) Compliance Analysis of the consumption of water and energy.

2.1 Definition of current supply scenario and future based on IPCC

The scenarios of use and supply of water indicate further reductions in the availability of water for industry. Water users (agriculture, livestock and population) are messy and work with distribution losses that can reach up to 34% [9]. Thus, the industrial sector need to care about increasing the supply of water and reduce their consumption at the same time through demand management and local source control programs.

The discussion about the energy consumption is linked to the greenhouse emission. The rise of GHG into the atmosphere depends on the burning of nonrenewable fuels or the lower conversion efficiency producing CO. The environmental impacts result in climate change and consequently alterations in water availability for society and for the economy as well as impact on biodiversity.

According the annual report about the inventory of CO_2 in the atmosphere, the amount of emission should be minimal to reverse climate change. Thus, the planet earth must send less and less CO_2 into the atmosphere, gradually, even with the growth of economies. The value target is 14 GTY in 2050 [10] and [11]. The forecast on the growth of the economy, with no GHG emission reduction action is to increase the current 50 GTY to above 62 GTY in 2050 (value stated in 2009) [10] - indicating the irreversibility of the greenhouse effect.

This paper aims to demonstrate how an analysis model on the availability of water and energy in the chemical industry and a behavior adjustment model may cause a large reduction in water consumption and energy in order to meet the expectations mapped by future supply of these inputs. The reduction in water consumption and energy depends on: management decisions; switching procedures; behavior changes in the short term; application reuse techniques and reprocessing of waste water and energy in the medium term; and development and deployment of technologies to increase the supply of water and energy and to reduce demand for water and energy in the long run.

2.2 Identification of the actual and future demand scenario to compare with the offer

A analysis of production and shutdown reports allow to build, in a preliminary way, the knowledge about the cause of losses and the behavior type to deal with these losses mainly of time, materials water, and energy. A rapid assessment of water and energy restrictions can give an idea of the scenario in the company. The production growth rate indicates if the demand will increase with the view to forecast the water consumption in the future, without including the preventive and corrective actions to demand management. When the supply and demand are compared, we can estimate the gap and which information is required to improve the analyses.

Water is classified in industry in various levels to meet the demand of the production plant and the administration. Large volumes of water are used in operations such as washing, transportation, ion exchange and others. One of the largest industrial water uses are in the cooling towers that meets the warm currents of the separation unit operations in the chemical process industry. Clarified water is the type of water used in cooling towers and represent 70% of the raw water inlet to the industry [12].

The measurement of water consumption is a legal obligation in order to inform the government environmental agency who granted a license with a utilization limit. This grant lets to use a quantity of surface or underground raw water and returns it (in effluent format) a quality better than the shot. Thus, it is necessary knowing the use of water, qualities (restrictions of equipment and processes), disposal and the behavior of the team and the manager regarding this precious resource to reduce the current water consumption and future need.

Regarding to energy consumption, classifying the type of energy used by industry is the first step to establish a demand profile. The electrical or thermal energy are primary energy and they can be converted into chemical energy and work. Thus, the main inputs in chemical industry are electric power and fuel. The measurement of consumption allows calculating the use of energy per unit of product giving an idea of the efficiency of its use. When a baseline and target (benchmark) is established and these are compared with the requested reduction of emission of GHG it is possible to establish the programs and investments in medium and long term.

2.3 Analysis of the availability of water, energy and competence

The starting point of the availability analysis is building the graph indicating water consumption and how much in terms of grants is authorized consume (immediately). In the case of supply, it is necessary to build an algorithm to increase the supply due to new sources or the reduction of losses. Another algorithm must be built when the source of water decreases due to climate change, increasing urban population, rise of irrigation area to agriculture, and so on. However, demand for water could consider two possibilities: (1) any action to reduce demand during the time (2016-2035) indicating only increase in the forecast increase in production or (2) actions and investments in process design, control routine and team behavior to reduce water consumption.

When the annual supply is projected for the future (including economic changes in society and the climate) and compared to future demand without considering corrective action, there is an indication of interruption of industrial operations for lack of water. Thus, actions to reduce water demand is indicated.

The availability of energy analysis actually compares the possible amount of CO_2 emission (which changes with time due to more restrictive situations in the future) from estimates by the International Energy Agency, with the current and future amount of CO_2 emission in different situations as the action taken or not to prevent irreversible climate change.

The initial approach of water/energy supply and demand needs to be confirm by mapping the thermal systems, equipment controls, water and energy saving programs, and analysis of sources and water and energy uses. The demand mapping is confirmed and compared with the preliminary data, they are made of the demand adjustment exercises in the timeline considering new best-case scenarios or more critical every five years.

Thus, the master plan for water and energy is builds up with the demand adjustment regarding the supply by technologies, programs, and technologies to be installed gradually. It is considered to predict supply by 2035 criteria related to industry growth, population, issues related to agriculture and irrigation, rainfall forecast, and the probable faults keeping the same demand.

This initial analysis should be reviewed after a comparison with external benchmark and improve the quality of information from direct and not secondary measurements.

Among the technologies are: reverse osmosis, geothermal, flexible cooling towers for reuse, process intensification, and others. Among the techniques are: water balance, energy balance, pinch, reuse of wastewater, Exergy analysis, analysis of material losses and energy, and others.

2.4 Analysis of Culture and People for behavioral change

A series of procedures and attitudes in the levels of implementation, management and strategy are based on increasingly complex decisions [13] and [14]. It is necessary to know how these decisions occur based on local cultural characteristics [15] and likely rules that influence not reduce water and energy consumption. It is important to analyze the influence of equipment failure, processes and operations of the direct losses and indirect materials to the reprocessing [16]. The decision to adjust the quality of the product can impact on the increase in the consumption of water, energy and the rework of workers.

An analysis of people and cultures will be able to identify indicators that evidence the inappropriate behavior that enhance water loss and energy. This allows building a program to control losses from these resources. With this identification, it is possible setting a program of training or behavioral adjustment to be implemented in the company.

The cultural influence is discussed for building a robust safety culture and is based on the dynamic behavior of the staff. This behavior is the result of a field of cultural forces: organizational desire, local culture and features of global culture. Some trends should be strengthened and other inhibited and this action are based on archetypes analysis [17].

The human factors analysis allows to establish programs to increase human reliability. Human factors can be [18]: conflict between policies and practices, the conflict between production and safety priorities, control facilities, managerial aspects, skills analysis, and preparation procedures are part of human factors that should be analyzed and monitored.

The moment of decision is then influenced by biases of culture and human factors causing cognitive impairment [19] occurs following human errors mechanisms as the default [20]. In addition to the analysis of cognitive impairment seeks to reduce the risk of accident through the joint analysis of technical factors and not technical. This strategy allows us to discuss what are the aspects that can impact the security processes in an organizational change [21].

2.5 Analysis of Routine and Integrated Management

After identifying the behavioral constraints or characteristics that lead the employee and the manager to start faults that cause loss of energy and water, it is important to analyze the routine. Check the risks of system stop on a daily basis allows you to work in a review of equipment and critical tasks. The knowledge of water systems, heat exchangers, energy integration, wastewater allows for maximum use of energy by increasing the energy efficiency. The water and waste energy reuse will rationalize the consumption reducing dependence on bulk supply facilitating sustainability of the company.

2.6 Analysis of the availability of water, energy and competence

The establishment of comparison criteria for use of water, energy and behavior that avoids the technical and human error adjusts techniques, technologies, decisions, behaviors and knowledge. So this step enables the continuous improvement of the risk management program in Energy and Water. (ISO 50001) [22].

3 Demand management

Some global factors have to force the industry to make efforts to increase productivity and reduce the consumption of basic materials as raw materials, water and energy. The water is already being alert reason for society and for the industrial sector, where specific events of lack of water can turn into long periods of shortage impacting on the economy and the population.

Energy is a key input to ensure the economic and social development and the rationalization of its use is an essential strategy to avoid losses during periods of restrictions resulting from the emission of CO_2 and the scarcity of natural resources. Energy rationalization might be to change procedures, behavior change, control techniques and technologies to enhance the use of energy and thus improve efficiency and improved thermal performance of the equipment and systems.

The water consumption in the industry is directly linked with energy demand, for the use and distribution of water in a production process, the energy has a fundamental role to keep the equipment operating. Similarly, the electrical and thermal energy coming from a unit of basic inputs need the aid of water for its existence.

Industrial activities that involve the consumption of basic inputs has suffered deployments due to significant consumption and existing waste. Cycles of concentration in cooling towers, water use outside of specified boilers, mass loss and energy in steam distribution systems, heat losses in electric power generators, discarded wastewater without prior analysis, are some examples that show opportunities for deployment techniques or technologies that contribute to the reduction of utilities, waste and therefore environmental impact.

However, to reduce the consumption of water and energy are proposed a number of solutions management, various techniques used by the STAFF during routine operations and technologies.

3.1 Management

The management plays a key role in the implementation of programs to control routine related to the behavior of the staff and control of deviations. Each technique has suggested goal, steps and impacts are described below.

The control program at the source [23] is to identify equipment and pathways of contamination to the final disposal, liquid (effluent), solid (waste inventory) and gas (emissions or losses fugitive emissions). Understanding on which the carrier fluid and how it is possible to establish measures to avoid losses. The additional technical material balance loss is the mass calculation of losses and the impact on fixed costs, variable and billing.

Control of critical task, named as POS, Procedures and Operational Standards [24], aims to identify, in this case of water and energy availability, which tasks are that they can produce losses and analyze the environment, planning, execution, control and the failed task.

The analysis of environmental deviations [25] want based on the fault tree method and field observation, identify deviations that lead to contamination of everyday life. When this technique is well understood in the organization it enables a behavior change operation.

This research is in development and is about identifying what the proper conduct of the operation team to prevent energy loss. Follows the line of CCPS in Operational Discipline [26] and includes aspects of the company culture. It takes into account that energy is intangible and less noticeable to control losses.

The calculation of the reliability of socio-technical systems [27] depends not only on engineering and includes the social and managerial area of the company. For this reason it will be led by the manager. When the reliability of each functional part of the system (process, equipment, operation and human) is identified, after the process of the complexity of the classification, the calculus of the system reliability level is possible. The system reliability is then corrected through social attractiveness that indicates the strength of culture, the level of cooperation (communication) and adaptive leadership to new realities. Through this analysis it becomes possible to more policy action on production controls.

By calculating the reliability of systems and identified the impact of human reliability begins research on each performance factor indicated by SPARH [28] and discussed by API770 [18]. Thus, it becomes possible to define critical areas for action by the manager. These analyzes made by the management has focused on reducing energy losses!

3.2 Engineering

The engineering staff has the role to develop adaptations to current processes and equipment and provide new equipment to avoid or mitigate energy losses. These solutions require equipment yield calculation, energy efficiency and thermal performance.

Initially the operational control equipment and thermal systems adjusts the best operating point through procedures and control systems. It is therefore important that the process engineering and equipment to track and control the income and performance of furnaces, boilers, cooling towers, electric motors, power contracts, distribution of steam, condensate distribution, reflux in the distillation, and balance energy in the process equipment and utility in general.

This control can identify the need to revise the performance calculation methodology to be open systems, such as cooling towers. In furnaces and boilers may be a need to review the maintenance procedures (installation of refractory) or operating as purging. In collecting and recovering the condensate, on the other hand, several factors may impact on live steam loss to the atmosphere, it depends on the thermal load of the plant, steam quality, operation of steam traps, and quality of thermal insulation.

The effluent mapping to identify the loads of contaminants requires the establishment of networks of sources and uses of water by continuous or intermittent processes procedures. It is needed when discussing energy sources and uses, in this case, less visible, and thus requiring additional technical field, such as using pyrometer surface of hot equipment.

The mass balance and energy are more detailed technical and require methodologies for utilities and process equipment. simplifications may be needed, where process streams with low heat capacity variation are classified to 80% balance of the impact, reducing the number of streams.

For combustion systems are used to heat the lower and higher power. For open cooling tower systems we use the useful heat exchanged. In the case of energy balance that includes the axis of work it is important to check the income and losses of energy through heat. The energy balance is important to compare the fuel consumption with the issue of greenhouse gases.

3.3 Technology

In terms of technology we will discuss those involving new criteria for design of industrial plants and new equipment that reduce energy losses and water or increase its offer. Among the technologies involving new standards for process design and its control: modularity, intensification, integration and exergy. Among the technologies involving new equipment to reduce losses and increase in supply: desalination, heat pump, cogeneration, trigeneration and combustion bomb.

4 Application in the Chemical Industry

Applying this analysis methodology on the availability of water and energy it is based on the experiences in research projects to rationalize water use in the chemical industry of linear alkyl benzene (LAB) and the fertilizer industry (UREA) [29]. Other research has enabled the discussion of opportunities to reduce energy losses in the petrochemical industry from energy balances in processes and utilities [30] AND [31].

The experiences in the LPG industry and own fertilizer industry [24] allowed the present discussion of cultures and people. And finally, the argument based on ISO31000 and ISO50001 enabled compliance analysis.

The implementation of these steps during the period of two years will have as a result the establishment of control programs in source and strategic plans for water, energy and skills. Every five years is necessary to readjust the supply condition and analyze the indicators that demand occurred in accordance with the master plan. Organizational changes can alter the demand for water, energy and time.

The intended application for two years will be addressed through the following topics: contextualization, supply management, demand management, strategic analysis of the availability, culture and people analysis, routine management and compliance analysis.

4.1 Contextualization

Audits for use and water sources that used the effluent mapping technique and performance analysis in cooling towers indicated the high water consumption in the industry LAB mainly to feed the cooling towers [32]. It was noted also that occurs excessive contamination of waste water with several points of discharge of effluents and products. The LAB production process is complex, with the possibility of reprocessing with handling high toxicity products such as hydrofluoric acid, but a well-structured staff to decisions simply to achieve better control, access to appropriate techniques for controlling consumption energy and water.

The energy and water balance techniques, due to the complexity of petrochemical processes, indicates that more rigorous methods require mathematics reconciliation data. In other cases seeking the simplification based on the impact of the amount of energy involved, where the reduction of line processes is necessary, within the volume control for mass balance.

In the petrochemical industry of basic inputs where the type of naphtha defines differences in production costs, the equipment, the process parameters and the quantity to be produced of each product (olefins or BTX), it is important to review the energy balance. Flow rates in these processes are very high, so errors in the way of operating equipment can cause the plant to have high energy losses, indicating a need for high fuel consumption and high or make-up water for cooling towers.

The demand by up to 30% water cuts and agreements of pressure climates, and briefly change the intensifying industries in energy use. The lack of assertive decisions or appropriate behaviors have removed the goals in relation to the consumption of energy and water in the routine. This indicates that the transformation requires greater care in the industrial segment!

4.2 Supply management: control and planning for 2035

Petrochemical industry of cases and fertilizer supply of real data and water demand and actual data supply and energy demand are discussed.

4.2.1 Water supply management: premises for the analysis of supply management in the fertilizer industry in northeastern Brazil

Assumptions: (a) average climate change scenarios A1B with rainfall reduction impacts in Bahia (2070-2100) in about 60% on the coast and 35% in semi-arid with 2-5°C temperature increase [33]. (b) increase in supply due to improvements in the operation of river basins and due to climate change by 2018; (c) reduction due to climate change estimated at 28% to 60% of the current offer in surface and groundwater respectively, for 2035; (d) reduction of reliability in service of the Paraguaçu River Basin [34] for 2011 to 2040, mainly for irrigation, industry and power generation; (e) Increase in the number of months during which there were failures in care:

the Savior supply (1% to 5%) and industrial demand (1% to 9%) with reduced power generation by 66% in Dam of Horse Stone [35].

Supply: the water supply to the industrial sector in the basin of Paraguacu and Salvador city between 2012 and 2034. **Offer 2012:** Surface water = 20.9 m^3/s, groundwater = 3.9 m^3/s, total = 24,8 m^3/s. **Offer 2018:** Climate change with increased rainfall: 24,8 to 78,8 m^3/s (3.2 x). **Offer 2034:** Reduced rainfall = 25% in 2018, 19.7 m^3/s.

Fertilizer unit: in the specific case of fertilizers unit [29] is considered that the initial flow supplied (2012) is equal to the demanded flow rate, that is, 700 m^3/h. **The Offer in 2018:** since part of the overall offering for the industry shifts to supply the city (an increase of 2 X instead of 3.2 X), that is, 1400 m^3 / h. **Offer 2034:** decrease due to climate change by 33%, that is, 462 m^3/h. The demand for water in the fertilizer unit should be less than this flow.

4.2.2 Power supply management: premises for the analysis of supply management in the petrochemical industry in northeastern Brazil

Considers that energy supply is limited by the capacity of the atmosphere of the planet receive GHG to prevent climate change. Thus, the supply limit is based on climate studies by the IPCC. On the other hand, the demand for energy in industry depends on the activities carried out by thermal and electrical systems. Thus, the balance between environmental restriction is sought (supply) and the actual greenhouse gas emissions by industry (demand).

The power availability analysis will be handled by the comparison between supply and demand. Supply management will have an upper limit maximum environmental capacity to receive GHG emissions to prevent climate change (14 GTY in 2050), this is the blue scenario. In this scenario emissions are monitored since 2005 with 28 GTY emissions and tend to grow up to the value of 62 GTY when the climate change process becomes irreversible. This emissions growth occurs if no action is taken towards sustainable economic growth of BRICS and developed countries [10].

Global targets need to be resized to establish regional or industrial sector or by enhancing industry in energy consumption targets. This schedule allows you to define local action strategy seeking to guarantee the reduction of GHG emissions.

In the case of a petrochemical in Camaçari (Brazil): since 1000 there are on planet earth petrochemical same capacity and the global industry emits 100 times the amount of CO_2 (based on fuel consumption) of the global petrochemical segment. The proportion of the global emission reduction for local equivalent to a reduction of 10-5 Industrial overall goal [10] for this local industrial. In the case of switching to renewable matrix, the overall goal is 9 GTY after the local design amounts to 0.09 MTY for this petrochemical industry.

Offer in 2050: (a) replacement by renewable energy → 10 gigatons year of global emissions or 0.1 mega ton year local issue; (b) reduction in CO_2 consumption by effective use of thermal energy resulting from the burning → 11,5 gigatons year of global emissions or 0.115 mega ton year local issue [10].

4.3 Demand Management: control and planning for 2035

Petrochemical industry of cases and fertilizer supply of real data and water demand and actual data supply and energy demand are analyzed.

4.3.1 Water demand management: assumptions for demand management analysis on the fertilizer industry in northeastern Brazil

Assumptions: demand management analysis in a fertilizer industry in the Northeast of Brazil: (a) average growth of 2,5% per year of urban demands of the cities of the RMS (increased population projection and reduction of per capita consumption) [36] and (b) growth in consumption by industrial expansion and irrigation.

Urban and irrigation demands will reflect the reduced availability of water for industry.

Demand: the demand for water for industrial thread in the basin of Paraguaçu and nearby cities including Salvador city between 2012 and 2034 considering that there is no demand tuning activities. **Demand in 2012:** 15,5 m^3/s city and 6,2 m^3/s for industry, total = 21,7. **Demand in 2034:** 40% of consumption by the population increase and reduction of losses of approximately 25% due to investments in the operation of water supply equivalent to 5% increase in the city, 16,3 m^3/h. New industry and new units with 15% increase indicating 7,2 m^3/s for the industry. Total 23,5 m^3/s. Indicating that the offer will be lower (19,7) than the demand (23,5 m^3/s). That is, if no action is taken to adjust to demand, the industry will be at high risk of interruption.

Fertilizer unit: in the specific case of fertilizers unit [29] is considered that the initial flow supplied (2012) is equal to the demanded flow rate, that is, 700 m^3/h. **Demand by 2034 without setting activities,** will increase by 15% in water consumption = 805 m^3/h. This unit of fertilizer also have problems which need to consume less than 462 and will be consuming 805 m^3/h.

4.3.2 Energy demand management: assumptions for demand management analysis in the petrochemical industry in northeastern Brazil

The suggestion by the International Energy Agency (IEA) is that there are different ways to reduce GHG emissions: energy efficiency (electrical and thermal), renewable energy (wind, solar, biomass), nuclear energy and carbon capture. The IEA also shows that the main actors for this reduction are: processing industry, power industry, transport and buildings.

Avila *et al.*, 2011, showed that a petrochemical site:

(a) replacing 20% of energy consumption based on fuel oil with biomass energy. Following the assumption that this biomass, which compose a forest previously cut off from the atmosphere by photosynthesis, the same amount of CO^2 emitted in their use. This petrochemical be taking 20% of CO^2 emissions initially issued.
(b) 50% reduces the loss of live steam (estimated at 30%) in the condensing system, steam systems, boilers and furnaces.
(c) recover 50% of losses due to non-use of cooling water recirculation flow in heat exchange in cooling towers (ideal thermal performance (dti) = 95%; dti real = 75%; 20% difference). Thus, an original temperature difference of 10°C, with higher

energy efficiency reaches 11°C, a reduction of 1°C in cold water → bringing as positive impact improving the profile of the separation columns and reduction of fuel demand thereof magnitude and CO^2 emissions as a result, 10%.

Note that:

1. The amount of biomass used (petrochemical-based local → global petrochemical → global industry) inhibits the emission of CO^2 with the change of 20% for renewable matrix 8,45 GTY, compatible with the demands of the International Energy Agency.

2. Increased energy efficiency in thermal systems (petrochemical-based local → global petrochemical → global industry) reduces burn and CO_2 emissions by 12,5 GTY, compatible with the demands of the International Energy Agency.

4.4 Analysis of availability: strategic plan

Exercises for analysis of the availability of water, energy and skills

The program to control wastewater at the source and master plan provided operational activities and new projects to achieve the demand reduction of 750 to 310 m3/h, so below the 462 m3/h.

Scheduled actions are [29]: (2013-2018) (1) pattern of change in the final product plant shutdown; (2) stability in the operation of purification and reuse; (3) reprocessing the waste water treatment, reuse in cooling tower; (4) increase in operational reliability; (5) reuse of disposal with a high level of contaminant; (5.1) purifier contamination reduction - reduction of product load; (6) condensate polishing removal of organics. (2019-2020) (7) energy conservation project, condensate (I); (8) control work in the steam generation (II); (9) condensate return for demineralized water; (10) the entry of new plants with improved design pattern. (2020-2024) (11) of rainwater use project; (12) improvements in the use of cooling towers; (13) Water reuse for fire and utilities; (14) improvements in water treatment plant; (15) rationalize water use in the area and the administrative use. (2025 and 2028) (18) effluent reuse in cooling towers; (19) advanced technologies to reduce vapor loss - cogeneration and heat pumps. (2029 and 2034) (20) new techniques for heating and cooling: intensification; (21) new cooling technology projects: geothermal energy and application of the products.

The availability of energy and competence [8] are discussed from Figures 1 and 2. Figure 1 is considered the emission of CO2 in a temporal analysis which compares the issue (with and without adjustment of shares) with the offer is the maximum capacity of the atmosphere to process GHGs without climate change. In Figure 2 we analyze the defendant competence for technology with the expertise offered by people with and without adjustments regarding the offer.

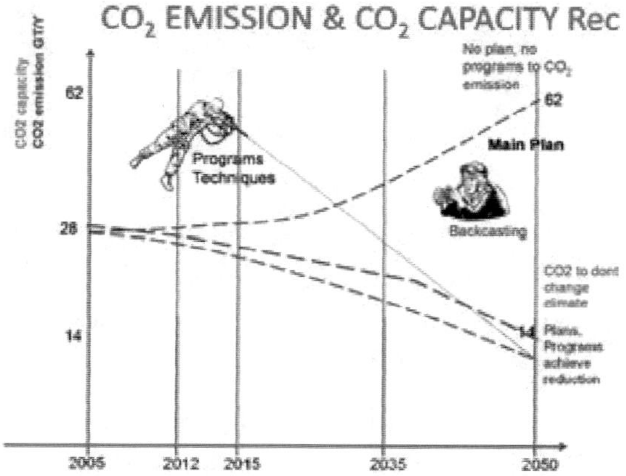

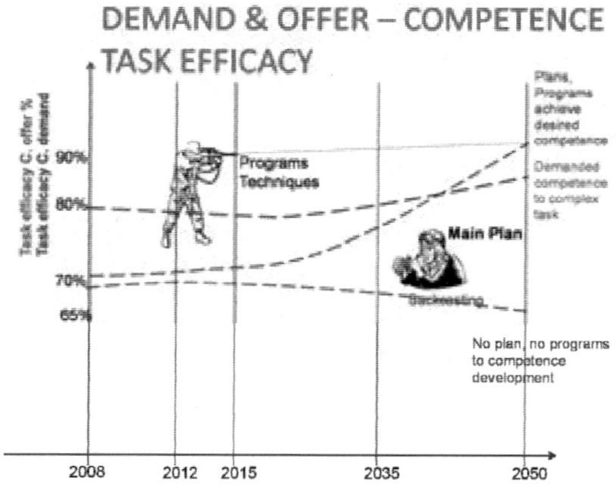

Figure 1 - Supply and demand GHG [8] Figure 2 - Supply and demand Competence [8]

<u>Master Plan for Water and Energy [8]</u>

The Master Plan is a strategic tool that enables meet the challenges of climate change and its consequences for the industry. Planning for adjustment and availability of water & energy for the industry is complemented with technical and management tools for the implementation of the Master Plan.

The Master Plan discusses the legal and institutional aspects of the laws dealing with water, energy and environmental impacts. supply management depends on legal issues that allow water and energy management in industry. The current and future availability is performed by comparing supply of water resources in local basins with the demand of these resources to the industry seeking the necessary adjustments. This discussion should also is made to address the constraints of the Brazilian energy matrix with the conditions for water use and CO2 emissions.

Analysis of water and energy classes and specifications of equipment and processes enables us to diagnose the possibility of water and energy reuse. The identification of classes used in the industry indicates their effective use to have qualities in different sources. The water treatment and effluent depends on the water balance enabling reuse projects.

Concepts and applications are detailed for adaptive management of supply and demand and which proposals for long-term strategic adjustments as credit demand reduction, industrial symbiosis concepts and opportunities, the importance of monitoring indicators, and the risk of lack of water for 10 and 40 years.

The ways to manage supply take into account: availability of water sources, climate change, hydro-geological perspective, costs and energy content. A calculation procedure is suggested for the adjustment of supply relative to the projected demand. In case of power must calculate the CO2 emission target for the specific case of the installation. The management of future demand includes the control program in the source material balance losses and others.

In the Annex of this plan are presented technical and managerial and operational methods. The intention is to keep demand and controlled supply. The main technologies to maintain demand and controlled supply for Industry are classified and detailed.

Control Equipment and Thermal Systems

The thermal control equipment and systems, process control and reliability of the systems are extremely important to regulate the demand for water and energy. The demand adjustment requires lower value to water supply (based on state water policy) and energy (based on local and global constraints of GHG emissions).

The equipment that has high impact on fuel consumption and the emission of greenhouse gases are: (a) cooling towers; (B) heat exchangers that utilize utilities; (C) generating steam boiler; (D) heating stoves or pyrolysis reaction; (E) steam distribution and condensate energy recovery (including traps); (F) current of high enthalpy process; (G) operating point and balance of separations such as distillation; and (h) electrical systems in general.

At this point will be discussed energy efficiency in some of these systems: (a) Deten-water research project in 2003, Ávila noted the importance of working with clean cooling water circuit to avoid blockages in the upper pools and highlighted the potential losses if maintaining the cooling tower structure (and water quality) is poorly made, with losses that may cause heating of water up to 1 ° C. By maintaining one should prevent the cooling water flow paths follow without heat and mass transfer with the dry air. (b) This project was highlighted the importance of working with cycle highest concentration in the towers to reduce water consumption. This control increases the risk of corrosion in order to be header exchangers and hinders water reuse in cooling towers [37] [38].

(c) There are several opportunities to recover waste heat in the boilers that generate steam in enhancing industry in energy consumption [30] [31] [38]. Since preheating of the water, control of stoichiometric combustion, the specification and maintenance of refractories, the utilization of heat of flue gas (new boiler, flue pump). (e) the design assumptions for steam distribution, steam traps installation and condensate energy recovery (including traps) should be based on the expansion possibilities of production and operating limits of the systems [30] [31] [38] . Thus, due to the limitations of operating with traps and thermal systems, search is the best point and with new needs, install new production modules.

Water balances, effluent, fuel, GHG and energy

In research projects to rationalize water and energy were used mathematical techniques mass balance (FAFEN-WATER, DETEN-WATER) and enthalpy (Ecobraskem) with multiplier weights varying according to quality information and then closing balance by data reconciliation. This technique was used for LAB production plants, Olefins, urea and petroleum [30] [31] [38]. As for the balance of utilities in equipment such as cooling towers, boiler and furnace energy balance it was based on ICPs and the formulas simplified the treaters water that is based on the wet bulb temperature, the flow and the temperature difference [38].

Sustainable projects: new techniques and technologies

The need to reduce energy consumption requires changes in equipment design and process. The main reason for these changes is to reduce the gaps, allowing a reduction in the consumption of utilities through direct heat transfer without the need for thermal fluid design adjustments considering the external environmental changes and changes in thermal load.

The spare capacity of operations and equipment is adjusted to increase the capacity of industrial plants. In industrial processes it is possible to carry out this adjustment due conservative design factors that enhance the capacity of the equipment, but in the case of utilities is not always possible. Thus, you can use the spare capacity of reactors, pumps, but it is not possible to trap and electric motors. With restrictions on the use of energy and knowledge in human reliability, it is intended to modularize the projects through more accurate system design [38].

Another option in the project area is to reduce the use of utilities and size of equipment by intensifying operations and processes where the heat necessary to the achievement of operations is transferred directly by the equipment wall. Where the heat is released from certain operation and received in another operation without the need for use of thermal fluids. If energy balance in petrochemical is noted that the use of exchangers with hot thermal fluids (steam) and cold (cooling water) consumes 4000 GJ / h compared effectively with the necessary heat (80 GJ / h) to unit operations work without utilities. With intensive use of energy can significantly reduce power consumption and size of equipment [38].

The rationalization of energy and water can be conducted using techniques which integrate the process with the utility and waste. The identification of water users with the material restrictions for equipment allows the water and energy savings through its reuse in equipment. This technique is applied in sewage networks and networks of heat exchangers.

In the case of wastewater networks (pinch), the contamination levels of waste water streams are analyzed to compose a quality goal that meets the requirements of equipment that can use this water. If exchanger network (pinch), the residual heat through the cooling water reuse allows a better use to reduce energy losses.

The main equipment that consumes water and has high potential for use of waste water is the cooling tower [32]. Thus, it is the integration of processes (sources) with the effluent (waste water standard), and returns to the utilities (water cooling).

Energy recovery from the best thermal performance in cooling tower and application of process integration (pinch) has as a result the reduction of water loss and the reduction of $CO2$ emissions. To account for this loss reduction in $CO2$ emission area is seeks to study the integration of the cooling tower, heat exchangers, separation operation (reflux), the heat demand (steam oven), burning fuel and $CO2$ emissions.

The decrease in water supply from climate change requires research on the feasibility of industrial desalination including comparing other solutions involving energy supply and water (eg lifting deep brackish water). The current technology of reverse osmosis membrane desalination is still not profitable for water treatment industries in the private sector.

4.5 Culture analysis and people

Private and public institutions recognize the importance of investigating the matter culture and people for sustainability, but do not yet know what priority to this issue. To establish an organizational and strong safety culture is required: to know the tasks and risks; analyze the behavior through the routine signals; and establish metrics to identify the stage of safety culture.

An onshore oil industry indicated that the image directly affects the profitability of the company, so often the actions prioritize avoid independent accident effect on production and on revenues. A proposed action was suggested for intervention in three years: (1) Task: To establish indicators of efficiency, formalizing forms of communication in organizational levels to review the inadequate standards establishing best practices; (2) Behavior: analyze routine deviations, establish a causal link with the process losses, see how minor emergencies are handled (decisions) and indicate gaps in relation to skills, knowledge and standard of acceptance; (3) Culture: Establish indicators for technical culture (bottom-up) and measure the difference in perception between the entrepreneurial vision and practices. Set organizational practices (rituals) to neutralize or mitigate biases culture.

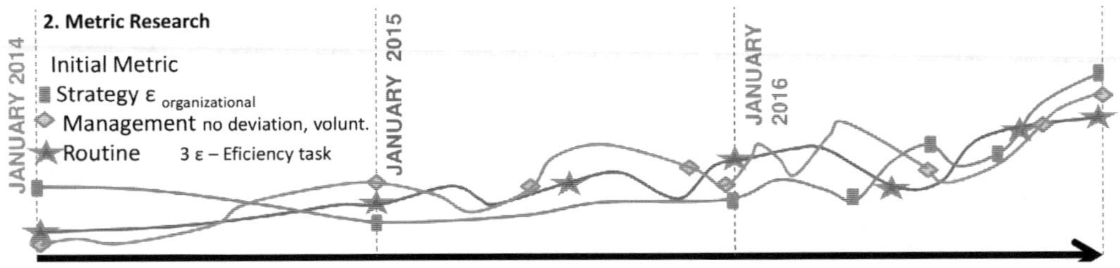

This initiative for the onshore site has been replicated through intervention in the fertilizer industry to analyze events in the routine that involves the effluent, where old vices of operation contribute to the process losses. The suggestion to change behavior and neutralization of decisions that led to contamination of effluent was through the educational program "Friends of emergency pond." The tools used in this program were: lectures in the control room and field, event analysis and hypothesis of the root cause, exchange of information through informative texts and making changes in the decisions of supervision and management.

The process that involves loss materials are more easily perceived as contamination of wastewater, fugitive emissions or the generation of solid waste. Large losses that affect the company's image as an accident or leaks has immediate impact on the stock market. Since energy losses are less visibility difficult behavior change to achieve targets in reducing GHG emissions. So it is important to consider cultural aspects and behavior to take more assertive decisions.

Analysis of elements or human factors [24]

In this GCPS organizational factors related to regional-global culture that are entered in 20 events in the chemical industry were discussed. These factors can initiate failure mode with final impact on energy losses. Thus, conflicts between policies and practices, economic difficulties, poor planning of the task, cultural traits that induce inappropriate decisions, risk management that does not include human factors are important factors to analyze the operational routine which leads to energy losses. The routine in deviant behavior difficult to establish operational discipline to avoid accidents and loss of energy [26].

Cultural analysis based on Archetypes [17]

Cultural phenomena do not exactly follow a logical order and are based on ancient customs rooted through tribal and new rituals introduced by the media. In an attempt to develop cultural barriers to the accident and GHG emissions seeks to analyze the phenomena resulting from the action of primary and derivative archetypes in the organization's tasks. This kind of technique indicates which new rituals should be deployed to vaccinate cognitive impairment.

Analysis of Cognitive Failure [19]

The cultural environment induces undue behaviors in individual or group during the task. Understanding the type of cognitive failure depends on the analysis of executive function in the various stages of cognitive processing: memory operations, environmental awareness for the construction of knowledge and action, attention calibration, job planning, job control and socioeconomic utility affective. It is important to identify the type of human error in the energy losses to prevent the ineffectiveness of action taken.

Reliability Analysis Sociotechnical [27]

The integration of production technology (product, process and control) with its operations on critical equipment indicates the importance of joint analysis of reliabilities. The operation achieves success if the understanding of the task objectives and knowledge of the process conditions (standard and fault). Thus, reliability of operations, processes and equipment depends on human reliability that receives the direct influence of cultural environments.

A propylene pumping operation in the refining industry was analyzed and it was concluded that the redundancy pumps is not enough when cognitive functions are weakened in conflict between policies and practices environment.

4.6 Routine management

The quality brought about the revolution of the administration [39] but did not ensure the sustainability of their results. The complex decisions and the investigation of systems failure mode indicate that to avoid losses should measure the efficiency and effectiveness of the task should be controlled thermal systems, critical procedures and also analyze the level of material losses and energy.

In the area of utilities and processes which circulates high amount of energy it is important to perform measurements and balances to prevent high losses of energy and water industry. The control of thermal systems involves critical steps such as maintaining water quality, purge tower and boiler, the level of control and the use of cells in cooling towers [31].

The production control by type of raw material defines the economic quantities of schedules based on variable cost of energy. In the petrochemical industry this issue has great importance in the definition of production programs knowing input naphtha type (weights and average).

Depending on the input naphtha occur demand for changes in processes, for example, the recycle gas will have another quality and will need to review the burners in the process gas boilers reuse [38]. To carry out the control of these tasks with high impact for the loss of energy is important to apply task analysis techniques in complex processes and have control over the realization schedule these tasks [40].

The mass balance and energy (water, materials and supplies, steam, fuel) enables the detection of losses and control at source [35] [23]. The balance of goods and inputs is closed with the quantification of losses: solid waste; liquid contaminants to the effluent; and fixed and fugitive emissions to the atmosphere.

4.7 Compliance Analysis of water and energy consumption

ISO 50001 [22] are described important procedures for the management and control of energy. It is important to identify sources (equipment and systems) lose water and energy in the production process. It is important to define if the losses are isolated or continuous, which the physical state and properties, the transport fluid and the impact receiving body. The output of thermal systems is measured by measuring the consumed energy and the comparison with expectations. Part of the energy loss and water is recoverable and the cost outweighs the benefit. compliance analysis indicates where the main losses and what the distance from the expected default behavior.

The fixtures and fittings that consume fuel or indirectly affect this consumption are extremely important: boiler, furnaces, heat exchangers, steam traps, cooling towers, water and steam distribution pipes, flash vessels, safety valves, and distillation columns .

It is important to define the consumption profile (baseline), the expected pattern of consumption, benchmark-based target, the difference between the measurement and the expected pattern (loss) and the percentage of impairment. It is also important to describe the operational procedures that can impact on the loss of water and energy consumption, expected standard and goal. These losses may have been caused by misconduct or lack of knowledge.

The criteria for equipment performance analysis are based on rules established by institutions such as ASME (boilers and furnaces) and CTI (cooling tower). Other criteria related to behavior and procedures may be based on API770 and operational discipline guidelines CCPS.

5 Conclusion

Initially considered the normal trend of events without intervention in the provision of management or demand management. On the one hand occur climate change altering the supply of water and energy, on the other hand is the population and industrial growth by increasing demand for water and energy. In this case the industry will be trapped in a situation of lack of water and energy across the demand relative to supply. In our case this will occur in 2022 (for water).

According to the problems already experienced shortages, so there is a balance between supply and water demand and energy are already demonstrated the means and methods that need to be taken into consideration and applied in the productive sectors and society. A case of the fertilizer and petrochemical industry have been discussed.

Already have the knowledge that technologies and techniques have been implemented in the industrial sector considerably reducing the waste of raw materials and fuel and thus environmental damage, but their implementations are not prioritized the way it should, as a general in contrast, there is a cultural influence, behavioral, economic and management that becomes increasingly slow the process of improvement.

With the projection of a future scenario and possible interventions mainly developing countries GHG emissions planned for the year 2050 is approximately 14GTano, very difficult task to achieve.

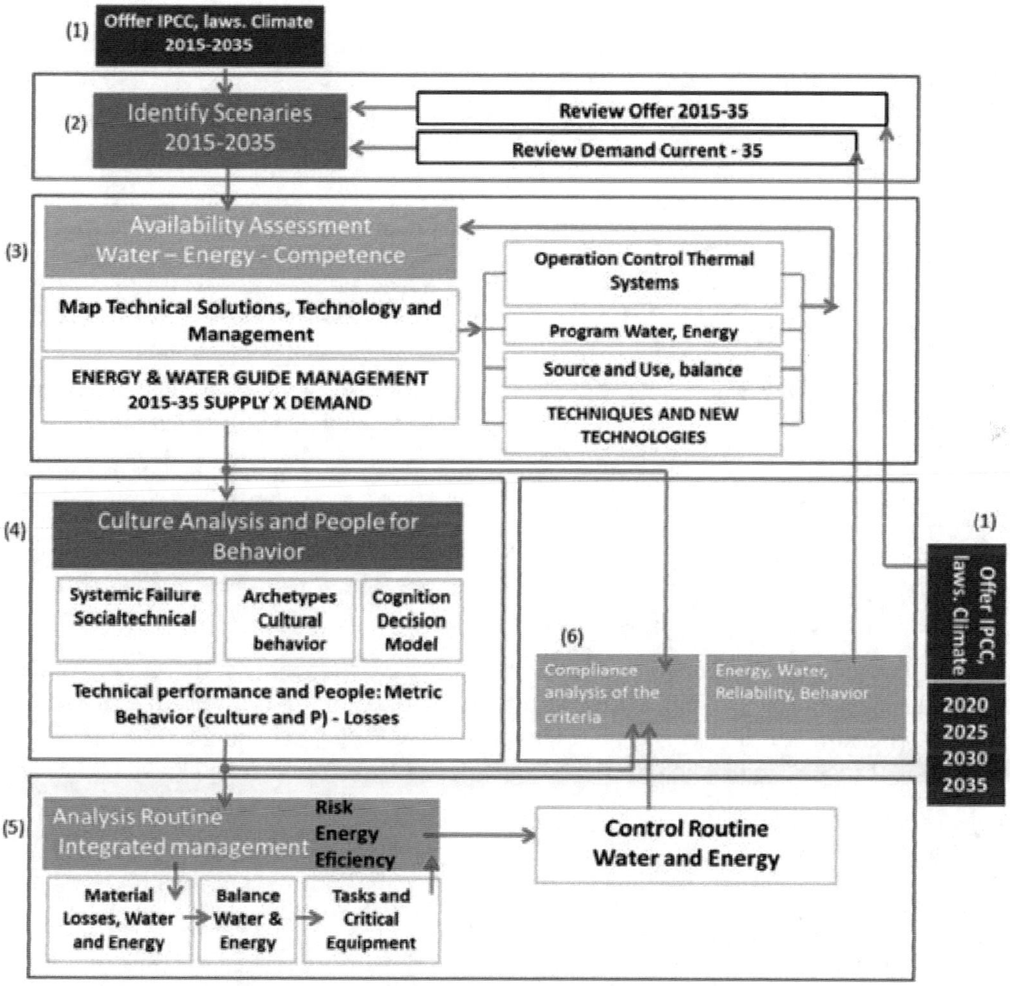

Figure 1. Analysis of Water Availability, Energy and behavior.

6 Bibliography

[1] CARMO, R. L.; DAGNINO, R. S.; FEITOSA, F. F.; JOHANSEN, I. C.; CRAICE, C. População e consumo urbano de água no Brasil: interfaces e desafios. In: SIMPÓSIO BRASILEIRO DE RECURSOS HÍDRICOS, **Anais...** Bento Gonçalves, 2013.

[2] WWDR – The United Nations World Water Development Report 2015: Water for a Sustainable World. Paris, UNESCO, 2015.

[3] FUNDAÇÃO SISTEMA ESTADUAL DE ANÁLISE DE DADOS – SEADE. Disponível em: < http://www.seade.gov.br/pibtrimestral/analise/>. Acessado em: 06 fev. 2016.

[4] PQU. SABESP. Sabesp e Polo Petroquímico lançam o maior projeto de Água de Reuso do País. 2008. Disponível em: <http://www.agenciagabc.com.br/noticias.asp?id=1045>. Acesso em: 06 fev. 2016.

[5] ROCKSTRÖM, J. *et al.* Planetary Boundaries: Exploring the Safe Operating Space for Humanity. Ecology and Society 14(2): 32. 2009. Disponível em: <http://www.ecologyandsociety.org/vol14/iss2/art32/>. Acessado em: 06 fev. 2016.

[6] ANEEL – Agência Nacional de Energia Elétrica. Matriz de Energia Elétrica. 2016. Disponível em: <http://www.aneel.gov.br/aplicacoes/capacidadebrasil/OperacaoCapacidadeBrasil.cfm>. Acesso em: 06 fev. 2016.

[7] MIERZWA, J. C.; HESPANHOL, I. Água na indústria: uso racional e reuso. São Paulo: Oficina de Textos. 2005.

[8] ÁVILA, S. F.; LOPES, J. R. N.; SOUZA, C. R. O. Strategies for industry sustainability. In: WREF 2012 - WORLD RENEWABLE ENERGY FORUM. 2012, Denver.

[9] BONALUME NETO, R. Perda de 30% da água durante distribuição contribui para crise em SP. **Folha de São Paulo**, São Paulo, 05 de jun. 2014. Fórum de Sustentabilidade.

[10] DOORNBOSH, R.; GIELEN, D.; KOUTSTAAL, P. 2008. Mobilising Investments in Low-Emission Energy Technologies on the Scale Needed to reduce the Risks of Climate Change, OECD SG/SD/RT (2008) 1, Paris, France.

[11] IPCC (2007), Metz, B.; Davidson, O. R.; Bosch, P. R.; Dave, R.; Meyer, L. A. Technical Summary in: Climate Change 2007: Mitigation. Contribution of Working Group III to the Fourth Assessment Report of the Intergovernmental Panel on Climate Change, Cambridge University Press, Cambridge, United Kingdom and New York, NY, USA.

[12] FURTADO, M. Água – Clientes exigem resfriamento de baixo custo. Revista Química e Derivados. São Paulo, pp. 32 – 34., 2009.

[13] PARRY, M. L. et al. (Ed.). Climate change 2007: impacts, adaptation and vulnerability; contribution of Working Group II to the Fourth Assessment Report of the Intergovernmental Panel on Climate Change. Cambridge (UK): Cambridge University Press, 2007f. 976p.

[14] ÁVILA, S. F. Análise de falha em processos complexos. In: CONGRESSO BRASILEIRO DE ENGENHARIA QUÍMICA, 19,, 2012. **Anais...** Búzios, Rio de Janeiro.

[15] ÁVILA, S. F.; PESSOA, F. L. P. **Proposition of review in EEMUA 201 & ISO Standard 11064 based on cultural aspects in labor team, LNG case.** Procedia Manufacturing, vol. 3, pp. 6101 – 6108, 2015.

[16] ÁVILA, S. F. **Reliability analysis for socio-technical system, case propene pumping**. Engineering Failure Analysis, vol. 56, pp. 177 – 184, 2015.

[17] ÁVILA, S. F.; MENEZES, M. L. A. Influence of local archetypes on the operability & usability of instruments in control rooms. In: EUROPEAN SAFETY AND RELIABILITY CONFERENCE – ESREL 2015. 25., Zürich, Switzerland.

[18] AMERICAN PETROLEUM INSTITUTE. API 770.2001: A manager´s guide to reducing human errors. Improving Human Performance in the Process Industries. Washington, DC., 2001.

[19] ÁVILA, S. F.; COSTA, C. Cognitive deficit analysis in routine tasks, as a strategy to reduce accidents and increase of industrial production. In: EUROPEAN SAFETY AND RELIABILITY CONFERENCE – ESREL 2015. 25., Zürich, Switzerland.

[20] REASON. J. Human error. New York: Cambridge University Press, 2003. 302p.

[21] ÁVILA, S. F.; DANTAS, E. T.; DUARTE, J. The Requirements & tools to treat process safety risk as result of Organizational Change. In: GLOBAL CONGRESS ON PROCESS SAFETY. 11., 2015. **Anais...** Austin, Texas.

[22] ABNT, Associação Brasileira de Normas Técnicas. **NBR ISO 50001:2011: Sistemas de Gestão de Energia: Requisitos com orientações de uso.** Rio de Janeiro: ABNT, 2011.

[23] ÁVILA, S. F.; KALID, R. A.; KIPERSTOK, A.; CARVALHO, V. S. Program to Avoid Product Leakage at Source, Fertilizer Case. Chemical Engineering Transactions. vol. 26, 2012.

[24] ÁVILA, S. F. Assessment of Human elements to avoid accidents and failures in task perform, cognitive and intuitive schemes. In: GLOBAL CONGRESS ON PROCESS SAFETY. 7., 2011. **Anais...**Chicago, Illinois.

[25] ÁVILA, S. F.; KIPERSTOK, A.; KALID, R. A.; SANTOS, A. C.; CARVALHO, V. S. Environmental Deviation Assessment, Record, Failure, Probability and Investment. Chemical Engineering Transactions. vol. 26, 2012.

[26] CCPS (Center for Chemical Process Safety), Conduct of Operations and Operational Discipline: For Improving Process Safety in Industry, American Institute of Chemical Engineers, New York, NY, 1985.

[27] ÁVILA, S. F. Reliability analysis for socio-technical system, case propene pumping. Engineering Failure Analysis, vol. 56, pp. 177 – 184, 2015.

[28] GERTMAN, D.; BLACKMAN, H.; MARBLE, J.; SMITH, B. C. The SPAR-H Human Reliability Analysis Method. Idaho National Laboratory, Washington DC, 2005.

[29] ÁVILA, S. F.; LOPES, J. R.N.; KALID, R. A.; KIPERSTOK, A. F. Gestão da disponibilidade de água e energia através do Plano Diretor. In: RIO OIL & GAS EXPO AND CONFERENCE 2014. Rio de Janeiro.

[30] ÁVILA, S.; KIPERSTOK, A.; BRAGA, B.; KALID, R. Avoiding loss of energy in a petrochemical industry, operation and design Congress 2011. In: WORLD RENEWABLE ENERGY CONGRESS 2011. Linköping, Sweden.

[31] ÁVILA, S. F.; COIMBRA, H.; SOUZA, C.; GOLÇALVES, A. Iniciativas de eficiência energética para sustentabilidade, oportunidades & desafios. In: RIO OIL & GAS EXPO AND CONFERENCE 2014. Rio de Janeiro.

[32] ÁVILA, S. F.; JESUS, Z. M. R.; KALID, R. A.; NASCIMENTO, V. Good practices in Cooling Towers: Comparison between industrial audits. In: Cooling Technology Institute Annual Conference. Corpus Christi, Texas, 2003.

[33] GENZ, F., TANAJURA, C. A. S., DE ARAÚJO, H. A. Impacto das Mudanças Climáticas nas Vazões do Rio Pojuca sob o Cenário A2 do IPCC - 2070 a 2100. In.: I CONGRESSO BAIANO DE ENGENHARIA SANITÁRIA E AMBIENTAL. Salvador. 2010.

[34] SILVA, S. F. Análise da disponibilidade e demanda para o sistema de abastecimento de água de salvador frente a cenário de mudanças climáticas. Dissertação (mestrado). Programa de pós-graduação em Meio Ambiente, Águas e Saneamento. UFBA. Salvador BA. 2012.

[35] KIPERSTOK, A.; BRAGA, E. A.; CESAR, M.; ÁVILA, S. F.; SAMARA.; KALID, R. A. Eds: 2012. Base de Informações para Plano Diretor de Água Fafen-BA, TECLIM RESEARCH FAFEN – UFBA. Camaçari, Brasil.

[36] SEDUR, SECRETARIA DE DESENVOLVIMENTO URBANO DA BAHIA. Estudos de Aproveitamento dos Mananciais da RMS – Revisão e Atualização das Proposições do Plano Diretor de Abastecimento de Água, Incluindo o rio Pojuca, e Visando o Uso Racional dos Recursos Hídricos Regionais - Estudo de Demandas. Relatório 1100.11-REL. EST. DEMANDAS-00. 73p. Salvador. 2010.

[37] SANTIAGO, P. S., & SANTOS, R. P. Major Improvements in cooling system performance at an ammonia/urea plant. IWC – International Water Conference, Proceedings: Pensylvania: 2000.

[38] ÁVILA FILHO, S., NASCIMENTO LOPES, J.R. Cooling Water Systems: An Overview, in: Z. Amjad, K.D. Demadis (Eds.), Mineral Scales and Deposits: Scientific and Technological Approaches, 2015, pp. 499–531.

[39] DEMING, W. E. Qualidade: A revolução da Administração. Rio de Janeiro: Editora Marques-Saraiva. 1990. 367 páginas.

[40] ÁVILA FILHO S., SOUSA C. R., CARVALHO A. C. Assessment of complexity in the task to define safeguards against dynamic risks. 6th International Conference on Applied Human Factors and Ergonomics (AHFE 2015) and the Affiliated Conferences, AHFE 2015. Procedia Manufacturing 3 (2015) 1772 – 1779.

442403 A Novel Open Source CFD Code for Explosion Modelling

Abstract:

The porosity parametrization of complex geometry is very important in explosion simulation. This project proposes a novel approach to calculate the mesh porosity (volume and area) based on the Gilbert–Johnson–Keerthi (GJK) distance algorithm. The GJK algorithm can be used to check the collision between convex objects. This concept is used in the current work to check the collision between an element of the mesh and an object of the geometrical model. The parametrized geometry is combined with the combustion model based on flamelet formulation in accordance with BML (Bray-Moss-Libby) model. The typical chemical libraries are replaced by the integral length scale of wrinkling based on semi-empirical formulation. One simple step reaction mechanism is considered. The turbulence closure problem is addressed via Boussinesq formulation using two equations model for turbulent kinetic energy and the rate of dissipation of turbulent kinetic energy.

The CFD (Computational Fluid Dynamics) solver relies on an explicit finite-volume code based on a structured Cartesian grid. The classical fourth order Runge-Kutta time marching approach is considered. The PDR (porosity distributed resistance) approach is used to model all geometry under analysis (including the large-scale objects and the small ones). The central differencing scheme is applied to discretize the governing equations and artificial viscosity is used to smooth out numerical wiggling. Additional resistance terms are considered in the momentum equation. Additional generation source term for turbulence is considered in the turbulent kinetic energy equation. The simulation of standard flow across different geometries presented consistency with the literature. Good agreement is found for similar simulations using a commercial CFD software. The developed open CFD code seems to be promising as an alternative to solve complex explosion problems from the engineering view point.

442442 Compliance with Process Safety Management Elements

Abstract:

OSHA promulgated 29 C.F.R. § 1910.119 regulation in 1992 and the fourteen Process Safety Management (PSM) elements. The main objective of defining these elements is to prevent unwanted discharge of highly hazardous chemicals exposing personnel to serious hazards. The elements are linked to each other and have intricate interdependencies. Every element provides information to other element(s) or uses information from other element(s). Given the interdependencies of these elements, an effective Process Safety Management Program (PSM) has to be in place which considers process design, technology, procedures (both normal and emergency), emergency response plans, training programs etc. The presentation talks about all these fourteen elements in details with main emphasis on the relief systems design.

442454 A Fuzzy Logic and Probabilistic Hybrid Approach to Quantify the Uncertainty in Layer of Protection Analysis

Abstract:

Layer of Protection Analysis (LOPA) is a widely used semi-quantitative risk assessment method. It provides a simplified but less precise method to assess the effectiveness of protection layers and the risk of an incident scenario. The outcome frequency and consequence are intended to be conservative by prudently selecting input data, given that design specification and component manufacturer's data are often overly optimistic. There are many influences, including design deficiencies, lack of layer independence, availability, human factors and wear in testing and maintenance, which are not quantified and are dependent on type of process and location. This makes the risk usually overestimated. So, there are different sources and types of uncertainty in the LOPA model that need to be identified and quantified. In this study, a fuzzy logic and probabilistic hybrid approach was developed to quantify the uncertainty of frequency of initiating event and the probabilities of failure on demand (PDF) of Independent Protection Layers (IPLs) based on the available data and expert experience. The method was applied to a distillation system with a capacity to distill 40 tons of flammable n-hexane. The outcome risk of the new method was proven to be more precise comparing to the results of the conventional LOPA approach.

442474 Quantitative Risk Analysis of Nitrogen Fertilizer Plant, of Moron Petrochemical Complex

Abstract:

Quantitative Risk Analysis of Nitrogen Fertilizer Plant, of Moron Petrochemical Complex

Considering the possible risk scenarios associated with new Nitrogen Fertilizer Plants (Ammonia, Urea and Utilities) of the Moron Petrochemical Complex, in general, are common to most petrochemical complexes in the world and are a result of materials hazardous that are manufactured, processed, handled and stored at the facility; the various processing systems and conditions; the different procedures used for the operation and maintenance of the facility; as well as the response of the detection and mitigation of hazards. Previous experiences in similar plants indicate that the most likely hazards are those who identify with the physical-chemical properties of the materials used and the process conditions.

For petrochemical facilities, the most common consequences are:

- Toxic Gas Effects
- Jet Fire
- Flash Fire
- Pool Fire
- Bleve
- Vapor Cloud Explosions.

Based on the premise of "Only what is known can be controlled," This paper aims to develop the Quantitative Risk Analysis (QRA) for new Nitrogen Fertilizer plants (Ammonia, Urea and Utilities) of Moron Petrochemical Complex, which it is located in the jurisdiction of the Municipality Juan Jose Mora, Carabobo State from Bolivarian Republic of Venezuela, considering for it the most severe scenarios identified previously by HAZOP analysis, SIL studies and previous plannings studies for the control and emergency response and contingency, then determining the levels of social and individual risk based on the probability of occurrence of the events of greater impact and its consequences and impact on people, equipment and buildings, comparing the risk levels obtained with tolerance criteria established by the corporation.

The Quantitative Risk Analysis (QRA) are developed from the standard and guidelines of Petroleos de Venezuela, (PDVSA); Center of Chemical Process Safety (CCPS), and The Netherlands Organization (TNO), through the computational tool PHAST RISK 6.7 . from Det Norske Veritas (DNV).

Management of Plant IPL's

Richard Stanley
Tehama Control Systems Consulting Services
Consultant to Mangan Software Solutions
26741 E Clifton Drive
Aurora, CO 80016
rick@tehamacontrolservices.com

Prepared for Presentation at
American Institute of Chemical Engineers
2016 Spring Meeting
12th Global Congress on Process Safety
Houston, Texas
April 11-23, 2016

Management of Plant IPL's

Richard Stanley
Tehama Control Systems Consulting Services
Consultant to Mangan Software Solutions
26741 E Clifton Drive
Aurora, CO 80016
rick@tehamacontrolservices.com

Keywords: IPL, LOPA, Independent Protection Layer, Safety Life Cycle Management, Safety Instrumented System, Safety Instrumented Function

Abstract

Independent Protection Layers (IPL) are usually identified in a Layers of Protection Analysis (LOPA) during a Process Hazards Analysis. The IPL portion of a LOPA identifies Functions or Design Features within a processing facility that serve to prevent or minimize the consequences of a deviation from the expected processing conditions. Many times a LOPA may identify various IPL's, but the facility then has difficulty in clearly associating the functions identified by an IPL with real plant assets. They then have difficulty in assuring that those assets have the necessary operational and maintenance visibility, and are operated and maintained to assure their availability. This paper discusses these issues and describes the characteristics of an effective Management Program for IPL's.

1 Introduction

During Process Hazard Analysis (PHA) for processing facilities, a number of Independent Protection Layers (IPL) may be identified as having the capability to reduce the risk of a given consequence occurring. A given scenario may require that one or more of these IPL's be in place to reduce risk to tolerable levels. Typically, these IPL's are identified during a Layers of Protection Analysis (LOPA), but other methods can be used.

Once these IPL's have been identified, the process of managing them has just started. Designation of an IPL adds Operational and Maintenance requirements that often exceed typical practices. This presents an operating facility with the challenge of adapting Operation and Maintenance practices to assure that the IPL's that are credited are well understood, that their protective functions remain available and meet the performance requirements that were defined for them.

This paper discusses the various type of IPL's and the requirements and methods for assuring their availability. The basic ownership requirements are described along with observations relative to the difficulty that real facilities have had in addressing these requirements. This particularly applies to the less visible IPL types such as those IPL's that are not associated with Safety Instrumented Functions (SIF).

After discussion of the ownership requirements and obstacles to meeting those requirements, the use of a Safety Life Cycle database to overcome these obstacles is discussed. The effective application of a Safety Life Cycle database can enable an operating organization to manage those requirements with maximum effectiveness and value with minimum impact of existing organizational structures and accountabilities.

2 Independent Protection Layers

A key concept in managing Process Hazards is that of an Independent Protection Layer. The formalized concept of an IPL was developed along with the LOPA process. During a PHA the potential consequences of various deviations in plant operation are evaluated, and the potential consequences of those deviations if no preventative measures are taken are defined in qualitative terms.

When an unmitigated risk identified by the PHA is higher than an organization's tolerance, the probability of the event occurring may be further evaluated to better understand the risk and the effectiveness of the various potential safeguards that could prevent consequence from occurring.

The most common technique for performing this analysis is the Layers of Protection Analysis (LOPA). In a LOPA, functions or design characteristics that are credited as being able to prevent the consequence are called Independent Protection Layers (IPL).

2.1 IPL Identification

The measure of effectiveness of identified IPL's is generally assessed against an organization's risk policies using a Target Mitigated Event Likelihood (TMEL). TMEL is usually based upon the severity of the LOPA scenario unmitigated consequences. For example, a severity level may have a required TMEL of 1e-04, or one occurrence of the unmitigated consequence in every 10,000 years.

During the LOPA, various Initiating Causes for a Deviation scenario are identified and each is assigned an expected frequency of occurrence. Some organizations may choose to apply additional frequency mitigation factors, such as the probability of personnel being in the area at the time of the Initiating Cause. Some organizations will choose to sum Initiating Causes, while others do not.

The product of expected frequency of the initiating causes, any additional frequency mitigation and Probability of Failure Upon Demands (PFD's) of the credited IPL's is used to calculate the value of the Total Mitigated Event Likelihood ($TMEL_{act}$). If the $TMEL_{act}$ is below the threshold specified for the consequence level associated with the event, the number of IPL's is considered to be sufficient. If the $TMEL_{act}$ exceeds the threshold the LOPA team recommends corrective actions, often recommending that additional IPL's be added. However, other recommendations such as process modifications to reduce the consequences or reduce the initiating causes may also be recommended. An example of a LOPA TMEL analysis is shown in Figure 1.

Figure 1 – Sample LOPA TMEL Analysis

LOPA TMEL Calculation Example

Consequence TMEL = 1.0E-05 (1 consequence Event per 100,000 years)

IC1 - 0.1 (once per 10 years)
IC2 - 0.1 (once per 10 years)
MF - 0.1 (Personnel present 10% of the time)

IPL1 PFD = 0.1
IPL2 PFD = 0.01

TMELact = (IC1+IC2) * MF * IPL1 * IPL2
TMELact = (0.1+0.1) * 0.1 * 0.1 * 0.01 = 2.0E-06

Actual TMEL (TMELact) is less than TMEL, so the IPL protection is acceptable. No additional protection is required

2.2 What is an IPL?

An Independent Protection Layer is an instrumented or physical function which, if effective, would prevent the unmitigated consequences of a scenario identified in a PHA or LOPA from occurring.

The primary characteristics of an IPL are:

- The proper functioning of the IPL would prevent the consequence associated with a scenario from occurring
- The IPL is independent of any of the scenario initiating causes and is independent of other IPL's cited for the scenario.
- The IPL is reliable enough to perform the expected function with the probability of failure being with the range with which the IPL is credited.

- The physical and functional design of the IPL meets the required independence and availability requirements.
- The IPL is Operated and Maintained in a manner that assures the credited reliability is achieved during plant operation.

CCPS [1] provides guidelines as to what functional, design or physical aspects of a process installation can qualify for being credited as an IPL, and describes the limitations for where multiple IPL credits may be taken. However, it is up to the facility owner or operator to determine specific local requirements. It should be noted however, that a 2015 US OSHA interpretation [2] appears to extend the concept of Recognized and Accepted Good Engineering Practices (REGAGEP) to publications that are not necessarily consensus documents. CCPS publications, such as the LOPA publication cited above, fall into this category. Furthermore, the OSHA interpretation notes that internal standards must represent REGAGEP and will reject standards that do not comply.

2.3 IPL types

Some of the more common IPL types are:

- Operator response to an Alarm
- A Basic Process Control System Automatic Action
- A Safety Instrumented Function (SIF) implemented in a Safety Instrumented System (SIS)
- A non-SIF interlock
- Pressure relief devices
- A physical barrier that would contain process fluids after a loss of primary containment. An example is a tank dike or a drainage system
- Physical equipment that would prevent a scenario from causing the unmitigated consequence associated with it. For example, this could be a valve limit stop, a restriction orifice or other flow restriction that prevents excessive inflow.
- Inherently Safe Design

2.4 IPL's and the Safety Life Cycle

The concept of the Safety Life Cycle is often considered to be associated with Safety Instrumented Systems (SIS) because its prominence in IEC 61508 and IEC 61511. However, the concepts are applicable to any protective functions or systems and are used in other contexts.

An important factor to recognize is that the risk reduction provided by IPL's is the product of their RRF's, and that if any IPL fails to meet the performance credited by a LOPA, the associated risk reduction of that IPL is lost. CCPS recognizes this by requiring that any IPL be Operated and Maintained in manner that assures that the credited Risk Reduction is effective.

The IEC 61511 and ISA 84.00.01-2004 Standards provide a framework for managing of Safety Instrumented Systems (SIS) and the Safety Instrumented Functions (SIF) that were identified by the Hazard Analysis process. However, these Standards are silent on managing of other non-SIF IPL types.

This has led some operating organizations to conclude that there are no corresponding requirements relative to the non-SIF IPL's. This is not the case: non SIF IPL's require sufficient care to assure their availability and cannot just be run to failure or be treated like non-IPL functions. The sections in the remainder of this paper discuss the practicalities of IPL implementation and ownership, the risk gaps that have been observed, and suggested management practices to prevent these gaps from developing.

3 What Should an IPL Management System Look Like

One of the fundamental gaps observed in practice is that there is often a perception that once an IPL is credited in a LOPA, and something exists in the facility that appears to perform the function defined by the LOPA, no additional action is required. This results in a system in which the effectiveness of an IPL is left to chance and the assumed risk reduction becomes an illusion.

An effective IPL Management System must be able to assure that IPL's are visible, that their design meets the functional and independence requirements for an IPL, and that the IPL's are Operated and Maintained to assure the required availability.

IPL's types and their implementation can vary considerably. While SIF IPL's may be paid more attention because of standards such as IEC 61511 and ISA 84.00.01-2004, non-SIF IPL's are usually shared with other basic plant systems such as the Basic Process Control System (BPCS) or blend in with other mechanical systems. Without an effective IPL Management System, non-SIF IPL functions often are either not effective to begin with, or atrophy to the point were they do not deliver the risk reduction with which they were credited.

Table 1 below identifies some of the gaps in IPL Management that have been observed in real operations. Unfortunately, these gaps are more common than not.

Table 1 – IPL Management Gaps

IPL Management Gap	Cause	IPL Management Requirements
Improper IPL credits are taken	Personnel are not properly trained on what the requirements and limitations for IPL qualification are, or local standards do not comply with REGAGEP	LOPA facilitators must be well trained on IPL requirements and advise participants on IPL requirements including IPL Types allowed and restrictions taking multiple credits. LOPA's should be reviewed by qualified and empowered individuals to assure validity.
IPL's as implemented do not meet the effectiveness, independence and reliability requirements defined in the LOPA's	Plant assets are identified of being IPL's, but are not evaluated to verify that they meet the IPL requirements.	Checklists for assessment of IPL qualifications are available and all IPL's are assessed against those requirements.
IPL functions are not understood by Operations and Maintenance personnel	IPL's are identified in LOPA documents and are not clearly identified in documents that are used by Operations and Maintenance personnel. LOPA documents are not readily available to non-PSM personnel and then are difficult to correlate to actual plant installations.	A system is needed that clearly identifies plant assets that are used to implement IPL's. The data should identify real plant assets that perform IPL functions, the Operational and Maintenance expectations for the IPL's, including operator actions, restrictions on disabling or bypassing the IP and Maintenance and Testing Requirements.
IPL's are bypassed or disabled	Assets credited as IPL's are not clearly identified. Operating and Maintenance procedures for IPL's are not prepared, available or enforced	Operating procedures need to exist that define criteria for bypassing or disabling of IPL's and requirements for management or operations while bypassed or disabled.
IPL's are modified or removed without proper MOC	The IPL list is not available, or is out of date. Projects making modifications do not recognize devices which are in IPL service and treat their modification like modification of non-IPL services	A continuously up to date IPL list is provided and linked to all LOPA, IPL design and operations and maintenance data. Projects evaluating proposed modifications have all the data needed to perform a proper MOC available in one place.
IPL's are not tested at the intervals required to maintain performance	IPL's are not identified in maintenance systems or do not have test procedures available. When procedures are available, management is ineffective because of many documents in many places, many of which may not be consistent or up to date	Test intervals, maintenance procedures and testing procedures are integrated in one database. Test records are captured in the Operations and Maintenance Module to allow for identification of next required test interval.

3.1 Why Is Compliance So Hard?

Many organizations have demonstrated difficulty in incorporating the requirements of IPL ownership and operation into their day to day practices. The basic reasons for this are that the importance of IPL's is not always recognized, and that effective management of IPL's generally does not fit within traditional plant management structures.

Effective IPL management requires that information cross traditional organizational boundaries. In traditional operating facilities, these boundaries usually result in many parallel information and documentation systems that have limited accessibility or provide only the functionality that a particular group requires and does not necessarily support the needs of other groups. Some examples of this are:

- Process Safety Groups have proprietary software that often has highly restrictive incensing terms that effectively lock up HAZOP and LOPA data access to just a few people.
- Maintenance Management Systems have historically been developed for management of major rotating equipment by and for rotating equipment personnel. Other functions are focused upon parts inventory management and work order tracking. These systems perform poorly when other systems are involved. They are particularly bad at handling Instrumented Systems. Resistance to making them effective for Instrumented Systems, or handling safety related non-rotating equipment has typically been quite high.
- Much of other traditional information management is done with discrete paper files or electronic files that are often not coordinated or readily available. These records are held by various organizations and are difficult coordinate and keep up to date. Often these records do not survive changes in personnel.

Management of IPL related data across these multiple platforms and formats is extremely inefficient and easily broken. Most facilities do not have the personnel or skill sets required to manage these problems. The Systems, even when well intentioned, tend to fail under the burden the highly labor intensive requirements of maintaining this data.

3.2 Safety Life Cycle Management

Effective management of IPL's across the entire Safety Life Cycle requires Safety Life Cycle Management systems that address the cross discipline nature of the data required and supports the following types of information:

- Defining relationships between Hazard Identification, Functional Requirements and Design of Protective Functions and Equipment, and Operation and Maintenance requirements.
- Capturing the Operational and Maintenance Events that are related to the IPL functions and equipment and using that data to provide Operation and Maintenance Management a clear picture of compliance and performance.

- Allowing users from various organizations to readily access IPL related data including the Hazard associated with IPL Assets, IPL functional requirements and equipment specifications and IPL Asset events and performance

Figure 2 illustrates a high level view of the relationships that should be defined and maintained and their general relationship to the Safety Life Cycle.

- HAZOP scenarios that have a high consequence are related to LOPA's
- LOPA's and the IPL's that were credited by the LOPA are identified
- Plant Assets that are used to perform the functions identified by a LOPA IPL are identified and their relationships to to the LOPA IPL's are shown. Note that in some cases, multiple LOPA IPL's may refer the the same Plant Asset.
- The Plant Asset data contains definition of the functions required, performance requirements, testing requirements, expected operator responses and physical equipment used to implement the functions.
- Operation and Maintenance Events are recorded and associated with the Plant Assets or Plant Equipment.
- KPI's are derived from the Operation and Maintenance Events and used to manage such things as period testing and assess performance such as demands, failures, bypasses.

Figure 2 – IPL Asset Data and Relationships

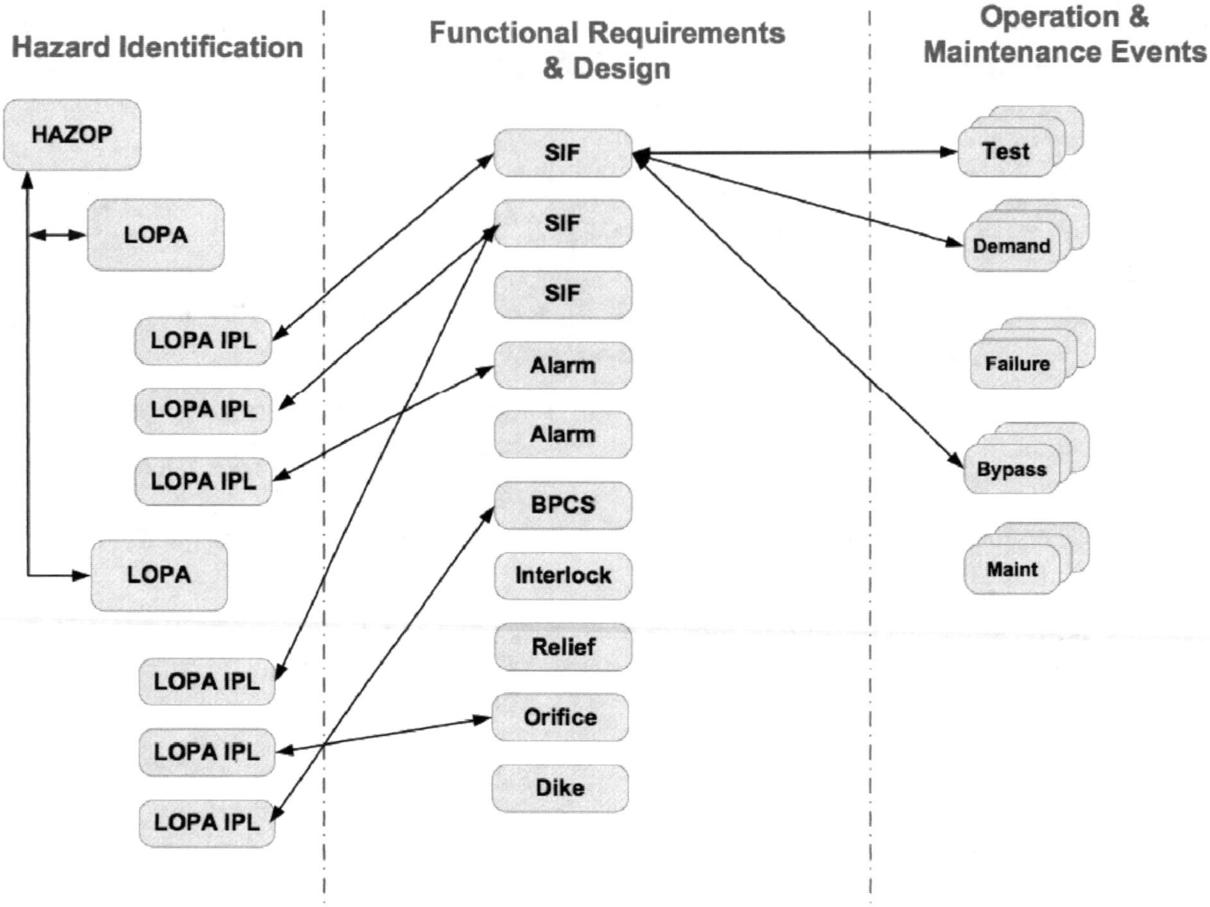

Figure 3 shows a more detailed set of data and relationships that focuses on one Plant Asset that is used to perform IPL functions.

- The Asset is related to one or more LOPA IPL's and to the equipment used to perform the Asset functions
- Plant Equipment that is used to implement an IPL Asset, such as instrumentation or other physical equipment is associated with the IPL Asset
- Assessment checklists and results are associated with the LOPA IPL's. Design Assessment checklists are associated with each Plant Asset and its associated equipment.
- Important documents such as Operating and Maintenance procedures or design documents are associated with the Asset for quick access.
- Operating and Maintenance Events are recorded and associated with their related equipment or Plant Assets

Figure 3 – IPL Asset Data and Relationship Details

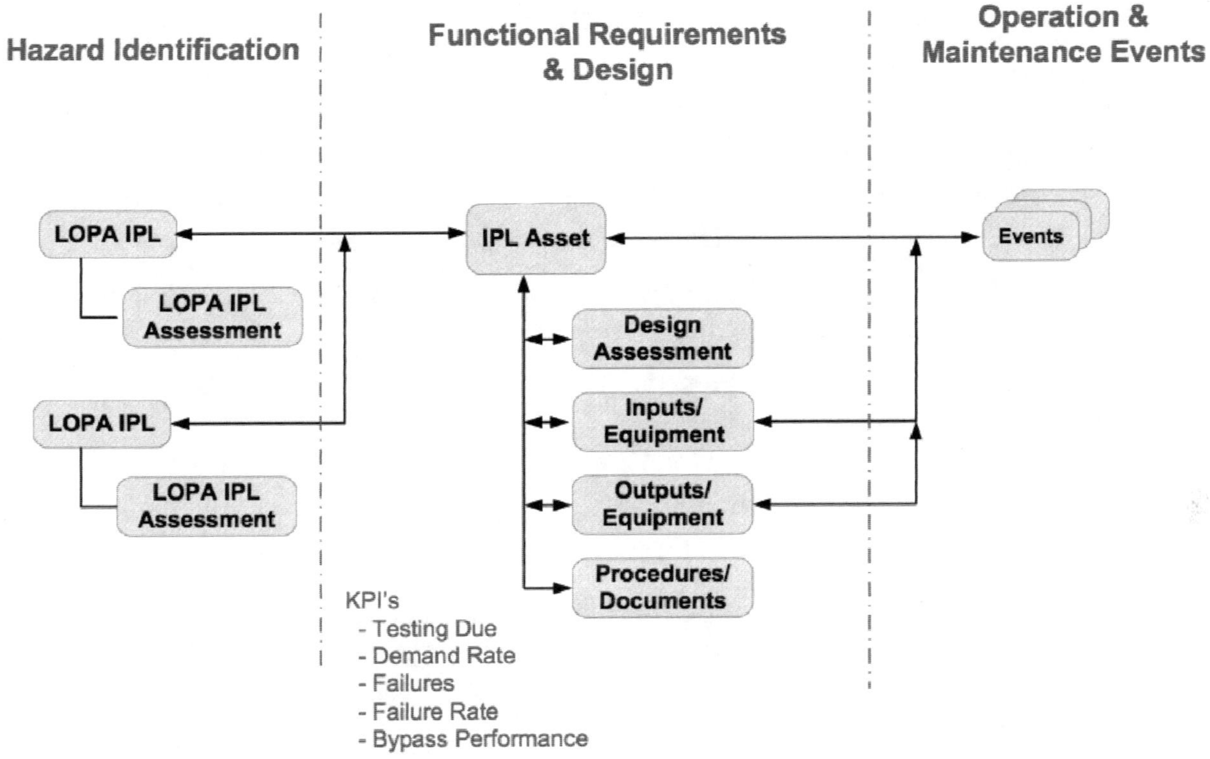

4 Using a Safety Life Cycle Management Data Base

A Safety Life Cycle Management System is best implemented in a database that is constructed specifically to manage the data and events that apply to the Safety Life Cycle. The data should be highly linked, so that a User can readily access related data. For example, an operator would be able to view the functional requirements for an IPL Asset, using a tag or other ID that is used within the Unit, and then directly access the Hazard data associated with the IPL. Appropriate Operations, Maintenance or Engineering personnel should then be able to associate Events with the IPL Assets and view KPI's associated with those Assets.

4.1 Data Relationships

A Safety Management Life Cycle Management Data Base should contain data objects that provide relationships among the various items shown in Figure 3. Among these are

- LOPA IPL's that define basic definitions of the IPL function and performance requirements and relate the LOPA IPL to the appropriate LOPA and HAZOP scenarios

- IPL Assets that define the actual plant assets that are used to implement the IPL functions. These should be identified using the same identifiers used in other plant documentation and by which plant operators, engineers and maintenance personnel know the asset. The IPL Asset data should define the design requirements such as function, expected response, accuracy and test and maintenance requirements
- Assessment Check Lists that define the items relative to performance, design, documentation or testing that is required of the IPL
- Input and Output Asset and Device objects that define the functional requirements for the Asset, Inputs or Outputs that are used to implement the IPL function and the specification data for the Devices that are installed in those Asset locations
- If IPL Performance is to be tracked, Event data objects that allow for tracking of events such as Demands, Failures, Testing, Bypasses, etc.
- Reference Documents such as procedures or design or specification data

4.2 Data Content

The data for each of the objects shown in Figures 2 and 3 should capture information necessary to support IPL Management. Some examples are:

- LOPA IPL's define the type of IPL, the performance requirements - typically in terms of a Risk Reduction Factor (RRF), and a functional description of the IPL. If operator response to an alarm is involved, the expected operator action and response time should be specified
- SIF Asset data should define the functional requirements, performance expectations and testing and maintenance requirements. In some implementations, data for a SIF could include all data necessary for definition of the Safety Requirements Specification (SRS) as defined in IEC 61511 and ISA 84.00.01-2004.
- Non-SIF IPL Asset data should define define the functional requirements, performance expectations and testing and maintenance requirements appropriate to the type of IPL.
- A LOPA IPL Assessment is a check list of items and their assessment results necessary to validate that the IPL will be effective and independent of other IPL's credited for the LOPA scenario
- An IPL Design Assessment is a check list of items and assessment results necessary to verify that the IPL Asset as designed meets the functional, performance and independence requirements for the IPL.
- Input/Equipment and Outputs/Equipment should define the functional requirements for Inputs and Outputs used by the IPL. The Equipment sections should contain or be linked to specification data for the specific equipment performing the functions.
- Procedure/Documents data should identify the specific Operating Procedures, Maintenance and Test Procedures or other Design Documents that apply to the IPL.

- Events – Events are used to describe the date, time and results of various types of Operation and Maintenance Events that occur during the IPL Life Cycle. The data for each event should be appropriate to the Event Type, such as as-found and as-left data for Functions and Equipment, start and end of bypasses, etc.

5 AN EXAMPLE IPL

The Figures below illustrate how a Safety Life Cycle data base application can be used to manage an IPL. An Alarm IPL is shown as an example.

5.1 IPL Functional Requirements

Figure 4 shows an example of an IPL Functional Specification for an Alarm with Operator Action IPL. This allows a User to define the functional requirements for the Alarm including such data as

- The Alarm ID – Typically this is the Tag Number of the Alarm as it appears on P&ID's, Operator Interfaces and other documentation
- Alarm functions, set points and basic implementation requirements
- Expected operator response and the time within which the response is expected
- Alarm operating restrictions such as when the alarm may be disabled or restrictions on changing set points or other configuration data
- Alarm Testing Requirements – type of testing and frequency of testing.
- LOPA Associations including the LOPA scenario for which the Alarm was cited as an IPL, the LOPA IPL and other IPL's that were cited by same LOPA scenario
- Associated Documentation such as Test Procedures, Operating Procedures

Figure 4 – Alarm IPL Functional Requirements Data

Alarm O-M Data | Alarm Functional | Alarm Devices | Revisions

Print | 1 Document | Attach Tasklist | Subscription | Bookmark | Admin Tools

Alarm Functional Requirements - FCC-FAL-138, Feed to East Riser

Alarm Requirements

Alarm ID:	FCC-FAL-138
Alarm Description:	Feed to East Riser Low Flow
Functional Description:	Feed flow to the East Riser is indicated on the BPCS by the analog input poing FCC-FI-138. This point has a low flow alarm confirgued to alert the operator of low riser feed flow
Asset Protection Category:	IPL
Operator Repsonse Time:	10 Min
Operator Repsonse Time Notes:	Nominal time to responsd to a high priorty alarm
Defined Corrective Action:	The operator shall place the East Riser Assist Steam flow controller FCC-FC-282 into automatic with a set point of 1100 PPH. Verfiy that the controller is properly proving steam to the riser

Protective Category:

Health and Safety	Environmental	Commercial	Reputation
Yes	No	No	No

Performance Requirements:	RRF 10 PFD 1.0 E-01
Alarm Operating Requirements:	Alarm must be enabled at all times when feed is in the unit. If the alarm must be bypassed or disabled prepare Safety Bypass Authorization. Set point is not to be changed without an MOC being performed
Alarm Response Time:	5 sec
Alarm Response Time Notes:	BPCS must activate alarm within 5 seconds of when the transmitter senses low flow
Alarm Location:	FCC Operator's Console in Main Control Room
Alarm Implementation:	Soft alarm on analog input tag
Alarm Priority:	High
Alarm Priority Rationale:	The risler low feed flow alarm indicates that conditios may exist the contribute to a riser flow upset but the situation is no yet crtical. Additional SIF layers of protection exist
Alarm Testing Requirements	Test transmitter, calibrate and test alarm activation and verify set points at turnarounds
On-Line Testing Interval	N/A
Off-Line Testing Interval	60 Months

Alarm Set Point (Software Based)

Set Point ID	Value	Type	Tolerance	Time Delay	Basis
FCC-138-SPL	115	Decreasing	5%	1 sec	Minimum HC flow w/o assist steam

Alarm Set Point (Device Settings)

Set Point ID	Value	Type	Tolerance	Time Delay	Basis
N/A	N/A	N/A	N/A	N/A	N/A

LOPA References

LOPA IPL ID	LOPA ID	LOPA IPL Description	LOPA Hazard	LOPA Consequnces	Required RRF
FCC-LOPA-15-1	FCC-LOPA-15	Low Feed Flow Alarm	Increase in riser catalyst density and loss of flow stability. Potential loss of flow and flow reversal	Hydrocarbon in regenerator. Very high regenerator temperataures and potenial hydrocarbon in precipitator. Potential explosion	10

Associated IPL's

ID	Type	Description
FCC-SIF-02	SIF	East Riser Emergency Steam
FCC-SIF-05	SIF	FCC Catalyst Reversal Protection System

Alarm Testing Procedures

Document ID	Title	Description
CE-TEST-7764	Alarm Testing Procedure	General Refinery Alarm Testing Procedures

5.2 IPL Operating and Maintenance History

Figure 5 shows an example of a view of the Alarm IPL's Operating and Maintenance history for the same Alarm as illustrated above. This view repeats the basic Alarm Functional Requirements data and the provides data specific to the Alarm operating status and the history of the events which have been entered into the database for the Alarm. This view provides the following types of data.

- Basic Alarm Function Requirements
- Alarm Testing and Performance Data – this section lists things such as the testing status for the alarm, the last test date and next test data, commissioning data, hours in service and a summary of performance statistics such as the number of demands and failures and the in-service failure rate
- Lists of the events that have been entered into the database for the alarm listed by type of event. In the example, this includes the Alarm Service Status history, Test Events, Demand Events, Bypass Events, etc.

In this view, many data fields are linked so the User can click on things such as Event ID's to go directly to the event detail view.

Figure 5 – Alarm Operation and Maintenance Data

Alarm O-M Data | Alarm Functional | Alarm Devices | Revisions

Print | 1 Document | Attach Tasklist | Subscription | Bookmark | Admin Tools

Alarm O-M View - FCC-FAL-138, Feed to East Riser

Alarm Data

Alarm ID	FCC-FAL-138
Alarm Description	Feed to East Riser Low Flow
Functional Description:	Feed flow to the East Riser is indicated on the BPCS by the analog input poing FCC-FI-138. This point has a low flow alarm confirgued to alert the operator of low riser feed flow
Operator Repsonse Time:	10 Min
Defined Corrective Action:	The operator shall place the East Riser Assist Steam flow controller FCC-FC-282 into automatic with a set point of 1100 PPH. Verfiy that the controller is properly proving steam to the riser

Testing and Performance Data

Alarm Testing Requirements	Test transmitter, calibrate and test alarm activation and verify set points at turnarounds
On-Line Testing Interval	N/A
Off-Line Testing Interval	60 months
Test Status	Current
Last Test Date	3/26/12
Next Test Due Date	3/26/17
Due Date Based Upon	Turnaround Schedule

Determining Devices	ID	Type	Description
	FCC-FT-138	Flow Transmitter	Feed to East Riser

Alarm Service Status	Commissioned
Service Hours	108,624 hrs
Alarm Commission Date	4/28/08
Alarm Commission Notes	None
Alarm Decommission Date	N/A
Alarm Decommission Notes	N/A

Alarm Performance	Life of Plant	5 Years	1 Year	Year to Date
No. of Failures	1	1	0	0
No. of Demands	6	6	0	0

In Service Failure Rate 7.0 E-02

Alarm Service Status History

Event ID	Date	From	To
FCC-FAL-138-Serv-001	4/28/2008	Created	Commissed

Test Groups

Test Group ID	Test Group Description	Date Last Run	Test Event ID
FCC-Alarms-001	FCC Safety Related Alarm Testing at Turnaround	4/4/2008	FCC-Alarms-001-Test-002
FCC-Alarms-002	FCC Safety Related Alarm Testing at Turnaround	3/24/2012	FCC-Alarms-001-Test-001

Functional Test Events

Event Date	Event ID	Parent Event ID	As Found	As Left	Event Notes	Event Status
4/4/2008	FCC-FAL-138-Test-001	FCC-Alarms-001	Failed	Passed	Pre-Commissioning Test	Approved
3/4/2012	FCC-FAL-138-Test-002	FCC-Alarms-002	Passed	Passed	Routine Turnaournd Tst	Approved

Demand Events

Demand Date	Demand Event ID	As Found	As Left	Event Notes	Event Status
4/17/2008	FCC-FAL-138-Demand-001	Passed	Passed	Startup	Approved
9/8/2008	FCC-FAL-138-Demand-002	Passed	Passed	Feed Pump Trip and Spare Start	Approved
12/17/2008	FCC-FAL-138-Demand-003	Passed	Passed	Feed Pump Trip and Spare Start	Approved
7/22/2011	FCC-FAL-138-Demand-004	Passed	Passed	Feed control valve sticking	Approved
10/9/2012	FCC-FAL-138-Demand-005	Passed	Passed	Feed Pump Trip and Spare Start	Approved
11/30/2013	FCC-FAL-138-Demand-006	Passed	Passed	Feed Pump Trip and Spare Start	Approved

5.3 IPL Assessment

As mentioned above, a critical part of the Safety Life Cycle is verification of an IPL's qualification to perform as an IPL. Check Lists can be created and tracked to document this process. Examples for an IPL Independence Check List and an Alarm Design IPL Check List are shown below.

In Figure 6 illustrates a Check List that would be completed near the time an IPL is cited in a LOPA is shown. This task list is intended to verify that the basic requirements to qualify the function as an IPL have been satisfied.

Figure 6 – LOPA IPL Check List

Item No.	Description	Finding	Comments
1.1	Verify that the IPL function, if effective, will prevent the consequence of the initiating cause from occurring	Passed	
1.2	Verify that the IPL is independent of the LOPA scenario imitating cause	Failed	The Alarm appears to use the same transmitter whose failure was the initiating cause.
1.3	Verify that the IPL is independent of any other IPL's that have been credited for the LOPA scenario	Passed	

Figure 7 shows a Check List summary that appears in the LOPA IPL view. This provides the user with a quick overview of the verification status and typically appears on the IPL Functional Requirements view.

Figure 7 – LOPA IPL Check List Status

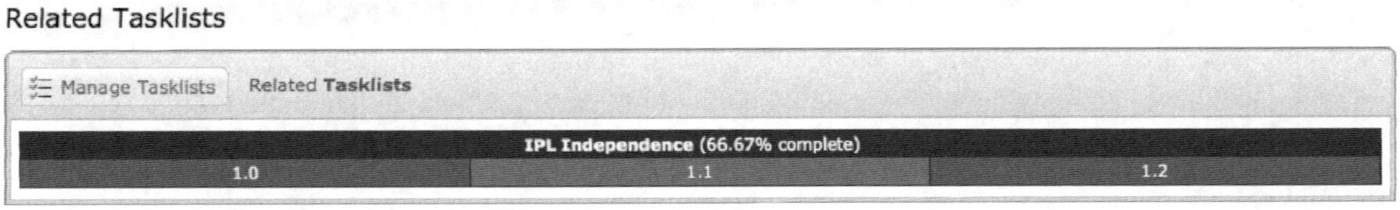

Figure 8 illustrates an IPL Design Check List for an Alarm. This check list addresses design requirements and operational preparedness items such as operation and maintenance preparedness items.

Figure 8 – Alarm IPL Checklist

Item No.	Description	Finding	Comments
3.1	Verify that the Alarm and its associated operator actions are capable of preventing the consequence for which the alarm has been cited as an IPL	Passed	
3.2	Verify that the alarm location is continuously attended	Passed	
3.3	Verify that the initiating or sensing device is independent of the initiating cause of the LOPA scenario	Failed	This alarm is derived from FCC-FT-138 whose failure is the LOPA scenario IC. Need an independent transmitter
3.4	Verify that the initiating device or sensor is independent of other protection layers claimed as IPL credits for the same LOPA scenario	Passed	
3.5	Verify that an operator action has been defined for this alarm, that the action can be accomplished and the result effective within the time between the initiating cause and the consequence occurring	Passed	
3.6	Verify that a written procedure for the operator response to this alarm has been prepared, that operators have been trained to locate the procedure and that the procedure is readily available to the operator	Failed	Operating procedures require updating.
3.7	Verify that this alarm has a suitable priority and it is identified as an IPL Alarm on Operator Graphics and Alarm views	Failed	Alarm needs to be added to operating graphics and IPL Alarm graphic

Item No.	Description	Finding	Comments
3.8	Verify that the settings for this alarm and disabling, inhibiting or bypassing of the alarm is under control of a Bypass Control Procedure and suitable approvals are required for the above actions and that return to operation is managed by this procedure.	Failed	The alarm needs to be added to the list of critical items that require Safety Bypass Authorization.
3.9	Verify that modifications to this alarm are under Management of Change Controls	Passed	Site Standard MOC Procedure
3.10	Verify that Maintenance and Testing Procedures have been prepared which are applicable to this alarm, and that the Procedures include capture of all testing records	Passed	Site Standard Maintenance procedure for testing of alarms
3.11	Verify that the P&ID; correctly shows this alarm	Failed	P&ID has not yet been updated to show this alarm as an IPL
3.12	Verify that the Alarm has been included in the Unit IPL List	Passed	

5.4 KPI's from Data Base

Once IPL data is captured in a data base, KPI's associated with the IPL Operation and Maintenance can be readily created and displayed. Some of the KPI's and summary data that can be developed are

- Identification of IPL Testing that is upcoming or overdue
- Identification on IPL's that have had high Demand or Failure rates
- Calculation of In-Service IPL Failure rates
- Reporting on IPL Assessment status and and completion of open items
- IPL quantities for Units, Sites or Enterprise-wide
- Identification of LOPA IPL's that are not linked to a Plant IPL Asset
- Identification of Plant Assets that are associated with multiple LOPA IPL's

Figure 9 shows a few examples of the types of IPL KPI data that can be provided to Users

Figure 9 – Typical IPL KPI's

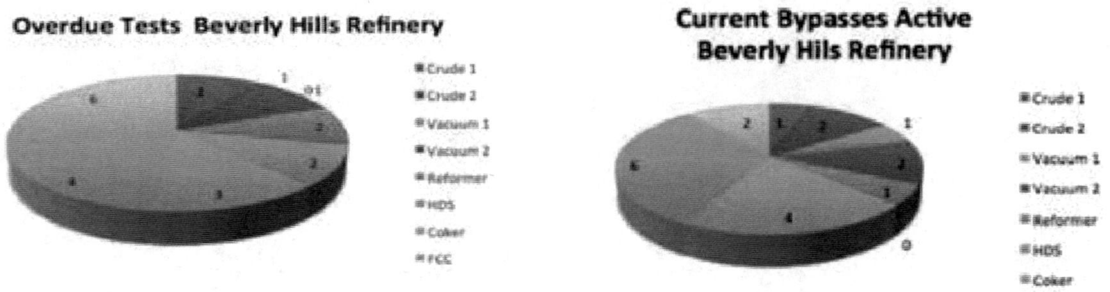

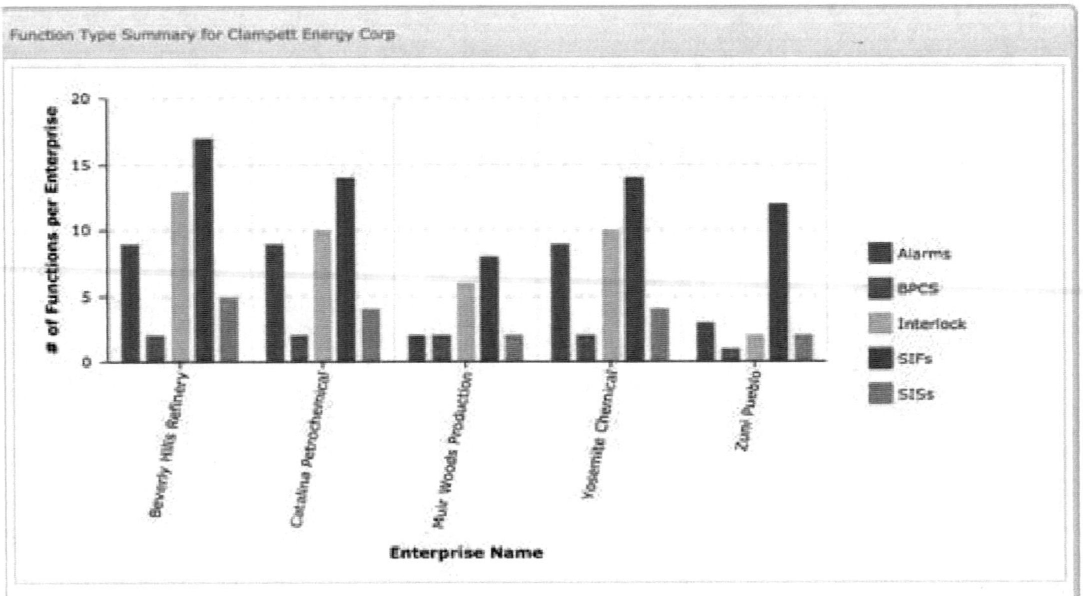

Enterprise Name	Alarms	BPCS	Interlock	SIFs	SISs
Beverly Hills Refinery	9	2	13	17	5
Catalina Petrochemical	9	2	10	14	4
Muir Woods Production	2	2	6	8	2
Yosemite Chemical	9	2	10	14	4
Zuni Pueblo	3	1	2	12	2

6 Conclusions

The Management of IPL's within an Operating Facility presents distinct challenges to Facility Management and trickles down to many organizations within the Facility. Many organizations are having signification problems in managing these requirements due to lack of the clarity of requirements and the cross discipline aspects that do not fit within the practical restrictions of management tools traditionally used. Force fitting IPL Safety Life Cycle management requirements as an overlay to these tools has generally been highly inefficient and generally ineffective.

The compliance gaps that have been observed as operating facilities struggle with management of IPL's and the Safety Life Cycle are a result of these problems. A practical and effective means of compliance is to apply a Management Process that is designed specifically for support of the Safety Life Cycle. This Management Process can maximize effectiveness of compliance and assure that the risk reduction that has been assumed for a facility is actually realized.

7 References

[1] – Layer of Protection Analysis, Center for Chemical Process Safety, ISBN 0-8169-0811-1, 2001

[2] – OSHA Memorandum – June 5, 2015 – Thomas Galassi, RAGAGEP in Process Safety Management Enforcement.

Rick Stanley has over 40 years' experience in Process Control Systems and Process Safety Systems with 32 years spent at ARCO and BP in execution of major projects, corporate standards and plant operation and maintenance. Since retiring from BP in 2010, Rick has consulted with Mangan Software Solutions (MSS) on the development and use of MSS' ProSys SLM Safety Lifecycle Management software and has performed numerous Functional Safety Assessments for both existing and new SISs.

Rick has a BS in Chemical Engineering from the University of California, Santa Barbara and is a registered Professional Control Systems Engineer in California and Colorado. Rick has served as a member and chairman of both the API Subcommittee for Pressure Relieving Systems and the API Subcommittee on Instrumentation and Control Systems.

442704 Jack Rabbit II: Experimental Large-Scale Chlorine Release Trials

Abstract:

In August and September of 2015, the U.S. Department of Homeland Security (DHS) Science & Technology (S&T) Chemical Security Analysis Center (CSAC) conducted the Phase 1 field trials of the Jack Rabbit II (JR II) program, a series of 5 large-scale outdoor chlorine release experiments with a collaborative team of partners from government, industry, and academia. Sponsored by the CSAC, the Defense Threat Reduction Agency (DTRA), and Transport Canada, these field trials were performed to fill critical knowledge, data, and capability gaps to support and improve chemical release modeling, industrial safety, and emergency response procedures. This paper summarizes the execution of the 5 successful chlorine release trials ranging from 5 to 9 tons, the instrumentation deployed, the data collected, and preliminary analyses of the high-definition video footage, cloud movement, and behavior and concentration/time profiles as a function of distance from the release site. This paper will also present plans for the JR II Phase 2 field trials scheduled for late 2016 which will include additional chlorine releases up to 16 tons.

445083 Novel Methodology for Inherent Safety Assessment in the Process Design Stage

Abstract:

Various physical and chemical hazards are intrinsic to the material itself or its manner of storage and use. Inherently safer design of the chemical process and equipment has been increasingly used by the design engineers to minimize or eliminate these hazards. This approach aims to mitigate hazards by design instead of add-on measures. This presentation describes a new design approach, introducing comprehensive inherent safety index (CISI), for use in the early stages of process development. Chemical, process and connectivity scores are calculated for individual units in the process, based on which the CISI assigns safety scores for the equipment. Each chemical in the unit has a weighted severity score. Reactivity score is calculated for mixture of chemicals in that unit. Connectivity score in introduced because the units can be highly interconnected, leading to compounding of hazards. Biodiesel and methyl methacrylate processes were studied to demonstrate the new safety assessment methodology. The results of the assessment are used to clearly identify key areas for improvement in a root-cause analysis.

Cost and Benefit Considerations of Designing for Higher Pressure

Juan P. Lopez
Smith & Burgess LLC
7600 W. Tidwell Rd., Ste 600
Houston, TX 77040
juan.lopez@smithburgess.com

Guomin Wang
Smith & Burgess LLC
7600 W. Tidwell Rd., Ste 600
Houston, TX 77040
guomin.wang@smithburgess.com

Prepared for Presentation at
American Institute of Chemical Engineers
2016 Spring Meeting
12th Global Congress on Process Safety
Houston, Texas
April 11-23, 2016

Cost and Benefit Considerations of Designing for Higher Pressure

Juan P. Lopez
Smith & Burgess Process Safety Consulting
7600 W. Tidwell Rd., Ste 600, 77040
juan.lopez@smithburgess.com

Guomin Wang
Smith & Burgess Process Safety Consulting

Keywords: inherently safer design, overpressure protection by system design, UG-140, maximum achievable pressure, process design.

Abstract

There are multiple changes within a processing plant that can trigger the need to change the design of equipment, flare, or availability of Safety Instrumented Systems (SIS). For example, a rate increase in the facility may lead to a specific relief valve being undersized, at the same time it may also impact the adequacy of the flare, and lastly in order to adequately manage risk a new SIS may be required. These three action items are due to the necessity of relieving fluid in order to prevent overpressure from a vessel that is not rated for the maximum possible achievable pressure. The proposed approach of designing the equipment for a higher pressure in the design phases, so it is able to contain the fluid without need for relief, could result in a decrease in future expenses related to upgrades of relief systems as well as lower risk. This paper will look into the disadvantages of designing equipment for higher pressure versus the benefit that may come from doing so, in accordance with ASME Section VIII – Division 1 UG-140: Overpressure Protection by System Design.

1 Introduction

The safety of a plant has become one of the foremost priorities due to the increasing pressure from government regulatory bodies and communities. As a result of several high profile incidents, the focus of the chemical process industry and regulatory bodies has shifted towards *Inherently Safer Design* (ISD). Incidents like Bhopal, India, in 1984 emphasized the need for eliminating hazards whenever possible and reducing or minimizing the risk when the hazard could not be completely eliminated through ISD.

Designing process equipment that is strong enough to contain the maximum achievable pressure is just one example of ISD. The option of designing process equipment for higher pressure can bring about many advantages when other ISD strategies may not be feasible. For example, refineries carry out many processes at high pressure (such as hydrotreating reactors). The ideal solution of ISD would be to eliminate the hazard of high pressure inventories by utilizing different processes carried out at lower pressures; however, based on the current technology, the proven effectiveness of these reactions combined with the availability of other forms of mitigating the risk such as process controls, instrumentation, or Safety Instrumented Systems (SIS) present a compelling case for managing the risk associated with having a large inventory of high pressure material. The risk of loss of containment due to overpressure in a hydrotreating reactor could also be lowered by designing the process equipment for the largest achievable pressure produced by a runaway reaction, without relying on SIS, operator intervention, or process controls and cutting the cost of maintenance of SIS or the need for a large emergency relief system.

ISD provides a different approach to process safety: instead of accepting hazards and then controlling the risk by adding layers of protection, the designer is challenged to provide a method to eliminate the hazard altogether. If the hazard cannot be eliminated, then the challenge becomes finding a method to make the process inherently safer. This paper proposes to design process equipment for higher pressure as a viable method to make the plant inherently safer. It also challenges the notion that designing for higher pressures is not cost effective; through careful design it may possible to make processes safer and economically feasible if ISD is considered throughout the early stages of the lifecycle of the facility. The paper will also provide some case studies of operating facilities where designing for the maximum achievable pressure could have helped in lowering the future expenses related to inflexible design.

2 Overpressure Protection by System Design within ISD

Inherently safer design is only one method for managing risk. The strategies for managing risk, according to the book *Inherently Safer Chemical Processes*[1], are classified into four categories:

"Inherent	Eliminating the hazard by using materials and process conditions that are non-hazardous; i.e., substituting water for a flammable solvent;
Passive	Minimizing the hazard through process and equipment design features that reduce either the frequency or consequence of the hazard without the active functioning of any device; .e., providing a diked wall around a storage tank of flammable liquids;
Active	Using controls, alarms, safety instrumented systems, and mitigation systems to detect and respond to process deviation from normal operation; i.e., a pump which is shut off by a high level switch in the downstream tank when the tank is 90% full. These systems are commonly referred to as engineering controls, although human intervention is also an active layer;
Procedural	Using policies, operating procedures, training, administrative checks, emergency response, and other management approaches to prevent incidents, or to minimize the effects of an incident; i.e. hot work procedures and permits. These approaches are commonly referred to as administrative controls."

All four strategies of risk management (or a combination of them) can present an acceptable and safe design. Inherent safety attempts to avoid the hazard completely. For situations in which removing the hazard may not be feasible, accepting the hazard and attempting to manage the risk with *passive*, *active*, and *procedural* strategies may produce a more desirable result. A good example is transportation: travel by airplane is more hazardous than traveling by car; however, there are a lot of safeguards that allow air travel to be ultimately safer than ground transportation and therefore, the benefit of fast air transportation outweighs the hazards associated with it because the risk can be controlled. In the same manner, a more hazardous procedure may be more desirable if the risk is managed appropriately through the utilization of *passive*, *active*, and *procedural* risk management strategies.

Overpressure protection by system design falls under the category of *Passive* risk management as it requires no active functioning of any device. It does not fall under *Inherent* category because it does not eliminate the hazard altogether, but it simplifies the process enough to make it *inherently safer.*

Among inherently safer design there are multiple strategies to eliminate or reduce hazards. According to *Inherently Safer Chemical Processes*[1], the four major strategies of ISD are explained as follows:

"Minimize:	Use smaller quantities of hazardous substances (also called *Intensification*).
Substitute:	Replace a material with a less hazardous substance.
Moderate:	Use less hazardous conditions, a less hazardous form of a material, or facilities that minimize the impact of a release of hazardous material or energy (also called *Attenuation* and *Limitation of Effects*).
Simplify:	Design facilities which eliminate unnecessary complexity and make operating errors less likely, and which are forgiving of errors that are made (also called *Error Tolerance*)".

The magnitude of the Bhopal incident occurred as a result of storing a large amount of an extremely toxic material. All four strategies could have helped in preventing the incident or reducing its consequence. Some inherently safer methods to reduce this hazard would have been to *minimize* the inventory of toxic material, *to substitute* it with a less hazardous substance, to *moderate* the impact of the hazard with better emergency response, or *simplify* the process to avoid the opportunity of misoperation. After the incident occurred, it was clear that ISD was required.

ASME Section VIII Div. 1 UG-140 "Overpressure Protection by System Design" states that "A pressure vessel does not require a pressure relief device if the pressure is self-limiting (e.g., the maximum discharge pressure of a pump or compressor), and this pressure is less than or equal to the MAWP of the vessel at the coincident temperature[2]".

Overpressure protection by system design falls under the *Simplify* strategy of ISD. The idea is to reduce the potential for overpressure and reduce the probability of loss of containment by simplifying the process and allowing for larger margin of error. Note that designing process equipment for the maximum achievable pressure does not completely eliminate the hazard, since the possibility of loss of containment of high pressure material still exists by means other than overpressure. However, the hazard is greatly lowered since the probability of loss of containment due to overpressure caused by process upsets, human error, or instrumentation failure is inherently reduced.

This paper focuses on designing for the maximum achievable pressure as a means to make processes inherently safer. There are applications today where this strategy could feasibly be implemented. The following sections discuss some examples on how potential hazards can be avoided if the systems are designed for higher pressure ratings. The examples will demonstrate the potential benefits in terms of relief valve sizing, disposal systems (flare) sizing, and cost reduction, as well as the challenges faced by overpressure protection by system design.

3 Designing for the Maximum Achievable Pressure

Inherently safer design is valuable for process risk management. Regulatory bodies, process safety experts, and policy makers have already embraced the proven concepts related to ISD over the last few years. However, there are still challenges to be faced in the industry.

In the book *Process Plants: A Hand Book For Inherently Safer Design*, the following is stated about equipment and operator failure: "In all industries, equipment failures and errors by operators and maintenance workers are recognized as major causes of accidents, and much thought has been given to ways of reducing them or minimizing their consequences.[2]"A procedure known as PRIM (Process Related Incidents Measure) is used to analyze incident data annually by the Canadian Chemical Producers' Association (CCPA) member companies. The 2014 PRIM analysis indicates the six Process Safety Management (PSM) elements shown in the figure below contribute to 85% of the total incidents.

PSM Element No.	*PSM Element*	*% of Incidents*
6	Process and equipment integrity	23.8
2	Process knowledge and documentation	21.2
4	Process-risk management	16.8
7	Human factors	8.9
5	Management of change	7.3
3	Capital project review and design procedures	6.5

Figure 1. Statistic of Incidents related to PSM Elements

According to PRIM, the largest percentage of incidents fall into the category of process and equipment integrity. Some of this may come from corrosion failure, but it is likely that some of these events may come from overpressure and loss of containment. In fact, out of the six elements of PSM listed in the figure, all of them could be resolved by inherently safer design and potentially through overpressure protection by system design. If the process vessels are designed for the maximum achievable pressure in accordance with ASME VIII Div.1 UG-140, the risk associated with equipment and operator failures could be greatly reduced, since the vessels would be able to contain the pressure without compromising equipment integrity.

Carefully establishing a design pressure with a combination of procedural or active layers of protection may provide the same risk reduction and may be cheaper than designing for the minimum pressure and then installing expensive SIS or complicated layers of protection to manage the risk. Designing for higher pressure does not necessarily mean that every system should be designed for the highest achievable pressure. It is easily recognizable that there are situations in which designing for the maximum achievable system pressure is not economically feasible; however, there are processes in which marginally increasing design pressures in combination with other cheaper active layers of protection (for example relief devices instead of SIS), may provide the same risk reduction level while still being a cheaper option.

If the concepts of Overpressure Protection by Design are applied to an entire facility, the need for large relief systems could potentially be eliminated. If every system within the plant was designed to be able to contain the inventory of material with no need to relieve it, the result would be that the relief systems required to handle the relief loads would be a lot smaller than the actual. Smaller flare systems could be easily achieved if enough process knowledge and research is spent in the early stages of the life cycle of the facility.

Additionally, designing vessels for higher pressure would reduce the risk associated with human or instrumentation error, and the conditions necessary for vessel failure would also be less likely to occur. The cost associated with maintenance of SIS's and training would go down, as well as the expenses resulting from testing and inspection requirements. The loss of product associated with venting requirements also needs to be included in the estimation; designing for the maximum achievable pressure would greatly reduce unnecessary product loss.

Along with designing for the maximum achievable pressure comes the flexibility of being able to increase the throughput of the facility without having to modify the active or procedural layers of protection. This is a common problem for operating facilities, where the change in throughput is increased over the years and this results in having to upgrade, modify or maintain additional layers of protection over the years.

In summary, a passive design is not only safer but it may also prove to be cheaper throughout the lifetime of the facility.

4 Why Not?

The ultimate goal of inherently safer design is to eliminate hazards. However, it has been established that a safe design can be achieved with a combination of passive, active and procedural layers of protection. It is necessary to find a balance between acceptable risk and economic factors using a combination of the risk management techniques. Even though the probability for loss of containment is greatly reduced, the hazard of maintaining large inventories of high pressure hazardous material through overpressure protection by system design is not eliminated.

There is a perceived notion that designing process vessels for higher pressure results in unnecessary cost or that the economic burden would drive capital projects to not be feasible. The reality is that some engineers design process equipment based solely on the operating conditions of process vessels just to have the cheapest and most "cost-effective" design possible. This approach results in design that is not flexible to process upsets caused by human error, controls or utilities failure. Because the risk associated with this design is commonly high, relief systems or SIS are necessary to adequately manage the risk. Designing for the maximum achievable pressure is an inherently safer strategy since there is no need to rely on additional layers of protection in order to reduce the risk. The upfront cost of design would be higher if every process vessel was designed for the highest achievable pressure, but this methodology would not be a fair cost and risk evaluation.

Instead, when considering the cost associated to different process safety alternatives, a truly fair analysis must also consider the cost and risk over the entire life cycle of the facility. It is likely that for some situations simplifying the design by careful consideration of the process and designing for the highest possible pressure would result in a more cost effective option in comparison with the upfront cost of active layers of protection (SIS, relief systems) plus the cost of operation and maintenance of the necessary active and procedural layers of protection. The perception of higher expenses is still an obstacle to the development of inherently safer techniques.

Another challenge is the fact that old designs are already established and there is very little room for innovation. It is easier to just recycle the last design because it has already been proven to work; there is no need to re-think and carefully analyze the process and search for inherently safer strategies for design. Knowledge of the process outside of the design conditions is required in order to fully utilize the concepts of overpressure protection by system design, which adds another challenge when competing with old and proven designs.

5 Case Study (PSV Sizing)

A system with higher design ratings can help minimizing the need for installing additional active layers of protection. Often times, the required relief rates due to certain overpressure scenarios are so large that they cannot be easily mitigated. The figure below represents one example to of the difficulties to mitigate the risk when the chosen design relies solely on overpressure protection.

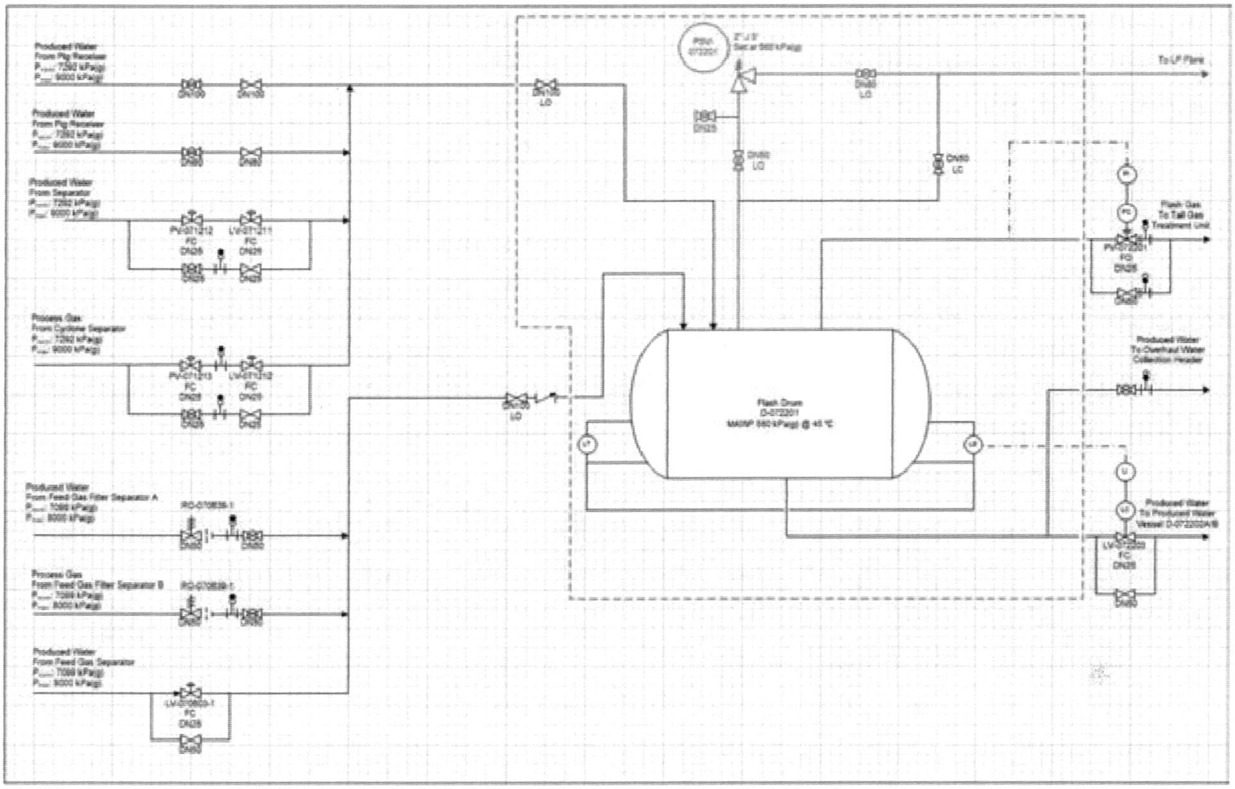

Figure 2. Flash Drum Example

The protected system is the Flash Drum. The controlling scenarios for the system shown in the above are instrumentation failure and the inadvertent valve opening of the normally closed valves. The normal operating pressure upstream of the control valves are around 7200 kPag, which exceeds the MAWP of the system (580 kPag). This risk of loss of containment could not be mitigated by the addition of relief devices (active layer) since addition of relief devices large enough to handle the area was unfeasible. Another possible solution was to install restriction orifices (procedural layer). Due to the large pressure differential between the normal upstream pressure and the PSV set pressure, the required restriction would need to be less than 7mm. An orifice area of 7 mm can potentially be plugged; therefore, this was not a feasible solution either. The only option was to install high integrity SIS in order to shut the feed in the event of low liquid level in the upstream system in order to prevent the possibility of gas-blowby. This approach can mitigate the risk to an acceptable level for one scenario; however, the SIL rated system has to be regularly maintained and tested to ensure its integrity. Note that the relief device still required an increase in size in order to mitigate the liquid overfill scenario. Due to this size increase, the inlet and outlet discharge piping had to be increased in size.

Note that the mitigation could have been simplified if the design pressure of the vessel was higher. The reason for having to install so many additional layers of protection is that there was a real risk of loss of containment since the upstream pressure for this system was approximately 10 times as large as the vessel hydrotest pressure. If the vessel had a higher design rating from the beginning of the life cycle, the entire procedure could have been avoided. It is not necessary in this case to design the vessel up to the maximum achievable pressure, there are situations in which this is economically not feasible; however, designing the vessel for a high enough feasible design pressure in combination with a relief device could produce the same risk reduction with lower maintenance cost when compared to the actual design in combination with SIS systems.

6 Case Study (Flare Design)

With additional process knowledge and carefully analyzed process safety information, designers have the ability to create processes that are safer and more cost-effective over the lifetime of a facility. Due to environmental regulations, a processing plant had action items from the authorities which included routing several relief valves currently relieving to the atmosphere to a closed system. The analysis included considerations of global scenarios (such as Refinery Power Failure) resulting in multiple systems simultaneously relieving to the flare.

Based on the original design pressures for each of the systems, the resulting global loads were larger than any combination of flare headers could handle. The calculated backpressure based on the generally accepted methodology for flare system analysis was so large that the simulation calculated accumulation levels high enough to lead to loss of containment if all the atmospheric relief valves were routed to the closed systems.

The end result of this analysis led to a combination of increased risk (which was accepted by the facility) and the installation of multiple additional active layers of protection to maintain the risk to an acceptable level. In addition, these modifications imposed a large upfront cost as well as future maintenance costs that could have been initially prevented if additional risk considerations were taken into account in the early stages of the lifecycle of the plant.

An advantage of carefully specifying design pressures and overpressure protection by system design is that relief loads can be reduced or eliminated altogether. Through careful analysis the, the designer could specify the maximum temperature of the heating medium to not exceed the bubble point of the working fluid at relieving pressure, this is called reboiler pinch-out. If a similar methodology of reducing or eliminating loads through overpressure protection by system design is followed throughout the entire facility, the disposal system requirements could be reduced.

7 Conclusion

The process safety industry is constantly learning from past incidents. Inherently safer design alternatives are needed in order to reduce or eliminate the hazards resulting from necessary processes. Designing for the highest achievable pressure is one strategy that results in inherently safer designs. There are numerous benefits of designing for the highest achievable pressure in accordance with ASME Section VIII – Division 1 UG-140: Overpressure Protection by System Design:

- Designing for the highest achievable pressure will make any process inherently safer.
- Initial cost of implementation of active and procedural layers of protection as well as the cost of operation and maintenance of active and procedural layers of protection will not be necessary.

- Disposal systems would be smaller if the strategy is implemented in the early stages of the life cycle of the facility.
- Rate increases, debottlenecking, regulatory practices, and changes in technology would be less likely to result in changes to the necessary existing layers of protection due to the added flexibility of designing for the maximum achievable pressure.

In order to more carefully assess the viability of overpressure protection by system design, the assessment of risk and economic variables should be done during the early stages of the life cycle of the facility. Researching the process and carefully analyzing all inherently safer design strategies as options will ultimately lead to a safer design; however, overpressure protection by system design has real and practical applications today, when other inherently safer design strategies may not be feasible.

8 References

1. CCPS (Center for Chemical Process Safety). "Inherently Safer Chemical Processes: A Life Cycle Approach, 2nd Edition". December 2008.
2. Trevor Kletz and Paul Amyotte. "Process plants: a handbook for inherently safer design. 2nd Ed". May 17, 2010.
3. BPVC Section VIII-Rules for Construction of Pressure Vessels Division 1

441692 Practical Risk Based Approach to Pressure Relief and Effluent Handling System Design

Abstract:

Recent updates to pressure relief system design RAGAGEP include provision for risk based overpressure protection. Although risk based approaches are not new, the recent RAGAGEP provides additional guidance and criteria. This study presents a practical risk based approach to pressure relief and effluent handling system design and a supporting case study that results in safe and cost effective design.

Historically, pressure relief system design focused on qualitatively identifying a controlling scenario or design basis, the situation that places the highest demand on the pressure relief system, and calculated that the relief system could adequately provide overpressure protection. The associated design basis documentation was inadequate and was often only a relief device specification sheet.

The promulgation of OSHA 29 CFR 1910.119, the PSM standard, has significantly improved design basis documentation. Since the implementation of the PSM standard, typical steps in relief system design and design basis evaluation and documentation include identification of overpressure scenarios; quantification of the associated relief requirements and relief system performance; and evaluation of the hydraulic performance of the effluent handling system. This approach has improved relief system design and design basis documentation, but has also significantly increased the cost without a corresponding benefit in plant safety.

Risk based pressure relief and effluent handling system design also begins with identification of overpressure scenarios. However, the foundation of the risk based approach leverages existing PHAs and risk tolerability criteria already implemented at PSM covered facilities. When PHAs use LOPA, the consequences and likelihood of the scenarios are determined. The consequence of an overpressure scenario can be further evaluated with computational techniques, which include consideration for the pressure rise, potential release of toxic and flammable fluids, and resulting vapor cloud. Next, using risk tolerability criteria the risk of the identified overpressure scenarios is compared to risk tolerability guidelines. Scenarios with an intolerable risk are then mitigated through enhancements and alternatives including traditional pressure relief systems. The risk tolerability and cost of the mitigation measures are then balanced to ensure a safe and cost effective relief system.

Integrating the pressure relief system design into existing PSM practices reduces the effort and cost of developing and maintaining relief system design basis documentation and focuses the designers' intent on improving plant safety. This risk based approach to pressure relief system design ensures adequate overpressure protection systems and sound investment in human and capital resources focused on risk reduction.

441563 Overpressure Protection By System Design - Use of High Integrity Pressure Protection Systems

Abstract:

Overpressure Protection by System Design - Use of High Integrity Pressure Protection Systems

Use of a safety instrumented system (SIS) in lieu of a traditional pressure relief device is commonly referred to as a high integrity pressure protection system (HIPPS). With the approval of ASME code case 2211 in 1996, ASME Section VIII has been updated to allow for overpressure protection by system design in the BPV Code Part UG-140. A detailed analysis and thorough documentation is required to ensure that the vessel is protected by "system design", or in other words, the pressure is self-limiting, or there are no credible overpressure scenarios that result in the vessel pressure exceeding 116% of the maximum allowable working pressure (MAWP). API 521 Pressure-relieving and Depressuring Systems Appendix E allows replacement of a pressure relief device with an SIS that has a reliability consistent with the risk of an overpressure event. The API approach is similar to the approach outlined by CCPS, i.e. they are risk based accounting for both the probability of the overpressure event and the consequences of such an event. In contrast, the ASME guidance is based only on reducing the likelihood of an overpressure event.

The use of HIPPS to replace conventional pressure relief protection is growing. Some of the main drivers include the need to reduce flare system loads on large green field projects, limit and control the release of green house gases and also to allow for increased capacity in brownfield expansions.

This paper will review the differences in the ASME, API and CCPS approaches for overpressure protection using safety instrumented systems, the "credibility" criteria in ASME and the design of high integrity instrumented systems required to meet the code requirements. The paper will also include a discussion on how different users have implemented HIPPS in the offshore and onshore processes.

442373 RAGAGEP Considerations for Relief and Flare Systems

Abstract:

Abstract

Recent emphasis on RAGAGEP increased the scope of relief systems risk factors that require evaluation in order to develop complete and compliant Pressure Relief and Flare Systems (PRFS) documentation.

This paper discusses how RAGAGEP considerations now require evaluation and proper documentation of risk factors that are often overlooked including but not limited to: dispersion analysis, thermal radiation, noise, vibration risk, reaction forces and structural supports, metal cold temperatures due to expansion cooling and two phase flow, hot temperatures due to fire exposure and/or runaway reactions, PRV stability, and chemical reaction systems.

Important RAGAGEP considerations for these additional risk factors are highlighted and illustrated using several examples and case studies. Recommendations are provided on how to best address these factor in the evaluation and documentation of design basis.

12TH GLOBAL CONGRESS ON PROCESS SAFETY

Considering External Fire to be a Non-Applicable Overpressure Scenario

Jessye Palladino
Smith & Burgess: Process Safety Consulting
190 S. LaSalle St.
Ste. 550
Chicago, IL 60603
Jessye.Palladino@smithburgess.com

John Burgess, Ian McPike
Smith & Burgess: Process Safety Consulting
7600 W. Tidwell Rd.
Ste. 600
Houston, TX 77040

Prepared for Presentation at
American Institute of Chemical Engineers
2016 Spring Meeting
12th Global Congress on Process Safety
Houston, Texas
April 11-23, 2016

Considering External Fire to be a Non-Applicable Overpressure Scenario

Jessye Palladino
Smith & Burgess: Process Safety Consulting
190 S. LaSalle St.
Ste. 550
Chicago, IL 60603
Jessye.Palladino@smithburgess.com

John Burgess, Ian McPike
Smith & Burgess: Process Safety Consulting

Keywords: External Fire, Overpressure, Relief Valve, Pool Fire

Abstract

An external fire is usually considered an applicable overpressure scenario as the necessary elements to fuel the ignition of a fire are almost always present within a refinery, gas plant, or petro-chemical plant. This article will demonstrate the situations and conditions when it is acceptable to not consider external fire as an applicable scenario. The primary circumstances that will be covered include: surrounding chemical inventory, ignition source distance, and equipment elevation. In the event that these conditions are such that fire is considered an applicable scenario, the required flammable liquid pool conditions should be evaluated to ensure that an external fire can both be created and sustained to result in an overpressure occurrence.

1. Introduction

In the recent past, there have been multiple catastrophes in the chemical industry as a result of fire. In 1988, explosions devastated a plant owned by the Pacific Engineering and Production Company of Nevada (PEPCON) and again in 2002 a manufacturing facility for Third Coast Industries was completely destroyed. Fire is a dangerous possibility in the chemical industry, and as such, should be considered when discussing and identifying hazardous scenarios for plant operations. After initial consideration, deeper investigation into the mitigating factors and potential root causes of a hypothetical pool fire event may prove beneficial in determining an external fire to be a non-applicable scenario for certain cases. Conservatively accounting for exterior pool fire for every piece of equipment can result in issues such as large disposal requirements or unnecessary modifications to existing relief systems and flare headers. The site layout, the availability of flammable liquid to pool, and facility history can be used to determine if the conditions exist such that a pool of flammable liquid can be created. Society of Fire Protection Engineers (SFPE) literature and American Petroleum Institute (API) standards are useful in quantitatively analyzing if sufficient chemical inventory is present to sustain an exterior fire event. Guidance from these sources can be used to determine the size and burn time of a given pool of flammable liquid. By considering the above criteria, the possibility of overpressure to equipment as a result of pool fire can be determined.

2. Why is Exterior Fire Important to Consider?

The PEPCON facility in Henderson, Nevada manufactured Ammonium Perchlorate, a powerful oxidizing material utilized in solid propellant rocket boosters. On May 4, 1988, workers were performing repairs on a building that had been damaged in a recent sandstorm. Welding led to ignition of the structure as it was covered in Ammonium Perchlorate residue from a nearby drying process. The conflagration eventually spread to the storage building directly adjacent, where several thousand 55-gallon drums of Ammonium Perchlorate were being stored, resulting in two major explosions. The second explosion ignited a high pressure natural gas line located under the facility which burned for approximately 1200 hours until authorities could stop the flow of fuel [1].

More recently, on May 1, 2002, a petroleum products facility owned by Third Coast Industries in Friendswood, Texas was completely destroyed by a fire that consumed approximately 1.2 million gallons of flammable liquid and burned for over 24 hours [4]. The Third Coast Industries facility blended and packaged various petroleum lubricants such as motor and hydraulic oil. A small fire outside one of the packaging warehouses caused nearby small storage drums to fail, contributing additional fuel and growing the fire to affect other larger containers in the area. Eventually, the pool fire consumed the entire facility as more storage tanks failed and released their flammable contents. The U.S. Chemical Safety and Hazard Investigation Board concluded that one of the major contributing factors to the severity of this disaster was that no containment or drainage was available to control the spread of flammable liquids at the site of the initial fire [4]. Additionally, onsite fire water was insufficient to quell the flames once discovered. This allowed sufficient time for the fire to cause numerous storage tanks to fail and release enough flammable liquid to fuel the fire for over 24 hours [4].

These examples highlight the importance of considering factors such as liquid drainage and burn time when determining risk associated with an external pool fire event. In each case, investigation determined that the effect of the fire could have been mitigated by elements that affect the formation and behavior of the flammable liquid pool [1] [4]. For facilities where these elements already exist, it may be possible for credit to be taken when determining the appropriate protection for equipment. Accounting for fire loads for every vessel and equipment can result in very large relief loads to the flare or disposal header. Designing for this basis may be problematic and unnecessary, especially if the flare requirements for other site-wide relief scenarios such power or utility failure are much smaller comparatively. Examining the conditions necessary for a pool fire to occur is useful in analyzing the possibility of overpressure.

3. Determining Applicability of Pool Fire

When determining the applicability of an overpressure event as a result of fire, conditions necessary to create a pool of flammable material may be considered. The formation of a pool of flammable liquid is contingent upon the availability of liquid fuel from nearby process equipment. These flammable pools typically result from a liquid release from tanks, vessels, and processes due to inadvertent valve opening or failures in vessel and piping integrity [5]. Operating conditions for certain processes and areas of a plant may be such that little to no liquid is available in the process equipment. Similarly, the chemical inventory in some units may be of a non-flammable nature. For each of these cases, a loss of containment due to one of the aforementioned initiating events may not pose a risk of exterior pool fire because no liquid would be available to pool during a release or the pool that would be formed would not be flammable.

The applicability of an external pool fire as an overpressure scenario depends on the amount of the flammable liquid inventory and the burn rates of flammable material. This quantitative analysis begins by evaluating the dimensions of a flammable liquid pool that could result from a loss of containment. The size and burn time of a fire is correlated to the area of the spill and the volume of the spill. A larger spill area will result in a larger fire circle and a larger spill volume results in a longer burn time for a fire. Furthermore, the main dimension for determining if a liquid-controlled flame spread pool fire can be sustained is the depth of the liquid pool. Experiments have shown that in fuel spills with depths less than 1.5 mm, the flames do not spread outwardly from the ignition source and thus the pool is not fully ignited [2]. A possible reason for this lack of flame propagation is due to the dampening of liquid recirculation at these shallower depths [2]. By losing this liquid recirculation, the flammable liquid material burns out before the circulation can aid in igniting the surrounding fuel. Further experiments concluded that the minimum depth of a pool fire necessary to support flame spread is two millimeters; however, different molecular weights may result in a deeper depth of liquid needed for a pool fire [2]. For example, 2.0 millimeters is needed for decane versus 2.3 millimeter for dodecane [2]. Since 2.0 millimeters is considered the minimum depth to sustain liquid-controlled flame propagation it is used as the required depth in this analysis. Additionally, minimum fuel depths are only applicable for fires that are in the liquid-controlled spread regime. By using this minimum depth and the spill area, the minimum required volume of flammable liquid can be determined from equation 1 [2].

$$V_{min} = A_{spill}\delta_{min} \tag{1}$$

$$V_{min}(m^3) = A_{spill}(m) \cdot (0.002m) \tag{2}$$

$$V_{min}(ft^3) = A_{spill}(ft) \cdot (0.00656ft) \tag{3}$$

Where V_{min} is the minimum volume of flammable liquid required, A_{spill} is the area of the liquid pool, and δ_{min} is the minimum depth of the pool. However, to calculate the minimum volume, an area of the spill has to be defined. The area of a spill can vary depending on if the spill is confined or unconfined. Confined spills have defined areas as there are physical barriers (open containers with fuel in them, walls, dikes, etc.) that prevent the liquid from spreading. However, free flowing spills are not constrained and can spread laterally depending on several factors such as the spill momentum, surface material, and surface tension of the spreading fluid. The greater the initial spill momentum a larger area is expected to be covered. The surface a spill is occurring on has many factors that influence the spreading of a liquid. A smooth flat concrete flow has less surface friction and thus allows for the liquid to spread easier. Whereas, a more porous rougher surface, such as dirt, limits the lateral expansion of a spill resulting in a smaller spill area with a greater liquid depth. Similarly, a fluid with less surface tension is expected to spread more easily than one with a higher surface tension. These multiple factors make the area of an unconfined spill difficult to define. If there is limited data about the area of a spill, then general assumptions can be made. Conservatively, API STD 521 §5.3.2 states "in the absence of any other governing factors, consideration of a fire-impact area is frequently limited to a ground area of 230 m^2 to 460 m^2 (2,500 ft^2 to 5,000 ft^2)" [3]. Using this range of area for a pool fire, the minimum volume equations are simplified to equation 4 and 5:

$$V_{min}(m^3) = 230m^2 * 0.002m = 0.46\ m^3 \tag{4}$$

$$V_{min}(m^3) = 460m^2 * 0.002m = 0.92\ m^3 \tag{5}$$

$$V_{min}(ft^3) = 2500ft^2 * 0.00656ft = 16.40\ ft^3 = 122.7 \text{ US Gallons} \tag{6}$$

$$V_{min}(ft^3) = 5000ft^2 * 0.00656ft = 32.81\ ft^3 = 245.4 \text{ US Gallons} \tag{7}$$

resulting in a range of liquid volume of 122.7 to 245.4 US gallons. By calculating the minimum volume needed to sustain a pool fire for the API suggested area (2,500 ft^2 to 5,000 ft^2), the flammable liquid of a plant can be inventoried to verify that there is a sufficient volume present to sustain a pool fire. If the fire area does not contain at least 122.7 gallons of flammable liquid then an external pool fire due to an uncontained spill may not be considered an applicable source for overpressure. Similarly, if the area of a spill can be confined then the minimum volume of fuel needed for the known area can be determined using the same method as above with the minimum depth of 0.002 meters (0.00656 feet). Additionally, the inventory of the confined spill area can be used to determine the applicability of a pool fire in that region. By using the minimum depth and either the API suggested area for an unconfined spill or a known area for a confined spill, the minimum volume to sustain a pool fire can be calculated. If the inventory in the defined area does not meet this minimum volume then the flames are not expected to spread as the depth of the pool would be less than 0.002 meters (0.00656 feet) and therefore the external pool fire is not expected to result in overpressure.

If sufficient flammable liquid inventory is present in the fire area, then further investigation of the burn time of the flammable liquid can determined if the pool fire can be considered a non-applicable overpressure scenario. While the minimum volume of liquid to sustain a pool fire may be present, the length of the fire may prove to be too short to result in overpressure. The time needed for a pool fire to raise a piece of equipment from operating conditions to relief pressure is another factor in the determining the possibility of an external pool fire resulting in overpressure. A transient fire calculation can be done for a vessel in the fire circle to obtain the time needed to reach overpressure. Combining the required time for an external fire exposure to cause overpressure with the burn rate of the liquid fuel, provides another condition the flammable liquid inventory of a system's surroundings that impacts the applicability of external fire. In order to evaluate the duration of a fire, the burn rate of the fuel is needed. There have been multiple studies performed on the burn rate of different materials that have resulted in various equations for the burn rate. But empirical data for a wide range of fuels have been found to correspond closely with a ratio of the heat of combustion to the heat of vaporization as given in equations 8 and 9 and this relationship is used for this analysis [2]:

$$\dot{m}''_{max}\left(\frac{kg}{m^2 s}\right) = 1 \times 10^{-3}\frac{\Delta h_c}{\Delta h_{v,sen}} \tag{8}$$

$$\dot{m}''_{max}\left(\frac{lb}{ft^2 s}\right) = 2.05 \times 10^{-4}\frac{\Delta h_c}{\Delta h_{v,sen}} \tag{9}$$

Where $\dot{m}''_{max}$ is the maximum mass burning rate per unit area, Δh_c is the heat of combustion of the material, and $\Delta h_{v,sen}$ is the heat of vaporization of the material. Note that the mass burn rate for uncontrolled spills have been shown to be significantly lower than the mass burn rate of controlled fires. Therefore, the NFPA suggests the burning rates of unconfined spills be estimated as 20% of the maximum mass burning rate [2]. Once the maximum mass burning rate for a liquid is known, the burn time of the fire can be determined. The units of the burn rate are given as mass per area-time. In order to isolate the time factor, the mass of the material must be divided by the mass burn rate that is multiplied by the area of the spill. Since the liquid inventory is typically recorded as a volume and not as a mass this numerator mass factor can be converted to a volume multiplied by the density of the fluid in order to obtain equation 10:

$$t_{bcontrolled} = \frac{V \cdot \rho}{\dot{m}''_{max} \cdot A} \tag{10}$$

$$t_{buncontrolled} = \frac{V \cdot \rho}{0.2 \cdot \dot{m}''_{max} \cdot A} \tag{11}$$

where $t_{bcontrolled}$ is the burn time of a controlled fire, $t_{buncontrolled}$ is the burn time of an uncontrolled fire, V is the volume of the flammable liquid, ρ is the fuel density, $\dot{m}''_{max}$ is the maximum mass burning rate per unit area, and A is the area of the fire. Overpressure due to an external pool fire may occur if the burn time of fire exposure exceeds the time it takes for a vessel to reach overpressure. Therefore, in addition to having the depth of the spill at least 2.0 millimeters, the volume of liquid inventory must also be able to sustain a fire long enough to result in overpressure for a vessel.

An example of an uncontrolled spill of m-xylene pool fire and a vessel containing water is used for analysis. A simplified transient fire calculation on the water vessel is done to obtain a time for the water vessel to reach overpressure. Then the maximum burning rate will be calculated along with the required volume of m-xylene to fuel pool fire to last long enough to cause overpressure in the water vessel. A simplified transient fire approach was taken in Appendix B to determine the time needed for a vessel containing pure water to raise from operating conditions to operating pressure and dew point and then how much time is needed to fill the vapor space in the vessel due to an external fire. Additional analysis can be done to determine the time needed to raise the vessel to relief pressure but for this analysis the time for the first two steps is conservatively taken as the time to overpressure the vessel, 43.6 minutes.

Based on equations 1-7 the minimal volume for a 2,500 ft^2 fire of m-xylene is 122.6 US gal, which translates to 889.5 lb_{mass} of m-xylene at atmospheric pressure and 60°F. Equation 9 combined with the heat of combustion to heat of vaporization ratio, obtained from the graph in appendix A, yields the maximum burn rate of m-xylene for an uncontrolled spill:

$$\dot{m}''_{max}\left(\frac{lb}{ft^2 s}\right) = \left(2.05 \times 10^{-4}\frac{\Delta h_c}{\Delta h_{v,sen}}\right) * 0.20 \tag{12}$$

$$\dot{m}''_{max}\left(\frac{lb}{ft^2 s}\right) = 2.05 \times 10^{-4}(70) * 0.20 \tag{13}$$

$$\dot{m}''_{max}\left(\frac{lb}{ft^2 s}\right) = 0.002867\frac{lb}{ft^2 s} \tag{14}$$

Once the maximum burn rate is acquired, the time of a fire can be calculated. First, the minimum volume required to sustain a 2,500 ft^2 is evaluated to see if that minimum volume (122.6 US gallons) can create a fire that will last long enough to result in overpressure. The time of this minimum volume m-xylene fire is calculated below,

$$t_b = \frac{V \cdot \rho}{\dot{m}''_{max} \cdot A}$$

$$t_b = \frac{\left(\frac{122.6\ gal}{7.48\ gal/ft^3}\right) \cdot 54.23\ \frac{lb}{ft^3}}{0.002867\frac{lb}{ft^2 s} \cdot 2500\ ft^2}$$

$$t_b = 124.1\ seconds$$

This fire is expected to burn out in less than 3 minutes. This calculation is assuming the entire pool is burning at its maximum rate for the whole duration of the fire and does not take into account the initial time it takes for the flames to spread to the whole pool from the ignition source. However, the time it takes for the flame to spread is substantially less than the burn time and therefore does not need to be considered in the duration calculation [2]. This minimum volume of m-xylene to have a pool fire that covers an area of 2,500 ft^2 results in a fire that is predicted to burn out well before overpressure in the water vessel is expected to occur due to the external fire. The volume required for an m-xylene pool fire of 2,500 ft^2 to burn long enough (43.6 minutes) to result in overpressure is calculated below:

$$V_{req} = \frac{t_b \cdot \dot{m}"_{max} \cdot A}{\rho} \quad (15)$$

$$V_{req} = \frac{(43.6 * 60)s \cdot 0.002867 \frac{lb}{ft^2 s} \cdot 2500\, ft^2}{7.177\ \frac{lb}{ft^3}} \quad (16)$$

$$V_{req} = 2613\ gal \quad (17)$$

$$Volume\ of\ m - xylene = 2{,}613\ gal \quad (18)$$

This volume of m-xylene is the minimum amount to have a sustained pool fire of 2,500 ft^2 for at least 43 minutes. The volume needed for a 5000 ft^2 fire is double the 2500 ft^2 volume, 5,226 gallons. The flammable liquid within a plant or fire circle can be inventoried and the possibility of a lasting pool fire determined. If a pool fire is determined to be an applicable overpressure scenario given the available inventory, the effects of site specific layout and features can be considered on the severity of the possible fire scenario.

4. Factors Affecting Severity of a Pool Fire

Pools of liquid fuel also are strongly influenced by the layout and features of a facility, such as grading, trenches, and diversion walls. In the aforementioned Third Coast disaster, the investigative boards determined that the destruction caused could have been drastically reduced by adequate drainage and diking at the plant [4]. Pools will naturally tend toward the low points of drainage ditches in a plot, and often these features are designed to carry the flammable liquid away from equipment and processes. The Occupational Health and Safety standards for flammable liquids state "the route of [a] drainage system shall be so located that, if the flammable liquids in the drainage system are ignited, the fire will not seriously expose tanks or adjoining property" [6]. If a pool fire were to occur in this case, the pool may be isolated away from processes and the risk of overpressure due to the impingement of flames upon equipment would be significantly reduced. Similarly, grading, drainage, and sloping in a plot may be enough to guide pools of liquid toward a specific area of the plant resulting in a lowered pool fire risk in other areas of the facility. Layer of Protection Analysis (LOPA) refers to this type of protection as "Postrelease Protection" and states that it "provide[s] a high level of protection if maintained properly" [7]. Dikes and drainage systems are specifically listed as an effective means to reduce the frequency of large consequences resulting from an exterior pool fire [7]; however, these measures require an extra layer of scrutiny when they are being considered to reduce the risk of exterior fire. Although walls are useful in diverting flammable materials away from particular vessels and processes, they also tend to increase the effects of fire on the side the pool fire is isolated. According to overpressure relieving standards set forth by API, a pool fire confined by walls or embankments can result in higher heat fluxes than an open pool fire [3].

Finally, the incident history of a facility serves as strong indicator of potential sources of leaks and other problematic issues that can lead to a pool fire. Conversely, plant history can also be used in conjunction with the aforementioned criterion to identify areas that have a low risk of exterior fire relative to other areas. If certain procedures have been identified multiple times to result in a release of flammable liquids, then the factors mentioned previously, such as grading and vessel inventory, can be considered in an attempt to predict the behavior of the hypothetical flammable pool and the relative risk of overpressure in the event that the pool is ignited. LOPA literature suggests that sufficient historical data is very effective at adjusting failure rates to reflect local conditions and situations [7].

5. Conclusion

Fires at refineries, chemical plants, or gas plants are serious and dangerous events. There have been numerous fires that have resulted in extreme damages, injuries, and even deaths. The devastating fires at the PEPCON and Third Coast Industries facilities are prime examples of just how dangerous an external fire within a facility can be. These fires resulted in extensive damages and highlight that external fire should not be overlooked and that proper precautions should be taken to prevent similar incidents from occurring. The risk associated with an external pool fire should never be taken lightly; however, in certain cases overpressure protection in case of an external fire may not be needed as an external pool fire is not feasible for the protected equipment. Certain facilities may have overly conservative relief loads for the external fire scenario that could result in a large, calculated flare relief that may not be plausible for the facility. These relief loads from completing external fire analysis on a facility could cause a facility to design for large disposals systems that are not needed or make unnecessary modifications to existing facilities, incurring unnecessary expense. To ensure overpressure protection is being implemented correctly in the case of an external fire, there are multiple conditions that can be evaluated which, if met, could cause the external pool fire scenario to be not applicable as a source of overpressure. The surrounding chemical inventory, site grading and drainage, and facility history are all factors that dictate whether or not external fire will be a potential source of overpressure. Additionally, the volume and burn rate of the surrounding chemical inventory can be quantified to determine the size and burn time of a potential fire from a loss of containment. The size and burn duration of a fire based on a volume of a flammable liquid can determine if a pool fire could be sustained in a plant. By evaluating each one of these circumstances carefully, it may become acceptable to exclude external fire as a source of overpressure.

6. Appendix A

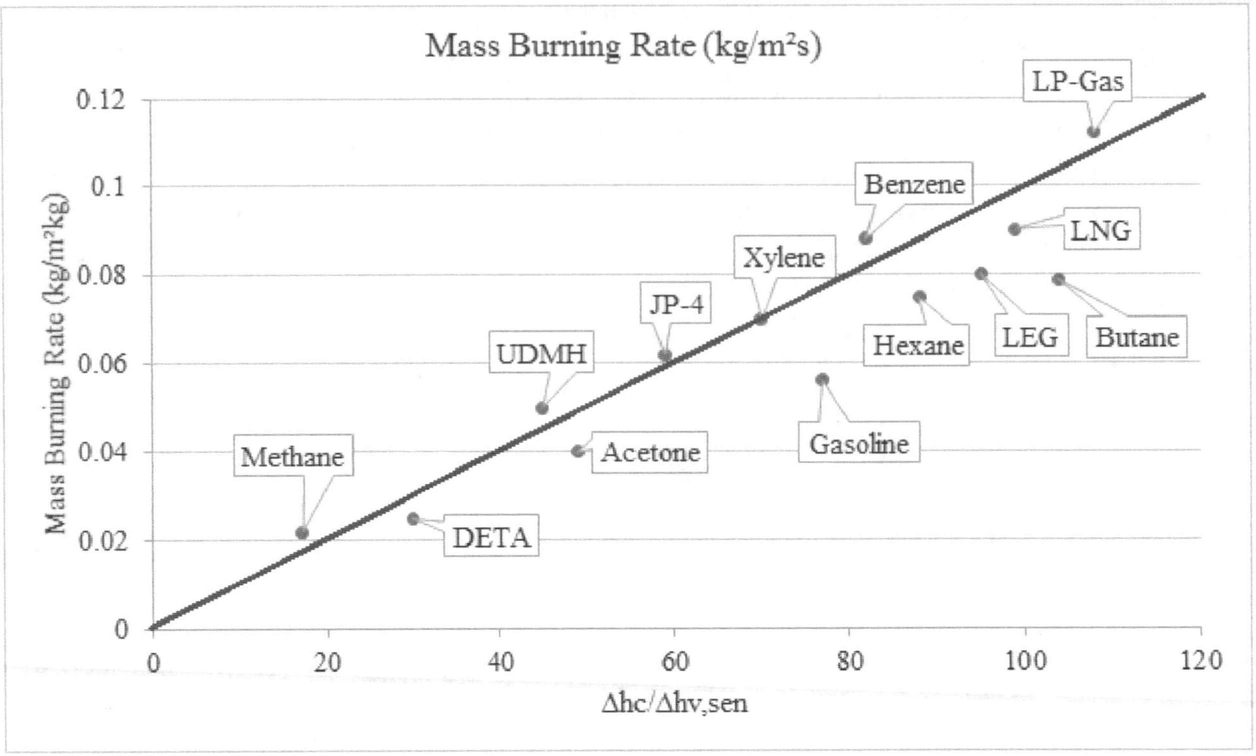

Figure A.1 Mass burning rate per unit area versus the thermochemical property of files burning as pan fires as a function of line 1.A [7]. The xylene value of 0.07 kg/m²s was used in the quantitative analysis.

7. Appendix B

This is the simplified calculation to determine the time taken to overpressure a water vessel for the example analysis.

Table B.1. Geometry and operating conditions of water vessel used in analysis

Description	**MAWP (psig)**	**MAWT (°F)**	**L -T/T (ft)**	**Dia (ft)**	**%LL**
Water Vessel	100	450	15	8	25%
	Composition	**Operating P (psig)**	**Operating T (°F)**	**Relief Pressure (psig)**	
	100% Water	80	275	121	
	Vessel Type	**Btm Head**	**Top Head**		
	Vertical	Elliptical	Elliptical		
Volumes	**Total (ft^3)**	**Liquid (ft^3)**	**Vapor (ft^3)**		
	888.0235	318.3481	569.6722		
Phase Density (lb/ft^3)		57.2873			
Mass of phase (lb)	18237.30	18237.30	0.00		
Molecular Weight (lbmol)	18.02	18.02	18.02		
Mass Heat of Vaporization (Btu/lb)		901.94			
Surface Areas	**Total (ft^2)**	**Liquid (ft^2)**	**Vapor (ft^2)**		
	515.7	165.0	320.7		
		Aws= 1.1 Aliq (ft^2)			
		181.5			

Using the API 521 4.4.13.2.4.2 eq (6) [3] the total heat input due to an external fire can be determined using the wetted surface area calculated above, 181.5 ft^2.

Table B.2. API 521 heat input due to external fire table

Heat Input Due to External Fire $Q = C * F * A_{wst}^{0.82}$	**Adequate Firefighting/ Drainage?**	**API Fire Constant**	**Environment Factor**	**Q (Btu/hr)**
API 521 4.4.13.2.4.2 eq (6)	Yes	21000	1.0	1,494,798.75

The first time step was determined as the time needed to heat the residing liquid to bubble point at operating pressure.

Table B.3. Step 1-Time to Heat Residing Liquid to Bubble Point

	Liquid (ft^3)
Volume (ft^3)	318.3481
Pressure (psig)	80
Temperature (°F)	75.00
Phase Density (lb/ft^3)	57.29
Mass of phase (lb)	18237.30
Molecular Weight (lbmol)	18.02
Mass Heat of Vaporization (Btu/lb)	901.94
Heat to Raise to Bubble Point (Btu, HYSYS)	974195.8
Time to Raise to Bubble Point (hr)	0.65
Time to Raise to Bubble Point (min)	**39.10**

The next time step is the time it takes to vaporize enough of the liquid to fill the vapor space of the vessel, after this point the pressure in the vessel will begin to rise but in this analysis the time taken from these first two steps are used as the time to overpressure, this is a simplified and conservative method as there are many factors that impact a transient fire calculation.

Table B.4. Step 2-Time to Vaporize Liquid to Fill Vapor Space @ Op Pressure

Volumes	**Vapor (ft^3)**
	593.8
Phase Density (lb/ft^3) @ OP Pressure and dew point	0.21
Mass of Vapor (lb)	125.16
Molecular Weight (lbmol)	18.02
Mass Heat of Vaporization (Btu/lb)	901.94
Heat to Fill Vapor Space (BTU)	112,885.47
Time to Fill Vapor Space (hr)	0.08
Time to Fill Vapor Space (min)	**4.53**

Total time needed to fill vapor space: 39.10 mins + 4.53 mins = **43.63 minutes.**

8. References

[1] Routley, J. Gordon, *Fire and Explosions at Rocket Fuel Plant – Henderson, Nevada,* United States Fire Administration Technical Report Series, May 4, 1988

[2] P.J. Dinenno, D. Drysdale Ph.D., C.L. Beyler Ph.D, and W.D. Walton, *SFPE Handbook of Fire Protection Engineering 4th Edition*, National Fire Protection Association, 2008.

[3] American Petroleum Institute, *Pressure-relieving and Depressuring Systems – API Standard 521 – 6th ed.,* January 2014

[4] *Investigation Report – Petroleum Products Facility Incident – Third Coast Industries*, U.S. Chemical Safety and Hazard Investigation Board, Printed Mar. 2003

[5] National Fire Protection Association, *NFPA 30 – Flammable and Combustible Liquids Code, 2015 Edition,* May 19, 2014

[6] Occupational Health and Safety Administration, *Occupational Health and Safety Standards 1910.106 – Flammable Liquids,* United States Department of Labor, Accessed 2/11/2016, Web.

[7] Center for Chemical Process Safety, *Layer of Protection Analysis: Simplified Process Risk Assessment,* New York, NY: Wiley, 2001. E-book.

441525 *Quantification of Double Contingencies in Overpressure Design Philosophy By Fault Tree Analysis*

Abstract:

The design of relief systems, specifically relief valve sizing, remains a crucial factor in mitigating the consequences of overpressure accidents in chemical plants. General criteria from engineering standards (API and ASME) and established engineering practices of operating companies and engineering contractors have provided guidelines in sizing relief valves. Sizing relief valves using standard and independent scenarios such as external fire, blocked outlet, and cooling water failure is a straight-forward matter however in instances where double contingencies or double jeopardy would apply, it is increasingly more difficult to assess the scenarios.

The paper will provide the general pathway to determine quantitatively if a double jeopardy scenario should become credible by using a deductive failure analysis method called Fault Tree Analysis. Fault Tree Analysis can analyze the overpressure state of the system using Boolean logic to combine lower level events such as two separate failures in a double contingencies scenario. A case study for seven permutations of control valve failure and its bypass valve positions will be compared and discussed with reference to Fault Tree Analysis results. Three of the permutations are based on engineering practices from several of operating companies and an assessment is made as to which approach is more realistic for design.

Fault Tree Analysis has been selected for study since it is considered an appropriate mathematical tool for analysis. By taking into account operation and design experience, Fault Tree Analysis can assist design engineers to incorporate or dismiss certain double jeopardy scenarios.

441463 PSM/RMP Modernization Programs in California - New Developments and Correlation to Evolution at the Federal Level

Abstract:

Paper 396953, presented at the 2015 Global Congress on Process Safety, provided a snapshot of "Changes in the PSM/RMP Regulatory Framework (California and National)". Post-publication, additional significant changes have been proposed by the agencies involved in the previous initiatives. On May 26, 2015, Cal/OSHA issued a comprehensive update to its proposed regulation for "Process Safety Management for Refineries," and CalEPA/CalOES issued an update to the California Accidental Release Prevention (CalARP) Program, creating a new program level "Program 4" that applies only to petroleum refineries. Both of these regulations coalesced as part of an overall initiative resulting from the efforts of the California Interagency Refinery Task Force. Subsequently, on September 24, Cal/OSHA made some minor changes and released its updated "PSM for Refineries" for public comment.

The final language for both of these new California petroleum refinery requirements is anticipated by December 2015, and promulgation is anticipated for Summer 2016. In parallel, Federal OSHA and U.S. EPA are creating updates to Process Safety Management (PSM) and Risk Management Plan (RMP) requirements as part of the PSM/RMP Modernization/Expansion Programs designed to address significant accidents that occurred in the past few years.

The changes being proposed for these programs are extensive, currently being formulated, and are not synchronized between the various agencies. The focus of this paper will be to provide relevant background information, a summary of the proposed changes, and tips for addressing key technical and regulatory challenges, as well as providing the reader with a "roadmap" for effective implementation.

The timing of GCPS-2016 is ideal for unveiling the final regulatory requirements in California and for identifying strategies for addressing these desynchronized requirements. The content of this paper will be important to communicate to GCPS attendees in the April 2016 time frame, and key topics include:

- Background – Pre-2012 PSM/RMP Universe
- Overview of Key Safety Management System Program Modernization Initiatives
- Key Elements of the Various SMS Regulatory Initiatives
- SMS Modernization Status and Potential SMS Program Impact
- What Should I Be Doing Now?

441338 PSM: Keeping Abreast of Legal Developments in Order to Stay Compliant

Abstract:

Compliance with PSM is an ongoing challenge. To stay ahead of the curve, industry needs to be aware of how the PSM Standard is being enforced by OSHA and interpreted by the Courts. Attendees will learn about the most recent legal developments involving the PSM Standard and where the next compliance challenges will be.

442487 Feasibility Study for Hydrogen Sulfide (H_2S) Optical Gas Imaging

Abstract:

Infrared (IR) cameras have been used extensively to detect hydrocarbon leaks from process equipment for both process safety and environmental applications. Recent advancements in the field have allowed for continuous and autonomous detection as well as quantification of hydrocarbon emissions. While these advances have led to new applications for hydrocarbon leak detection, optical gas imaging to date has not be utilized for detection of H_2S leaks. H_2S is a very toxic chemical commonly found in oil and gas streams and some process streams. Due to the toxic nature of the gas, it is highly desirable to remotely image and detect the presence of H_2S. Unfortunately, due to its IR spectral characteristics there has been no demonstrated method to achieve IR imaging of H_2S plumes.

A series of tests were conducted to evaluate the feasibility of imaging H_2S and to determine the sensitivity of an infrared imager in terms of ppm-m. Various spectral regions were examined for suitability of H_2S imaging. The results of the test show that it is feasible to image H_2S in the SWIR region. The results of these tests and the corresponding sensitivity of an H_2S imager are presented, along with potential applications for this technology.

Design Considerations for Oily Water Lift Stations (OWLS)

William J Celenza
KBR Inc
Houston, TX
William.celenza@kbr.com

Janani Sankar, AIChE Member

Prepared for Presentation at
American Institute of Chemical Engineers
2016 Spring Meeting
12th Global Congress on Process Safety
Houston, Texas
April 11-23, 2016

Design Considerations for Oily Water Lift Stations (OWLS)

William J Celenza
KBR Inc
Houston, TX
William.celenza@kbr.com

Janani Sankar
AIChE Member

Keywords: Oily Water Sewer, Pump Station, Sump, Gravity Drain

Abstract

Oily water, a mixture of hydrocarbon residues and water, is contaminated water released during normal operations at an oil refinery. Process wastewater flows are collected by an oily water or industrial sewer system that can be combined with area drains to collect potentially contaminated surface runoff from the on plot process paved areas as well as areas where spillages and wash water may occur. Surface runoff includes precipitation and fire water which is drained to catch basins, or area drains, through a network of gravity drainage piping and manholes. A clean storm water system collects precipitation from unpaved areas, roofs, roads and other areas not likely to be subject to contamination for direct offsite discharge. An oily water system needs to safely transfer collected flows to treatment or storage facilities to prevent contamination from impacting human health or the environment.

Refineries that segregate the oily water collection system from storm or fire water flows may drain runoff by gravity directly to one influent lift station at the entrance to the treatment plant, without using any intermediate lift stations. However, with combined systems the reliance on gravity conveyance can lead to excessive excavation depths and increased sewer construction complexity. The use of lift stations strategically placed throughout a combined collection network can simplify underground sewer piping networks, accommodate high water tables, and provide functional isolation for complexes or units within a refinery.

The design of an Oily Water Lift Station (OWLS) is distinct from that of municipal or highway related storm water lift stations due mainly to the industrial nature of the inflows, which typically are non-flammable but can include limited amounts of flammable and combustible hydrocarbons. The potential for explosive atmospheres in oily water lift stations, typically sealed and vented for VOC emission compliance, presents what is to some an unacceptable risk and has led to a design philosophy of minimizing or even eliminating the use of lift stations in this service, regardless of the increase in installation complexity. Also system flexibility can be increased and environmental risk associated with operating an Oily Water Drainage System (OWDS) can be better managed by including OWLS. Even though relying on lift stations presumes lower system reliability, appropriate pump selection, use of an emergency power supply, and automatic control and alarm systems have been successfully used to reduce this risk.

This paper describes methods for managing risks associated with operating OWLS through design considerations and related operational and maintenance procedures. Compliance with current US codes and standards for fire, electrical, and flammable vapor emission issues are discussed for gravity drain systems with lift stations, as well as the use of complementary closed drain pressurized collection systems, chemical sewers, and dedicated collection systems for constituents not compatible with atmospheric gravity drain systems. However, specialized systems such as Benzene Waste Operations NESHAPs compliant collection networks are not included in this discussion.

1 Background and Purpose

Modern refineries segregate direct contact process wastewater, such as desalting washes, process condensates and sour water, as well as solvent discharges in dedicated collection systems. Segregation allows for pre-treatment near the point of generation and potential cascading or reuse onsite with the aim of reducing end-of-pipe treatment loads. Wastewater is then typically pumped directly to the Wastewater Treatment Plant (WWTP) headworks. Also, flammable, toxic, corrosive, reactive or high pour-point liquid materials are segregated into closed drain systems (CDS), consisting of dedicated piping and vessels connected to selected drains that collect these liquid types for recovery or safe disposal. Hydrocarbons from tank drains, solvents or sour water from sulfur recovery or tail gas treatment units, and amines from acid gas removal units are examples of process streams which should use CDS. Typically only residuals or cleaning washes from these operations would be drained to a gravity Oily Water Drainage Systems (OWDS).

Oil-free wastewater such as boiler blowdown, cooling system blowdown, and demineralizer waste if not needed or suitable for reuse is typically collected then pumped around the headworks and primary treatment section of the WWTP as shown in Figure 1.

Not many refineries have segregated OWDS and process area storm water drainage systems possibly due to local regulations. But if segregated systems exist, the water is directed to separate treatment systems and the storm water system is likely to follow first flush principles subject to local regulation.

The combined OWDS will collect potentially contaminated storm water runoff from the on plot process paved areas and any areas where spillages and wash water may occur. This system will also collect and drain all fire water. Collecting these flows ensures that any contaminated runoff will not be discharged to the environment, or infiltrate into the groundwater. The areas being drained to the OWDS system are usually concrete paved.

Sanitary sewer systems should be entirely segregated from an OWDS. Sanitary wastewater may eventually be commingled with the oily water if the mixing may prove beneficial for the treatment of the combined wastewater at the WWTP.

As schematically shown in Figure 1, only some wastewater generated from refining operations is conveyed to the site WWTP via an underground gravity OWDS. The wastewater streams presented in Table 1 when only nonvolatile, nonflammable, and non-toxic components are present are typically managed in a refinery gravity OWDS. Since these streams are either continuously or potentially oil contaminated, treatment through the primary treatment section of a refinery WWTP is always recommended. Also as indicated in Table 1, the process area storm water is collected through the OWDS as a combined system. When compared to the typical influent for a refinery API Separator treatment system, the OWDS pollutant levels would generally be lower in chemical oxygen demand (COD), phenols, sulfides and ammonia but higher in oil & grease and total suspended solids (TSS). The API separator influent includes the OWDS combined with other wastewater streams at the WWTP headworks.

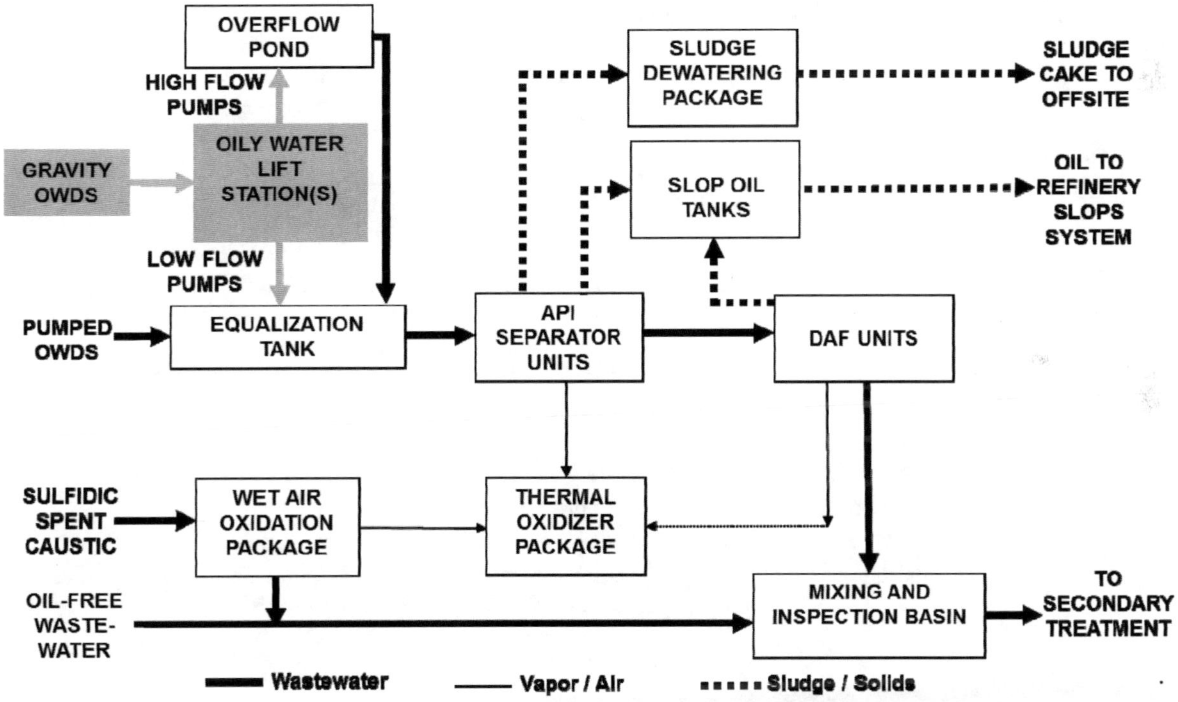

Figure 1 Refinery Wastewater Collection and Primary Treatment System

An OWLS, as part of the OWDS, receives flows that are incidental in nature and thus highly variable in flowrate as well as pollutant loading. In addition to storm water and fire water events, incidents such as hydrocarbon spills, hose station wash-downs, testing of fire water sprays, process vessel washes, pump baseplate working loses, etc. need to be anticipated.

Table 1 Wastewater typically managed in a refinery gravity OWDS (derived from US EPA 1980)

Wastewater Source	Typical Conventional Pollutant Levels (mg/l)						
	Oil & Grease (1)	COD	TSS (2)	Phenolic Cmpnds	Ammonia Nitrogen (3)	Sulfides (3)	pH
Hydrocarbon Equipment Drains (tanks, vessels, etc.) for dewatering, cleaning, hydrotesting, etc.	< 1,000	< 300	< 1,500	< 5	Trace	<10	5 to 9
Pump Station Drains (casing, base plate, etc.)							
Utility Drains (sample points, level gauges, instruments, etc.)							
(Un)loading Station Drains							
Pipe Trench Drains							
Area Drains (Wash, Storm, Fire Water)							

Notes:

(1) Oily Water is a mixture of non-flammable hydrocarbon residues and water, released during normal operations and is not expected to create hazardous conditions if discharged directly to a gravity OWDS.
(2) Can include fine dust/carbonate type solids that have been observed to produce a sticky paste when mixed with water and oils at some sites.
(3) For instance, Sour Water which contains more than 10 mg/l of hydrogen sulfide with other components such as ammonia, phenols, and benzene would not be permitted as discharge to an OWDS.

2 OWDS SYSTEM DESCRIPTION

The basic types of gravity OWDS are “dry” or “flooded”. Dry systems consist of free-draining sewer networks interconnecting catch basins and manholes. A typical catch basin or manhole has a sand trap and a liquid seal as shown in Figure 2. The sewers free drain to an OWLS that transfers the wastewater to a site WWTP for treatment or temporary storage.

Flooded systems maintain continuous wastewater gravity flow within the sewer piping and drain directly over a weir located at the entrance to an OWLS or to an API Separator at a WWTP. The system is flooded by the weir at these locations; and the catch basin and manhole components are the "dry-box" type with sewer inlet and outlet lines located flush with the flat bottom (Mason 1984). Water entering the OWDS will cause the level to rise and water will eventually flow into the OWLS by overflowing the weir. The weir needs to be designed to manage the peak instantaneous flow. Flooded systems may be impractical in locations sufficiently dusty where excessive solids may cause clogging (Lees 2012a, Mason 1984).

Compared to a flooded OWDS, a dry OWDS requires deeper excavation to accommodate free draining sewer lines and operation of the OWLS nearly empty. Thus a dry OWDS typically requires more intermediate OWLS for the same topographic conditions, such as available terrain slope, and water table elevation.

Per the OWDS in Figure 3, raw wastewater streams normally are introduced into the collection system through individual or area drains, which can be open to the atmosphere or sealed to prevent wastewater contact with the atmosphere. Individual drains may be dedicated to a single source or piece of equipment.

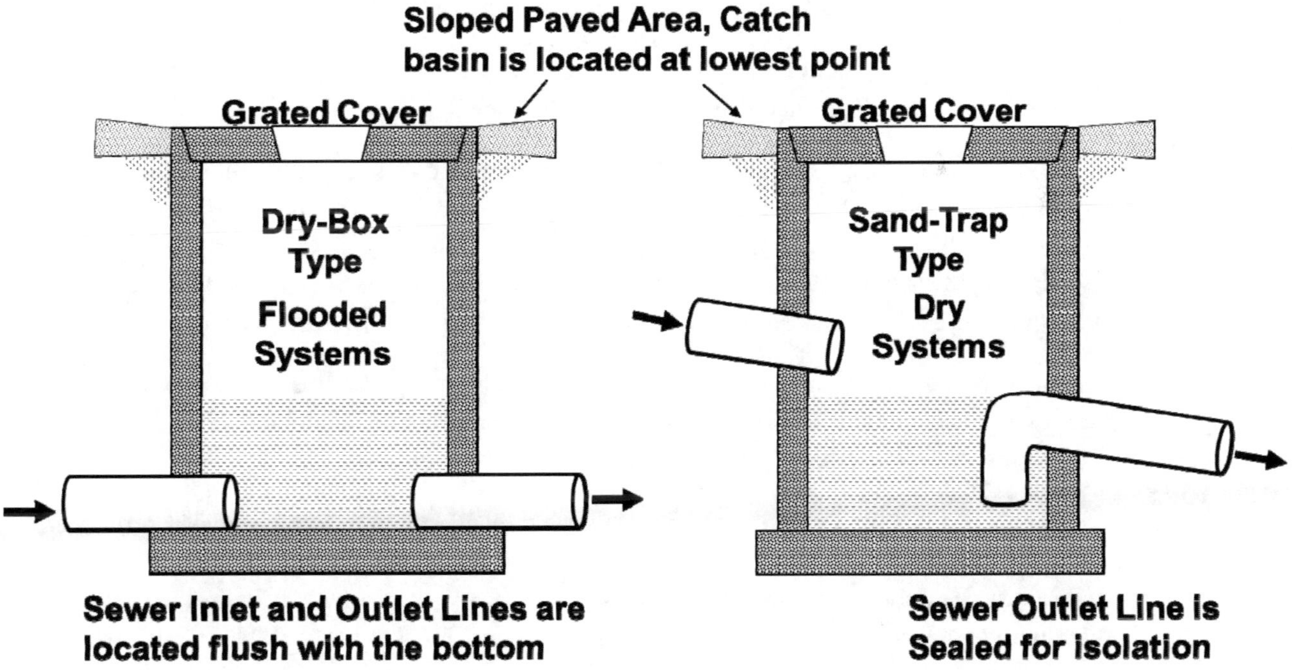

Figure 2 Dry-box "Flooded" vs. Sand-Trap "Dry" Type Catch Basins

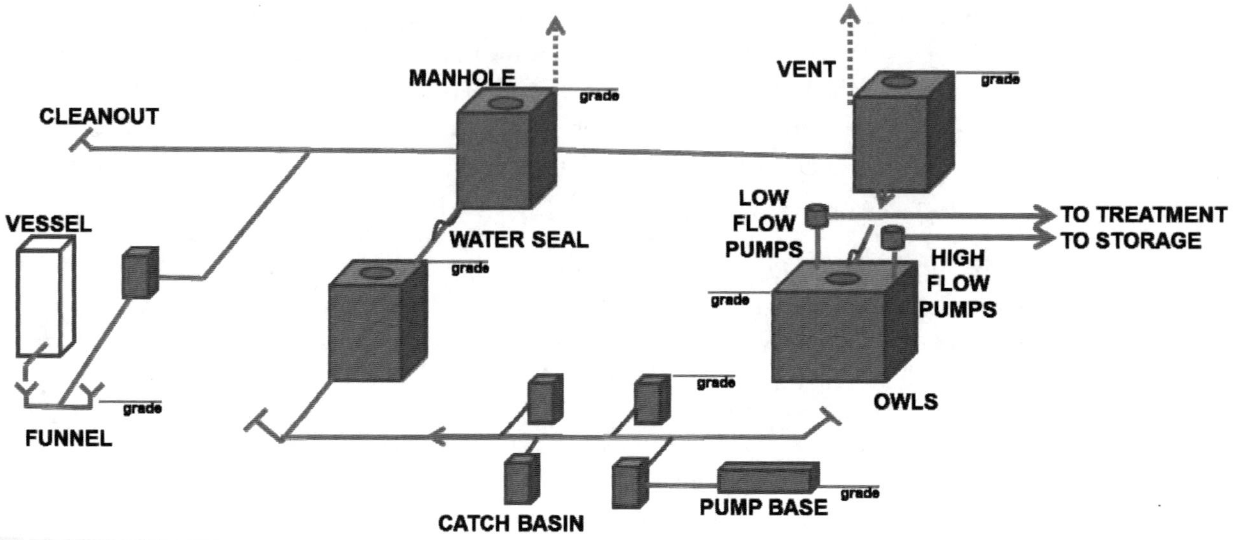

Figure 3 OWDS, Local Oily Wastewater Lift Station (OWLS) System

Drain Pipe / Funnels are outlet pipes that collect oily water from points such as pump bases, drain points, or vessel drains and discharge through a catch basin or manhole into the OWDS.

Area drains, or catch basins, serve several sources and are located centrally among the sources or pieces of equipment that they serve. Catch basins have a grate on top, an integral sediment trap and outlet liquid seal for dry systems and also serve as a collection drain for surface flows. Individual catch basin drainage areas are typically limited to approximately 5,000 square feet to ensure rapid surface runoff and to isolate groups of equipment for fire protection

Manholes in sewer lines permit service, inspection, and cleaning of a line. They may be located where sewer lines intersect or where there is a significant change in direction, grade, or sewer line diameter. Manholes include an outlet liquid seal for dry systems.

An OWLS, as part of the OWDS, is usually the last collection unit before transfer for treatment or storage, accepting wastewater from one or several sewer lines. An OWLS is a retaining structure, usually under atmospheric pressure, located below the normal grade of the gravity OWDS network. The collected wastewater is normally pumped to the WWTP.

An OWLS is sized to manage the larger of the following design flow rate cases:

- Normal process operation flows plus the peak fire water flow from the local risk area; or

- Cumulative storm water flow, typically based on a 5-year or 10-year return frequency, to prevent flooding that would disrupt operations or prevent a safe shutdown.

In some regions, the allowance for fire water may be more than the volume allowed for a 5-year or even 10-year storm water event; while in other regions the storm water event sizes the system. In those cases where fire water dominates, the normal process flows would be added to the storm water for the sizing analysis. This analysis also depends on the size of the collection area, since storm water flows are proportional to the ground surface area while fire water flows are not.

An OWDS with a network of intermediate OWLS is shown in Figure 4. The OWLS pumps typically would discharge into common headers, with one header dedicated to convey the wastewater to the WWTP for treatment and another header to storage.

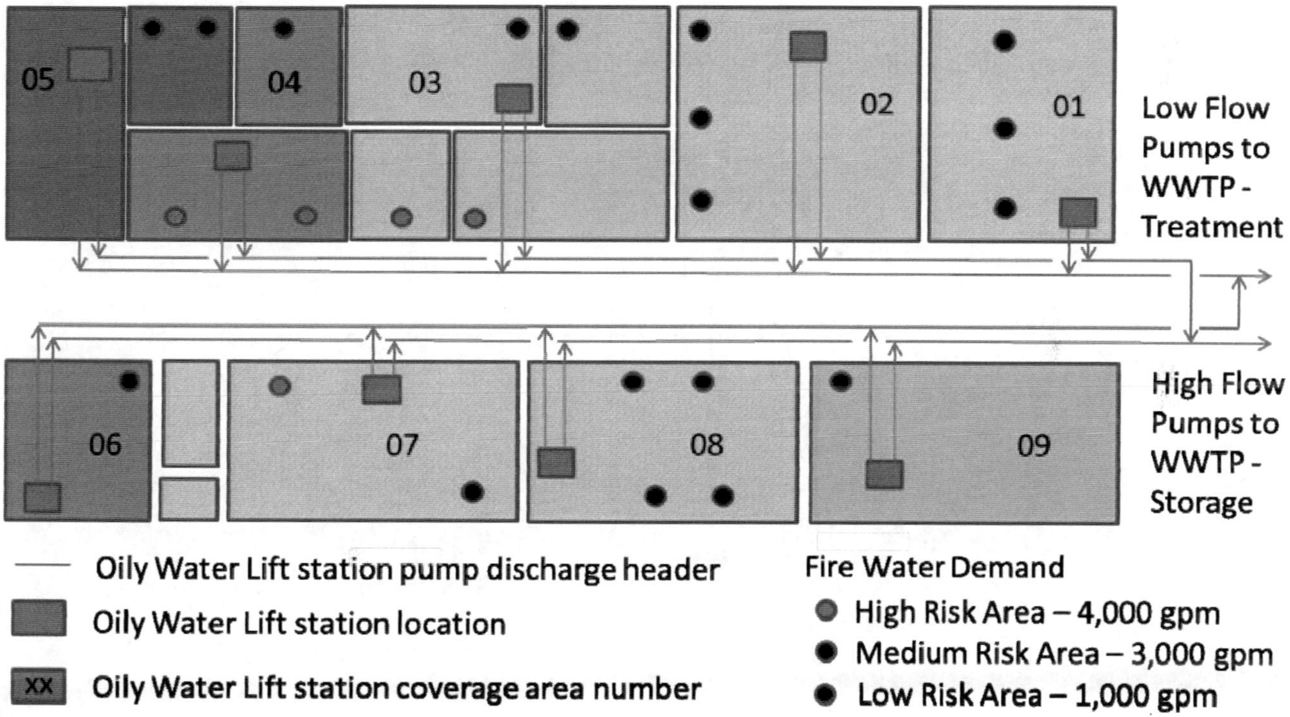

Figure 4 OWDS With A Network of Intermediate OWLS

Depending on the difference between normal and peak or high inflows, an OWLS may have one or two sets of pumps. If the high flow is less than 10% more than normal flow, a single set of pumps sized for the high flow condition can be used efficiently. If these flows vary by more than 10%, two sets of pumps, one for the normal and one for the high flow condition may be more efficient. For a two pump set system, a single or dual pumping chamber OWLS may be considered. Figures 5 and 6 show plan and section views of a dual chambered OWLS. Normal flows are managed in a smaller chamber with the larger chamber of the same OWLS reserved for the high flows.

All OWDS wastewater initially enters the low flow or process flow chamber and the high flow or wet weather chamber receives overflow from the low flow chamber. The minimum required working volumes of the chambers are related to their pump capacities, with the low flow pumps typically around 50 gpm or less and the high flow pumps 500 gpm or more each. However as indicated in Figure 6, the low flow chamber must also accommodate an underflow baffle and overflow weir rated at the design inlet flow. These weir dimensions often dictate the footprint shape, and often result in a working volume larger than the required minimum.

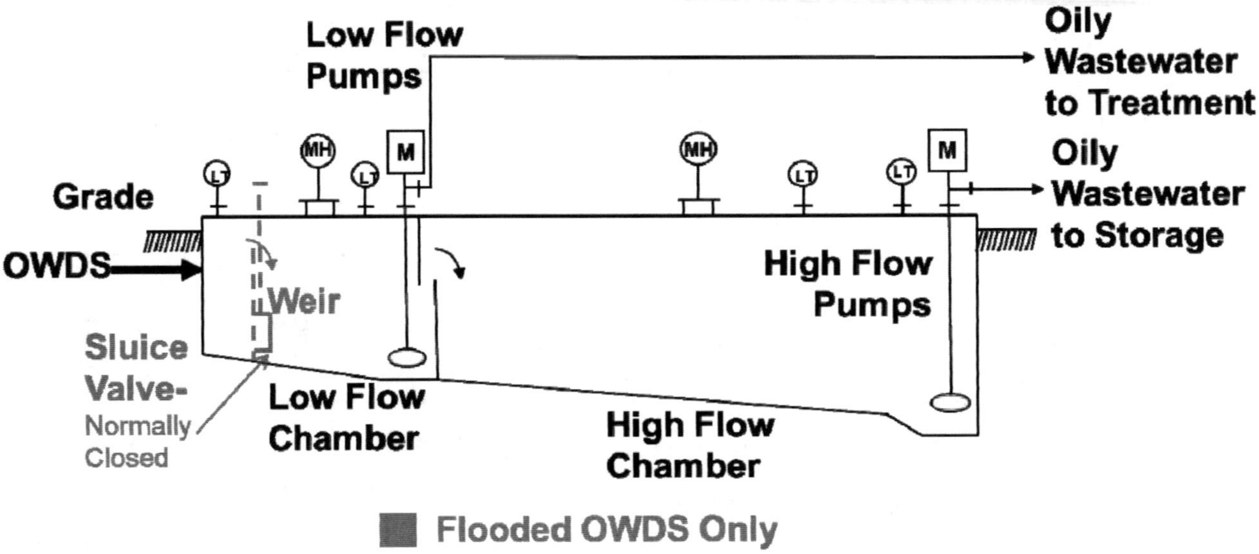

Figure 5 Dual Chamber Oily Water Lift Station (OWLS) – Section A-A

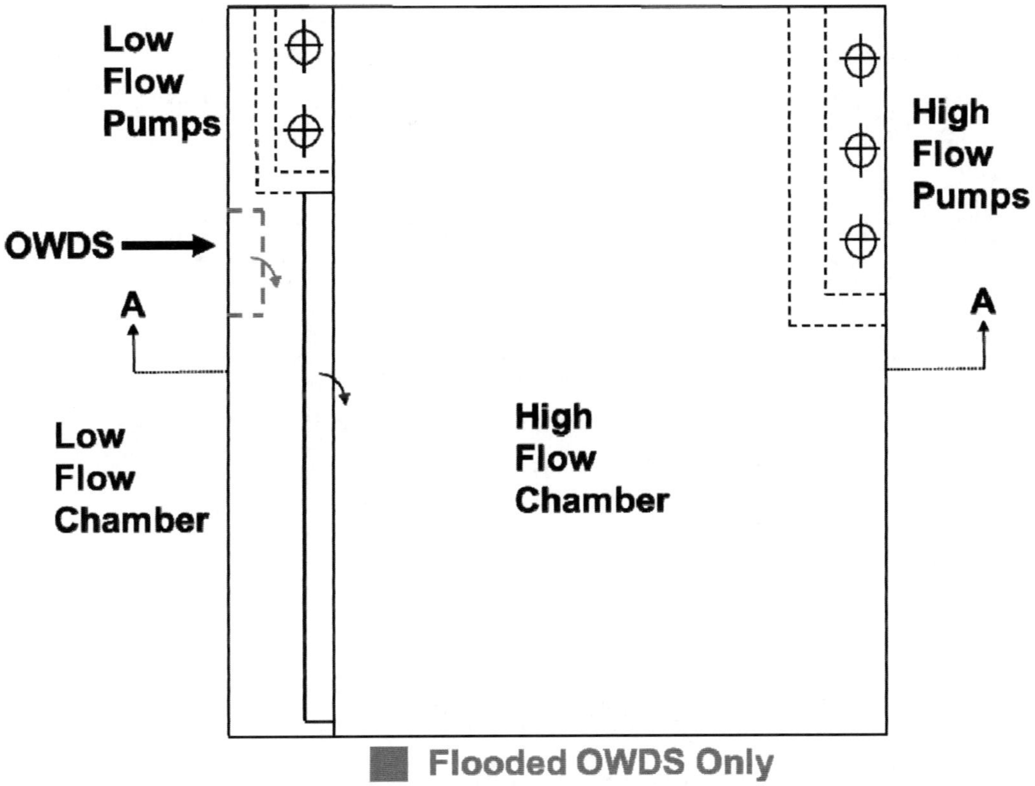

Figure 6 Dual Chamber Oily Water Lift Station (OWLS) – Plan View

3 SAFETY CONSIDERATIONS

A majority of safety incidents occur during startup or in auxiliary areas such as transport, storage, waste management, and recovery of raw materials or products (Lamberti 2011). Flammable liquids in wastewater discharged to an OWDS can cause fires / explosions in collection systems-particularly near the point of discharge, where high temperatures promote evaporation into a relatively fixed volume of air and form vapors that are trapped within the collection system. As the heated gases expand within the confined space, they create underground pressure seeking an opportunity to vent. Due to the vapor space confinement within OWDS and OWLS, when a flammable vapor cloud is ignited, the combustion probably will give rise to an overpressure resulting in a vapor cloud explosion (VCE) rather than a fire (Lees 2012b). This pressure will blow off at the weakest point, such as a manhole cover, a liquid seal trap, and / or an OWLS cover. Once dislodged, more oxygen may be introduced triggering additional explosions. A VCE can be ignited by electric sparks, friction, surfaces such as manhole covers heated by the sun, or heat generated by chemical reactions. Often workers use metal tools that can accidentally strike against a concrete surface or manhole cover creating sparks. (US EPA, 1992). Also, potential pump-related ignition sources include defective pumps, dry running pumps and faulty cables on submersible pumps (Bloch 2015).

Twenty years of historical data from a refinery and terminal facilities operator recorded a total of three dozen OWLS incidents in dry type systems (confidential 2008). Forty percent of the incidents involved a fire / explosion, while the rest were overflows. The main contributors to the incidents were:

- Presence of flammable liquids that formed an explosive vapor mixture – fire incident,
- Malfunction of level instrumentation - overflow,
- Failure of pumps - overflow.

Table 2 summarizes methods for managing risks associated with operating the components of an OWDS, preventing system failures, and increasing reliability.

Table 2 Methods for Managing Risks Associated with oily water lift stations

Design Considerations	
Electrical Area Classification	OWDS are classified as Class I, Zone 0 locations for manholes, catch basins, OWLS, underground piping, and vents. Zone 0 is defined as an area where an explosive gas mixture is continuously present or expected to be present 1,000 or more hours/year or 10% of the time (API RP 505, reaffirmed 2013).
Seals / Isolation	All flows into a dry OWDS enter through sealed catch basins or through a sealed connection to a manhole. All sealed connections, including sections of a drainage line located between liquid seals, from upstream of the first sealed manhole in an OWDS to downstream of the farthest catch basin, need to be vented to direct the release of any flammable or hazardous gases and to prevent a vacuum which could drain out the water seal connection. A water seal typically consists of an elbow or a tee with an outlet extending downward to provide a minimum of a 6 inch seal.
	In exceptionally hot and dry locations the seals in dry OWDS may require water addition to avoid evaporation; and a flooded OWDS may require a water purge to maintain a level of working liquid (Mason 1984).
	The use of a network of independent OWLS permits the isolation of individual process areas, units or complexes during site maintenance shutdowns. Whereas a site-wide sewer network increases the risk of a fire spreading and may lead to effluent stagnation during emergency situations.
Managing	Closed Drain Systems (CDS) are needed to manage flammable liquids and toxic materials from process water draw-offs, hydrocarbon contaminated

Flammable Liquids and Corrosive or Toxic Materials	cooling water, de-inventorying equipment drainage, and streams which could otherwise create hazardous conditions if they were discharged directly to OWDS. Even though no continuous flammable liquid source connections should exist, flammable liquid hydrocarbon content less than 50 mg/l may be considered for drainage. A combustible liquid at a temperature within 15 F (8 C) of its flashpoint would be considered flammable.
	Examples of toxic materials include: spent caustic, benzene, phenol, sour water that contains H2S (>10 mg/l) and ammonia (> 0 mg/l). Corrosive materials have a pH < 3 (acid) or pH > 12 (base).
Venting	Since flammable gases may accumulate in an OWDS, manhole covers and OWLS covers should be sealed or gasketed and vented to a safe location away from any potential ignition sources.
	End-of-Line Deflagration Flame Arresters may be used on vent lines to prevent flames from entering an OWLS, such as if a flammable gas mixture flowing from the vents were to encounter an ignition source.
Piping line depths	The OWDS piping in onplot unit areas is located under reinforced concrete slabs. Thus the configuration of the standard catch basin or funnel / drain pipe assembly vs. Minimum Cover Over Top of Pipe typically sets the minimum depths. Since a connection at the extremity of an OWDS sets the network depth, drain lines at these locations may be embedded in concrete to limit depth if the required soil-bearing loads are high.
	In a dry OWDS the slope for a free draining line is usually 0.6% to 0.8%. A flooded OWDS line slope would only require a nominal 0.1% slope (Mason 1984).
	For a dry OWDS, all gravity flow lines entering an OWLS must be above the maximum water level that will be maintained during operation. This design requirement ensures that the line will not become overfilled and not work properly. A 6-inch elevation difference is typically required to not impede incoming flow and cause a backflow and accumulation of sediment (confidential report 2008).
	Typically, when gravity OWDS lines are installed in trenches deeper than 20 feet, the installation cost increases significantly due to more complex excavation and trench shoring techniques (US EPA 2000a)
Siting	When locating OWLS, consideration for the distance to major process equipment should be considered. Offset, such as 150 to 250 ft. from process equipment, 30 to 70 ft. from occupied buildings, and 10 ft. from pipeways; may limit options and even lead to consideration of off unit plot locations.
	If a high water table or large distances are an impediment to the design of an OWLS, even after considering converting from a dry OWDS to a flooded and / or segregation of the OWDS from the storm / fire water, then additional OWLS may need to be installed.
	The number of local OWLS is determined by the following:

	• Type of OWDS, dry vs. flooded. • Site topography, available elevation change. • Minimum gravity flow line velocity, • Maximum invert depth to prevent the OWLS bottom elevation from being located below the water table, • Prevention of gravity flow lines from crossing large underground site utility lines, • Prohibition of locating near process equipment, inhabited buildings, drainage swales, diked areas, or designated pipeline corridors. • Avoidance of the pump lift station head from exceeding 20 ft. to allow the use of self-priming pumps.
	For OWLS vent locations, a safe location should be considered as no closer than 50 ft. horizontally from a furnace or other ignition source; and if within 100 ft. consider ending the vent line at least 50 ft. above grade.
<u>Downstream impact</u>	As the number of OWLS pumping to an API Separator increases, the turbulence impact from on-off pumping on the oil / water separation process should be considered. Either installation of a break tank with equalization or ensuring that the API Separator design is robust enough to manage the potential fluctuations and turbulence is suggested.
	Due to potential disruption of the pump(s), such as by mechanical or power supply reasons, the WWTP design may need to include quick operational turn down.
<u>Pump Selection</u>	Due to a Class 1 Zone 0 Electrical Area Classification, pumps with submersible motors would not be permitted. Bloch (2015) further discusses pump types for Zone 0 service.
	Several types of centrifugal pumps are frequently used in OWLS applications. Straight-flow centrifugal pumps are suitable for low-flow / high head conditions, angle-flow are appropriate for low to moderate heads, and mixed flow for large flows at low head (US EPA 2000c). Concerns for causing emulsification that would impact downstream treatment may also guide pump selection.
	Self-priming pumps have been found to be more reliable in OWLS service (confidential report 2008). If the required pump flow / discharge pressure is within the operating range of a self-priming pump, and the OWLS depth does not exceed a 20 ft. suction lift then the use of a self-priming pump may be an option. Use of a piping foot valve or anti-siphon valve would be recommended in this application. Or an option would be a vertically suspended line shaft type pump.
	US EPA (2000b and c) states that the number of pumps should be chosen to correspond as nearly as possible to flow fluctuations. In OWLS with peak inflows less than 700 gpm, two pumps are customarily installed with each

	able to meet the peak influent rate, subject to flow variability concerns. For instance, if the peak flow is less than 10% more than normal flow, a single set of pumps sized for the peak flow condition can be used efficiently. For larger OWLS or OWLS with greater variability, strategies include: installing pumps of different sizes, using parallel pumps, trimming impellers, installing variable speed drives or two-speed motors in order to prevent frequent starting / stopping and to not require excessive OWLS storage capacity. However for OWLS with rapid flow increase or high head losses, a single peak pump may be favored because two pumps operating together yield only slightly higher flows than one pump. Parallel peak pumping is favored for large OWLS with relatively flat system head curves, where multiple pumps deliver substantially more flow and provide more flexibility.
	Ursino (2014) states that adding a redundant pumping element in standby is less effective than increasing the number of pumping units that may function in parallel. Thus a larger number of smaller online pumps may prove more reliable than a less number of larger pumps, despite provisions for standby operation.
	Overall pump efficiency depends on the type of pump, the control system, solids loading, and the fluctuation of influent. Well-designed pump systems may operate at up to 75 to 85 percent efficiency most of the time (US EPA 2000c); however, discharge header conditions may limit efficiencies to closer to 50 percent at times (US EPA 2000a). Also, the mechanical efficiency of the pump impeller would normally decrease with increased solids handling ability, in relation to the solid size to be passed.
	Installing an inlet OWLS screen with means to clean it, will provide for filtering out incoming debris to protect the pumps from blockage.
<u>Power Supply</u>	Including separate transformer feeds to all pump systems can help to ensure a continuous source of power is provided. Emergency generators whether onsite standby or portable with quick connection can also be used as a backup power source. However, in the case of a power failure, options to de-inventory an OWLS would be: a) Use of Vacuum Trucks with suction hose access through a sump cover manway and, b) Use of portable diesel pumps with suction hose access through a sump cover manway or by way of dedicated connections on pump piping lines.
<u>Instrumentation and Controls</u>	Instrumentation to ensure overflows or running pumps dry does not occur is recommended; however, faulty or complex instrumentation should not hold up pumps from automatically operating when needed for instances such as removing fire water. The confidential report (2008) states: 1. Provide two radar level transmitters (LTs) for redundant level measurement in the OWLS. Level indication shall be in the control room. LTs should be configured to fail low such that pumps will not start on

	transmitter failures. LTs are preferred over switches because switches are blind devices that can fail in an undetected manner. Typically one LT should be designed to control the sequenced start-stop of pumps. Each LT should indicate its level on a DCS display in the control room and be provided with low and high level alarms. The two associated LTs should be compared and alarm on deviation, to provide more confidence that the level signal is true. 2. Maintain the water seal within an OWLS by having the incoming flow line termination point at 6 inches below the pump shut off level. Also the OWLS high-high level shall be reset below the incoming flow line. 3. A flow or pressure transmitter should be provided on each pump to trip on low flow. An orifice based flowmeter is recommended, but a pressure transmitter may be considered if the pump discharge pressure will reliably drop to a known pressure on loss of flow. Flow transmitters (FTs) are preferred over switches because switches are blind devices that can fail in an undetected manner. FTs should trip its associated pump on low flow and be configured to fail low. Low flow tripping should be active when the pump is on automatic or manual operation. However, consider low flow alarming only for pumps likely to be used for fire water service, to allow for enhanced availability. 4. Local controls should be located at a safe distance from the edge of the OWLS – 10 ft. away or more - to keep operators away from potential explosion exposure. Consider a local analog flow indicator from each pump FT and a local analog level indicator for each sump. Each pump should also have a local hand-auto-off (HOA) switch and push button start. However, since operating the pumps in manual mode increases the risk of pumps dry running, this switch should be locked or car-sealed in auto mode. Also consider providing the control room DCS with an indication of HOA switch position. 5 The control room DCS should be provided with a running and stopped indicator for each pump. Also consider DCS logic that will monitor both the OWLS level and pump flow / running conditions, so to provide an alarm if a pump is running with acceptable flow but the level is not moving.
Operational and Maintenance Procedures	
Spills	Operating procedures will need to stress the importance of detecting and isolating oil spills.
	The dual chamber version of the OWLS contains an overflow / underflow weir system to segregate free oil in the inlet portion of the system. However, an automated oil removal system is usually not installed. As a result, oil will collect in this first chamber and needs to be removed

	manually. The main function of the underflow / overflow weir system is to avoid oil carryover into the second or high flow chamber since only the low flow or dry chamber discharges directly to the WWTP's oil removal system.
	During vessel turn-around after depressurization, equipment inventory is drained into a CDS. Only residual inventories of flammable and toxic liquids, less than 26 gallons from equipment should be drained to an OWDS under special procedures. Flushing the sewer with water for dilution after draining until the percent lower explosive limit (%LEL), determined by lowering an LEL tester sample hose near the liquid level, is reduced to < 10% is suggested. Before draining any small amounts of flammable liquids or hazardous materials to the OWDs, also consider: (a) wind direction, (b) positioning of any ignition source beyond 50 ft., (c) H2S or ammonia content, and (d) use of proper personal protective equipment (PPE) including breathing apparatus.
Removal of Surface Oil	For lighter hydrocarbons, the insoluble oil tends to float on the surface. When the OWLS pumps down, the oils often are retained when the low level is reached. Special pump designs are available to allow for removal of much of this oil, and / or vacuum truck access can be provided to remove surface oil.
By-pass for maintenance	Facilities for bypassing an OWLS should be considered, such as emergency pump connections on the underground gravity inflow line and on the outlet pump discharge line. In addition, provisions for placement of an isolation sluice gate on underground inflow lines to assist in a temporary pump-around may be practical.
Pump Failure	Confidential report (2008) states that OWLS pump failures are mainly due to: • Dry running: from dirt plugging the pump strainers or faulty level instruments. • Lack of lubrication: poor dry packing design or poor maintenance practices. • Vertical pump application: more prone to failure than horizontal self-priming pumps because of their suspended long shaft, which is more likely to wear against pump bushings and packing if dirt or vibration increases. Also more difficult to detect problems as components are below the surface and not easily accessible. • Lack of effective preventative maintenance (PM): even if an OWLS pump PM program exists; confirm that all pumps are covered.
	Maintenance goal should be to have Mean Time Between Failures (MTBF) for pumps above two years.
Temperature	From a materials of construction consideration, the temperature of any drainage should not be higher than the temperature rating of the drainage piping. Hot steam condensate is not recommended to be discharged to the

	OWDS.
	In the case that Potentially Contaminated Condensate (PCC) has been detected from a leaking exchanger, cooling would be required before any PCC would be approved for draining to an OWDS subject to flammability concerns. Collecting PCC in a CDS for pumping directly to a WWTP based dedicated cooler should be considered. Cooling to at least 140F is recommended in either case. For operational input: 1. In the event of any contamination of steam condensate with hydrocarbons, identify the source of contamination and isolate for correction. 2. If for any reason, hydrocarbon-contaminated hot steam condensate is directed to OWDS, then it must be cooled by dilution with utility or fire water.
<u>Solids</u>	Recognize that EPA HAZARDOUS WASTE CODE F037 pertains to Petroleum refinery primary oil/water/solids separation sludge –including any conveyances; sumps; and storm water units receiving dry weather flow. Also, in some regions fine dust/carbonate which produces a sticky paste when mixed with water and oils can be expected in an OWDS. Any sand or grit that collects in process areas and gets washed into an OWDS may mix with hydrocarbons and create sticky oily solids. One pound of dry solids can create ten pounds of oily sludge (IPIECA 2010). Solids issues favor the use of non-clog pumps, typically open impeller trash pumps, since they have been demonstrated to be reliable for handling oily wastewater and provide protection against oily waste causing pump failure.
<u>Firewater</u>	The OWDS function to remove fire water is essential to firefighting activities, so OWLS pumps must be set-up to operate when required. The primary concern is to contain firewater runoff onsite. Firewater within unit battery limits should gravity drain through the OWDS and collect in a local OWLS. Both the OWDS and OWLS in a combined system should be designed for firewater loads. For dual chamber OWLS, the excess inflow will overflow to the high flow chamber. When the level in the high flow chamber reaches the high flow pump start level, the high flow pumps will start and the low flow pumps stop. However, an operator can shut off the low flow pump and eventually rely on only the high flow pumps without loss of containment. The high flow pumps will discharge to storage, thus isolating the collected firewater from the WWTP treatment system. Note, the low flow pumps discharge directly to the WWTP treatment system but their relatively small flow rate will limit the impact of firewater on the WWTP. Once collected in storage, firewater can be analyzed and even pre-treated before discharging through the WWTP for treatment at a controlled rate to limit potential impact. Any significant free oil layer should be manually skimmed off before discharging back to the WWTP.

	The first flush approach while applicable to storm water, may not necessarily apply to firewater runoff which can remain contaminated throughout a fire fighting event. Therefore, the disposition of any collected firewater runoff will have to be determined on a case-by-case basis, and should be expected to depend on the specific firefighting scenario. Firewater can carry environmental contaminants, with a scenario including flammable and/or combustible materials, such as petroleum products, foam solution containing perfluorooctane sulphonates, dissolved and particulate materials from combustion processes, materials generated through quenching, as well as heavily contaminated solids and grit. Since the hazards and associated risks, such as unknown health effects and the impact on a biological treatment system are difficult to estimate in practice, both the liquids and solids should be considered as potentially hazardous wastes. Also, if the analyses indicate that the detained solids / sludge in an OWDS / OWLS after a fire event may not be recommended for treatment at the onsite WWTP then arrangements for offsite treatment and/or disposal via licensed waste contractors may be necessary. Note, even if the firewater runoff liquid is determined acceptable for the onsite WWTP, the solids may require a different disposal route.

4 CONCLUDING REMARKS

The management of oily water involves managing process safety hazards. Since all hydrocarbons can form an explosive atmosphere under the right temperature and pressure conditions; methods for mitigating and controlling risks associated with operating OWLS should be firmly established.

Safety hazards associated with OWLS can be mitigated through proper design, permits and procedures. The "safety culture" must extend to control auxiliary areas such as wastewater management systems; and should consider the OWLS as a hydrocarbon process facility in regards to HAZOPs and process safety analysis. The most labor-intensive task for OWLS is routine preventive maintenance (US EPA 2000c); however regardless of the excellence of a maintenance program, equipment use causes wear and ultimate failure or breakage of parts should be accounted for in planning.

5 REFERENCES

API RP 505, (Reaffirmed 2013), Recommended Practice for Classification of Locations for Electrical Installations at Petroleum Facilities Classified as Class I, Zone 0, Zone I, and Zone 2.

Bloch, H.P. (August 2015). Consider Zone 0 vertical pumps for slop pits. Hydrocarbon Processing. Vol. 94 No. 8.

Confidential Report, (2008).

IPIECA (2010), Petroleum refining water / wastewater use and management. Operations Best Practice Series. Refinery Water Management Task Force. Storm water and sewerage section.

Lamberti, P.G., JL Woodward. (March 2011). Learning from Events: Major Process Safety Incidents from Wastewater in Process Chemical Plants, Or Take Care of Your Wastewater or It Will Take Care of You. Process Safety Progress Vol. 30, No. 1.

Lees, F.P. (2012). Lees' Loss Prevention in the Process Industries (4th Ed) Elsevier. (a)Chapter 10 – Plant Siting and Layout, Section 10.17 Drain Systems.

(b)Chapter 17—Explosion, Section 17.28 Vapor Cloud Explosions

Mason, G.S., C. Arnold. (1984) Contain Liquid Spills and Improve Safety with a Flooded Storm Water Sewer. Chemical Engineering V 91 (Sept 17th), pp 105-109.

Ursino, N. (2014), P. Salandin. Combined effects of mechanical availability and storage capacity in pumping station risk assessment. Procedia Engineering V 70, pp 1679-1687.

US EPA, (July 1980) Treatability Manual Volume II Industrial Descriptions, Petroleum Refining Category Section14, EPA-600/8-80-042b

US EPA, (June 1992) Guidance to Protect POTW Workers From Toxic And Reactive Gases and Vapors, Section 3.1. EPA 812-B-92-001.

US EPA, (September 2000a) Wastewater Technology Fact Sheet – Sewers, Force Main, EPA 832-F-00-071.

US EPA, (September 2000b) Collections Systems Technology Fact Sheet – Sewers, Lift Station, EPA 832-F-00-073.

US EPA, (September 2000c) Wastewater Technology Fact Sheet – In-Plant Pump Stations, EPA 832-F-00-069.

12TH GLOBAL CONGRESS ON PROCESS SAFETY

Critical Considerations in the Pressure Relief Analysis of Refinery Sour Water Systems

James (Jim) R. Bridgett
james.bridgett@siemens.com

Robert (Bob) Siml
robert.siml@siemens.com

Gabriel Gandhi Martiniano Ribeiro de Andrade
andrade.gabriel@siemens.com

Siemens Energy, Inc.
4615 Southwest Freeway, Suite 900,
Houston TX 77027

Prepared for Presentation at
American Institute of Chemical Engineers
2016 Spring Meeting
12th Global Congress on Process Safety
Houston, Texas
April 10-14, 2016

UNPUBLISHED

Critical Considerations in the Pressure Relief Analysis of Refinery Sour Water Systems

James (Jim) R. Bridgett
james.bridgett@siemens.com

Robert (Bob) Siml
robert.siml@siemens.com

Gabriel Gandhi Martiniano Ribeiro de Andrade
andrade.gabriel@siemens.com

Siemens Energy, Inc.
4615 Southwest Freeway, Suite 900,
Houston TX 77027

Keywords: safety, pressure relief, two-phase flow, inherently safer design, sour water

Abstract:

Sour water systems are an integral part of all refineries. They are designed to collect sour water (i.e., water containing hydrogen sulfide, other sulfur compounds, and trace amounts of hydrocarbons) from numerous sources for treatment prior to disposal. Due to the complexity of these systems, the pressure relief analysis is inherently difficult. This paper discusses critical considerations in the pressure relief analysis, including the impact of light hydrocarbons entering the sour water system, design flaws that set traps for operators, deficiencies in operating procedures, and some critical considerations for the pressure relief analysis. Also presented will be potential mitigations for existing systems and recommended best practices for the design of new systems.

1. Introduction

Sour water collection and treatment systems are an integral part of all refineries. Although not a core production unit, the proper function of the system is important to the safe, reliable economical operation of a refinery. While it might be tempting to assume that non-core production units such as sour water systems are low risk and therefore don't require the same level of overpressure protection analysis and review as other parts of the refinery, the work recently done by Siemens to support the study of a process upset in a sour water system strongly indicates that is not the case. Since sour water systems inherently contain numerous high to low pressure transitions, the overpressure protection analysis is much more complex and difficult than might first appear. This paper will discuss in detail the lessons learned from our recent work and how those lessons can be used to strengthen the overpressure protection of sour water and similar systems. Improving the overpressure protection of these systems is a key part of reducing risks associated with operating these types of systems to as low as reasonably practicable (ALARP).

2. Initiating Event

A typical refinery sour water system consists of multiple sources, including streams that are pumped or pressured in on level control, as well as streams that are manually drained.

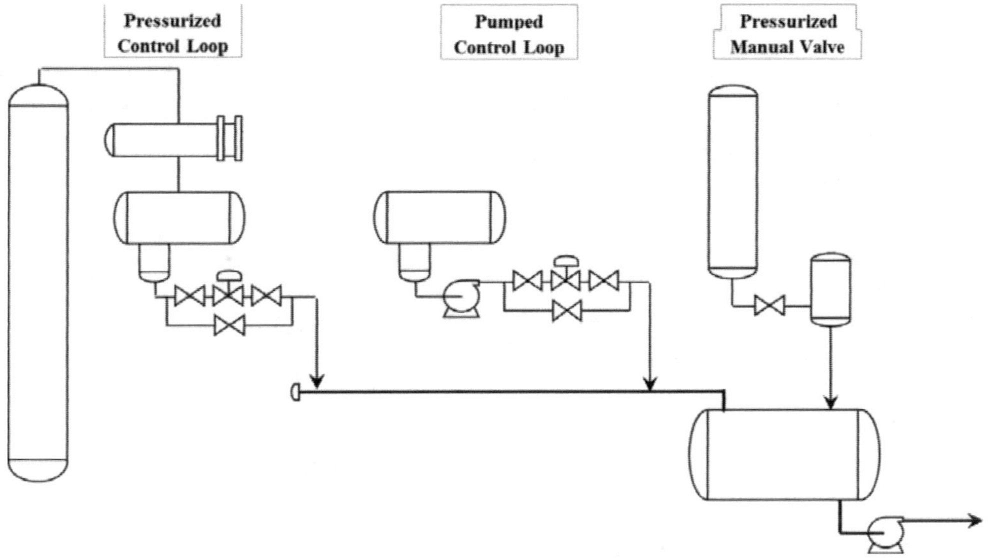

Due to a process upset, the feed to the column contained much more water than normal. Water was entering the overhead accumulator and separating out in the water boot faster than the interface level control could remove the water to the sour water system. Therefore the outside operator was asked to open the bypass valve around the interface level control valve, to reduce the water level.

It appears that the operator opened the bypass valve gradually, in steps, over a few minutes. The water level in the boot decreased in corresponding steps, and then sharply dropped to zero. At the same time the pressure in the sour water drum increased rapidly. Over about 30 seconds, the pressure increased from the normal operating pressure of approximately 10 psig to at least 55 psig. The actual pressure the vessel reached is not known because 55 psig was top of scale for the pressure transmitter on the sour water drum. As the pressure in the drum spiked, both the pressure control vent valve and the relief valve on the sour water drum opened as designed, releasing vapors and/or gases to the flare.

The pressure transmitter remained "pegged" at the maximum reading of 55 psig for another 30 seconds. The pressure decreased to approximately 15 psig over the next 4 minutes. As the pressure in the system dropped, the relief valve closed. Over the next 30 minutes or so, the pressure gradually dropped from 15 psig and stabilized at the normal system pressure of 10 psig. As the pressure dropped, the vent valve returned to normal operating position.

In response to the loss of interface level and the pressure increase at the sour water drum, the operator immediately closed the level control bypass valve. The interface level was re-established in 10 to 15 minutes.

3. Preliminary Review

The above information is based on both process data as captured by the data historian and information from operations. Siemens first conducted a preliminary evaluation to validate the process data and time line as described above. This evaluation was intended to answer the following questions:

a. What was the root cause of the pressure rise in the sour water system?
b. How much liquid hydrocarbon, if any, entered the sour water system?
c. What maximum pressure did the sour water system reach?
d. Did the time line make sense based on estimated flow rates through the bypass valve, etc.?

Since these questions can be answered using standard and well understood calculation techniques, detailed calculations will not be presented in this paper. Instead, the focus will be on the results and conclusions.

3.1 Root Cause of the Sour Water System Pressure Rise

Three potential root causes of the pressure spike in the sour water drum were considered:

a. Release of trace amounts of hydrocarbons, hydrogen sulfide and other dissolved gases from the additional sour water flow that occurred when the level control bypass valve was opened.

Based on the dimensions of the water boot, the maximum amount of sour water that could have been drained to the system was about 15 gallons (approximately 120 pounds). This volume of water is simply too small, and would have contained too little dissolved hydrocarbons, hydrogen sulfide, etc. to contribute significantly to the pressure rise in the sour water drum.

If the additional sour water was the only source of light gases and vapors into the system, this analysis indicates that the process vent valve would have had more than sufficient capacity to handle the event; as a result, the pressure would not have reached the relief valve set point.

b. High pressure hydrocarbon vapors from the overhead accumulator (i.e. "gas blow-by" scenario).

For a gas blow-by type scenario to occur, all of the liquid in the overhead accumulator would have to have been (drained) transferred to the sour water drum. Since the hydrocarbon level in the overhead accumulator did not drop to zero at any time during this upset, this ruled out the gas blow-by scenario as the root cause of the pressure spike in the sour water drum.

In addition, even though there was more than enough hydrocarbon in the overhead accumulator, the high level alarm in the sour water drum never activated, no frost line occurred, and the interface level was quickly reestablished.

c. *Liquid hydrocarbon from the overhead accumulator flashing as it entered sour water drum.*

Primarily by a process of elimination, the root cause of the pressure spike in the sour water system was determined to be a moderate amount of liquid hydrocarbon entering the sour water drum (see 3.2 for details). Due to the difference in pressure between the overhead accumulator and the sour water drum, the hydrocarbon flashes as it enters the sour water system. The process vent valve could not respond fast enough and/or did not have enough capacity by itself to vent the vapor volume generated to the flare, and as a result, the pressure in the system increased until the relief valve opened.

Additional evidence that supports this conclusion includes:

- Little to no change in the hydrocarbon level in the overhead accumulator during this process upset.

- The relatively short duration of the maximum pressure reading and the relatively quick return of the system to normal pressure.

- Interface water level quickly being reestablished.

3.2 How much, if any, liquid hydrocarbon entered the sour water system

Based on the available empty volume of the sour water drum and the documented pressure changes, this analysis indicates that about 100 pounds of hydrocarbon vapor was required to increase the pressure in the sour water drum from the normal 10 psig to 55 psig. This sets a lower limit on the amount of hydrocarbons that could have entered the sour water drum.

The total amount of hydrocarbon that entered the sour water drum, in addition to the amount required to pressure up the system, includes the following:

a. *Amount released to the flare through the process vent valve.*

Based on the flow coefficient (C_V) of the control valve and estimated vessel pressures plus the established time line, the best estimate of the total amount of material released to the flare through the vent valve is about 300 pounds. This estimate includes approximately 200 pounds over the 5 minutes as the pressure spiked and then decreased to 15 psig and at least 100 pounds over the next 30 minutes as the pressure dropped from approximately 15 psig back to the normal 10 psig.

b. *Amount released to the flare through the relief valve*

Since there was no indication that release was two phase or liquid, the flow through the relief valve was based on standard relief device vapor sizing equations using ASME certified parameters for the relief valve orifice area and coefficient of discharge. Based on the time line, the pressure relief device was open for less than 2 minutes. During this time, approximately 90 to 110 pounds of material was released through the relief valve to the flare.

Overall, it is estimated that approximately 500 pounds (110 gallons) of hydrocarbon drained from the overhead accumulator into the sour water drum during this upset.

3.3 What maximum pressure did the sour water drum reach?

Based on the recorded process data, it was known that the pressure in the sour water drum reached at least 55 psig. Since 55 psig was the maximum reading of the pressure transmitter, the actual pressure reached was not known. To address this question, a process simulator was used to plot the vapor pressure of the overhead accumulator hydrocarbon vs. temperature. Unfortunately, the temperature of the sour water drum was not available. Therefore, the temperature of the overhead accumulator, 85 °F, was used as the best available estimate of the temperature in the sour water drum. This temperature is consistent with cooling water temperatures expected during the time of year the upset occurred.

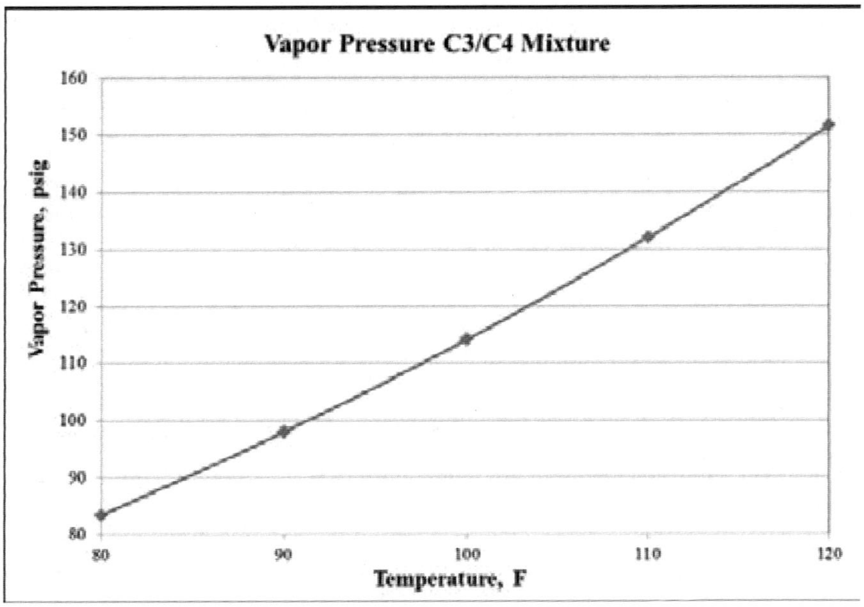

Using the estimated temperature of the sour water system, the maximum potential pressure in the sour water drum is approximately 90 psig when liquid hydrocarbon is present. Based on the capacities of the relief valve and process vent valve, the best estimate for the pressure in the sour water drum during this upset is 60 to 70 psig.

This pressure is just above the top of scale pressure for the pressure transmitter on the sour water drum and is an indication that the pressure transmitter was reasonably accurate.

3.4 Does the time line make sense based on estimated flow rates through the bypass valve, etc.?

As previously indicated, the estimated volume of sour water in the overhead accumulator boot was approximately 15 gallons. The interface control station consists of a 1 inch control valve and a 2 inch gate valve as the bypass. Although it is not unusual for gate valves to be used as a control valve bypass it is generally recognized that gate valves are not suitable for flow control. It is unusual to see an oversized gate valve as a bypass. Obviously, due to both size and flow characteristics, at the same pressure drop, the capacity of the gate valve is much greater than the control valve.

For sour water, the wide open capacity of the bypass was as high as 500 gpm based on the normal operating pressure in the overhead accumulator (185 psig) and the normal pressure in the sour water drum (10 psig). This estimate assumes little or no flashing occurs across the control valve.

For the hydrocarbon, the wide open capacity of the bypass was approximately 150 gpm based on the same pressure differential before the pressure in the sour water drum started to rise. This flow rate is much lower than the flow rate on water due to the lower density of the hydrocarbon and the flashing of the hydrocarbon liquid across the bypass.

In general, the estimated flow rates through the bypass valve are consistent with the rapid short duration of the pressure spike in the sour water drum.

Some time and effort was spent to investigate whether the flow characteristics of the gate valve also contributed to this process upset.

The number of turns from fully closed to fully open for a 2 inch gate valve varies between 11 and 13, depending on manufacturer. Definitive opening characteristics for gate valves were not readily available in the open literature, but it was generally noted that due to the geometry of the valve, the flow characteristics of gate valves are highly non-linear. There is some anecdotal evidence that gate valves may flow as much as 80% of capacity at only 20% open, but this is not well documented and should only be used with appropriate care and judgement as a relative indicator of the opening characteristics of gate valves.

While the evidence is not as definitive as desired it is doubtful that the bypass valve was ever fully open and was closed quickly once the interface level indication in the water boot went to zero (0) and the pressure spiked in the sour water drum. All indications are the 2 inch gate valve was grossly oversized and not suitable for control of flow out of the water boot.

4. Overpressure Protection Analysis

The following scenarios would typically be considered during the overpressure protection analysis of the sour water drum:

a. External fire
b. Blocked liquid outlet (overfilling)
c. Process vent control valve failing closed
d. Failure of automatic controls on each feed stream

Of these scenarios, a, b and c did not play a role in this process upset, and the analysis of these scenarios is fairly straightforward using standard overpressure protection assumptions and calculation methods appropriate to each type of scenario.

The failure of automatic controls cases were the primary focus for two reasons:

a. A failure of automatic controls and/or associated bypass played a central role in the process upset being studied.
b. Sour water systems inherently contain numerous high to low pressure interfaces at control valves or manual valves.

The failure of automatic controls scenarios typically include:

1. The control valve failing or being driven 100% open
2. The control valve bypass valve being 100% open
3. Some combination of control valve and bypass valve open, per client preference.

For all of these cases, the required relief rate calculation is based on:

a. Control valve flow coefficient (C_v) provided by the valve manufacturer
b. The pressure differential across the control valve. The typical assumption used in the calculation is normal upstream pressure minus downstream relief pressure.

When evaluating control valve and bypass scenarios the first step in the analysis is to determine if the liquid in the upstream system will fit in the downstream system.

If the liquid will fit, then the relief device on the downstream vessel should be sized for the vapor breakthrough or gas blow-by case.

If the liquid does not fit, then the relief device on the downstream vessel should be sized for the "liquid displacement" case. Liquid displacement is a phenomenon where vapor or two-phase material entering a liquid full system displaces the liquid. The relief device must be sized to handle the displaced liquid at the volumetric flow rate entering the system. Typically, sizing for liquid displacement requires a considerably larger relief device.

Liquid Displacement

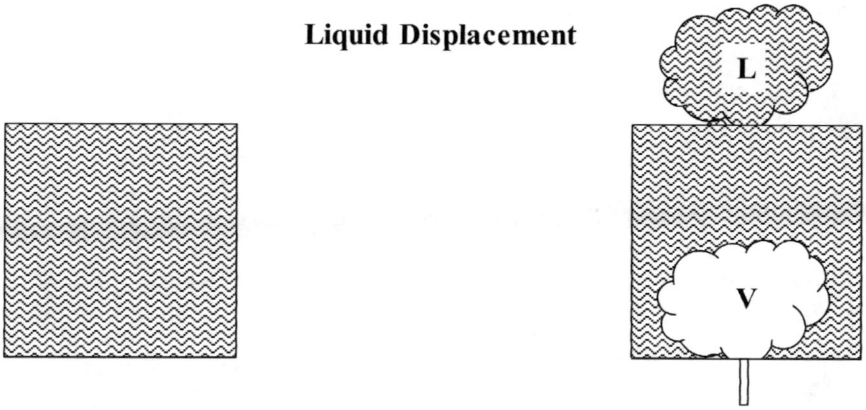

Liquid Full **Volumetric Flow In = Liquid Volume Flow Out**

In this particular case, the hydrocarbon is flashing at the lower pressure in the sour water system. Initially, the liquid will tend to separate out and accumulate in the drum. The flow through the relief device would only be vapor. Note that any liquid hydrocarbon that accumulates in the drum would continue to flash and generate additional vapor due to transfer of heat from the sour water. In general assuming all the liquid will vaporize is conservative. Evaluating the potential for liquid to collect and instantaneously generate a higher vapor rate due to mixing is very difficult to predict and is beyond the scope of this paper.

One of realizations of this analysis is as liquid separates out and begins to fill the drum, the space available for liquid-vapor disengagement is reduced. At some point, the vapor velocity in the drum will increase to the point that the liquid will not separate out and instead will be re-entrained out the relief device. This two phase release case can significantly impact the size of the relief valve. Therefore any process vessel with a source of two-phase flow must be evaluated for vapor-liquid disengagement and the relief system must be sized for two-phase flow if re-entrainment will tend to occur from the level rising.

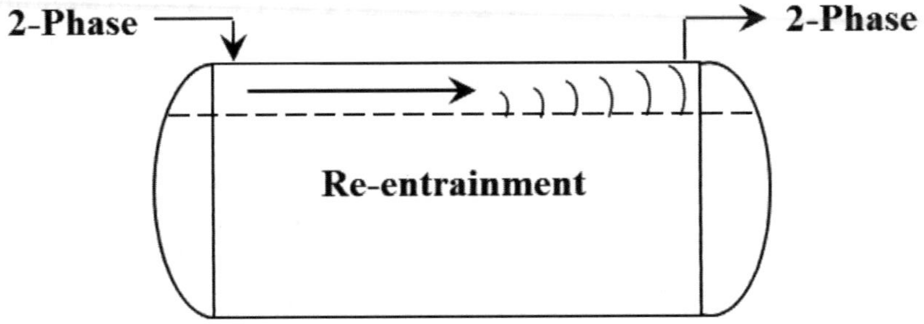

One of the most surprising realizations is liquid displacement will not occur due to the entry of two-phase flow when downstream vessel is not initially liquid full, even when there is enough liquid in the upstream system, because re-entrainment will occur first.

5. Conclusions

- Overpressure protection analysis of sour water systems in particular, and other similar collection systems in general, is not simple and straightforward. Particular attention must be paid to all high pressure to low pressure interfaces and ensure all credible scenarios have been identified.
- Liquid displacement will not occur when the flow entering is two phase if the vessel is not initially liquid full because re-entrainment will occur first.
- An oversized gate valve is not suitable for use as the bypass valve and can set a trap for operations.
- Any process upset is an opportunity for continuous improvement.

6. Recommendations

Corrective Actions for Existing Systems

- Check for oversized bypass valves:
  - Consider downsizing, installing a mechanical stop, or restricting orifice.
  - Consider replacing gate valves with globe valves.
- Review the overpressure protection analysis.
- Consider increasing the span of pressure transmitter.
- Consider recording process temperatures in the data historian

Inherently Safer Design for New Systems

- It's generally not practical to design the sour water system for max upstream pressure of every source. To the degree practical, design for the vapor pressure of the hydrocarbon at upper end of the sour water system operating temperature.
- In the design phase, balance the size of the water boot versus design of control and bypass valves to ensure the operator has adequate time to respond.
- Bypass valves for use in normal operation should generally be globe valves of similar size to the control valve.
- If oversized bypass valves are required for start-up, shut down, or other specific operating conditions, consider installation of a separate bypass valve that is car-sealed closed during normal operations.
- When reasonably practicable, the process vent valve should be sized to keep the relief valve from opening.

Disclaimer:

The information contained in this document represents the current view of the authors at the time of publication. Process safety management is complex and this document cannot embody all possible scenarios or solutions related to compliance. This document is for informational purposes only. Siemens makes no warranties, express or implied, in this paper.

For more information, please contact:

Siemens Energy, Inc.
Process Safety Consulting (SAF)
4615 SW Freeway, Suite 900
Houston, TX 77027
1 (800) 658-8809
+1 (713) 570-2940
Website: www.siemens.com/energy/oilandgasconsulting
Email: info.ogconsulting.energy@siemens.com

Practical Application of Advancing Process Safety Tools for the Petroleum Refining and Petrochemical Industries

Rodney Reibold
Marathon Petroleum Company
539 S. Main, Findlay, OH 45840
rrreibold@marathonpetroleum.com

Prepared for Presentation at
American Institute of Chemical Engineers
2016 Spring Meeting
12th Global Congress on Process Safety
Houston, Texas
April 11-13, 2016

UNPUBLISHED

Practical Application of Advancing Process Safety Tools for the Petroleum Refining and Petrochemical Industries

Rodney Reibold
Marathon Petroleum Company
539 S. Main, Findlay, OH 45840
rrreibold@marathonpetroleum.com

Keywords: AFPM, API, Advancing Process Safety, APS, Hazard Identification, Practices Sharing, Event Sharing, API 754, Human Reliability, Process Safety Site Assessments

Abstract

Over the past 5 years, the American Fuel & Petrochemical Manufacturers (AFPM) and American Petroleum Institute (API) have developed process safety tools for use by the refining and petrochemical industries to assist in improving process safety across the entire industry. The suite of programs are called Advancing Process Safety. This paper describes what industry is seeing in the way of process safety performance, the tools and programs available to these industries to help them improve performance, and practical ways companies are using the tools to improve performance.

1. Background and Development of Advancing Process Safety Programs

The petroleum refining and petrochemical manufacturing industry is committed to protecting the health and safety of its workers, contractors, neighbors, and the environment in which it operates. For decades, industry has worked together to develop safe work practices and standards aimed at reducing risks of major industry hazards and preventing process safety incidents. Despite these efforts, process safety incidents occur. As a result, industry, government, non-governmental organizations (NGO's) and the media have increased focus and attention on process safety.

In 2010, a group of refining and petrochemical industry executives belonging to the then NPRA (National Petrochemical and Refiners Association) now AFPM, tasked process safety professionals within their companies to develop a working group to explore ways to collectively improve process safety performance across the refining and petrochemical industries. The goal was to develop programs that would create a significant improvement in process safety performance. Shortly after work began, API was asked to join the effort to leverage the strengths and resources of both organizations to successfully develop and implement the programs. Over a two year period, the group developed a suite of six programs currently being utilized by the petroleum refining and petrochemical industries. These programs are Process Safety Metrics, Process Safety Event Sharing, Hazard Identification and Practices Sharing Program, Process Safety Regional Networks, Process Safety Site Assessment, and Focused Improvement. Together these programs are called the Advancing Process Safety (APS) program.

A key objective of the APS program is to develop tools and information, aimed at improving the industry's process safety performance, and to make these tools and information freely available to member companies. The APS program is able to use data streams unique to the program and specific to our industry to target areas for improvement.

This paper introduces and illustrates some of the tools and information available to member companies and how these tools and information can be used to raise process safety awareness and improve process safety performance.

2. Advancing Process Safety Data Collection and Analysis

AFPM and API members provide the following data on an annual basis to APS to be analyzed by industry experts:

- Facts regarding Process Safety Events (PSEs) per *ANSI/API RP 754 Process Safety Indicators for the Refining and Petrochemical Industries,*
- Tier 1 and 2 PSE summary reports to the Event Sharing database, and
- Results and observations from the API Process Safety Site Assessments.[1][2]

The following sections describe the data collected, the analysis process, and the resulting observations for industry improvement. This data analysis occurs on an annual basis. At the time of this paper, the most recent full year of data was 2014 process safety data, which was reported and analyzed in 2015. For the purposes of this paper, this will be referred to as the 2015 data analysis.

2.1 Data Analysis

Each group, Process Safety Metrics, Event Sharing, and Site Assessment individually analyze their data and present their observations during the annual Deep Dive Analysis meeting. The meeting participants consist of over forty process safety experts from member companies.

The results of the analyses are combined into one Deep Dive analysis to provide insight into opportunities for improvement that will be discussed later. It should be noted that data from each program (Process Safety Metrics, Event Sharing Database, and Site Assessments) are discrete data sets, and therefore some observations might vary from data set to data set. Trends observed in the 2015 data analysis are discussed in the following sections.

2.1.1 API 754 Process Safety Metric Data

The experts analyzed the 2014 Process Safety Metric data submitted by 59 companies representing 199 sites and were able to make several observations.[3] The data consisted of 921 events: 268 Tier 1 PSEs and 653 Tier 2 PSEs.

Figure 1 represents the four Tiers of performance indicators, as explained in API RP 754. For more information, please refer to Appendix A for API RP 754 Tier 1 and 2 definitions.

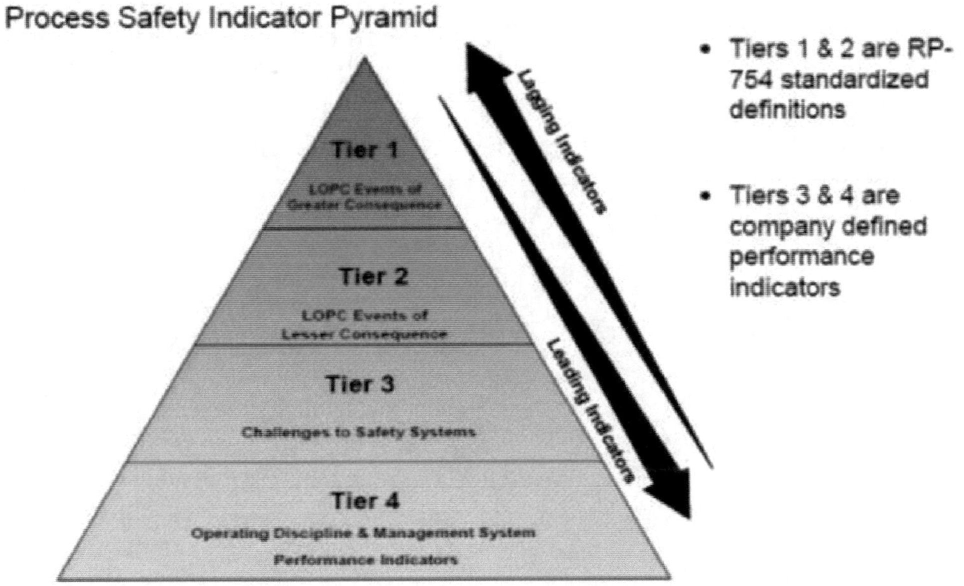

Figure 1. API RP 754 Process Safety Indicator Pyramid

Basic Observations:

- <u>Mode of Operations</u> – 70% of reported incidents occurred during Normal Operation with a majority of those events occurring during Steady State (72%). The next largest category of releases occuring during Normal Operation occurred with Loading and Unloading acvtivities (10%).
- <u>Point of Release</u> – Loss of Primary Containment (LOPC) from Piping Systems and Tanks accounted for 63% of the incidents reported (Piping Systems 49% and Tanks 14%). Figure 2 illustrates that the largest subset of releases from Piping Systems were caused by Piping and valve leaks.

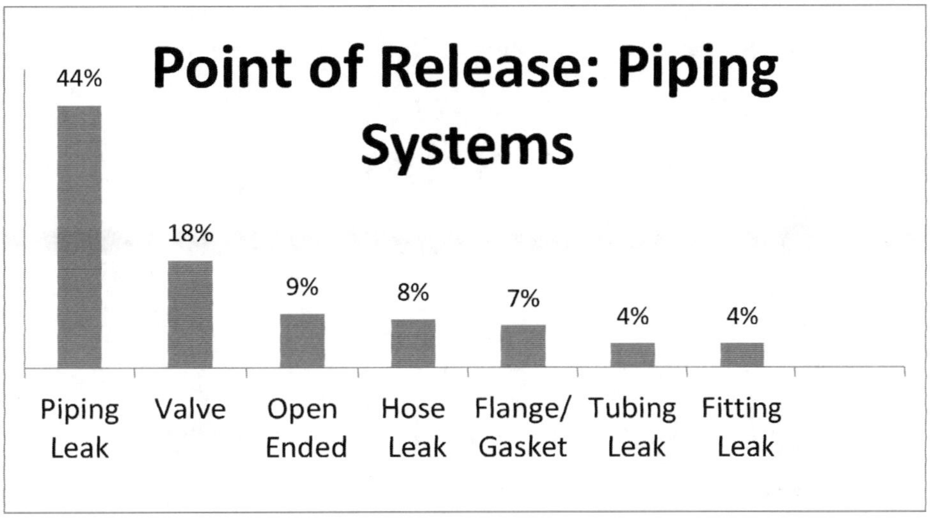

Figure 2. Points of Release from Piping System PSEs

- Piping Leaks – The top contributing factors to piping leaks were fixed equipment inspection, equipment reliability and design. Fixed equipment inspection issues are related to programs for detecting and mitigating metallurgical issues such as corrosion or cracking, before piping failure.
- Valves – Observations from incidents that involved releases from valves indicated that a majority of events were related to "Valves Left Open" and a majority of those events occurred during "normal operation". See Figure 3.

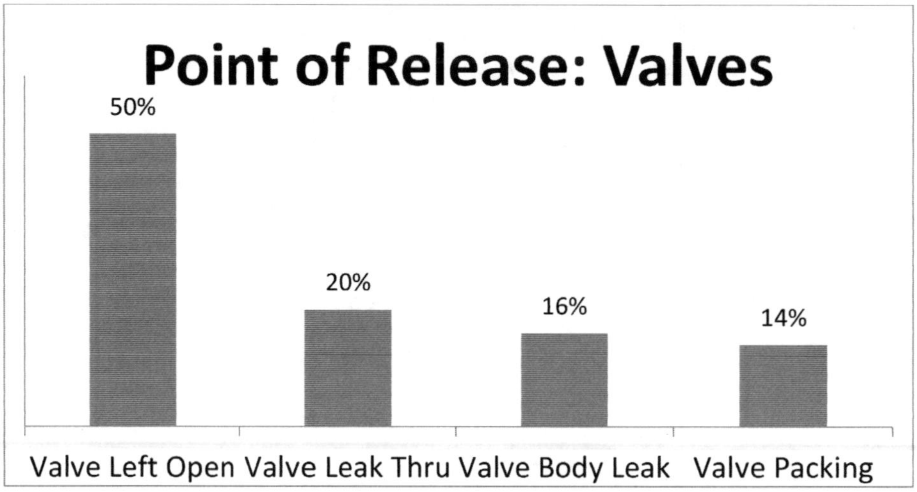

Figure 3. Valve Related PSE Subcategories

- Release from Atmospheric Tanks – Atmospheric tanks (including floating roof tanks) were the second largest point of release locations for PSE Tier 1 and 2 events after Piping Systems. A majority of the events occurred in the Tank Farm followed by the Sulfur Recovery unit. The largest group of those events were considered "overfill" events. Human factors and equipment reliability were the top causes for tank related events and human factors and procedures were the main causes contributing to the overfilling of atmospheric tanks. Figure 4 illustrates the types of releases from tanks.

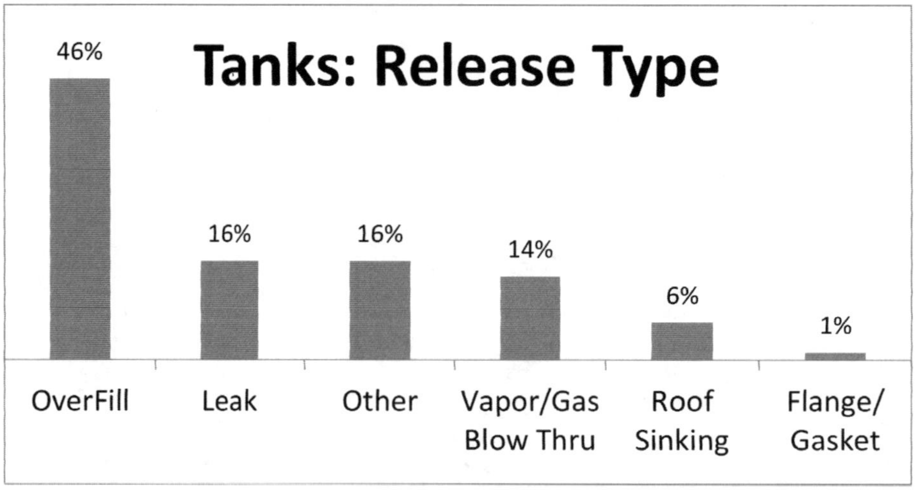

Figure 4. Release Type for Tank PSEs

- Causes – One of the limitations of API RP 754 statistics is that the data does not contain the cause of the incident as the member company determined during their incident investigations. The information provided under API RP 754 includes a synopsis of each process safety event that occurred. The team analyzing this data was able to identify some apparent causes for about 50% of the data, by reading the event description that was included. Although the analysis was based on the limited sample size, the team concluded that human factors, equipment reliability and inadequate fixed equipment inspection were the primary apparent causes of the events. Figure 5 shows the apparent causes of the PSEs, derived from the 2015 API RP 754 data analysis.

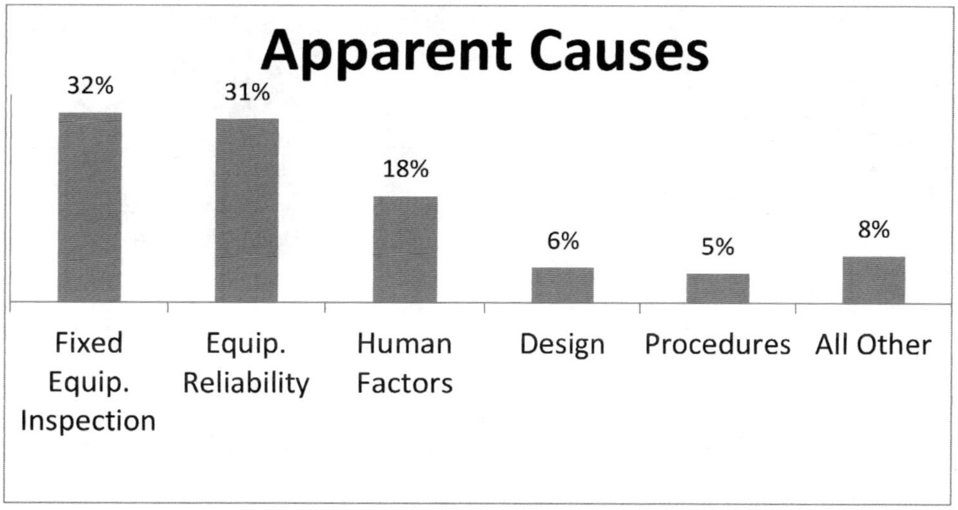

Figure 5. Overall Apparent Causes of PSEs

2.1.2 Event Sharing Database

The Event Sharing Database is a smaller subset of the API RP 754 data and provides more descriptors of the event, causes as determined by the member company incident investigations, and lessons learned.[4] At the time of the 2015 data analysis, approximately 428 process safety events were reviewed from the AFPM Event Sharing Database. Eighty percent of those events were API Tier 1 events that occurred from 1989 to 2014. Some general themes observed from the deep dive analysis include LOPC as a result of valves being left open or inadvertently opened, loss of primary containment originating from small bore piping and/or tubing (2" or less in diameter), and corrosion related LOPC events. Figure 6, illustrates the proportion of these Human Reliability events and how they break down into sub-categories.

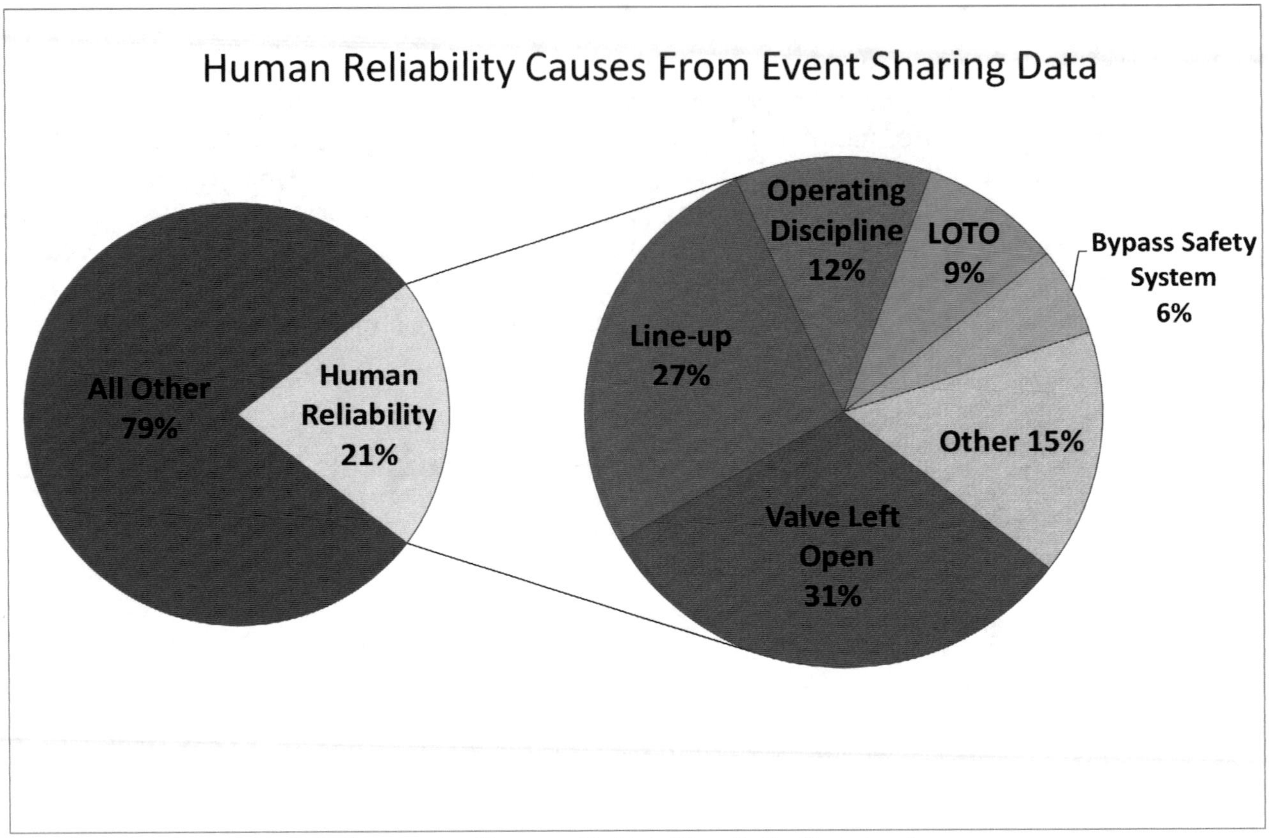

Figure 6. Human Reliability Sub-Categories

Open Valves – Twenty-one percent of the events from the database were identified as being caused by human reliability problems, including valves that were inadvertently left open to atmosphere or closed valves that were inadvertently opened. Nearly 1/3 of this subset involved bleeders, vents, and drains with 40% of the "open valve" events occurring during the startup of equipment.

Operating procedures and human factors (i.e., operating discipline) were identified in the database as the two most prevalent causes for these types of events. "Procedures" were often named as a cause because they did not mention the valve(s) in question or the specific task(s) being performed. In only a few cases were the procedures found to be unclear or just not followed.

Small Bore Piping – LOPC events involving small diameter piping continue to be prevalent in the 2015 data analysis with at least 19% of the reported events being related to piping or tubing 2" in diameter or less.

Piping Corrosion – Events associated with corrosion represented about 19% of events in the database. Corrosion continues to relate to a significant number of PSEs.

Winterization – Freeze-related events trended much higher in 2014 than historical rates. In 2014, 15% of events entered were associated with freeze-ups compared to 6% in the entire database.

2.1.3 Site Assessment Program

Observations from the Site Assessment program are shown in Figures 7 and 8.[2]

Basic observations:

- The lowest scores came from Mechanical Integrity (MI) and Process Hazards Analysis (PHA).
- When looking at the protocol categories, there appeared to be a wide distribution of scores for topics like shift turnover, operator rounds, and facility siting which indicates that some sites are doing those things very well and that others have an opportunity to learn from those sites.

A note about assessment scores: Higher is better, but the scoring isn't quite the same as a college exam. In the Site Assessment world, a score of 75% represents strong performance that meets the requirements of the assessment protocol. To achieve a 100% score, a program must show it meets or exceeds the requirements of the protocol, show a cycle of continuous improvement, must demonstrate sustainability, and show evidence of being embraced by the workforce.

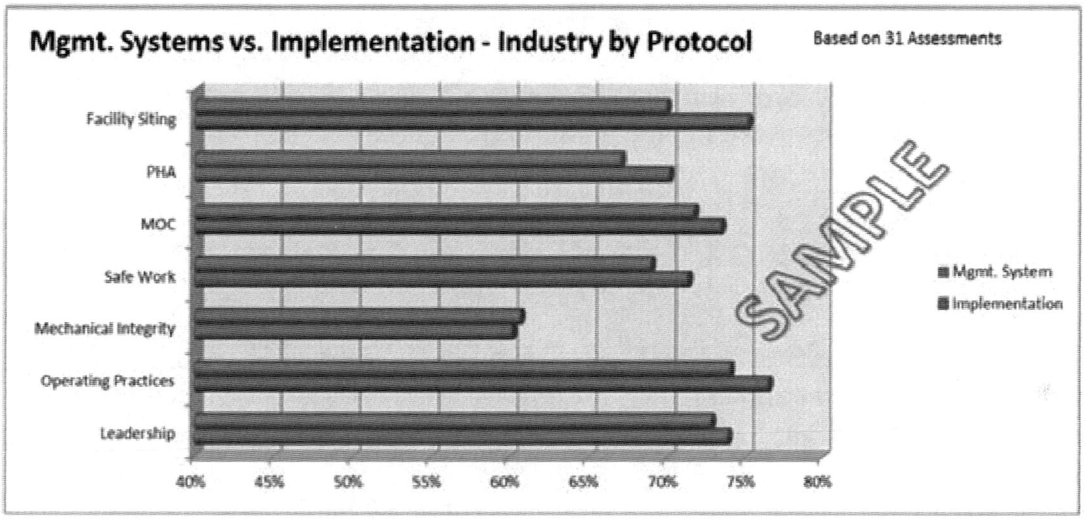

Figure 7. Management System versus Implementation by Protocol

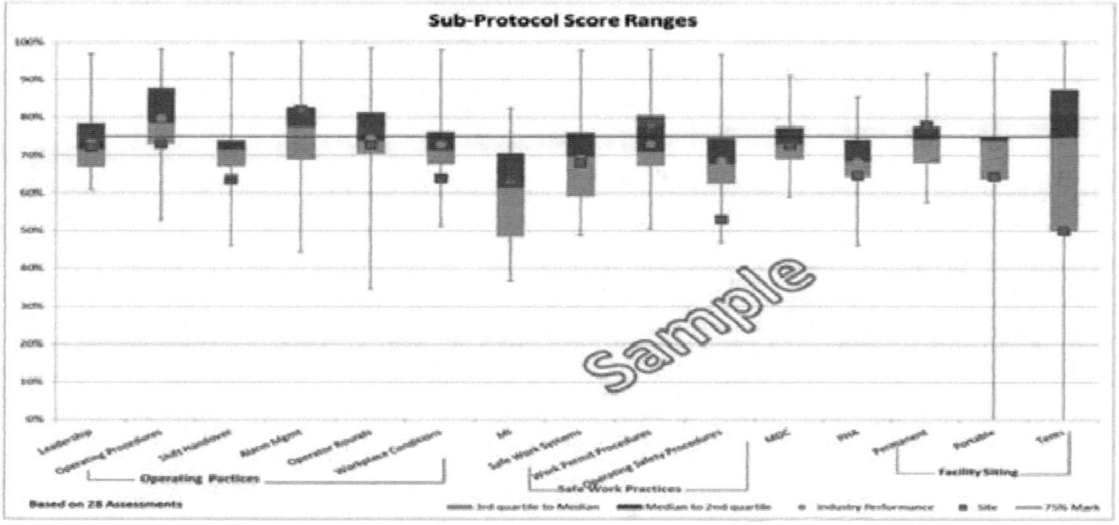

Figure 8. Site Assessment Sub-Protocol Scores

2.2 Using the Data to Improve

Does the data above look familiar? Are you experiencing similar types of events at your facility? It is likely that similar types of incidents ARE occurring at your processing facilities. The APS data indicate that these common themes are occurring over a broad cross-section of the industry; small plants and large plants, refineries and chemical plants, domestic and international. It is incumbent upon each company to collect, analyze, and make improvements based on their own data. The foundation of APS is to collect, analyze, and develop tools for improvement based on the industry's data.

The Advancing Process Safety program has used the data for industry improvement initiatives and to develop many tools for member company sites to use to improve their own performance. The following are a few actions that have been pursued based on the annual deep dive analysis:

- Developed and proposed a number of revisions to strengthen API 570 (Piping inspection code) and API RP 574 (piping inspection practices) in the areas of small bore piping.[5][6] The main idea is to increase awareness of small bore piping issues as part of a company's routine piping inspections procedures.
- Developed a number of Hazard Identification documents to address hazards associated with equipment, or work practices on equipment, uncovered in the data analysis. Examples include topics such as Piping Dead-legs, Winterization, and Small Bore Piping.[7]
- Developed a number of Hazard Identification documents and Shared Practices to address common personnel issues uncovered in the data analysis. Examples include topics such as Shift Change issues, Operator Communications, Operator Line-ups, and Operator Evaluation Rounds.[7]
- Developed a program called "Walk the Line" to address common Human Reliability issues. This includes prepackaged training materials, prepared tool box meeting outlines (scripts), informational newsletters, and videos.[8]

All of these materials and much more are available to member companies for use in their process safety improvement journey. The following section describes how information and tools developed by the APS Program can be used to improve your process safety performance.

3. Advancing Process Safety Tools and Ways to Use Them

The following sections outline the individual APS programs, tools, and information as well as how some companies are using them to improve their process safety performance. They are organized according the APS program where the information or tools are developed, and are freely available to member companies. It should be noted that this paper only discusses a few of the ways these tools can be used. In practice, there are many more ways that each of the tools discussed below can be used. The products described in these sections can be found on the AFPM Safety Portal website.[9]

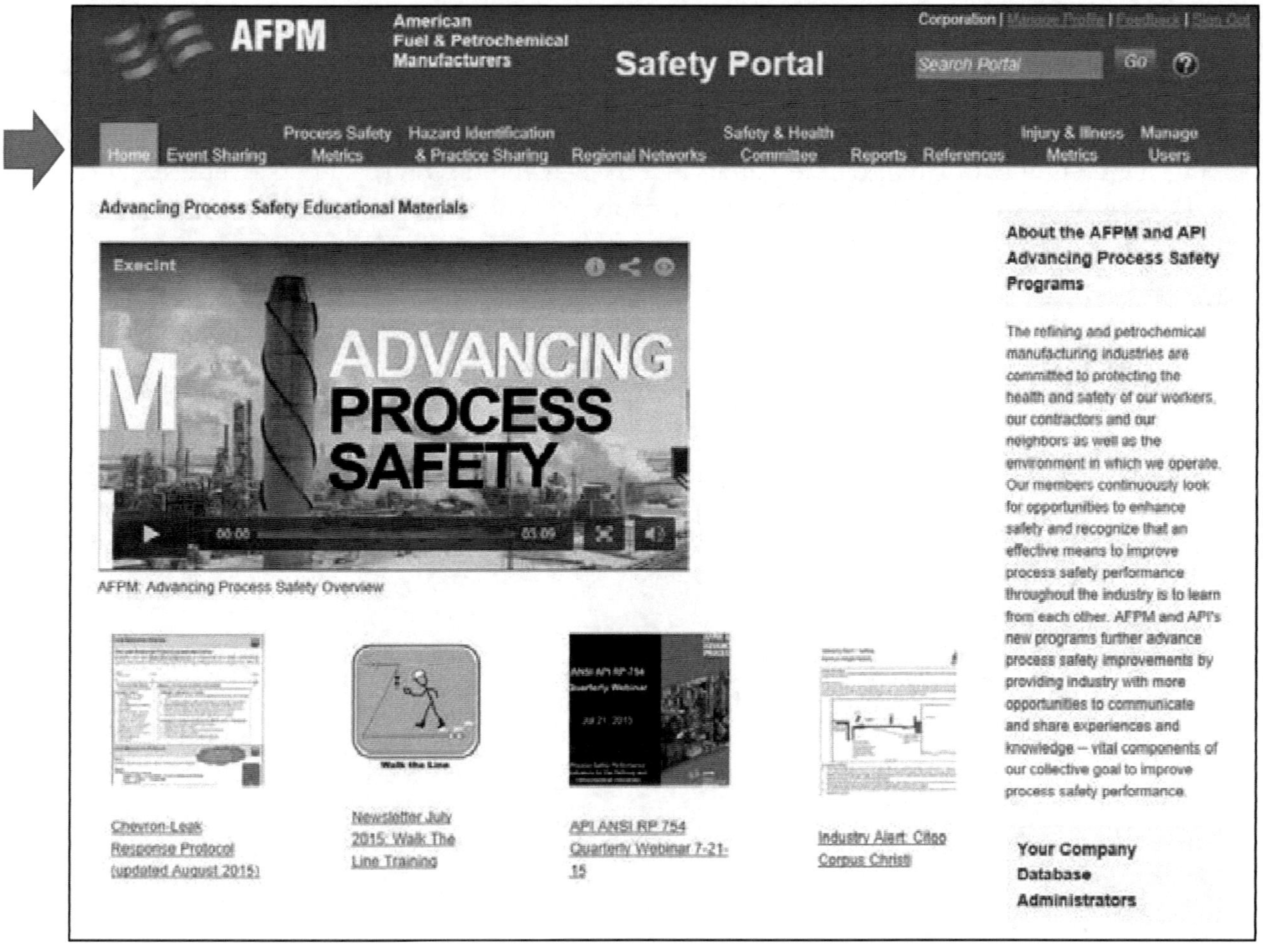

Figure 9. AFPM Safety Portal

3.1 API 754 Process Safety Metrics

The purpose of API RP 754 is to identify and track leading and lagging process safety indicators.[1] Knowledge of these indicators and taking appropriate action based on those indicators reduces the overall risk of major hazards in industry. On an annual basis AFPM and API publish a Process Safety Event Report and data analysis tool for both Tier 1and Tier 2 events.

The following illustrates a few practical examples of how APS member companies use Process Safety Metrics data.[3]

3.1.1 Benchmarking Against Industry

Figure 10 shows how benchmarking can be used by a company to compare its process safety performance to industry performance. Smaller companies, without a central corporate Process Safety Management (PSM) organization, may also use this tool to compare their sites to each other without need for a centralized compilation of data. Performance can be compared based on multiple criteria including the type of facility, size of facility, exposure hours, type of incident, incident year, or for specific process units. Comparisons can be done by both PSE count and PSE rate. In this example, a fictional company has four sites and is comparing its number of Tier 1 Process Safety Events over a three year period to other refining facilities of any size. The metric database is an extremely flexible benchmarking tool. By making simple choices from drop down menus, similar comparisons can be quickly created based on a wide variety of criteria. Figure 11 shows a comparison of Site D's performance on its fluid catalytic cracking (FCC) unit compared to other small facility FCC units. Figure 11 also shows that 44 of the FCC units at small facilities went the entire period without a Tier 1 or Tier 2 process safety event.

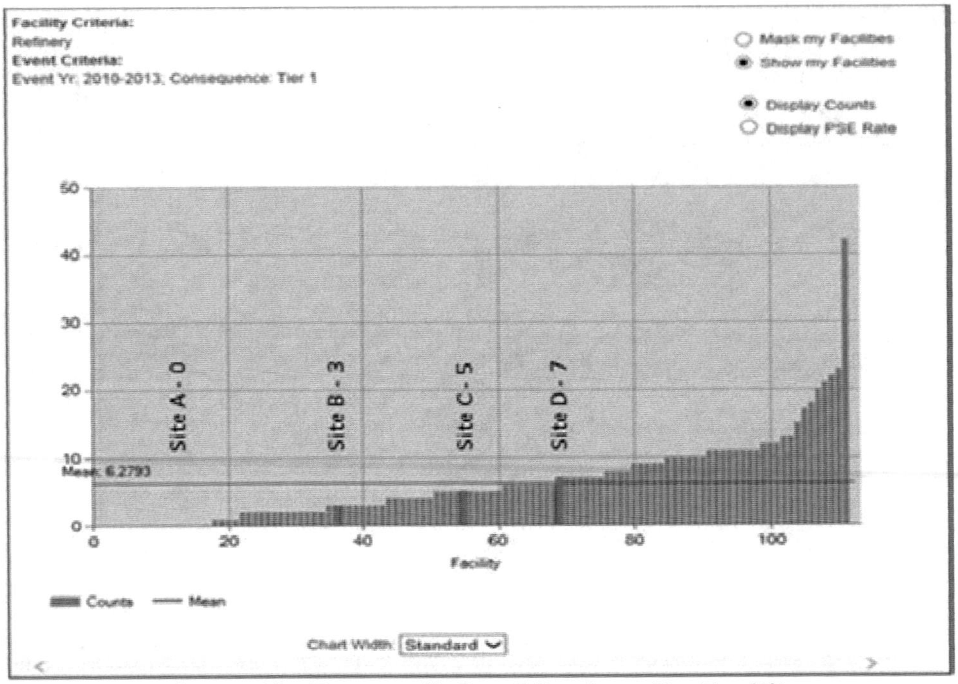

Figure 10. Process Safety Metrics – By Facility

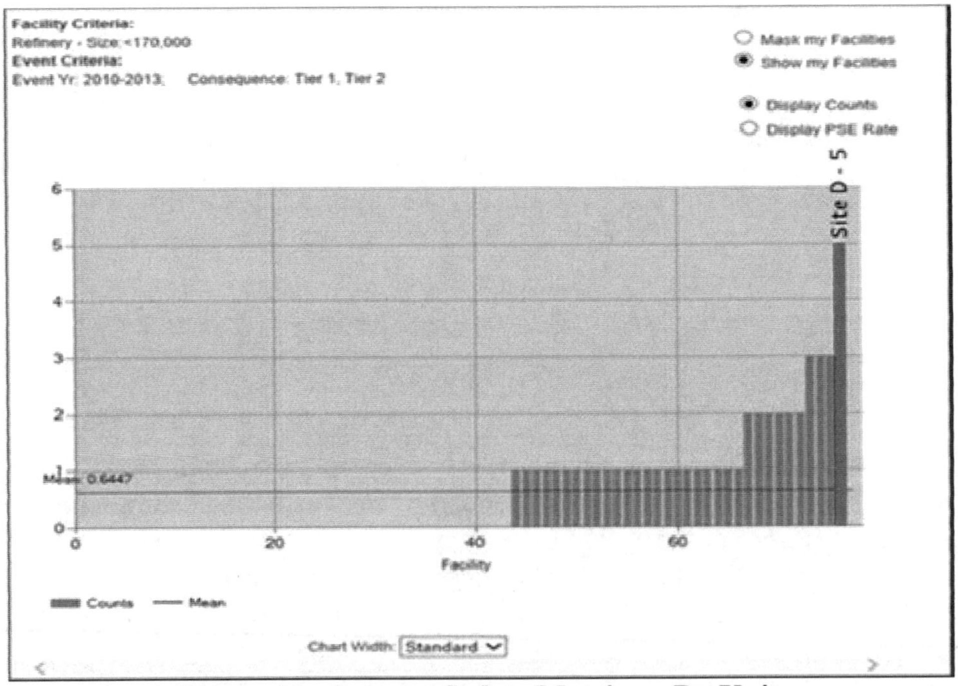

Figure 11. Process Safety Metrics – By Unit

- Goal Setting – Companies set process safety performance goals for their sites. This tool allows a company to identify what level of performance they will need to target to be an industry leader.

- Learning – The benchmarking tool can be used to identify successful practices within a company so that they can be shared across their sites. For instance in the example in Figure 10, Site D can use the tool to identify that Site A has implemented successful PSM programs and practices. Site D can then engage Site A to understand these programs and practices to implement at their own site.

- Prioritization of Resources – The benchmarking tool can be used to identify where resources need to be focused to improve PSM performance. For instance in the example in Figure 11, Site D would be able to identify that the large majority of their PSEs are occurring in their FCC unit, and that they are the worst performer in FCC process safety among similar sized sites. With this information, the site can target improvement efforts at the FCC. The site may also use this tool to identify which of their sister sites have strong FCC process safety performance, and direct resources to learning from the other sites and implementing existing successful practices to improve performance.

3.2 Event Sharing

The Event Sharing Database is a smaller subset of the API RP 754 Process Safety Metrics data and provides additional descriptors of the event, causes, and lessons learned. Where the main intent of API RP 754 data collection is **benchmarking**, the main intent of the APS Event Sharing program is **learning**.[4] Data submitted to APS includes a description of the event and equipment involved, as well as the cause(s) and learning(s) identified by root cause analysis performed by the member company. Individuals have the opportunity to search the database for specific information or export the database in a variety of forms. Key information about the event, cause(s), and learning(s) are clearly laid out, as is other information not shown here. See Figure 12. Other tools derived from the database are Safety Bulletins and the Annual Learnings Report.

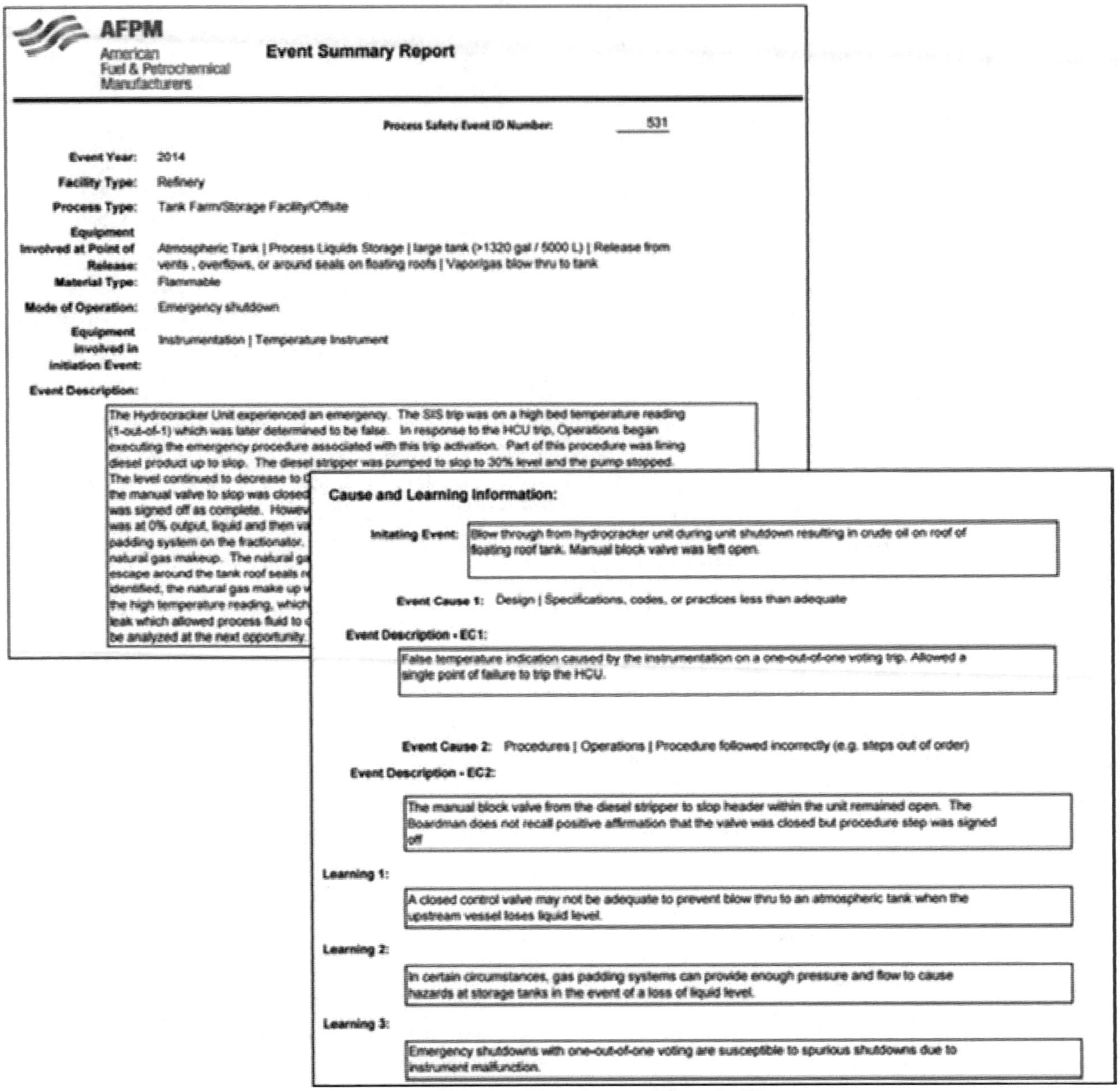

AFPM
American
Fuel & Petrochemical
Manufacturers

Event Summary Report

Process Safety Event ID Number: 531

Event Year: 2014

Facility Type: Refinery

Process Type: Tank Farm/Storage Facility/Offsite

Equipment Involved at Point of Release: Atmospheric Tank | Process Liquids Storage | large tank (>1320 gal / 5000 L) | Release from vents , overflows, or around seals on floating roofs | Vapor/gas blow thru to tank

Material Type: Flammable

Mode of Operation: Emergency shutdown

Equipment involved in initiation Event: Instrumentation | Temperature Instrument

Event Description:

The Hydrocracker Unit experienced an emergency. The SIS trip was on a high bed temperature reading (1-out-of-1) which was later determined to be false. In response to the HCU trip, Operations began executing the emergency procedure associated with this trip activation. Part of this procedure was lining diesel product up to slop. The diesel stripper was pumped to slop to 30% level and the pump stopped.
The level continued to decrease to
the manual valve to slop was closed
was signed off as complete. Howev
was at 0% output, liquid and then v
padding system on the fractionator.
natural gas makeup. The natural g
escape around the tank roof seals
identified, the natural gas make up
the high temperature reading, which
leak which allowed process fluid to
be analyzed at the next opportunity.

Cause and Learning Information:

Initating Event: Blow through from hydrocracker unit during unit shutdown resulting in crude oil on roof of floating roof tank. Manual block valve was left open.

Event Cause 1: Design | Specifications, codes, or practices less than adequate

Event Description - EC1:

False temperature indication caused by the instrumentation on a one-out-of-one voting trip. Allowed a single point of failure to trip the HCU.

Event Cause 2: Procedures | Operations | Procedure followed incorrectly (e.g. steps out of order)

Event Description - EC2:

The manual block valve from the diesel stripper to slop header within the unit remained open. The Boardman does not recall positive affirmation that the valve was closed but procedure step was signed off

Learning 1:

A closed control valve may not be adequate to prevent blow thru to an atmospheric tank when the upstream vessel loses liquid level.

Learning 2:

In certain circumstances, gas padding systems can provide enough pressure and flow to cause hazards at storage tanks in the event of a loss of liquid level.

Learning 3:

Emergency shutdowns with one-out-of-one voting are susceptible to spurious shutdowns due to instrument malfunction.

Figure 12. Event Sharing – Sample Event

There are a number of ways companies use information gathered through Event Sharing submissions including:

- PHA Preparation and Conduct – One tenet of conducting a process hazard analysis is to identify and evaluate incident scenarios which could have catastrophic consequences. The Event Sharing Database is a convenient tool for supplementing a site's own incidents to ensure hazard scenarios from the broader industry are identified and evaluated.[4] The database can be queried to extract events related to the process unit in question. The query can be further refined to pinpoint a specific type of equipment (e.g. Fired Heater) or a mode of operation (e.g. Start-up). The events can be prescreened by the PHA leader and appropriate events selected for review by the PHA team. The team can consider the database events as well as the causes and learnings as they evaluate hazard scenarios, causes, and safeguards.

- Procedure and Safe Work Practice Development – A company or facility can use the database when developing procedures or practices. The database can be queried to find events that relate to a specific unit and further refined to find events where procedures or safe work practices were listed as a cause. The company or site can then use the information in these events to ensure they have adequate measures incorporated into their procedures or practices to prevent a similar incident from happening.

- Mode of Operation Scenarios – In preparation for starting up a unit after turnaround, an operations foreman can query the database to find events on the same type of unit which occurred during start-up. The foreman can use the learnings from the event(s) to ensure the same incident does not happen during the pending start-up. For example: Many events during start-up are related to bleeder valves left open when feed is started to the unit. The Foreman could use this information to verify that a unit walk down occurs prior to start-up which looks for open bleeder valves.

- Design Reviews – When designing a new unit or making modifications to an existing unit, a design engineer can use the database to find events on the same type of unit where Design was listed as a cause. The engineer can then take appropriate measures to ensure their design will not be susceptible to a similar event.

- Tool Box Topics – The Event Sharing database is an excellent source of events to use as tool box topics or "What-if" drills. An event for a pertinent unit or task can be selected for discussion of what happened, how to respond to a similar event, and how to prevent a similar event from occurring in the first place.

3.2.1 Process Safety Bulletins

There are times when it is important to highlight key and/or urgent information to companies when a high learning value event or trend has been identified. Each Bulletin contains detailed information about events, causes, and lessons. Process Safety Bulletins are located in the Safety Portal under the Reference tab. Figure 13 is a sampling of recent bulletins which were issued to member companies.

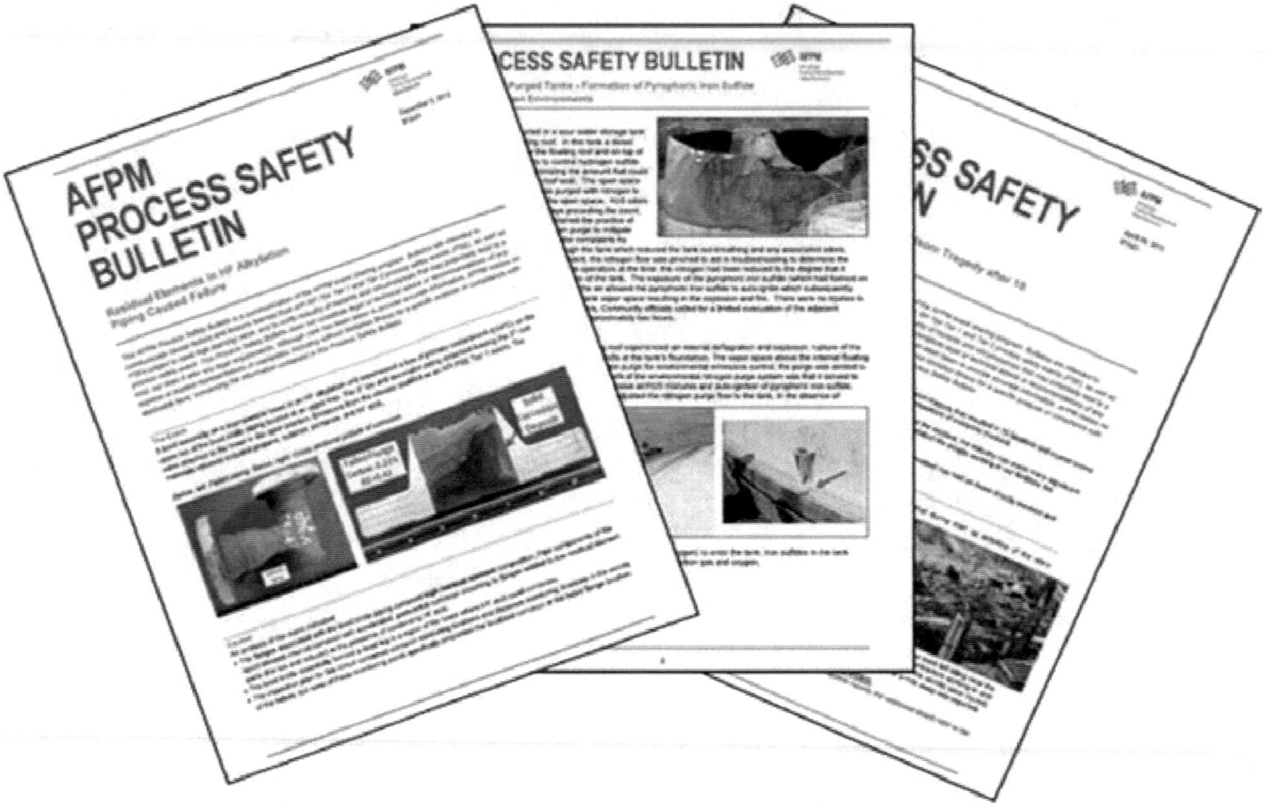

Figure 13. Process Safety Bulletin - Samples

Companies can use Process Safety Bulletins in a number of ways, including:

- Awareness – A company may use the bulletins to inform personnel of a serious incident that has occurred or a risk that exists, which could affect them.
- Standards, Specifications, and Practices – Subject matter experts can use the bulletins to evaluate the Standards, Specifications, and Practices for their company or facility to make sure that they have adequate requirements, features, and/or safeguards to prevent a similar event from happening at their company.
- PHA Preparation and Conduct – The process safety bulletins are appropriate for use by PHA teams to identify hazards and ensure adequate safeguards are in place.
- Leadership/Management Review – A facility leadership team can review the content of the bulletins to increase awareness of hazards and to determine if action at the facility is necessary to prevent future incidents. Any actions can be assigned and tracked to completion by the facility.
- Tool Box Topics – Like the Event Sharing database, the Process Safety Bulletins are an excellent resource to use as tool box topics or "What-if" drills. A pertinent bulletin can be selected for discussion of what happened, what the hazard is, how to respond to a similar event, or how to prevent a similar event from occurring at your site.

3.2.2 Annual Learnings Report

The Annual Learnings Report contains the results of an annual data analysis that identifies event trends in the industry. Figure 14 is a sampling of a recent Learnings Report.

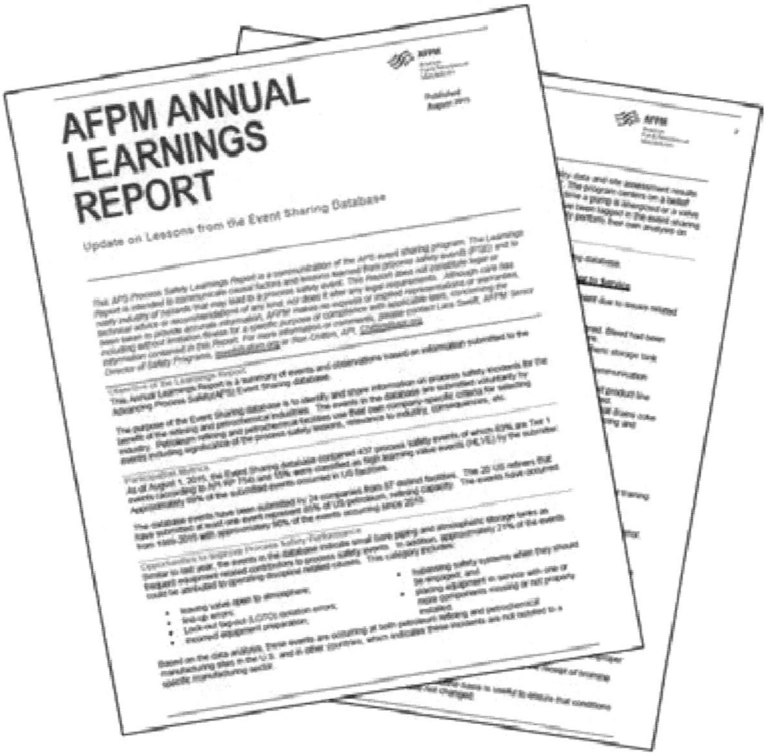

Figure 14. Event Sharing Annual Learnings Report

In addition to the same uses as outlined for the Process Safety Bulletins, the Learnings Report is also an opportunity for introspection and a chance to recognize "blind spots." A company can consider that "If this is an industry trend, is it also happening at our facilities?" Where appropriate, the company may choose to take measures or actions intended to prevent the events from happening at their facilities.

3.3 Hazard Identification

The AFPM Process Hazard Identification program supports the sharing of process safety hazards.[7] The program provides information and ready guides for potentially overlooked and not widely known process safety hazards in order to improve process safety awareness. The goal is to create tools that capture and share knowledge that could be used by other companies or sites when developing new process safety practices or improving existing ones. A sample Hazard ID document is shown in Figure 15. The Hazard ID documents are located on the AFPM Safety Portal.[7] Current Hazard ID documents are shown in Figure 16.

Each Hazard ID document contains sections with information on:

- Purpose and Use of the document,
- Scope,

- Examples of Inherent Concerns or Hazards,
- Detailed breakdown of potential causes for the hazards,
- A listing of available references and resources, and
- Links to known industry events related to the hazard.

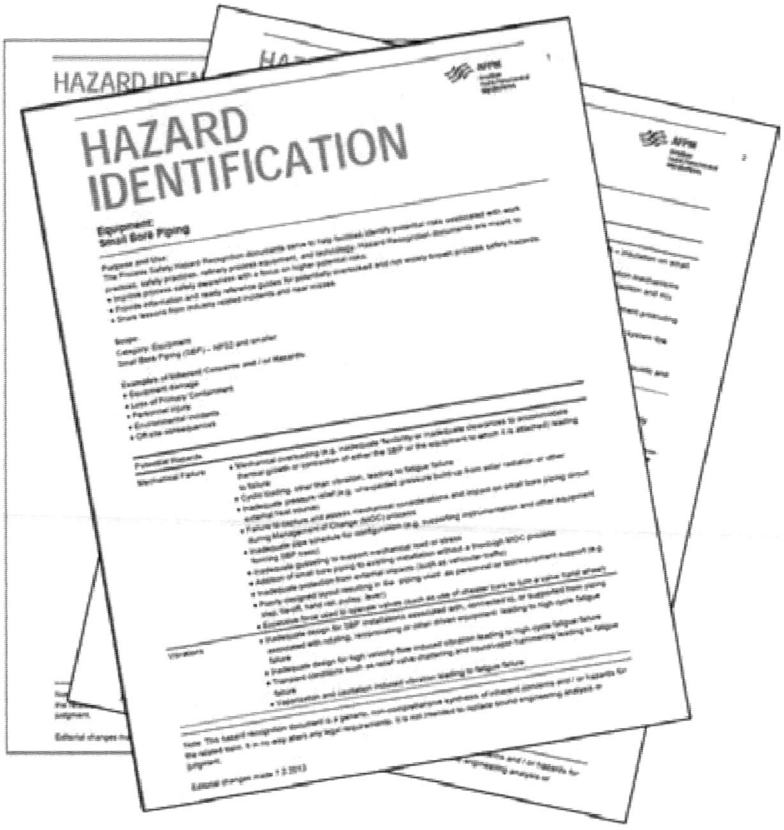

Figure 15. Sample Hazard Identification Tool

Hazard ID documents can be used in a number of ways, including:

- PHA – The Hazard ID documents are useful during hazard reviews as a check to make sure that no credible hazards were missed or omitted by the review team. The Hazard ID documents can be prescreened by the PHA leader and appropriate documents selected for review by the PHA team. The team can consider these hazard identification tools as they evaluate hazard scenarios, to ensure no credible hazards are missed, and to verify they have adequate safeguards or controls to address the identified hazards. The References and Resources section can be used to help the team determine what actions to take if they feel the existing safeguards are not adequate.
- Pre-Job Hazard Assessments – A number of the Hazard ID documents are associated with recurring tasks in a Refining environment such as vacuum truck operation, sampling, lighting or tuning a fired heater, live flare work, etc. These

documents are great tools for a pre-job hazard review to facilitate discussion and calibrate a technician before they go do a task. They are brief, to the point, and include a robust listing of potential hazards. During the meeting the Hazard ID document can be used to discuss how each potential hazard will be mitigated during the upcoming task.

- Procedure and Safe Work Practice Development – A company or facility can use the Hazard ID tools when they are developing procedures or practices. The Hazard ID portal can be reviewed to identify documents that are applicable to procedures or practices that are being developed, or are up for their periodic review. The company or site can then use the information in these documents to ensure they have adequate measures incorporated into their procedures or practices to mitigate the identified hazards. Another way to use these tools would be to stay abreast of new Hazard ID documents as they are created, and proactively complete a gap assessment with your existing management systems to ensure all identified hazards are adequately addressed.
- Design Reviews – When designing a new unit or making modifications to an existing unit, a design engineer can review the Hazard ID tools during the design to ensure hazards are identified and mitigated.
- Training Development – The Hazard ID documents can be useful tools when preparing training materials. One key aspect of refinery training is to include an overview of the hazards of the process. These tools are a great way to make sure all of the bases are covered in this portion of the training.

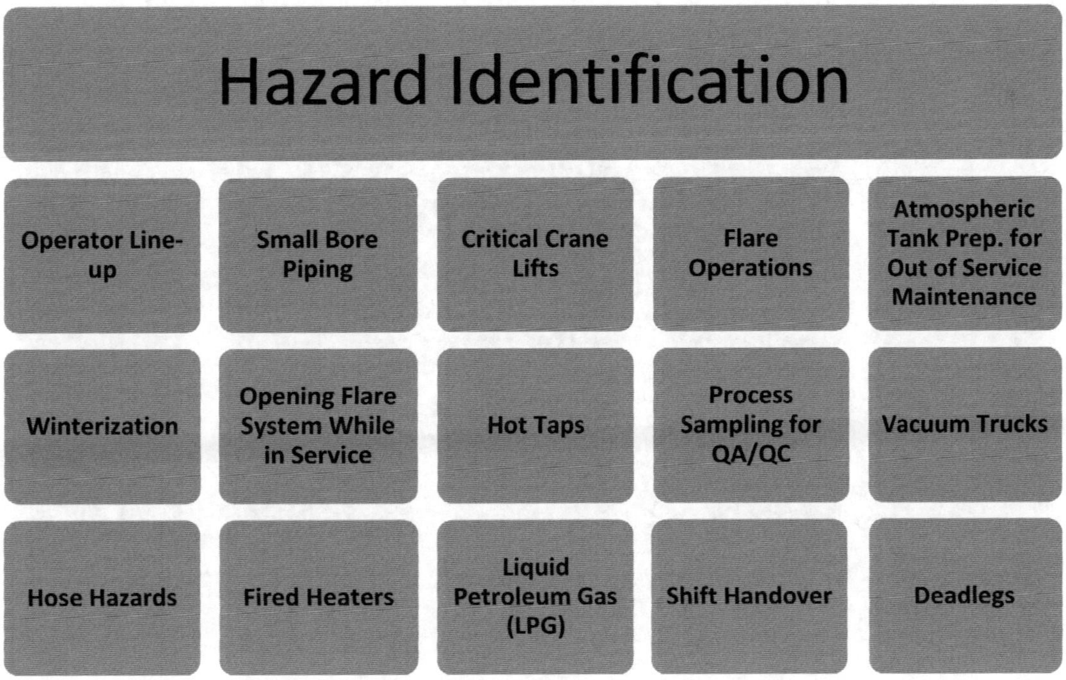

Figure 16. Currently Available Hazard ID Documents

3.4 *Practice Sharing Documents*

Practice Sharing Documents are designed to share information on process safety practices in order to help improve process safety performance and awareness throughout industry.[7] The goal is to capture and share knowledge that could be used by other companies or sites when developing new process safety practices or improving existing ones. The Practices being shared have been successfully used by an industry member, but this does not mean they are right for everyone, or that they will produce similar results at your site. Rather, they are an option to consider when implementing or adjusting programs and practices at a site. Figure 17 shows a sample Practice Sharing document.

Each Practice Sharing document contains sections with information on:

- Purpose and Use of the document,
- Scope,
- A description of the practice,
- Implementation suggestions, and
- References and Resources.

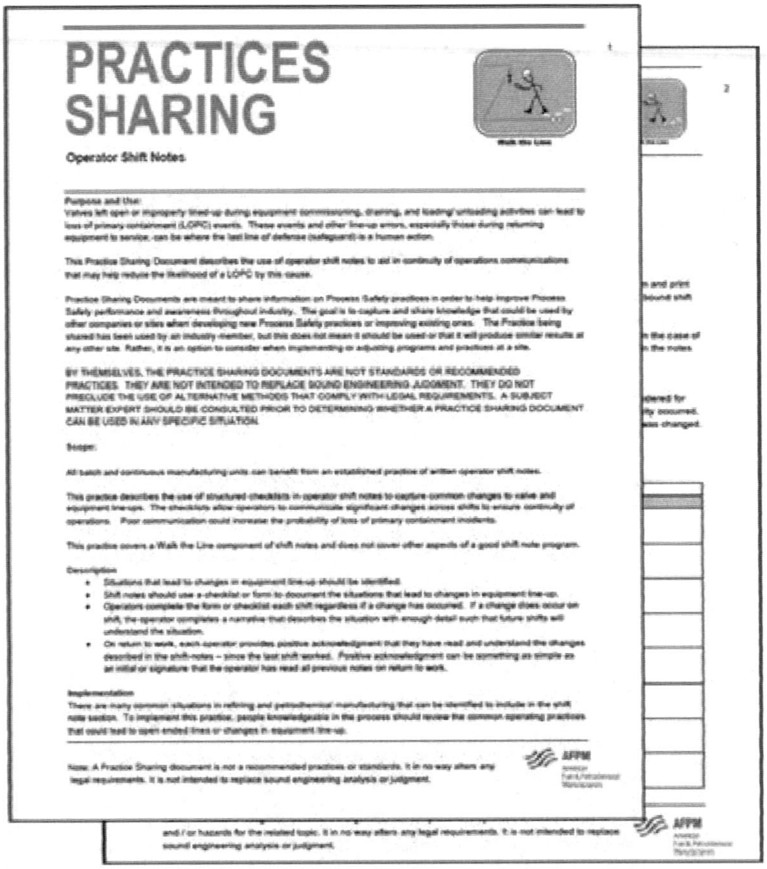
PRACTICES SHARING

Operator Shift Notes

Figure 17. Sample Practices Sharing Tool

Practice Sharing Documents can be used in the following ways:

- Procedure and Safe Work Practice Development – A company or facility can use the Practice Sharing tools when they are developing or updating procedures or practices. A shared practice can be helpful when a facility is developing a new practice, by providing related concepts so that the facility is not starting from scratch. Similarly, the practice shared could be consulted by a site developing a related practice, to see if there are any aspects in the shared practice that the site might want to implement. Where a company or facility has an existing practice of their own, the shared practice can be useful during the periodic reviews of the existing practice to determine if there are any useful concepts to incorporate from the shared practice.
- Incident Investigation – When a site has completed a root cause investigation and is looking for ways to address a gap in site practices, a relevant shared practice could be a useful tool for developing a new site practice or modifying an existing one.

Practices Sharing is a new program developed in 2015. Currently there are nine Practices Sharing Documents available:

- Critical Bleeder Valves,
- Threaded Lanyard Plug,
- Operating Instructions,
- Operator Shift Notes,
- Shift Change Meeting,
- Shift Change Communications,
- Process Safety Field Audit,
- Operator Evaluation Rounds, and
- Informal Unit Walk-Through.

More practices will be identified and shared as time goes on. These practices can be located on the AFPM Safety Portal.

3.5 *Process Safety Site Assessments*

The Process Safety Site Assessment (PSSA) Program is a collaborative industry effort managed by API, focused on higher risk process safety activities in petroleum refining and petrochemical facilities.[2] This is **NOT** a run-of-the-mill compliance audit! This program primarily involves the assessment of a site's Process Safety systems by a small team of independent and credible third-party industry-qualified process safety assessors. Each assessor is an industry veteran subject matter expert with decades of experience. The expertise of the assessors is the strength of this program and what makes it so successful. Through the use of industry-developed protocols that span over 1,000 questions in total, the site assessments evaluate both the quality of the written programs and the effectiveness of field implementation in eight focus areas: Leadership, Management of Change, Mechanical Integrity, Safe Work Practices, Operating Practices, Facility Siting, Process Hazards Analysis, and HF Alkylation/API RP 751. More information about the Process Safety Site Assessment program is available HERE.

There are two primary ways to use the Site Assessment Program

Actual Assessment

When you have a Site Assessment performed at your site, you will get the benefit of seasoned veterans, looking for soft spots in your critical programs, using an industry developed protocol. The site assessment protocols are not just a check-the-box verification of compliance with the PSM regulations, but far more. The site assessment team evaluates your written programs and how your facilities implement them, using over 1000 very specific protocol questions. The resulting report provides a scoring which shows your performance in overall programs, sub-programs, and also drills down to very specific protocol provisions. The team also provides the facility management with a benchmarking report that compares your performance against the industry or against sister sites within your company. Figure 18 shows a sample page from an assessment protocol. One of the many benefits of a site assessment is having access to industry experts in each area; it's "like going to school." While they are on-site, they are also available to answer questions and engage in discussions on topics pertinent to the protocols.

Benefits of a Site Assessment Include:

- Identification of soft spots in your programs or their implementation.
- Suggestions for how to improve.
- Access to hundreds of years of combined expertise from the assessors. It's almost a training opportunity for your people.
- Obtaining a scored report and benchmarking against the industry.
- Having multiple sites assessed and use the scores to benchmark them against each other, to identify strong programs that could be shared from one site to another.
- Satisfies API 751 audit requirements with an additional audit.

Self-Assessment

Conduct a self-assessment, using subject matter professionals from your company. The same protocols used by the expert assessors are available, to member companies to use in self-assessments. These free tools are an excellent way to evaluate some of your site's critical programs. You won't get benchmarking, but you can score your own assessment to find areas for improvement. It's as simple as this: (1) APS member companies request the protocols from the API PSSA webpage, (2) put a team of your own employees together to apply the protocols to your site, (3) identify areas where improvements could be made in either your management systems or their implementation, and (4) go make the selected improvements.

Benefits of a Self-Assessment Include:

- The protocols alone can serve as roadmap for a successful program.
- Assessing your programs against the items suggested in the protocols can identify weak areas in your programs or their implementation.

- The protocols can also be used when developing new management systems as a check to ensure key aspects are covered in the program or practice being developed.
- Provide a standardized way to compare multiple sites within a company.

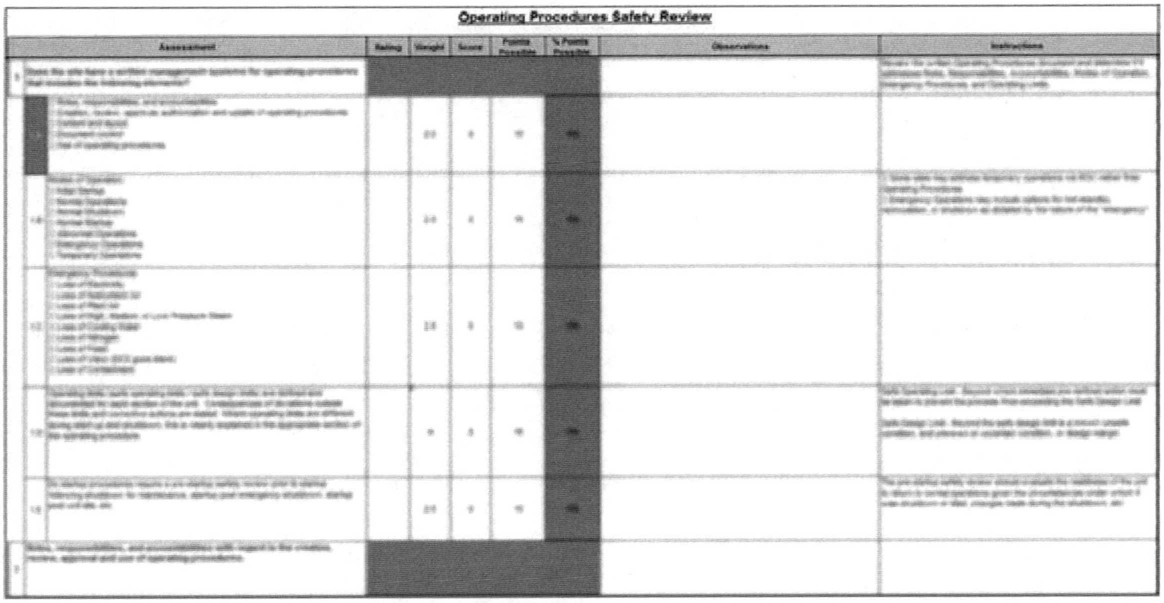
Operating Procedures Safety Review

Assessment	Rating	Weight	Score	Points Possible	% Points Possible	Observations	Instructions

Figure 18. Sample Site Assessment Protocol

3.6 *Human Reliability – "Walk the Line"*

Analysis completed by the APS Program has determined that approximately 20% of Tier 1 and Tier 2 process safety events are caused by human reliability issues. Types of events in this group include:

- Leaving a valve open to atmosphere,
- Valve line-up errors,
- Lock-out tag-out (LOTO) isolation errors,
- Incorrect equipment preparation,
- Bypassing safety systems when they should be engaged, and
- Placing equipment in service with one or more components missing or not properly installed.

"Walk the Line" is a program developed by the APS members to create a set of tools, free to member companies, to address the issues and cultures that lead to human reliability related events.[8] Walk the Line addresses the human reliability associated with conduct of operations and specifically, operating discipline by providing guidance for operations leadership and support to set the expectation to "walk the line" (to know exactly where a material will flow in a process). It reinforces the fundamental responsibility of operations to know with 100% certainty where a material will flow between any two points in a process. Walk the Line addresses the culture of operator line-up and provides tools that help the operator and front-line operations supervision become successful in meeting this responsibility.

The Walk the Line program is meant to generically address loss of primary containment in the refining and petrochemical industry caused by operator error when making a valve line-up to transfer process materials or operate a process. While there are many common causes and solutions, each company and site should have an understanding of their specific causes. Here is how you can use Walk the Line at your site or company:

- Set a Goal of Zero Walk the Line Incidents and commit to it.
- Analyze your LOPC Data.
- Identify Walk the Line Causes.
- Review and customize Walk the Line tools.
- Train employees using your site's customized Walk the Line tools.
- Repeat the discussions often in various meetings, tool box talks, and informal settings with process operations personnel.

The suite of Walk the Line tools are available to APS members HERE.

- Newsletters – The Walk the Line Newsletters are useful for increasing awareness about these Human Reliability issues, and for planning your sites effort to confront these causes and shift culture. The newsletters are suitable for top to bottom awareness and planning, from management level review to "boots on the ground" front line personnel.
- Tool Box Topics – The tool box topics offer short, scripted tool box meeting topics aimed at identifying common incident scenarios, engaging front line personnel in discussions, and providing suggested solutions to prevent incidents. There are 10 tool boxes related to Walk the Line topics, that are available for immediate use. While the tool box topics are fully complete, it is expected that each company or site might modify them to better align with their work practices or culture.
- Hazard ID Documents and Practices Sharing – A number of the tools being developed as part of the Walk the Line project are Hazard Identification Documents and Shared Practices. These have been previously discussed in this paper as has their suggested uses. Preparation of these documents for the Walk the Line project is on-going and these tools will be made available to APS member companies as they are completed. A development schedule is included on the Walk the Line page of the Safety Portal.
- Training Materials – Packages of "ready to go" training materials have been prepared and are available for off the shelf use. There are packages intended for both front line personnel and front line supervision.

4.0 Conclusion

The AFPM and API have been working together on an industry led initiative, Advancing Process Safety, with the overall objective of improving our process safety performance. Data obtained through the APS programs of Metric Reporting, Event Sharing, and Site Assessment identify high priority areas for improvement. Each area for improvement is being addressed through a APS work group, where tools and materials are developed and made available for AFPM and API member companies.

References

[1] API. "Process Safety Performance Indicators for the Refining and Petrochemical Industries," ANSI/API Recommended Practice 754, First Edition, April 2010.

[2] AFPM. AFPM Safety Portal Process Safety Site Assessment Programs, Available at http://safetyportal.afpm.org/SiteAssessment.aspx – Login credentials required. Accessed on February 10, 2016.

[3] AFPM. The AFPM Safety Portal Process Safety Metrics Database, Available at http://safetyportal.afpm.org/ProcessSafetymetrics-access.aspx – Login credentials required. Accessed on February 10, 2016.

[4] AFPM. The AFPM Safety Portal Event Sharing Database, Available at http://safetyportal.afpm.org/EventSharing-access2.aspx – Login credentials required. Accessed on February 10, 2016.

[5] API. "Piping Inspection Code: In-service Inspection, Rating, Repair, and Alteration of Piping Systems," API 570, Fourth Edition, February 2016.

[6] API. "Inspection Practices for Piping System Components," API Recommended Practice 574, Third Edition, November 2009.

[7] AFPM. AFPM Safety Portal Hazard Identification & Practice Sharing, Available at http://safetyportal.afpm.org/HazardIdentification-access.aspx – Login credentials required. Accessed on February 10, 2016.

[8] AFPM. AFPM Safety Portal – Walk the Line, Available at http://safetyportal.afpm.org/walktheline.aspx - Login credentials required. Accessed on February 10, 2016.

[9] AFPM. AFPM Safety Portal, Available at http://safetyportal.afpm.org/default.aspx – Login credentials required. Accessed on February 10, 2016.

[10] AFPM. AFPM Safety Portal – References, Available at http://safetyportal.afpm.org/References.aspx - Login credentials required. Accessed on February 10, 2016.

Appendix A: API RP 754 Tier 1 and Tier 2 Definitions

A Tier 1 or Tier 2 Process Safety Event begins with an unplanned or uncontrolled release of any material, including non-toxic and non-flammable materials from a process that results in one or more consequences described in the RP. The consequences that define a Tier 1 and Tier 2 process safety events are tabulated below.

Tier 1 Consequences	Tier 2 Consequences
An employee, contractor or subcontractor "days away from work" injury and/or fatality	An employee, contractor or subcontractor recordable injury
A hospital admission and/or fatality of a third-party	
An officially declared community evacuation or community shelter-in-place	
A fire or explosion resulting in greater than or equal to $25,000 of direct cost to the Company	A fire or explosion resulting in greater than or equal to $2,500 of direct cost to the Company
A pressure relief device (PRD) discharge to atmosphere whether directly or via a downstream destructive device that results in one or more of the following four consequences: ▪ liquid carryover; ▪ discharge to a potentially unsafe location; ▪ an on-site shelter-in-place; ▪ public protective measures (e.g. road closure); and, a PRD discharge quantity greater than the threshold quantities in Table 1 in any one-hour period.	A pressure relief device (PRD) discharge to atmosphere whether directly or via a downstream destructive device that results in one or more of the following four consequences: ▪ liquid carryover; ▪ discharge to a potentially unsafe location; ▪ an on-site shelter-in-place; ▪ public protective measures (e.g. road closure); and, a PRD discharge quantity greater than the threshold quantities in Table 2 in any one-hour period.
A release of material greater than the threshold quantities described in Table 1 of the standard in any one-hour period	A release of material greater than the threshold quantities described in Table 2 of the standard in any one-hour period

442500 Forge Bonding: A Safer Metal Joining Process

Abstract:

When planning in-service repairs or enhancements that involve the joining of metals; sparks, flames and even elevated temperatures are a concern for refinery and storage terminal owners and operators. Processes that generate uncontrolled elevated temperatures that exceed Auto Ignition Temperatures (AITs) published by American Petroleum Institute (API) or published Hot Surface Ignition Temperatures (HSITs) are often scrutinized and classified as "hot work" within these facilities. A metal bonding process has been developed that in inherently safe in these environments where flammable hydrocarbons are present.

Friction welding is a solid-state welding process that generates heat through mechanical friction between work pieces in relative motion to one another, with the addition of a lateral force to plastically displace and fuse the materials. Technically, because no melt occurs, friction welding is not actually a welding process in the traditional sense, but a forging technique. An advanced form of friction welding called forge bonding has been developed for portable entry into refineries, improving worksite safety by reducing the time at temperature and the hot surface area at the bonding site. In addition, forge bonding has a distinct advantage over welding in that it can join dissimilar metals that have largely different melting points.

An introduction to the forge bonding process followed by results from extensive ignition testing to determine the safety margin of multiple chemicals commonly found in refineries when in the presence of forge bonding, will be presented. The dependency of the ignition temperature of hydrocarbons on the hot surface area and the ignition lag time are discussed along with support from data from API 2216 and other well known publications. A presentation of the problem followed by an investigation of the forge bonding solution will prove that ignition of common hydrocarbons is not possible in the presence of the low energy levels created by the small hot surface area, short ignition lag time and moderate temperatures.

Real life implementations add further proof that forge bonding is an alternative to welding that is inherently safe in refineries, storage terminals and anywhere flammable hydrocarbons are present. This technology has far reaching implications for improved safety in other industries, such as pharmaceutical, pipelines, mining and marine to name just a few.

Best Safety Practice for Refinery and Petrochemical Plants During the Major T&I (Turnaround & Inspection)

Ayman Natto
Safety Department, Technical Safety Section, Rabigh Refining and Petrochemical Company (Petro Rabigh),
Rabigh, Saudi Arabia
ayman.natto@petrorabigh.com

Prepared for Presentation at
American Institute of Chemical Engineers
2016 Spring Meeting
12th Global Congress on Process Safety
Houston, Texas
April 10-13, 2016

UNPUBLISHED

Safety Practice for Refinery and Petrochemical Plants During the Major T&I (Turnaround & Inspection)

Ayman Natto, Process Safety Engineer
Safety Department, Technical Safety Section, Rabigh Refining and Petrochemical Company (Petro Rabigh),
Rabigh, Saudi Arabia
ayman.natto@petrorabigh.com

Keywords: Safety Practice during T&I, Safety in Major shutdown, Process Safety roles in shutdown, Pre-Start up Safety Review, Management of Change, Mechanical Completion Certificate.

Abstract

Planned shutdowns or Turnaround and Inspections (T&I) are normal activities for gas and oil companies the world over, but Petro Rabigh carried out a major shutdown at its huge refinery and petrochemical complex consisting of Refinery, Monomer, Polymer and Utility plants. This paper is aimed at the best Safety Practice implemented during the T&I which was completed at the end of December 2015. It explains the Safety Program developed and implemented by Safety Department to execute the T&I safely. Planning was a key element of the T&I preparation. A dedicated taskforce was established for the T&I planning with representatives from Manufacturing, Maintenance, Engineering, Safety, and other involved departments. The taskforce addressed various safety programs including Work Permit System, Critical Lift Plan, Confined Space Entry, Process Hazard Analysis (PHA) etc. Moreover, critical jobs were identified and Job Safety Analysis (JSA), Management of Change (MOC), Mechanical Completion Certificate (MCC) and Pre-Startup Safety Review (PSSR) were conducted for each of these jobs. Special training and presentations were given to both Petro Rabigh and Contractor Employees on the lessons learned from T&I incidents experience worldwide. A huge contractor workforce and the massive scale of activities were managed safely with the above innovative implementation of the practice, Petro Rabigh would like to improve in the 2019 T&I with many of its new Phase II plants.

1. Introduction

Most oil and gas companies and organizations worldwide have periodic planned plant shutdowns known as Turnaround and Inspections (T&I). They covers activities such as inspection, overhaul, modification and the installation of new parts or equipment. Activities that cannot be performed when the plant is in normal operation are also carried out during the shutdown period.

Rabigh Refining and Petrochemical Company (Petro Rabigh) is located on a 3,000 acre site in Rabigh, on the Red Sea coast of western Saudi Arabia. Petro Rabigh - a joint venture between Saudi Aramco and Sumitomo Chemical Company, began operations in April 2009, with the capacity to produce 18.4 million tons of petroleum based products and 2.4 million tons of ethylene and propylene-based derivatives per annum.

This integrated Refinery and Petrochemical complex has a 400,000 barrels per day Refinery with Distillation, Hydrotreating, Hydrogen Production and Recovery plants, and petrochemical plants consisting of the world's largest High Olefin Fluidized Catalytic Cracking plant, Ethane Cracker plant, Alkylation & Isomerization Unit, Propylene Oxide, Mono Ethylene Glycol, Polyethylene, Polypropylene, High Density Polyethylene plants, etc. These are supported by a vast network of utilities, such as a storage tank farm and marine berthing facility for the receipt of crude oil/raw materials and dispatch of finished products. Moreover, Petro Rabigh is currently constructing many new production plants - tied-in with existing plants during the 2015 T&I - for its Phase II project, known as Petro Rabigh II.

The first T&I at Petro Rabigh was successfully completed in 2011 with about 7 million man-hours logged, over seven weeks. With proper planning, preparation and perfect execution, Petro Rabigh was able to exhibit an enviable safety record, viz., two lost-time injury incidents of contractor employees, three medical treatment cases and five first-aid injury incidents. A huge contractor workforce and the massive scale of activities were managed safely with the above innovative practices and by team work, cooperation and dedication from all. Petro Rabigh would like to continue these practices in the 2019 T&I with improvement in the field of Process Safety, in addition to the hundreds of tie-in points for the Petro Rabigh II.

2. T&I Preparation

This was a major complex-wide turnaround, as it involved all the plants and utilities in the Petro Rabigh Complex. More than one hundred CAPEX works and Warranty Claim works were also included in the T&I scope of turnaround, adding to its daunting magnitude.

Based on the work scope and schedule, the peak contractor manpower in the Petro Rabigh Complex during the turnaround was over 13,000. This was in addition to the regular company employees, nearly 4000. On average the daily man-hours logged was 165,000 per day during the twelve weeks long turnaround period. At the end of the turnaround, the total man-hours logged were more than 15 million.

Petro Rabigh developed a T&I link in the company main web page which showed the day to day status of all Petro Rabigh T&I activities. Safety was one the main activities and included data such as daily safety observations, daily safety performance dashboards, safety flyers, incident flyers, lessons learned, near misses and safety suggestions. This link was updated daily by the Petro Rabigh Safety Department below is the screen print from the web page of that time:

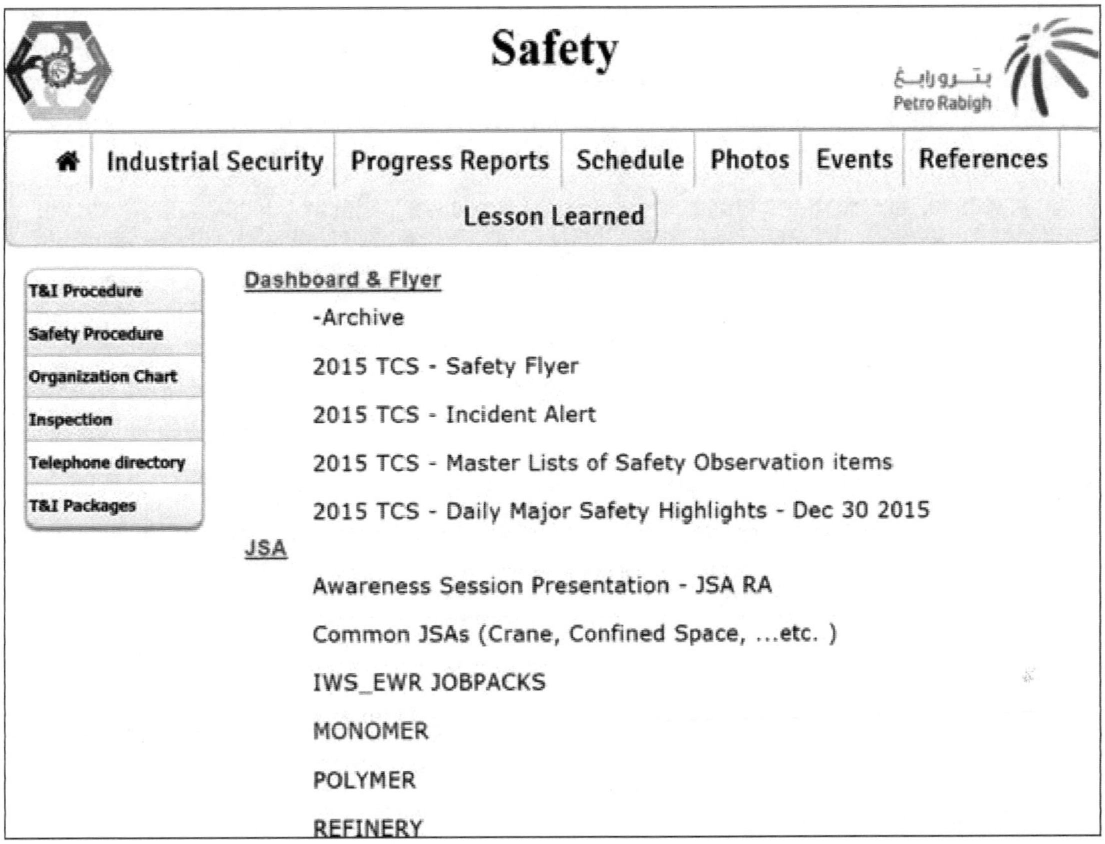

2.1. Contractor Safety Management

Planning is Good Business. A Good Plan ensures safe execution.

It was essential that all the contractor organizations were made sufficiently aware of Petro Rabigh' s minimum safety requirements. The following actions were implemented during the T&I:

No. of Contractor Companies	57
Approved Hazard Identification Plan (HIP)	66
Approved CVs Of the Safety Representative	181

2.1.1. Safety Orientation sessions

Petro Rabigh provided about 13,000 attendees for the Safety Orientation to all the contractor employees through a specially developed Safety Orientation video film. This film was produced in five different languages for the different nationalities who would be working in the Turnaround.

2.1.2. Turnaround Safety Forum Sessions

Turnaround Safety Forums for Contractors and Petro Rabigh staff were conducted before the Turnaround. The Forum included short presentations and discussions on the safe execution of work activities related to turnaround, lessons learned from the 2011 T&I and best industrial safety practices etc. to further enhance safety awareness and improve safety communication during the execution.

In addition, several sessions of Petro Rabigh Work Permit Receiver Training were provided to the selected, qualified Contractor Supervisors to understand the work permits system requirements and standards. This ensured alignment of safe work systems of the contractors with the Petro Rabigh Work Permit System.

2.1.3 Confined Space Standby Man Training

Thousands of confined spaces were opened during the T&I, with high risk especially in such a large refinery and petrochemical plant. Petro Rabigh trained over 1100 standby personnel to understand their roles and responsibilities upon confined space entry. Petro Rabigh issued a standby man certificate for the trained contractor employees to ensure the confined spaces were monitored and managed by a competent person.

2.2. Critical jobs

From among the listed jobs for the Turnaround, critical jobs were identified by a team of Petro Rabigh Operations, Maintenance, Engineering and Safety representatives and a JSA was conducted by the team for each such job. Similar review was done for all scaffold erection plans and critical lift plans.

Awareness Training was also given to Operations and Maintenance staff of Petro Rabigh covering JSA and Scaffold Safety.

2.2.1 Tools Inspection and Certifications

A dedicated team of Petro Rabigh Maintenance Department examined all tools and equipment of the contractors and issued approval sticker on only the properly functioning and safe tool / equipment. This prevented the contractors from using damaged tools and equipment in Petro Rabigh work site. Such inspection also covered Heavy Equipment – such as cranes, loaders, forklifts, low-bed trailers etc. through approved Third Party agencies.

Petro Rabigh established the certification requirement for the Scaffold Inspectors, Scaffold Supervisors and Crane Riggers.

All such measures in the planning stage collectively helped in managing the Turnaround safely.

3. T&I Safety Implementation and Monitoring

All the planning and preparation were converted into implementation, which ensured an excellent safety record during the Turnaround.

The Safety Department hired 30 contractor safety advisors to monitor the T&I in addition to dedicated Petro Rabigh safety Inspectors and Engineers who went around the contractor work areas to evaluate contractor safety performance, observing the activities, identifying unsafe actions, initiating talks with the workers and supervisors, establishing rapport, correcting unsafe practices at site and investigating incidents. Checklists were developed to record field observations.

The data provided by the Safety Inspectors was analyzed on a daily basis and transformed into a daily dashboard called the Turnaround Daily Safety Performance Dashboard. The dashboard indicated the Top Risk Category Items, the Top Three Unsafe Behaviors, the percentage of unsafe activities – areas-wise, etc.

Petro Rabigh Safety personnel conducted a total of 61,022 daily formal interactions with contractor employees and Petro Rabigh staff, in each area. Meetings were established with contractor management directly, to highlight important concerns to improve safety. Such meetings were held twice every week.

Daily Safety interaction meetings were held by Safety Department with contractor safety representatives on day shifts and night shifts.

The daily Tool Box Talks became very popular and served as a means of safety communication. Topics of the tool box talk sessions were selected based on the Top Risk Category of the dashboard for the previous day. This session was conducted in the beginning of each shift by each contractor.

Field safety observations and analysis were listed together and released in the form of Safety Flyers by Petro Rabigh. The lessons learned from the near miss and injury incidents were also released in the form of Safety Flyers.

3.1. Safety Observation Tours

Based on the statistics of observations shown in Table 1, "Sub-standard use of PPE" and "Unsafe Tools and Equipment" have appeared as our Top Risk Categories in turns. Top Three Unsafe Observations range from various categories related to "Housekeeping, Scaffold Inspection, Confined Space, Basic Safety Rules, and Barricading" etc.

The dashboard along with the Top Risk Category and Unsafe Behaviors generated constructive discussions in T&I meetings thereby providing focus on areas of concern for necessary improvement. Also, it included the percentage of observed people working safely, indicating the status of improvement in safety performance and reflecting the joint efforts of the whole organization.

The daily dashboard will continue to highlight the priority areas where everyone's efforts must be concentrated.

3.1.1 Safety Observation Statistics

Table 1. Safety Observation Statistics

Counselling meetings were conducted with low performing contractors top management to communicate safety concern and observations and action plan was demanded from contractors to reduce safety observations at site. Visible improvement was observed in contractor performance after these meetings.

Area	Total No. of Observations
Refinery	1439
Monomer	2781
Polymer	621
Utilities & Terminal	646
Shops and Laydown Yards	1560
Total	7047

3.1.2 Safety Flyers and Incident Alerts

In order to address Top Risk Categories and Unsafe Behaviors, safety flyers were developed and disseminated among all T&I concerned parties, including contractors.

The flyer information contents were discussed during Day and Night T&I meetings, and also during contractors' Toolbox Talks.

3.2. Near Miss & Safety Suggestions Reports

3.2.1 Lessons learned

Petro Rabigh encouraged contractors and employees to enter lessons learned whenever on the T&I web page visible to all Petro Rabigh employees to avoid recurrences and learn from it. About 550 lessons learned were received during the T&I which was already communicated and closed in the site.

3.2.2 Near Miss & Safety Suggestions

Petro Rabigh requested employees to submit near miss incidents and safety suggestions through the Petro Rabigh web page throughout the year, and this practice continued during the T&I as shown in Table 2.

Table 2. Near Miss & Safety Suggestion during the T&I

Total number of PRC submitted Safety Suggestions	01
Total number of PRC submitted Near Misses	10
Total number of Contractors submitted Safety Suggestions	06
Total number of Contractors submitted Near Misses	44

4. Process Safety through the T&I

Petro Rabigh had about 25 major modifications that were considered as a project which included designing, reviewing and constructing that had completed during the T&I in addition to the hundreds of tied-in points for the Phase II Project. The Safety Department had to ensure that all these projects complied with the process Safety regulation and standards. By implementing the Engineering and Process Safety elements smooth and timely transition from construction to start-up phase was achieved by implementing the Engineering and Process Safety elements smooth and timely transition from construction to start-up phase was achieved.

4.1 Management of Change (MOC)

Management of Change (MOC) is a proactive process with a two-fold objective: first, to challenge the need for change and to analyze its wider impacts on the business goals; and second to evaluate, authorize changes before they are made and ensure proper closure after the changes are completed. Its purpose is to establish the procedures necessary to ensure that the all the risks arising from proposed changes are decreased to an acceptable level. It is also an excellent communication tool to inform impacted groups of areas that a change is being considered and/or about to occur.

The core elements of the MOC process that are implemented in all modifications during the T&I:

- Agreement on the justification and reasons for the change by appropriate management,
- Risk analysis of the proposed change by a multi-disciplinary team of competent people who have required knowledge and experience, including vendors and specialists,
- Approval of complex changes or design deviations by engineering managers or technical authorities of sufficient experience and knowledge,
- Develop operating procedures to supplement or implement the change,
- Communicate the change to all concerned and train the workforce, those who are directly affected by the change,
- Conduct a pre-startup safety review to ensure system is ready for safe start, i.e. that all actions from the risk analysis are implemented, operating procedures are developed and staff are trained,
- Monitor the change during implementation period and capture any lesson learned for future reference, and
- Close out the change after specified time period otherwise initiate a permanent change process.

The MOC process is also critical in day to day operations in case of safe operating envelope excursion during process upset; it requires immediate action such as shut down or reporting of incident for further investigation if conditions were normalized quickly. Overall a robust MOC process and its effective implementation is an integral element of chemical process safety in order to prevent and control major incidents risks.

4.2 Mechanical Completion Certificate (MCC)

An important initiative during this T&I was the implementation of MCC through a Committee chaired by Proponent representative and comprised of Construction Agency, Inspection Organizations, Safety and other representatives as appropriate, depending upon the nature of the facility. The committee had to complete the following MCC steps: (1) accepting new facilities as mechanically complete, (2) commissioning and start-up, and (3) final performance acceptance and financial closeout. The Construction Agency was responsible for delivering a completed facility as defined in the project scope of work. After the facility was accepted as Mechanically Complete, the Commissioning, Start-Up and Performance Acceptance became the responsibility of the Proponent Organization with the Construction Agency providing necessary design and construction personnel, technicians and vendor representatives to assist in these activities.

For all facilities and especially where an existing facility was undergoing an upgrade or retrofit, both the Construction Agency and the Proponent were to assign personnel during the design phase to establish a Pre- Commissioning, Commissioning and Start-Up Plan.

4.3 Pre Startup Safety Review (PSSR)

Pre Startup Safety Review (PSSR) aids in the transfer of projects from the construction stage to the operations stage. It provides assurance that commissioning and start-ups can be carried out in a safe manner following the installation of new facilities. In particular, PSSR aims to confirm that what was built and installed conforms to the original design and standards, and that the hazards during start-up of the project have been properly addressed and controlled. Petro Rabigh dedicated a team to conduct about 25 PSSR for the projects taken up during the T&I, in all four Operation departments.

Nominated leaders from each Operation department led their respective PSSR teams (comprising of members from Operation, Engineering, Maintenance, Project Management Team, Safety and Environment) and started conducting PSSR for such modifications.

5. Conclusion

T&I execution is a major challenge that needs team work, cooperation, dedication, and flexibility in order to be successfully accomplished. With proper planning, implementation of safety programs, innovative practices and active involvement of contractors, Petro Rabigh completed the T&I scope safely, timely and reliably.

12TH GLOBAL CONGRESS ON PROCESS SAFETY

In-breathing requirement for a tempered tank exposed to prolonged sudden rainfall

Presenter/Dr.-Ing. Davide Moncalvo
Dr.-Ing. Michael Davies
Braunschweiger Flammenfilter GmbH (PROTEGO Germany)
Industriestraße 11, 38122 Braunschweig, Germany
davide.moncalvo@protego.com

Prof. Dr.-Ing. Roman Weber
Prof. Dr.-Ing. Reinhard Scholz
Institute of Energy Processes Engineering and Fuel Technology, TU Clausthal
Agricolastraße 4; 38678 Clausthal-Zellerfeld, Germany
weber@ievb.tu-clausthal.de

Prepared for Presentation at
American Institute of Chemical Engineers
2016 Spring Meeting
12th Global Congress on Process Safety
Houston, Texas
April 11-23, 2016

In-breathing requirement for a tempered tank exposed to prolonged sudden rainfall

Presenter/Dr.-Ing. Davide Moncalvo
Dr.-Ing. Michael Davies
Braunschweiger Flammenfilter GmbH (PROTEGO Germany)
Industriestrasse 11, 38122 Braunschweig, Germany
Davide.moncalvo@protego.com

Prof. Dr.-Ing. Roman Weber
Prof. Dr.-Ing. Reinhard Scholz
Institute of Energy Processes Engineering and Fuel Technology, TU Clausthal

Keywords: enter a few keywords here to aid searching

Abstract

In this paper a model is presented aimed to determine the air in-breathing requirements to compensate for collapse of a tempered tank when vapors condense inside due to the strong cooling brought by a prolonged intense rainfall. The effect of liquid product evaporation on tank cooling dynamics is also discussed. The presented model shows that vapor condensation strongly affects the in-breathing requirement and that liquid evaporation affects the cooling dynamics.

In order to verify this calculation procedure with measurements, a test program with a cylindrical API 12F tank will be initiated.

1 Introduction

Tempered tanks are an equivalent way to address liquid storage tanks. Protection of these tanks against corrosion of the bare metal surface due to rainfall is accomplished through either painting or applying an organic coating. Painting, though cheaper, does not protect the stored liquid from noticeable cooling in case of prolonged rainfall. Since liquid storage tanks have design pressures near atmospheric levels, a prolonged rainfall may result in vapor condensation, which may result in tank deformation and eventually collapse. In order to prevent this occurrence the general practice is the installation of vacuum breathers with a cumulative capacity larger than the so-called in-breathing requirement of the tank.

The principal standards governing breathing of low-pressure liquid storage tanks are ISO 28300 [1], API 2000 7th Edition [2], and EN 14015 [3]. These standards, as far as in-breathing requirements are concerned, vaguely refer to an earlier work of Förster et al. [4], where the rainfall was modeled as a uniform rain coat on the tank. Fullarton [5,6] evinced the effect of condensation on the in-breathing requirement of a tank filled with humid air saturated with steam and methanol vapors. On one hand, condensation releases latent heat to the walls thus dampening their cooling and therefore the in-breathing requirement. On the opposite hand, it is responsible for additional in-breathing to fill in the volume previously occupied by condensed vapor. A later work by Salatino et al. [7] tried to quantify the effect of condensation but he lacked proper modeling for the dampening of the wall temperature.

In this paper the authors are developing a model for an uninsulated liquid-filled tank exposed to a severe rain shower after a prolonged, uninterrupted solar exposure. Both vapor condensation and liquid evaporation will be addressed in the mass transfer dynamics. For simplicity the vapor space would be assumed to contain initially only liquid vapors.

2 Thermodynamical model

A sketch of a tank exposed to a rain shower of given intensity and impingement angle is shown in Fig. 1. In general, rain and ambient air temperatures differ. For a first analysis the rain film is assumed to be of constant thickness along the lateral walls and the roof. In the following model the bulk product inside the tank is assumed to have a uniform temperature and that the liquid level height is assumed to remain constant.

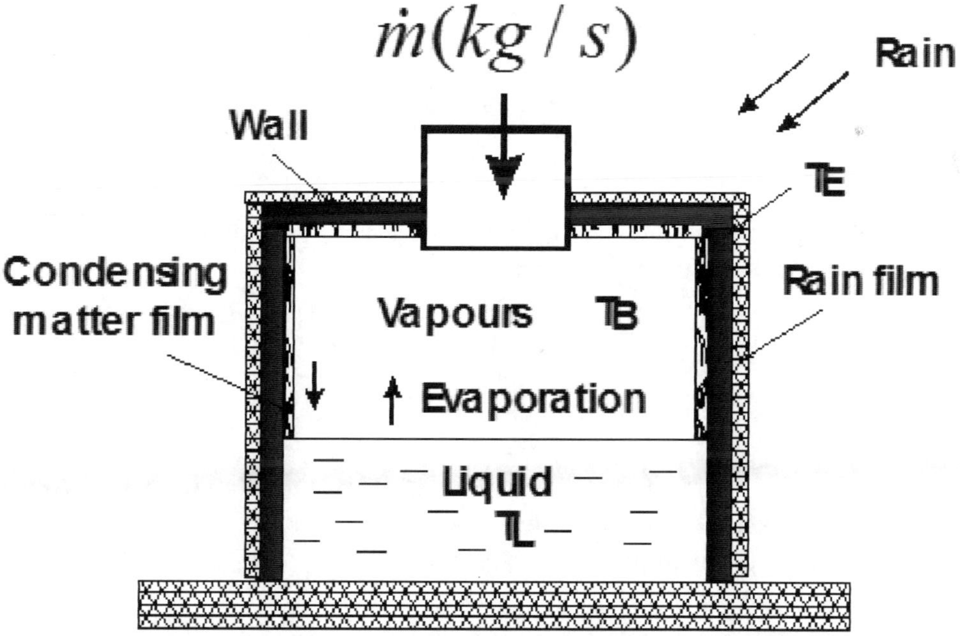

Figure 1: Model for the tank exposed to a rainfall

In this model the condensed vapors are assumed to form a film that is at the same temperature of the bulk product and condensation is expected to occur mostly at the cold tank walls, while evaporation mostly at the liquid free surface. Additionally, the eventuality of a non - uniform wall temperature distribution is neglected. Under these assumptions, it is possible to write a separate energy balance for the tank product and the tank walls. The energy balance for the tank product in Eq. (1) is derived by the authors. According to this equation, the rate of cooling of the gas bulk temperature is affected by both the internal heat transfer to the walls and the warming-up of the in-breathed colder air.

$$\left(M_{B\,liq}c_{B\,liq} + M_{B\,vap}c_{B\,vap}\right) dT_B/dt = -h_{int}A_{lat}(T_B - T_{walls}) + \dot{M}_{air}c_{air}(T_{amb} - T_B) \quad (1)$$

The energy equation for the tank walls in Eq. (2) relates the temperature change in the walls to the internal convective heat inflow from the tank product, the external heat transfer to the cold rain film and the heat release to the walls from the condensing liquid film. Its structure is similar to the one used by Fullarton [6], differing from it in the estimation of the rate of condensation in the tank.

$$M_{walls}c_{walls}\frac{dT_{walls}}{dt} = h_{int}A_{lat}(T_B - T_{walls}) - U_{rain}A_{lat}(T_{walls} - T_{rain}) + \dot{M}_{cond}\Delta h_{cond} \quad (2)$$

As long as no product breaths out of the tank, there is no loss in the mass of stored product but merely a transfer from one phase to the other. However, some air inflow is still needed to compensate for the volume loss of the bulk gas vapors due to rain cooling. In Eq. (3a) it will be assumed that the instantaneous change in the mass of the product vapors must be compensated by an ambient air flow of same magnitude. In Eq. (3b) the temporary change in the mass of the vapors is split into a term for the volume loss over time and the phase change mechanisms.

$$-\,dM_{B\,vap}/dt = \dot{M}_{air} \quad (3a)$$

$$dM_{B\,vap}/dt = d\left[p_B\,V_{B\,vap}/\hat{R}\,T_B\right]/dt + \dot{M}_{evap} - \dot{M}_{cond} \quad (3b)$$

The rate of phase change, either via condensation or evaporation, is directly related to the change in saturation concentration within the gaseous phase; see Eq. (3c) and (3d)

$$\dot{M}_{cond} = \rho_{mix}\beta[y(T_B) - y^*]A_{liq\,film} \quad (3c)$$

$$\dot{M}_{evap} = \rho_{mix}\beta[y(T_B) - y^*]A_{free} \quad (3d)$$

The overall heat transfer coefficient to the rain film is calculated as shown in Eq. (4a), considering that the rain mass flux wetting the tank $\dot{m}_{film}$ is less than the rain intensity on ground $\dot{m}_{rain}$ in Eq. (4b), following the formula of Förster et al. [4]

$$1/U_{rain} = 1/h_{rain} + 1/h_{free} = 1/\left(\dot{m}_{film}c_{rain}\right) + 1/h_{free} \quad (4a)$$

$$\dot{m}_{film} = \dot{m}_{rain}\ [\pi D^2/4 + DH\ cot\theta]/[\pi D^2/4 + \pi DH] \quad (4b)$$

For reference, in Continental Europe a rain of an intensity of 75 kg/(m2 h) occurs once a year, a rain of 150 kg/(m2 h) intensity occurs once every twenty years, while a pouring rain with an intensity of 225 kg/(m2 h) happens once in a century.

3 Results

For a given tank size, wall thickness and specified weather conditions, i. e. temperature and pressure, the above set of balance equations Eq. (1) to (3) can be integrated to calculate the temperature profiles of the bulk product and the walls, as well as the in-breathing flow rate as a function of time.

Fig. 2 and 3 show the in-breathing flow rates for a 2651 m^3 tank at an initial temperature of 49 °C containing water and heptane, for rain intensity levels from 75 to 225 kg/m^2h. Ambient and rain temperature are both equal to 15°C. Liquid level is below 1 %; at these low filling levels evaporation is negligible. These figures clearly show that when the rain intensity increases from 75 to 225 $kg/(m^2\ h)$, the in-breathing flow rate requirement rises by a factor of 1.7 for steam and 1.45 for heptane vapors. In each graph the maximum value of the cooling rate is added in parentheses and reported for completeness in Table 1. The maximum cooling rate coincides with the moment when the maximum air flow rate is breathing into the tank, which usually occurs shortly after the start of the rain shower (3 – 10 minutes). Both the maximum in-breathing flow rate and the cooling rate increase with the rain intensity, though not linearly. In a previous publication [8], the authors found that accounting for vapor condensation augments the calculated maximum in-breathing flow rates by approximately 25 % for the tank filled with steam and roughly 65 to 80 % for the tank filled with heptane vapors. As already observed by Fullarton [5], air in-breathing in order to refill the volume occupied by condensed vapors clearly dominates over any dampening effect due to latent heat release. This explanation should also work for partially filled tanks.

Furthermore in the latest editions of ISO 28300 [1], API 2000 [2] and EN 14015 [3] a constant cooling rate of 28 K/h is assumed for a tank volume of 2651 m^3. Table 1 clearly shows that even with a rain shower of 75 $kg/(m^2\ h)$ intensity this value is not at all conservative.

Table 1: In-breathing and cooling rates for the tank of 2651 m^3 volume filled with water and heptane vapors in function of rain intensity. Filling level is below 1 %.

Rain intensity ($kg/m^2/h$)	Maximum in-breathing (Nm^3/h)		Maximum cooling rate (K/h)	
	Water tank	Heptane tank	Water tank	Heptane tank
75	640	496	32	29
150	937	640	52	45
225	1097	721	67	50

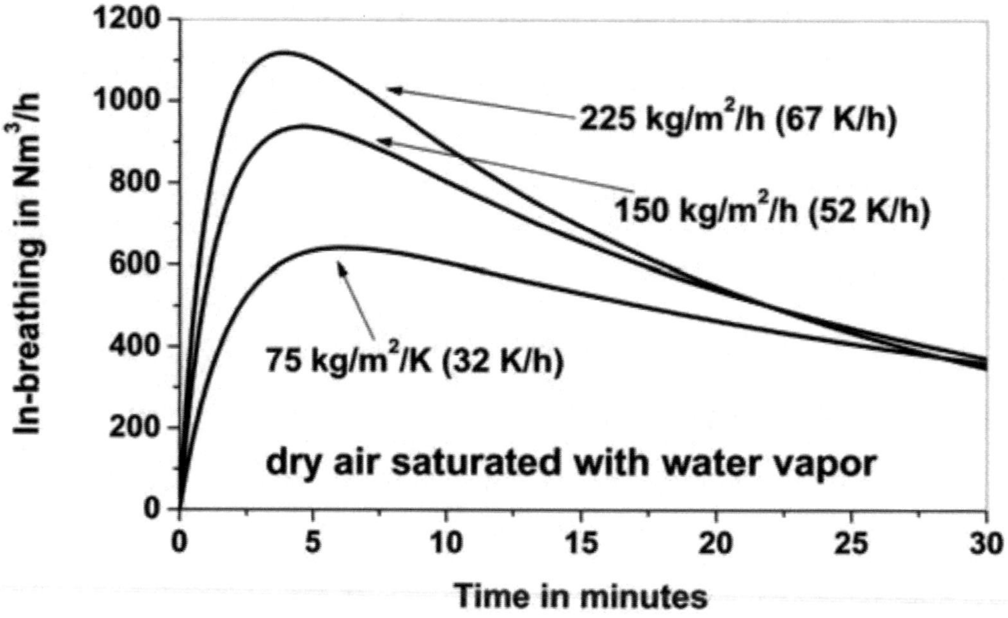

Figure 2: In-breathing as a function of time for a water tank of 2651 m³ volume under a rainstorm of 75, 150 and 225 kg/m2/h intensity. Filling level is below 1 %. Source [8]

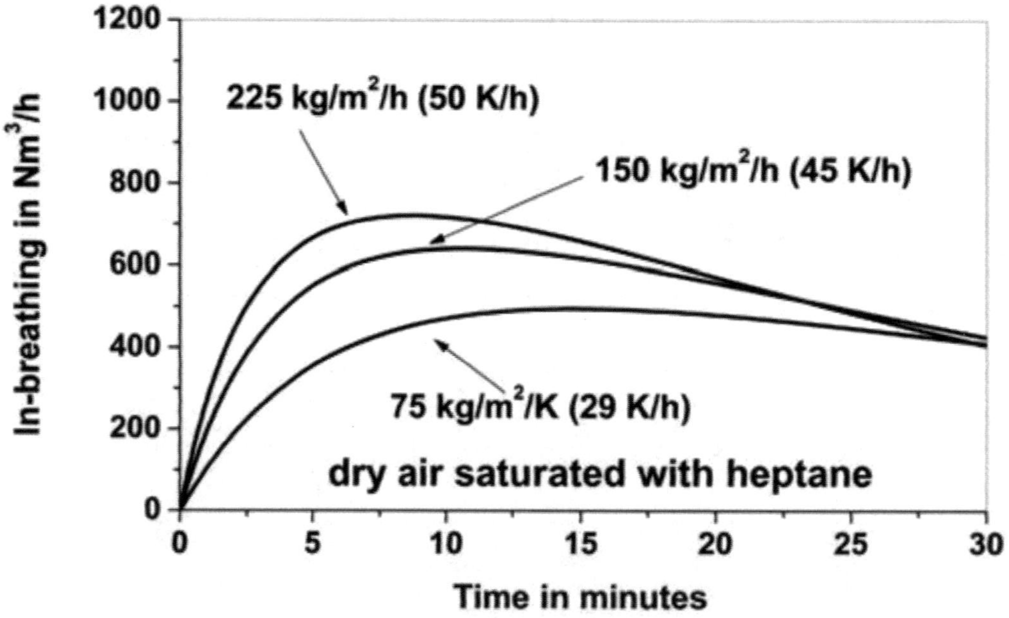

Figure 3: In-breathing as a function of time for a heptane tank of 2651 m³ volume under a rainstorm of 75, 150 and 225 kg/m2/h intensity. Filling level is below 1 %. Source [8].

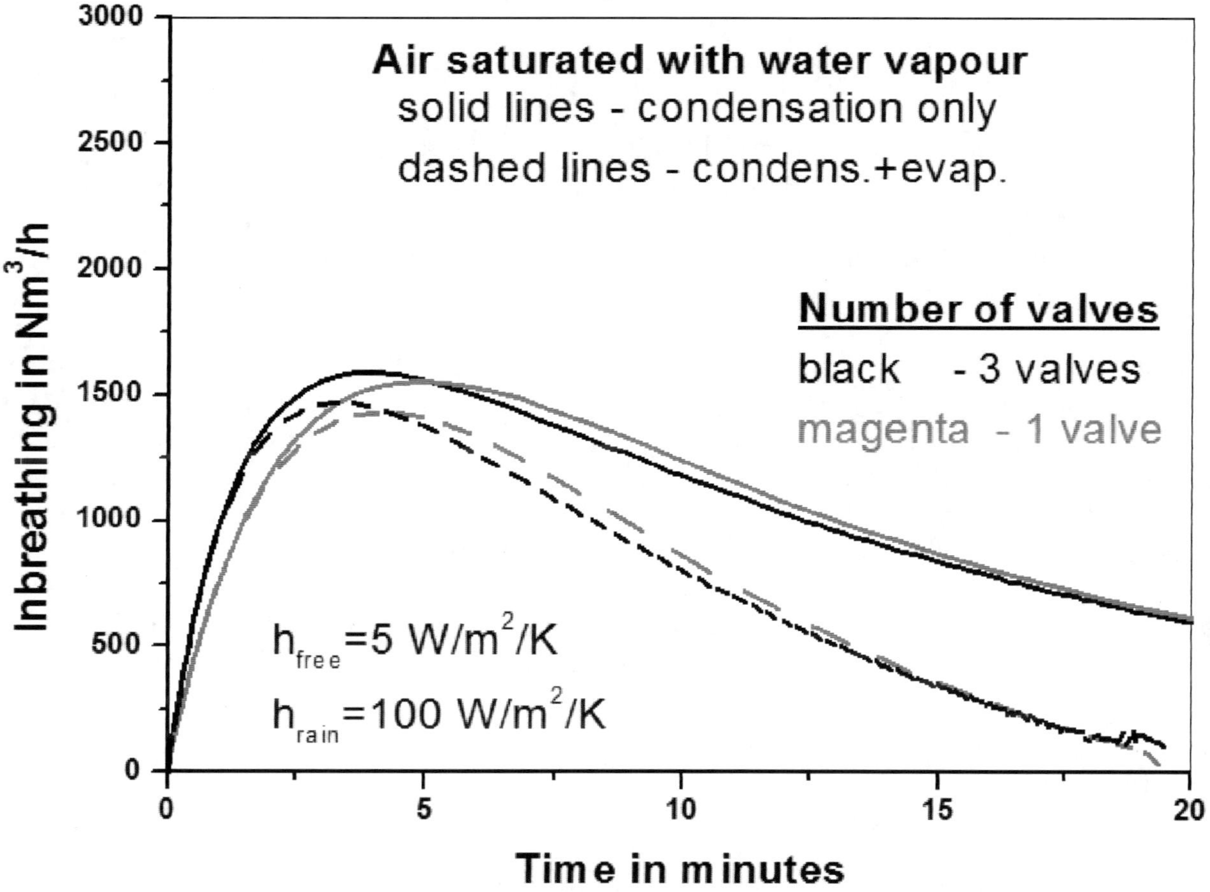

Figure 4: In-breathing as a function of time for a water tank of 4003 m³ volume with a free convection and rain equivalent heat transfer of 5 and 100 W/m²K. Filling level is 50 %-vol.

In Fig. 4 the in-breathing flow rate as a function of time is shown for a water tank with a filling level of 50 %-vol. From this picture it can be clearly seen that taking into consideration the evaporation of liquid strongly reduces the flow of in-breathed air over time, as expected. Additionally, it has been investigated how strongly the simultaneous in-breathing through several vacuum valves of given type and size is impacting the breathing requirements. For a near atmospheric tank the capacity of a breather valve can be assumed proportional to the instantaneous incompressible flow capacity though an ideal nozzle, as shown in Eq. (5). In this equation K_{valve} has a similar meaning to the discharge coefficient for safety relief valves as described in API 520 [9] and ISO 4126-1 [10] and is dependent on type and size of the device.

$$\dot{M}_{valve} = K_{valve} \cdot A_{throat} \cdot \sqrt{\rho_{air}[p_{amb} - p_{tank}(t)]} \quad (5)$$

On behalf of Fig. 4 it may be concluded that the valve throat area is large enough so that one device suffices, since the peak of demand for ambient air is uninfluenced by the quantity of vacuum breathers.

4 Measurements and future work

The above discussion clearly points out the importance of cooling rates on in-breathing. In this context, proper calculations of the heat transfer, particularly concerning the heat transfer coefficient, are essential. More precisely for a given rain intensity and impingement angle, the rate of cooling and the in-breathing flow rate have to be measured as functions of time in order to verify the proposed model.

The first set of measurements should involve an API 12F tank filled with air, that is saturated with steam and heptane vapors, in order to model the effect of condensation. In a second set of measurements the tanks should be filled with saturated water at several filling levels. The effect of precipitation may be recreated by ad-hoc showers in order to avoid random changes of rain intensity and impingement angle. Temperature and pressure inside the tank and at the walls as well as air flow rate shall be recorded.

Future extension of the test program, for instance to tanks of larger or smaller size or using other storage media, like methanol or ethanol, will be discussed on the base of the results obtained from the first round of tests.

5 Conclusions

A physical model for the calculation of the in-breathing requirement and the cooling rate of low-pressure liquid filled storage tanks exposed to a prolonged rainfall has been presented in this paper. The model relies in part on a previously published model, which has demonstrated a substantial effect of vapor condensation on ventilation requirements, which is not accounted for in the most recent editions of the standards ISO 28300, API 2000 and EN 14015. The model presented here includes in comparison with the previous one the effect of liquid evaporation. Our calculations showed that evaporation has a severe impact on the thermodynamics of the tank cooling. The quantity of top mounted vacuum breathers showed a rather negligible impact in the determination of the in-breathing requirements: nevertheless, this conclusion should not be generalized based on one calculation for only one scenario. Realizing the short-comings of the current venting models in the standards, among the other aspects an arbitrary specification of the medium cooling rates, the need for a comprehensive set of measurements dedicated to this important issue proves necessary. A series of experiments involving a cylindrical API 12F tank at several liquid filling levels will be started soon.

6 Nomenclature

A	Surface area	[m²]
c	Specific heat	[J/(kg K)]
D	Tank diameter	[m]
h	Heat transfer coefficient	[W/(m² K)]
Δh_{cond}	Specific latent heat of evaporation	[J/kg]
H	Tank height	[m]
$\dot{m}$	Mass flux	[kg/(m² s)]
K	Proportionality constant	[--]
$\dot{M}$	Mass flow rate	[kg/s]
M	Mass	[kg]
p	Pressure	[Pa]
t	Time	[s]
$\hat{R}$	Specific gas constant	[J/(kg K)]
T	Temperature	[K]
U	Overall heat transfer coefficient	[W/(m² K)]
V	Volume	[m³]
y	Mass fraction of the condensing vapor in the gaseous phase	[--]
y*	Mass fraction of the condensing vapor at saturation temperature	[--]
β	Vapor mass transfer coefficient	[m/s]
ρ	Density	[kg/m³]
Θ	Rain impingement angle	[--]

amb	Ambient	
B	Bulk product	
cond	condensation	
evap	evaporation	
film	liquid film (condenses or rain)	
free	Free or natural convection; free liquid surface	
int	Internal, i. e. between tank walls and bulk product	
lat	Lateral surface incl. roof	
liq	Liquid phase	
mix	Mixture of product vapors and in-breathed air	
throat	Throat, i. e. minimum surface area of a venting device	
vap	Vapor or gaseous phase	

7 References.

[1] ISO 28300, Petroleum, petrochemical and natural gas industries – Venting of atmospheric and low-pressure storage tanks, First Edition, 2008

[2] API 2000, Venting atmospheric and low-pressure storage tanks, Seventh Edition, 2014

[3] EN 14015. Specification for the design and manufacture of site built, vertical, cylindrical, flat-bottomed, above-ground, welded, steel tanks for the storage of liquids at ambient temperature and above, 2005

[4] H. Förster, K. Schampel, H. Steen, Atmungsvorgänge infolge von Witterungseinflüssen an Lagertanks für brennbare Flüssigkeiten, PTB Report W-22, ISSN 0341-6739, 1984

[5] D. Fullarton, Sicherheitstechnische Überlegungen bei Atmungsvorgängen in Lagertanks – Einfluss der Kondensations von Produktdämpfen, Chem. Ing. Tech. 58 (1), 1986, 38 – 40

[6] D Fullarton, E.U Schlünder, Näherungsweise Bestimmung der Austauschfläche bei der Kondensation von Gas—Dampfgemischen, Chem. Eng. Proc., 18 (5), 1984, 283- 292

[7] P. Salatino, G. Volpicelli, P. Volpe, On the design of thermal breathing devices for liquid storage tanks, Proc. Saf. Env. Prot. 77 (B), 1999, 354-359

[8] D. Moncalvo, M. Davies, R. Weber, R. Scholz, Breathing losses from low pressure storage tanks, J. Loss Prev. Proc. Ind., accepted

[9] API 520 Sizing, selection and installation of pressure-relief devices, Ninth Edition, 2014

[10] ISO 4126-1, Safety devices for protection against excessive pressure. Part 1: safety valves, Third edition, 2013

Methodology to Determine Relief Load for Trans-Critical Pool Fire

Craig Powers
Aspen Technology, Inc.
20 Crosby Dr, Bedford MA 01730
craig.powers@aspentech.com

Prepared for Presentation at
American Institute of Chemical Engineers
2016 Spring Meeting
12th Global Congress on Process Safety
Houston, Texas
April 11-23, 2016

Methodology to Determine Relief Load for Trans-Critical Pool Fire

Craig Powers
Aspen Technology, Inc.
20 Crosby Dr, Bedford MA 01730
craig.powers@aspentech.com

Keywords: supercritical pool fire pressure relief

Abstract

Adequate pressure relief design is of high priority in a variety of industries in order to maintain compliance with governmental regulations including OSHA 1910.119 and Pressure Equipment Directive 97/23/EC. Accurate relief sizing reduces the potential risk of damages to personnel, equipment, and production schedule. One of the most common overpressure scenarios for relief sizing is external pool fire. Among the most challenging types of fire analysis is a trans-critical pool fire, where the process fluid is subcritical at the onset of relief then progresses to become fully supercritical during the fire. Often, process simulation is required to get accurate fluid properties at different times during the fire, then a relief load is estimated using the 'alternative method' outlined in the 6th edition of API Std 521, Section 4.4.13.2.4.4, which examines the change in specific volume over time. This paper will examine, through process simulation, the effect of using several alternative methods to determine the required relief load for trans-critical fires.

1 Introduction

Pressure relief systems are an essential component of any process unit design. Relief systems must be designed to meet regulatory requirements, including 29 CFR 1910.119 and Pressure Equipment Directive 97/23/EC. Accurate relief sizing reduces the potential risk of damages to personnel, equipment, and production schedule. One of the most common overpressure scenarios for relief sizing is external pool fire. External pool fire scenarios are often calculated using the wetted vessel method outlined in the sixth edition of API Standard 521, section 4.4.13.2.4. [1] As an alternative, the Guidelines for Pressure Relief and Effluent Handling Systems suggests performing a batch distillation calculation. [2] For systems without phase change, API also offers a stepwise expansion calculation. Similar methods are also available in the literature. [3] [4] [5]

Upstream oil and gas production facilities present significant challenges in performing fire calculations. Because these facilities handle fluids at near-wellhead conditions, they operate at high pressures. The hydrocarbon liquids in the high pressure separators in these facilities may have relatively high fractions of methane, leading to critical pressures that are close to operating conditions. When performing a fire calculation, it may prove impractical to obtain a heat of vaporization to calculate the required relief based on a boiling calculation. Stepwise flash calculations are complicated by the potential for the liquid to disappear rapidly as the pressure passes the cricondenbar (the maximum pressure at which two phases are present, which may be higher than the critical pressure). Moreover, in three-phase systems, a water phase may remain, which will also influence relief calculations.

In this paper, we will explore the use of process simulation to calculate fire relief requirements for fluids operating near their critical pressure. The method that is presented involves a combination of spreadsheet calculations and process simulator unit operations. Limitations and potential extensions will also be discussed.

2 Calculation Method

2.1 Required Relief Load

Fundamentally, a vessel exposed to fire goes through a batch distillation process: the fire heats the vessel; the contents heat up, increasing the pressure until the opening pressure of the pressure relief device is reached. Then, as heat input continues, the density in the vessel will decrease as material is removed—in a typical vessel containing liquid, the material relieved is principally vapor evolved from the boiling liquid, but in vessels containing only vapor or supercritical fluid, the relief is from the changing density of the vessel contents.

To model this process, a single-stage batch distillation is performed in discrete steps. The system is maintained at the relief pressure (design pressure plus 21% allowable overpressure), and the required relief volume is the excess material due to vaporizing liquid and density changes in all phases. The calculation may be performed using a process simulator combined with a spreadsheet calculation.

The process simulator portion of the calculation involves two parts. The first part takes the vessel contents to relief conditions. First, the relative quantities of liquid and vapor are adjusted to agree with the specified liquid level in the system. This may be accomplished using a separator to split the liquid and vapor, a tee to remove the excess material (typically in the vapor phase), a mixer to combine the phases back together, and an adjustment calculator to achieve the volume ratio target. A conceptual diagram of this process is shown in Figure 1.

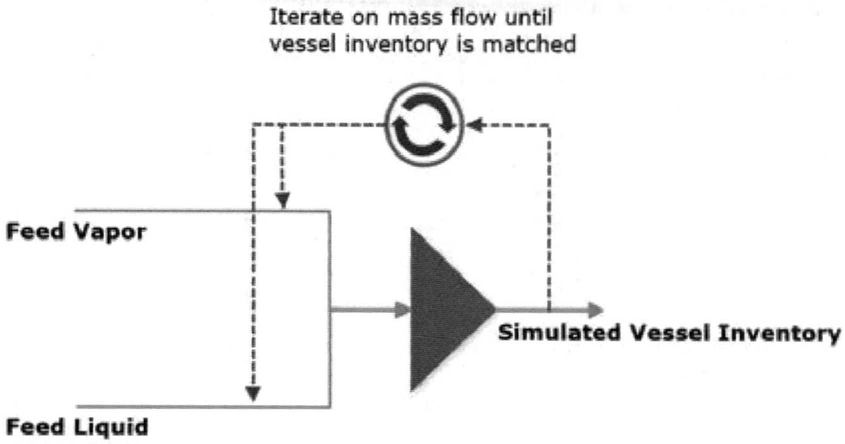

Figure 1. Feed Rebalance Operation

Then, the resulting stream is compressed to the relief pressure and heated until the bulk density matches the bulk density at operating conditions. This may be accomplished using a compressor or pump unit operation to increase the pressure of the material, a heater to increase the temperature, and another adjustment calculator to vary the temperature to attain the bulk density target.

The stream from the previous steps is then used as the input into the stepped fire calculation. Each calculation stage involves a heater to perform the step fire heat input, a separator to split the resulting liquid and vapor phases, a tee to pull the relief off of the vapor phase, and a mixer to produce the new vessel contents. A sample flowsheet of this process is shown in Figure 2. The output of the heater may be based on a temperature step, a target vapor fraction, or a duty step. The amount of vapor removed (the relief quantity) is obtained from the spreadsheet calculation detailed below. The following properties are used as inputs to the spreadsheet calculation: bulk specific enthalpy entering the step, bulk specific enthalpy leaving the heater, fraction of each phase leaving the heater, density of each phase leaving the heater, and molecular weight of each phase leaving the heater. These are then used in the spreadsheet to obtain the required relief on each step.

It is assumed that there is enough freeboard in the vessel so that the relief is entirely vapor. In cases with low freeboard or materials that tend to foam, a disengagement calculation is recommended. Methods for doing this calculation are available in the literature. [6]

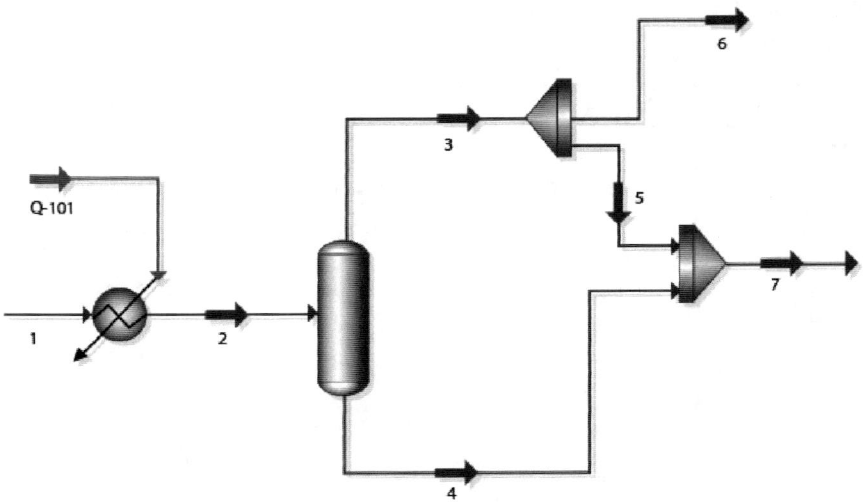

Figure 2. Flowsheet For Each Analysis Step

The calculations in the spreadsheet are as follows.

The volume taken up by the liquid after heat input in step (n) is:

$$V_\ell = \frac{m_\ell}{\rho_\ell}$$

Where: V_ℓ is the volume of liquid, m_ℓ is the mass of liquid, and ρ_ℓ is the mass density of liquid after heat input

The volume of vapor relieved in step (n) is:

$$V_{rel} = \frac{m_{total} f_v}{\rho_v} - (V - V_\ell)$$

Where: V_{rel} is the volume of vapor relieved, m_{total} is the total mass at the beginning of the step (both liquid and vapor), f_v is the mass fraction vapor after heat input, and V is the total vessel volume

The mass of vapor relieved in step (n) is:

$$m_{rel} = V_{rel} \rho_v$$

Where: m_{rel} is the mass of vapor relieved and ρ_v is the vapor density after heat input

The liquid and vapor properties may be taken from stream "2" in Figure 2.

The mass from this step is used to divide the vapor from this step into the part that is relieved (stream "6") and the part that is retained (stream "5"). The accounting in the process simulator may be easier if the initial flowrate is set to be the same as the initial total mass of the vessel contents (e.g. if the mass is calculated to be 5,000 lb, then the initial flowrate is set to 5,000 lb/h)—then, the amount of vapor removed in the simulator follows directly from the value of m_{rel} calculated above (e.g. if the mass relieved in step (n) is 250 lb, then the vapor relief stream in the simulator would have a flowrate set to 250 lb/h).

At this point, the process simulator is ready for the next calculation step, but the results obtained thus far are not sufficient for a relief calculation. To evaluate a pressure relief valve, a flowrate is required. As noted before, steps are typically taken by temperature, vapor fraction, or enthalpy, so an additional calculation must be taken to determine how much time passed.

Based on the volume of liquid, the wetted area of the vessel may be calculated. In this paper, the simple API heat flux equation is used to calculate the fire heat input into the vessel: [1]

$$Q = CFA_{ws}{}^{0.82}$$

Where: Q is the heat input into the vessel, C is a unit conversion constant, F is a factor to account for the presence of fireproof insulation (the value of F is 1.0 for an uninsulated vessel), and A_{ws} is the wetted surface area of the vessel

The amount of time that passed in the calculation step is:

$$\Delta t = (\hat{H}_{final} - \hat{H}_{initial})m_{total}/Q$$

Where: Δt is the amount of time passed in the calculation step, $\hat{H}_{final}$ is the specific enthalpy on step (n) after heating (but before material is removed for relief, stream "2"), $\hat{H}_{initial}$ is the specific enthalpy on step (n) before heating (stream "1"), and m_{total} is the total amount of material present

Then, the required relief flow rate is simply:

$$w = m_{rel}/\Delta t$$

Depending on the behavior of fluid, the maximum relief requirement may be on the first step of the procedure, or it may occur later. The maximum relief is best characterized based on the required pressure relief device area, as the literature as shown that this maximum may not correspond to the peak of either mass or volumetric flow. [3] Strictly, the system should be analyzed up to a projected vessel failure condition (or other standard end point, e.g. a time limit), although there may be an earlier step where it is reasonable to conclude that the peak relief has already passed.

As an alternative to the simple API heat flux equation, a more detailed treatment of heat absorption could be used. A combination radiation/convection Stefan-Boltzmann model is presented in Annex A of API 521 and in Appendix A of Scandpower report 27.207.291/R1. [7]

2.2 Required PRD Area and PRD Capacity

There are two methods offered for calculating relief device required area and capacity in API Standard 520 Part I that are applicable for vapor or supercritical systems. [8] One is for vapor sizing, recommended for use when the compressibility Z is within a range of $0.8 < Z < 1.1$. Kim and coworkers have studied the extension of this method to cases involving more extreme conditions, as might be seen in relief of a supercritical system. [9] The other method offered in API 520 Part I is the general direct integration method.

The unmodified API vapor method is unlikely to give satisfactory results for relief of a supercritical fluid as in this case. Both Kim et al.'s modification of the API Vapor and the direct integration method have been observed to give reasonable results. Kim et al.'s method is easier to use as part of a table calculation for screening purposes, while direct integration may be easily used in a process simulator that supports calculating a property table for a stream, which makes it suitable for a more exact calculation of required relief on a specifically-identified stream. In the cases that follow, the method of Kim et al. is used to screen for the step with the largest required PRD area, then the required PRD area for that specific relief stream is calculated using the direct integration method.

3 Applications

3.1 High Pressure Separator, Case (a)

The equipment to be analyzed is a high pressure separator operating at 90 °F and 1200 psig, with a design pressure of 1440 psig. The separator normally handles a three-phase fluid. For the purposes of this analysis, the separator is assumed to be a horizontal vessel with 2:1 ellipsoidal heads, a diameter of 5 ft, a tan/tan length of 10 ft, and an initial liquid level of 2 ft. To account for associated process piping, an additional 15% is added to the calculated wetted area. No credit is taken for adequate drainage and firefighting. The overall composition of the stream entering the separator is given in Table 1. Calculations were performed using Aspen HYSYS V8.8. Initial estimates for required orifice area were calculated using the method of Kim et al., while the final required orifice area was calculated using direct integration. A phase envelope of the system prior to relief is shown in Figure 3.

Table 1. Case (a) Composition

Comp.	Mole Fraction
N2	0.001
CO2	0.02
C1	0.73
C2	0.135
C3	0.057
iC4	0.013
nC4	0.02
iC5	0.007
nC5	0.006
nC6	0.003
nC7	0.002
nC8	0.001
nC9	0.0003
nC10	0.0001
H2O	0.0046

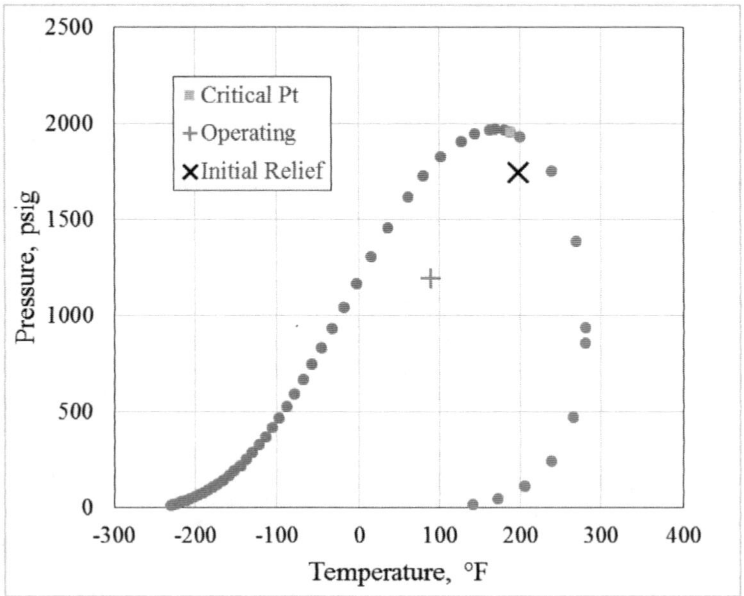

Figure 3. Phase Envelope for Case (a)

3.1.1 API Wetted Vessel / Boiling Liquid Calculation

The simplest form of the API wetted vessel / boiling liquid fire calculation requires application of the fire heat flux to boil the liquid contents of the vessel. The liquid is typically taken at operating conditions, then brought to its bubble point at the required relief pressure. In this case, with 21% allowable overpressure for a fire scenario, the relief pressure is 1742 psig.

This simplified calculation fails in this case, because the cricondenbar of the light liquid phase is below the relief pressure.

3.1.2 Stepwise Boiling Liquid Calculation

As an alternative to the simplified problem setup above, consider a more rigorous calculation of the vessel contents at relief conditions. The quantities of the liquid and vapor phases are adjusted to match the operating volume fraction, then an isochoric flash is performed to bring the fluid to the relief pressure, as discussed in the Calculation Method section above.

When this process is performed, at relief pressure, there is still a substantial amount of hydrocarbon liquid present. This may be surprising given the failure of the simplified boiling liquid calculation; however, the inclusion of the vapor in the constant density flash is very important. In this case, a significant amount of methane enters the liquid phase as the mixture is compressed; in fact, comparing the light liquid at relief to the light liquid at operating conditions, approximately 50% more methane is present on a mole fraction basis (0.47 vs. 0.29).

However, this calculation is still problematic, as there are several dramatic phase transitions that complicate the use of a boiling liquid calculation. First, around 225 °F, the hydrocarbon liquid phase ceases to vaporize, becoming a supercritical phase that the simulator identifies as vapor. Then, around 265 °F, the phase identification changes from liquid to vapor. Finally, around 620 °F, the water phase is vaporized. Stepping by temperature will give unrealistically large requirements in the steps where the large phase transitions occur. The two most likely alternatives both fail in this case: stepping by vapor fraction cannot work with the supercritical phase transition in the hydrocarbon phase, and stepping by enthalpy results in a flash failure. Moreover, the relatively small amount of liquid that vaporizes before the phase transition suggests that this method (based solely on liquid vaporization) is a poor choice.

3.1.3 Supercritical / Vapor Expansion Calculation

Given that the hydrocarbon liquid portion is a supercritical fluid at the relief pressure of 1742 psig, an alternative method of calculation would be to separately consider fluid expansion in the vapor/hydrocarbon portion of the system and boiling of the water. This method gives a required relief rate of 4,380 lb/h and a required orifice area of 0.03 in^2. The required relief rate for boiling the water portion of the feed is significantly smaller, 830 lb/h with a required area of 0.011 in^2.

3.1.4 Detailed Step Calculation

The method presented earlier in this paper is applied to the system, with heat input taken based on the simple API heat flux. To account for reduced heat transfer to supercritical / gas space, a correction factor of 0.3 was applied to the full-vessel heat transfer. [2] For any step, the larger of the wetted heat transfer (without correction factor) or total vessel heat transfer (with correction factor) was used. Maximum relief occurs on the initial step, with a required relief load of 4,370 lb/h and a required orifice area of 0.03 in² was obtained by a direct integration calculation. Although the water phase might be expected to flash suddenly, this was not observed in the simulator, as it disappeared gradually over a broad temperature range, with the vessel contents becoming entirely vapor around 375 °F.

Selected results from this calculation are presented in Table 2.

Table 2. Selected Results from Case (a)

T, °F	$\hat{H}_{initial}$, BTU/lb	$\hat{H}_{final}$, BTU/lb	VF, molar	HC frac, molar	ρ_v, lb/ft³	ρ_ℓ, lb/ft³	m_{rel}, lb	Δt, s	w, lb/h	A, in²
202.8	-1567.6	-1562.9	0.664	0.256	11.85	24.47	54.69	45.07	4369	0.0321
207.8	-1565.6	-1560.8	0.690	0.230	11.92	24.49	54.49	46.59	4210	0.0307
212.8	-1563.7	-1558.9	0.718	0.201	11.99	24.56	54.33	48.30	4049	0.0294
247.8	-1555.9	-1551.0	0.918	0	12.10	58.48	63.40	54.87	4160	0.0298
252.8	-1556.3	-1551.5	0.918	0	11.84	58.34	61.01	53.97	4070	0.0294
257.8	-1556.7	-1551.8	0.918	0	11.59	58.19	58.73	53.15	3978	0.0289

3.2 High Pressure Separator, Case (b)

The equipment to be analyzed is a high pressure separator operating at 5 °C and 64.5 bar, with a design pressure of 80 bar(g). The separator normally handles a two-phase fluid with no water present. For the purposes of this analysis, the separator is assumed to be a horizontal vessel with 2:1 ellipsoidal heads, a diameter of 2 m, a tan/tan length of 5 m, and an initial liquid level of 0.75 m. To account for associated process piping, an additional 15% is added to the calculated wetted area. No credit is taken for adequate drainage and firefighting. The overall composition of the stream entering the separator is given in Table 3. Calculations were performed using Aspen HYSYS V8.8. Initial estimates for required orifice area were calculated using the method of Kim et al., while the final required orifice area was calculated using direct integration. A phase envelope of the system prior to relief is shown in Figure 4.

Table 3. Case (b) Composition

Component	Mole Fraction
N2	0.0299
CO2	0.2183
H2S	0.0479
C1	0.5974
C2	0.0461
C3	0.0235
nC4	0.0148
nC5	0.0065
nC6	0.0048
nC7	0.0030
nC8	0.0032
nC9	0.0032
E-Mercaptan	0.0014

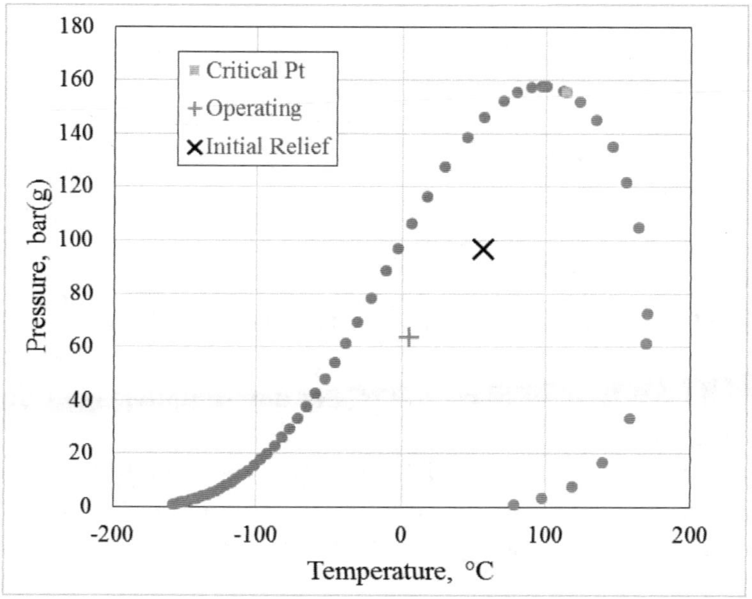

Figure 4. Phase Envelope for Case (b)

3.2.1 API Wetted Vessel / Boiling Liquid Calculation

Because the liquid portion of the separator feed is below its critical pressure at the relief pressure of 96.8 bar(g), a simplified boiling liquid calculation may be performed. Based on the liquid initially at its bubble point and finally at 10 mole percent vaporized, with sensible heat included, the heat of vaporization is 730 kJ/kg. The required relief rate is 3,840 kg/h with a required orifice area of 0.45 cm^2.

Note that using increasing spans of 10 mole percent vaporized (e.g. initially 10% vaporized, finally 20% vaporized) will decrease the heat of vaporization, increase the required relief rate, and increase the required orifice area, assuming that no changes in liquid level are considered. However, at higher fractions vaporized, the liquid level would be reduced, and less heat input to the fluid would be expected. This effect is part of the motivation for the step calculations.

3.2.2 Stepwise Boiling Liquid Calculation

The stepwise boiling liquid calculation was conducted as described under Case (a). As would be expected given the success of the simple boiling liquid calculation in this case, the stepwise boiling liquid calculation begins without issue. Similarly to the previous case, this calculation fails due to the phase behavior of the liquid as vapor is removed and its temperature increases. Around 170 °C, the liquid exceeds its cricondenbar. Phase identification in supercritical fluids is a difficult problem; in this case, the simulator continues to identify the material as liquid phase until it reaches 250 °C, at which point the phase identification changes to all vapor. For a boiling liquid calculation, this is a major concern, as all of the liquid turning into vapor over a short step requires an excessive amount of area for adequate relief. Alternatively, the initial step could be considered, but the required relief rate then is unreasonably low at 160 kg/h with a required device area of 0.02 cm^2.

3.2.3 Supercritical / Vapor Expansion Calculation

Because the liquid contents of the vessel are not supercritical at initial relief conditions, a supercritical / vapor expansion calculation cannot be conducted except by changing the initial state of the material in the vessel.

3.2.4 Detailed Step Calculation

The method presented earlier in this paper is applied to the system, with heat input taken based on the simple API heat flux. To account for reduced heat transfer to supercritical / gas space, a correction factor of 0.3 was applied to the full-vessel heat transfer. [2] For any step, the larger of the wetted heat transfer (without correction factor) or total vessel heat transfer (with correction factor) was used. Maximum relief occurs on the initial step, with a required relief load of 4,310 kg/h with a required orifice area of 0.52 cm^2. Note that the required area calculated here is 18% higher than the area requirement from the simple API wetted vessel calculation.

Selected results from this calculation are presented in Table 4.

Table 4. Selected Results from Case (b)

T, °C	$\hat{H}_{initial}$, kJ/kg	$\hat{H}_{final}$, kJ/kg	VF, molar	ρ_v, kg/m³	ρ_ℓ, kg/m³	m_{rel}, kg	Δt, s	w, kg/h	A, mm²
76.3	-4007.2	-3949.0	0.582	137.83	519.2	437.4	374	4207	51.9
96.3	-3843.1	-3784.7	0.614	135.16	491.0	365.9	356	3696	46.5
116.3	-3698.7	-3639.3	0.654	135.17	456.6	325.6	348	3367	42.8
196.3	-3290.9	-3230.2	1	153.16		294.9	373	2849	35.8
216.3	-3230.2	-3177.1	1	139.29		246.8	294	3017	39.3
236.3	-3177.1	-3124.6	1	128.36		194.7	265	2650	35.6

4 Conclusion

The detailed stepwise fire calculation method presented in this paper offers significantly more robust performance for fluids near their critical point. In particular, simple boiling liquid calculations often fail in these cases, and may be less conservative than the detailed method when they do not fail. The single-phase expansion method may give comparable results for some cases, but is not suitable for cases where two phases are present in the vessel at the initial relief conditions. As an additional benefit, the detailed method is suitable for incorporating more advanced heat input methods, such as the detailed convection-radiation Stefan-Boltzmann model.

5 References

[1] American Petroleum Institute, *API Standard 521 6th Ed.: Pressure-relieving and Depressuring Systems,* Washington, DC: API Publishing Services, 2014.

[2] Center for Chemical Process Safety of the American Institute of Chemical Engineers, Guidelines for Pressure Relief and Effluent Handling Systems, New York: American Institute of Chemical Engineers, 1998.

[3] R. Ouderkirk, "Rigorously Size Relief Valves for Supercritical Fluids," *Chemical Engineering Progress,* pp. 34-43, August 2002.

[4] R. C. Doane, "Designing for Pressure Safety Valves in Supercritical Service," *Hydrocarbon Processing,* pp. 63-67, January 2010.

[5] M. A. Grolmes, "Vent Sizing for Pressure Relief from the Super-Critical State - Revisited," in *DIERS User Group Meeting*, Las Vegas, April 2013.

[6] H. Fisher, H. Forrest, S. Grossel, J. Huff, A. Muller, J. Noronha, D. Shaw and B. Tilley, Emergency Relief System Design Using DIERS Technology, New York: American Institute of Chemical Engineers, 1992.

[7] Scandpower Risk Management AS, *Guidelines for the Protection of Pressurised Systems Exposed to Fire,* 2004.

[8] American Petroleum Institute, *API Standard 520 Part I 9th Ed.: Sizing, Selection, and Installation of Pressure-relieving Devices,* Washington, DC: API Publishing Services, 2014.

[9] J. S. Kim, H. J. Dunsheath and N. R. Singh, "Proper Relief-Valve Sizing Requires Equation Mastery," *Hydrocarbon Processing,* December 2011.

Dynamic Simulation Modeling to Protect Flare Headers from Air Ingress

Jesse Mumm
Bechtel Oil, Gas and Chemicals
Houston, Texas
jlmumm@bechtel.com

Satyajit Verma
Bechtel Oil, Gas and Chemicals
Houston, Texas
sverma4@bechtel.com

Freeman Self
Bechtel Oil, Gas and Chemicals
Houston, Texas
feself@bechtel.com

Prepared for Presentation at
American Institute of Chemical Engineers
2016 Spring Meeting
12th Global Congress on Process Safety
Houston, Texas
April 11-23, 2016

Dynamic Simulation Modeling to Protect Flare Headers from Air Ingress

Jesse Mumm
Bechtel Oil, Gas and Chemicals
Houston, Texas
jlmumm@bechtel.com

Satyajit Verma
Bechtel Oil, Gas and Chemicals
Houston, Texas
sverma4@bechtel.com

Freeman Self
Bechtel Oil, Gas and Chemicals
Houston, Texas
feself@bechtel.com

Keywords: flare, air ingress, dynamic simulation, flare header, make-up gas, vacuum protection

Abstract

A flare header system is subject to vacuum formation inside when a hot relief stops and the internal gases cool and/or condense. If make-up gas is not provided at an adequate rate, air may ingress from outside and potentially create a flammable air-hydrocarbon mixture inside the header. With the presence of a pilot as a ready source of ignition, explosions inside the header system could occur with potentially catastrophic consequences. This paper addresses six major considerations for selecting, sizing and specifying make-up gas requirements. The pros and cons of the locations and combinations of various types of sensors—such as differential temperature, actual temperature, header pressure, gas flow, and oxygen concentration—are discussed. Finally, recommendations are provided for instrumented controls to activate make-up gas addition, in addition to methods for determining the rate and duration of addition. Graphs and charts showing temperature and pressure profiles as well as vacuum formation as a function of time are presented to illustrate the concepts.

1. Overview of Vacuum Formation Mechanism

The pressure inside a flare header fluctuates depending upon the nature of relief events and ambient temperature conditions. On one hand, back pressure generated during a relief event may exceed the design pressure of the header, while on the other hand, severe vacuum may form inside the header after the relief has stopped and potentially exceed the design vacuum. In a header designed to mechanically withstand the extremes of high and low pressures, still another condition may exist. Because the flare stacks, when not equipped with a liquid seal drum or other means of preventing air ingress, are essentially open to the atmosphere, the flare headers are prone to air ingress from outside, even at small vacuums. This can potentially result in explosive air-hydrocarbon mixtures inside the header.

Air ingress into a flare header system is a distinct possibility when cooling or condensation of hot vapors happens after a release event is over and the header is stagnant. The resulting rate of reduction in volume can easily be larger than the constant seal purge flow to the flare stack. Under these circumstances, outside air will be drawn into the header and the resulting mixture of air and hydrocarbons in some sections of the header may fall within the flammability envelope. Because sources of ignitions are ubiquitous—the pilot being a guaranteed source—explosions are likely. The consequences of an explosion can range from damage to the flare header system to significant injury to plant personnel, lost production, and harm to company reputation.

In order to avoid the ingress of atmospheric air into the flare header, make-up gas must be injected into the flare header to compensate for the reduction in volume of vapors in the header.

2. Dynamic Simulation Model

Dynamic simulation enables one to rigorously model the process, equipment, and controls using differential and algebraic equations resulting from mass and energy balances, phase equilibria, heat transfer, and hydraulics. These equations are integrated over time to predict how the process and its equipment will respond to various disturbances.

A detailed computer model of a flare header system was prepared and divided into small segments. The model included pressure relief valves situated along the header, a flare knock-out drum, and a flare stack with a velocity seal that was open to atmosphere. The system did not contain a liquid seal drum at the base of the stack. Various reliefs, one at time, were introduced into the header system and the resulting flow conditions and heat transfer inside the header followed as a function of time and length of the header using appropriate heat transfer equations and thermodynamic properties of the released streams. Commonly available dynamic simulation software was used for this purpose.

For simplicity and speed of convergence, each header segment was considered to be a well-mixed region in terms of thermal and physical properties. The models were integrated long after relief was stopped to capture the cooling of relief gases, as well as the cooling of the header itself that took place due to the header being at a temperature higher than the ambient temperature. Included in the study were cases where the cooling vapor will partially or completely condense. The flow of gases through the flare stack was monitored as well as the temperature and pressure profiles along the length of the header. As a result of vacuum formation mentioned above, flow reversal is noted in many cases within a few seconds after the hot relief stops and cooling begins. The methods of initiating make-up gas injections, required make-up gas flow rates, and duration of the injections were investigated to determine the resulting header conditions and develop a control scheme that would prevent flow reversal at the stack that could otherwise result in air ingress into the header system. Cases studied included warm releases, cryogenic releases, and combination releases into the flare header system.

3. Establishing the Design Parameters

In order to provide a robust but economical design for instrumentation, piping, and make-up gas usage, the following six major design parameters should be considered:

1. Cooling/condensation mechanisms
2. Sensors to detect relief
3. Location of temperature sensors
4. Make-up gas flow rates
5. Location of make-up gas addition point
6. Starting and stopping of make-up gas

For existing systems with instrumentation and piping constraints, additional variations and simplifications will also be discussed.

3.1 Cooling/Condensation Mechanisms

There are two cooling mechanisms to consider when designing a make-up gas system:

- Cooling from ambient temperature or rain – The gas in the header is cooled by heat transfer through the pipe wall to the ambient air. If only cooling of the gas is involved, meaning no condensation occurs, the heat transfer is typically limited by the inside heat transfer coefficient. The occurrence of rain has minimal effect in these instances

  If condensation of the gas also occurs, the outside heat transfer coefficient becomes more important. Rain will increase the outside heat transfer coefficient, but the impact is small compared to other variables. Additionally, while a layer of ice on the header may provide insulation from cold ambient temperatures, it can also absorb heat from the gas. However, as with the effect of rain, the impact is not generally significant within the modeling capabilities.

- Cooling/Condensation from a cold header – Cooling and/or condensation of the flare gas may occur due to a cold header, which has been cooled below ambient temperature. In these instances, the header is cooled by a previous cold release. These cases are most common in cryogenic plants such as ethylene or LNG. The duration of the initial cold release has a large impact on the conditions of the header; the longer the initial cold release, the colder the initial temperature of the header.

3.2 Sensors to Detect Relief

When designing or configuring the control scheme for make-up gas injection, two important aspects to consider are how to detect the hot release and where to locate the sensors. This section will discuss possible methods of detecting relief as well as the importance of properly locating the sensors.

There are several methods to sense when relief occurs that, if cooled or condensed, will cause volume contraction in the header. This section will discuss advantages of each of the following detection methods:

- Temperature difference (gas minus ambient) – The temperature difference measures the difference between the gas temperature and the ambient temperature, under the assumption that the ambient temperature drives the cooling
- Actual temperature of gas being flared
- Pressure in header
- Oxygen composition in header

3.2.1 Temperature Difference (Gas Minus Ambient)

For cases where contraction is expected to occur through cooling, the rate of volume contraction is dependent on the rate of heat loss from the header to the ambient surroundings. Because the rate of cooling is driven by the temperature difference between the hot gases inside the header and the ambient temperature, the trigger to start make-up injection should be based on a temperature differential measurement. When this temperature difference exceeds a pre-determined criterion, make-up gas should be introduced. Once the temperature in the header drops back below this criterion, the addition of make-up gas can safely be stopped.

Measuring temperature along the entire header to ensure that all points are below the specified threshold before stopping the make-up gas flow would be ideal for reliefs where condensation or cooling are possible, however, installing many such measurements points may be infeasible. Specifying an additional time to keep injecting make-up gas after the threshold has been reached may be necessary to ensure the temperature downstream of the sensor has sufficiently cooled.

Identifying the minimum temperature difference criteria to inject make-up gas is crucial to ensuring the system can prevent ingress under a variety of conditions. Even at low temperature differentials a vacuum can form, however any seal purge gas injected upstream of the velocity seal of the flare stack may feed the vacuum up to a certain threshold ΔT before the addition of make-up gas is required.

This ΔT threshold that the seal gas can maintain will generally become the criterion to inject make-up gas, however to avoid any potential constant injection of make-up gas, additional consideration should be made to ensure the ΔT between the make-up gas and ambient temperature does not exceed this value.

3.2.2 Actual Temperature of Gas Being Flared

<u>Cooling cases</u>: Using actual temperature instead of differential temperature to detect hot vapor reliefs is generally not a recommended solution due to the difficulty of establishing an effective criterion separate from ambient temperature. The severity of contraction is dependent on the temperature differential and not the actual temperature of the gas, and because the ambient conditions can vary significantly due to changes in weather, a single temperature criterion may not be suitable for all conditions and relief cases. For example, the criterion should be chosen so that

it will effectively cover low ambient temperature conditions, but on a hot day the ambient temperature may approach the criterion and unnecessarily trigger make-up gas injection. Conversely, if the criterion is set too high, on a cold day the temperature differential between relieving gas and ambient might exceed the minimum threshold of the seal gas while the actual temperature remains below the criterion and therefore fail to trigger make-up gas.

Condensation cases: For cases where condensation can occur, actual temperature readings may be used because the dew point is independent of the ambient temperature and a condensable gas must necessarily be above its dew point initially in order for contraction from condensation to occur. Oftentimes the relieving temperature of a condensable gas is much higher than the minimum threshold for air ingress due to cooling only, so a higher temperature criterion may be used. Because contraction due to cooling may be still observed in addition to condensation, however, it may be advantageous to use a single temperature differential criterion that accounts for both cooling and condensation instead of actual temperature for purposes of simplification.

For cases when condensation can be expected to occur only if the header has been pre-cooled to below the dew point of a relief, the actual temperature of the header itself should be measured. For these cases, condensation can be expected at temperatures well below ambient, so the actual temperature criteria to begin make-up injection should be based on the dew point of the condensable gas. Because the header itself is the source of cooling in these cases and not the ambient temperature, the skin temperature of the header wall should be measured and gas should be injected for as long as the header is below the dew point of any condensable reliefs. Ideally the temperature along the entire header would be measured to ensure condensation is no longer possible, however it may be necessary to calculate the required time to warm up the header using make-up gas and use this data to set a minimum duration for make-up gas injection.

3.2.3 Pressure in Header

Pressure sensors are generally not sufficiently sensitive or fast enough to detect ingress and trigger make-up flow. If located near the flare stack, the pressure reading may be less than atmospheric pressure because of the buoyancy effect. Additionally, if the control action is not fast enough, the ingress of air may have already occurred and the header will pressurize as air fills the vacuum, thus fooling the control action.

3.2.4 Oxygen Composition in Header

Because the purpose of make-up gas injection is to prevent the formation of flammable air-hydrocarbon mixtures, the amount of oxygen in the header is of significant concern. However, using measurements of oxygen to trigger make-up gas injection is difficult for several reasons. First, the sensor will not sense the presence of oxygen until air has already started to ingress. Additionally, because such a sensor would need to be located near the flare stack in order to sufficiently detect the ingress of air, the potential distance between the relief and the measurement may introduce a significant time delay and even more air may ingress before the make-up gas can mitigate. Finally, the reliability and speed of such measurements is a concern, especially when compared to temperature measurements.

3.3 Location of Temperature Sensors

Improper choice of location for temperature sensors could introduce potential delays in taking preventive action. The preferred location to detect relief is on the main header immediately downstream of where the relief lateral ties into the header.

Sensing the temperature further downstream, such as at the knockout drum, can be ineffective for many reasons. First, the relieving gas will cool as it travels down the header and the temperature difference between the gas and ambient will begin to narrow and will not be representative of the temperature upstream. Secondly, as a result of this cooling and the holdup of the upstream portion of the header, a time lag in sensing the temperature may be introduced.

For the example of a cooling vapor relief case, Figure 1 compares sensing the temperature in the header downstream of the relief lateral tee to sensing it at the knockout drum. The set point to initiate make-up flow was 5°C above ambient ($\Delta T = T_{gas} - T_{ambient}$) and a hot vapor relief started at t=10 minutes. As shown in this figure, when the sensor to trigger make-up was located near the lateral, the temperature reached the set point seconds after relief began. When the sensor was located at the knockout drum, the temperature reached the set point after nearly 10 minutes. Because of this delay, if the duration of the relief was less than 10 minutes, the make-up gas would not be triggered and air ingress could occur.

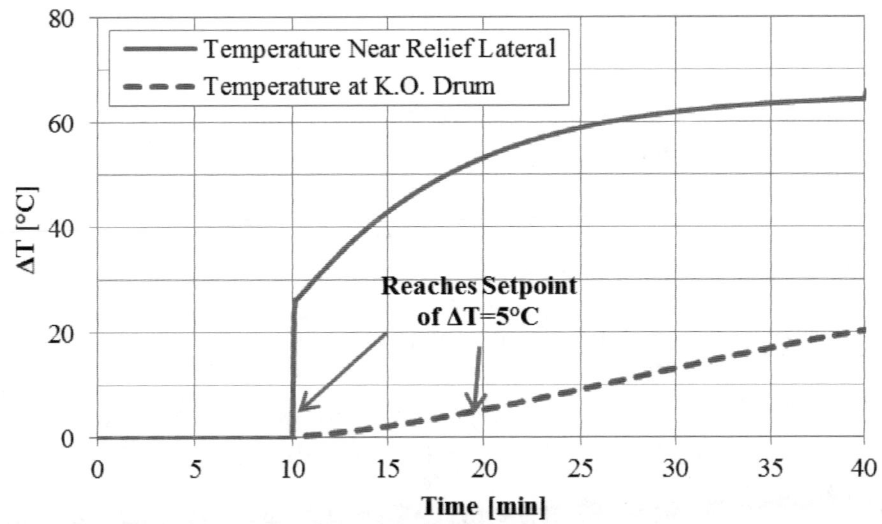

Figure 1: Sensing temperature at the relief lateral and at the knockout drum for a hot vapor relief.

Sensing the temperature inside the header is suitable for most cases. Sensing directly in the relief tailpipe or relief lateral may provide a faster response, but may also be infeasible if multiple reliefs must be considered. Additionally, locating the sensor in the lateral will not adequately measure the cooling of the header by the make-up gas, which is injected into the header rather than the lateral. This may unnecessarily extend the duration of make-up injection. On the other hand, sensing in the lateral could be used in conjunction with other sensors for special cases, such as individual cases that require separate higher make-up flows or are determined to require significantly faster response times.

3.4 Make-up Gas Flow Rates

Although the required make-up gas rate is a function of many factors, generalizations may be made to facilitate the establishment of the rate. These factors include the rate of cooling, location of the make-up gas injection point, potential for condensation, relief rate, seal purge rate, and duration of the relief. The following sections describe how these factors can impact the required rate of make-up gas.

3.4.1 Seal Purge Rate vs. Ambient Temperature

Seals are provided for flares open to the atmosphere, regardless of whether a pipe flare or a flare utilizing burners similar to industrial furnace burners (sometimes called sonic burners) are used. These seals are located on the open line upstream of the burner. The two most common types are velocity and kinetic seals. Seal purge gas is provided to increase the effectiveness. Husa (1) and API (2) offer recommendations for flow rates.

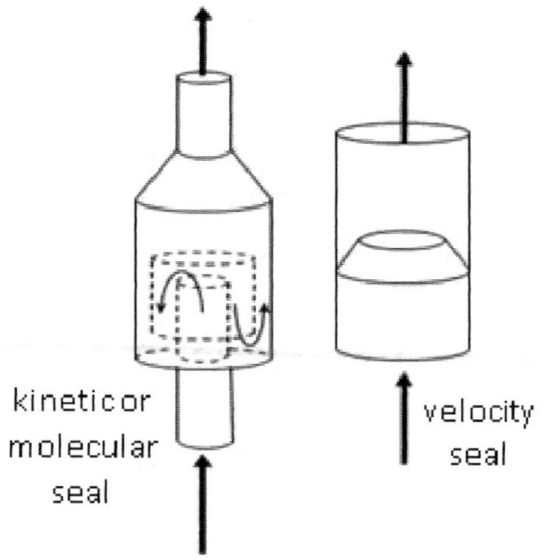

Figure 2: Velocity and kinetic seals.

However, even though these rates are not sufficient in most instances to protect against cooling and condensing of the flare gas, the seal purge does offer some minimal protection. This is dependent upon the header volume, the seal purge flow rate, and the rate of heat loss to the ambient (and therefore the differential temperature between the ambient and hot gases inside the header). The higher the seal purge rate, the higher the differential temperature range up to which the header is protected.

3.4.2 The Impact of Relief Duration on Make-up Gas Rates

The duration of relief has a significant impact on the formation of a vacuum. The temperature differential between the relieving gas and pipe wall is highest immediately after the relief begins, so relieving for short duration would yield the highest instantaneous rate of contraction. As relief continues, the pipe wall will begin to warm up with the gas in the header. This will decrease the heat transfer rate between the gas and the pipe wall but increase the overall total volume of gas that will contract.

The highest potential instantaneous rate of air ingress occurs simultaneously with the highest instantaneous rate of contraction. In order to eliminate any potential for air ingress, the make-up gas rate should be sufficient to prevent air ingress at this point. In order to determine the maximum temperature differential, and therefore highest rate of contraction, simulations should be conducted at the minimum ambient temperature.

The dynamic simulation example illustrates when the maximum rate of air ingress can occur. Figure 3 compares the volumetric flow rate measured at the base of the flare stack for a hot vapor release of methane at 130,000 Sm^3/h and 100°C at durations of 1, 5, and 30 minutes. A negative volumetric flow rate would indicate that a vacuum has formed and air has begun ingressing through the flare stack. As shown in the figure, the maximum rate of air ingress for this case occurred at a very short relief duration of 1 minute, but over time, longer relief durations exhibited a higher sustained rate of ingress because of the higher overall volume of relief gas present in the system.

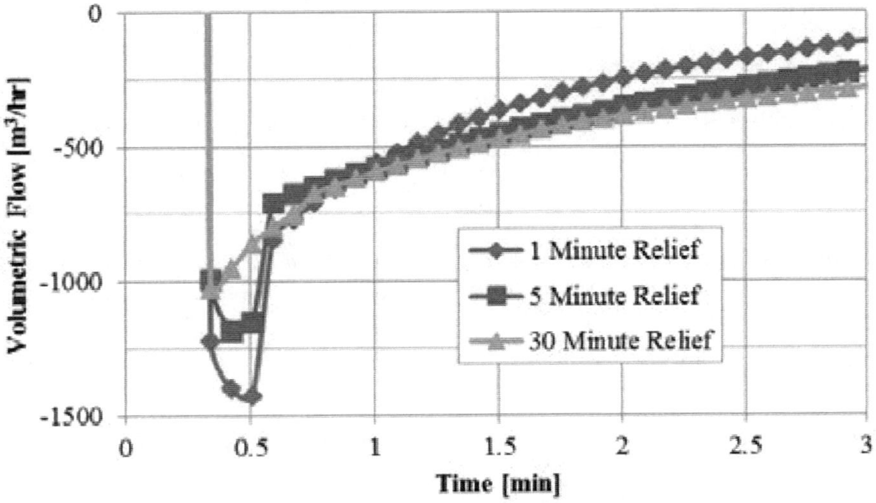

Figure 3: Volumetric flow rate at base of flare stack for hot vapor cooling case at varying relief durations.

3.4.3 Gas Composition and Cooling vs. Condensation

Hot releases of vapor that cool but do not contain condensable gases will exhibit slower and lower rates of volume contraction, with the heat transfer rate usually limited by the inside heat transfer coefficient. On the other hand, vapor containing condensable gases exhibit quick and high rates of volume contraction, with the rate of contraction strongly affected by the composition. In these instances the heat transfer rate will be limited by the outside heat transfer coefficient.

This difference is demonstrated by comparing Figure 3 for the cooling only case to Figure 4, which compares the volumetric flow rate measured at the base of the flare stack for a condensable vapor release of light naphtha at 30,000 Sm^3/h and 150°C at durations of 1, 2, and 5 minutes. As for the cooling only case, the maximum rate of air ingress occurred at a very short relief duration of 1 minute because the difference between ambient and the relieving gas remained the highest. At longer relief durations, the header may even warm above the condensing temperature of the gas, resulting in a slower rate of contraction because the header would need to cool before the gas could be condensed. This is demonstrated by the delayed negative flow rate peak for the 5 minute relief duration.

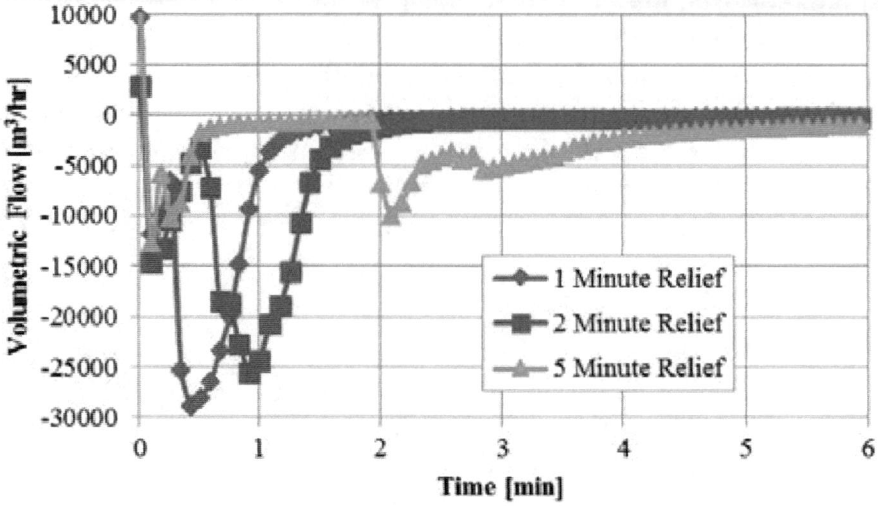

Figure 4: Volumetric flow rate at base of flare stack for hot vapor condensation case at varying relief durations.

3.4.4 Establishment of Make-up Gas Rate

Once the potential for vacuum formation and air ingress is understood, additional dynamic simulations can be performed to determine the flow rate of make-up gas necessary to prevent air ingress. The actual required rate is not necessarily equal to the maximum reverse flow observed, depending on the location of the make-up gas injection point and temperature and composition of the make-up gas. The make-up gas may compensate for volume contractions and displace the hot gas in the header and therefore reduce the overall rate of contraction. For cases with condensation, the make-up gas may also dilute the condensing gas and favorably affect the dew point of the mixture.

Figure 5 shows the volumetric flow rate at the base of the flare stack for a condensable relief with and without make-up flow. For this case, a hot vapor stream of light naphtha consisting of mostly C5-C6 hydrocarbons was released into the header for five minutes. At an ambient temperature of 20°C, this mixture will start to condense after an initial period of cooling. Without make-up, significant rates of contraction were observed immediately after relief subsided. A make-up flow rate of 6500 Sm^3/hr was required to prevent air ingress.

The condensing cases required a significantly higher make-up gas rate than the cooling only cases in Figure 3. For the cooling only cases shown in Figure 3, a make-up gas rate of only 975 Sm^3/hr was deemed sufficient to prevent air ingress.

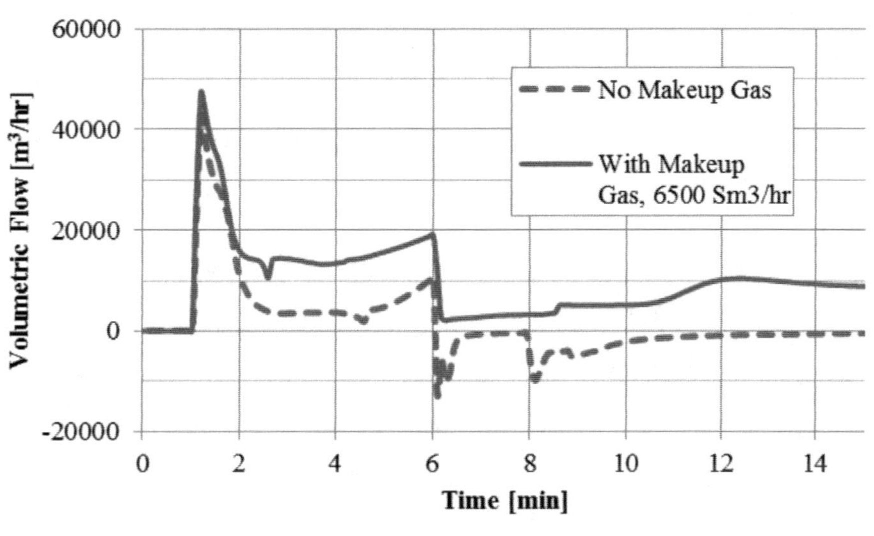

Figure 5: Volumetric flow rate at the base of the flare stack for a condensable release (light naphtha) with and without make-up gas flow.

3.4.5 Reliability of Make-up Gas

It is important to note that—considering the consequences of failure to provide adequate make-up gas to counter the potential vacuum formation—the source of the make-up gas must be highly reliable. Usually the source for the pilot gas, seal purge gas and the make-up gas are the same.

3.5 Location of Make-up Gas Addition Point

The preferred location of the make-up gas addition point is at the farthest end of the header, upstream of all reliefs, so that the flowing gas fills the hot spot first, compensates for the volume contractions, and displaces the remaining hot gases in the header. This method has an implicit advantage of reducing the total volume contraction requirement by pushing the hot gases outside of the header and potentially favorably affecting the dew point of a condensable mixture by dilution.

This location is the optimum regardless of how the other parameters are established. If the make-up gas addition point is located far downstream of the relief tie-in, such as near the flare stack or the knockout drum, the make-up gas would have to flow in the reverse direction to the relief point where the volume contraction is occurring.

Figure 6 demonstrates the impact that the location of the addition point has on the required make-up gas rates. The volumetric flow rates at the base of the flare stack when make-up gas is injected upstream of the relief and when it is injected downstream at the knockout drum for a condensable relief case are compared. For this case, a hot vapor stream of light naphtha consisting of mostly C5-C6 hydrocarbons was released into the header for five minutes. At an ambient temperature of 20°C, this mixture will start to condense after an initial period of cooling. For both cases, make-up gas was injected at a rate of 6500 Sm^3/hr. As shown in the figure, when make-up gas was injected near the knockout drum, negative flow was observed through the flare stack, whereas positive flow was maintained when the make-up gas was injected upstream of the relief.

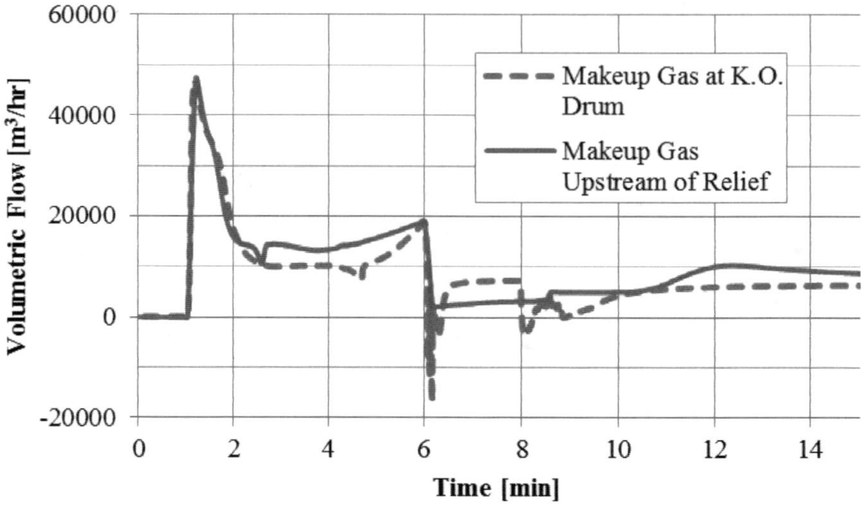

Figure 6: Volumetric flow rate at the base of the flare stack for a condensable release (light naphtha) when make-up gas is injected upstream of the relief and near the knockout drum.

3.6 Starting and Stopping of Make-up Gas

In the recommended control scheme, make-up gas is injected once relief is detected in the header via the appropriate temperature criteria. Valves and control schemes are provided to inject the make-up gas upstream of where the relief enters the header. Once a pre-set temperature condition in the header is satisfied, make-up addition can be safely stopped. Note that because the addition of make-up gas is initiated while the relief is occurring, the total backpressure at the discharge of the relief valve will be higher than the estimated backpressure when only the relief valve is discharging. Appropriate backpressure corrections should be made for the increased header flow rates.

The criteria for starting and stopping the injection of make-up gas depend on whether the expected reliefs will result in contraction due to cooling only, due to condensation and cooling by ambient conditions, or due to condensation and cooling by a pre-cooled header. The appropriate criteria for each condition are discussed in the following sections.

3.6.1 The Case of Cooling Only of Relief Gases – Starting and Stopping Make-up Gas

Starting make-up gas flow: As described in section 3.2.1, the difference between the temperature of gas in the header and the ambient temperature should be used to detect relief flow and to trigger make-up gas for cases where only cooling is expected.

To prevent excess injection of make-up gas, a flow override may be incorporated into the control scheme where make-up flow would not be injected until the flow rate close to the relief tie-in point drops below a specified minimum threshold. Both the temperature and the flow measurements should be located close to the relief point to minimize any potential delays that could result in air ingress.

Stopping make-up gas flow: After relief subsides, make-up gas should continue to be injected for as long as the temperature difference between the gas in the header and ambient remains above the minimum threshold value. Because the header will cool fastest near to the make-up gas injection point, once the temperature at the sensor drops below the threshold value, the header may still contain hot gas further downstream of the sensor location. To account for this, make-up gas should be programmed to continue to be injected for a defined duration to allow the gas temperature to stabilize.

Dynamic simulation can be used to determine how long it will take for the entire header to cool below the minimum ΔT at a given make-up flow rate. To determine the maximum time required, the relief simulation should initially be run long enough to establish the maximum steady temperature profile along the header before stopping the relief and allowing the header to cool.

Figure 7 shows the temperature profile along the example header for the hot vapor relieving case at various times after 30 minutes of relief ends. As shown in the figure, it took 90 minutes for the temperature along the entire header to fall below the threshold of 25°C (5°C above the ambient temperature of 20°C) assuming a make-up gas rate of 950 Sm^3/hr. This information can be used to determine how much additional time the make-up gas should be injected, depending on where the sensor is located relative to the make-up gas injection point and the relative rate of cooling at that location.

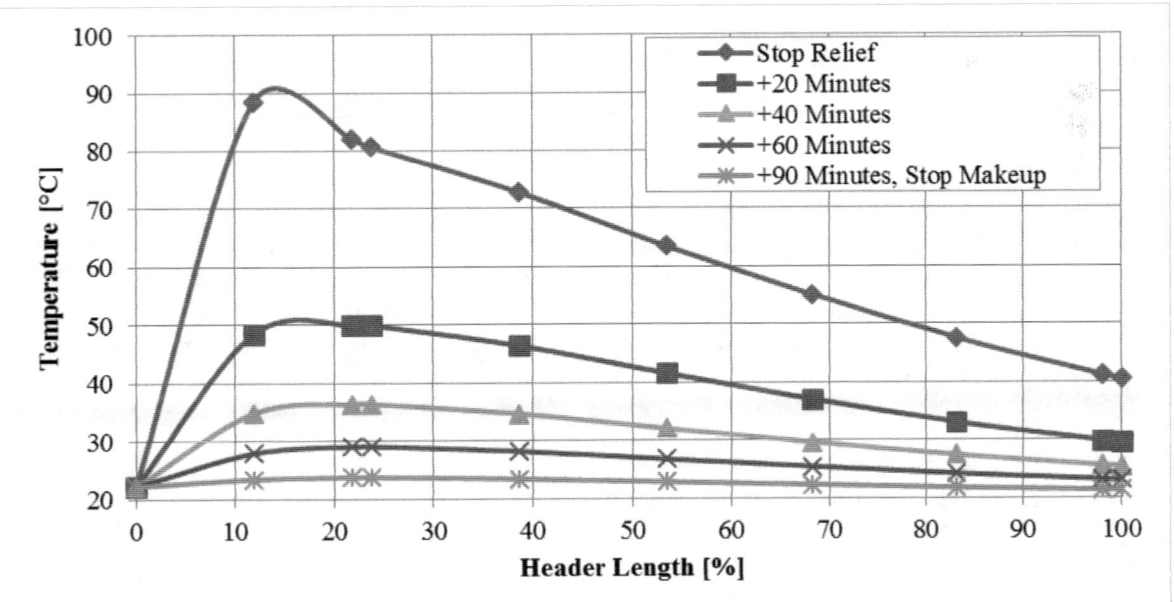

Figure 7: Temperature profile along header at various time intervals after hot vapor relief subsides.

3.6.2 Condensation of Relief Gases due to Ambient Conditions – Starting and Stopping Make-up Gas

Starting make-up gas flow: For cases where condensation can occur at or above ambient temperature, or as a result of rain or other weather conditions outside of the header, starting make-up gas should similarly be triggered by a minimum temperature difference criterion because the dew point of the gas will typically be well above the expected ΔT threshold for air ingress across all ambient temperatures as well as above the temperature of the make-up gas source. Otherwise, actual temperature may be used.

Unlike for cooling only cases, flow control may not be feasible when condensation is expected. Because the rate of contraction due to condensation is significantly faster than for cooling, a minimum flow override for injecting make-up gas only at high temperature and low flow may not provide a fast enough response to prevent vacuum formation and subsequent ingress of air.

Stopping make-up gas flow: As for cooling only cases, make-up flow should be stopped based on a minimum temperature difference threshold because the hot release may warm the header wall, resulting in a slower sustained contraction due to cooling after condensation begins to wane. Similarly, since it may be infeasible to measure temperature at all points along the header to ensure none exceed the minimum ΔT threshold, make-up gas should be injected for a pre-determined minimum duration.

3.6.3 Condensation of Relief Gases in a Cold Header – Starting and Stopping Make-up Gas

Starting make-up gas flow: A cold release in a warm header typically does not require injection of make-up gas because the header will warm and cause the gas to expand, which will result in a net forward flow out of the header. However, certain cases may exist where either a hot or cold release that is not normally condensable at ambient temperature occurs after the header has been cooled to below the dew point of the release.

Starting and stopping make-up gas for such a case requires a different approach than for the standard case of condensing that occurs at or above ambient temperature. Instead of injecting make-up gas once the temperature climbs above a specified threshold, it becomes necessary to inject make-up gas when the temperature of the header wall itself drops below the dew point of any potential subsequent reliefs.

Take for example a flare system that would experience releases of both cryogenic methane (boiling point of -162°C) and propane (boiling point of -42°C). There may be an instance in which a cryogenic release of methane may cool the header to temperatures to well below -42°C. If propane were to be released into the cooled header, it would condense and air ingress could occur. To prevent this, make-up gas would be injected once the temperature of the header drops below -42°C in order to warm the header. After the temperature along the entire header increases above this point, make-up gas would no longer be needed.

Stopping make-up gas flow: Ideally, after a cold release, the temperature at multiple points along the header would be measured to ensure that the entire header length has been warmed above the minimum condensing criteria. The make-up gas should continue for a specified duration to ensure that the temperature in the interim space between sensors has had adequate time to warm. This time can be established using dynamic simulation by calculating the time it takes to warm the entire header from the coldest state to above the threshold using only make-up gas.

Figure 8 shows the initial cooling by a cryogenic release of methane at three points along the header, followed by the subsequent warming of the header by make-up gas. A cold release of cryogenic methane at -130°C was released into the header for 3 hours, causing the entire header to cool to well below the boiling point of propane. Make-up gas was injected after the methane release subsided, and the header began to warm back up. As shown in the figure, it took nearly 40 minutes for the make-up gas to warm the entire header to above -42°C. This duration establishes how long the make-up gas should be injected once the cold release subsides.

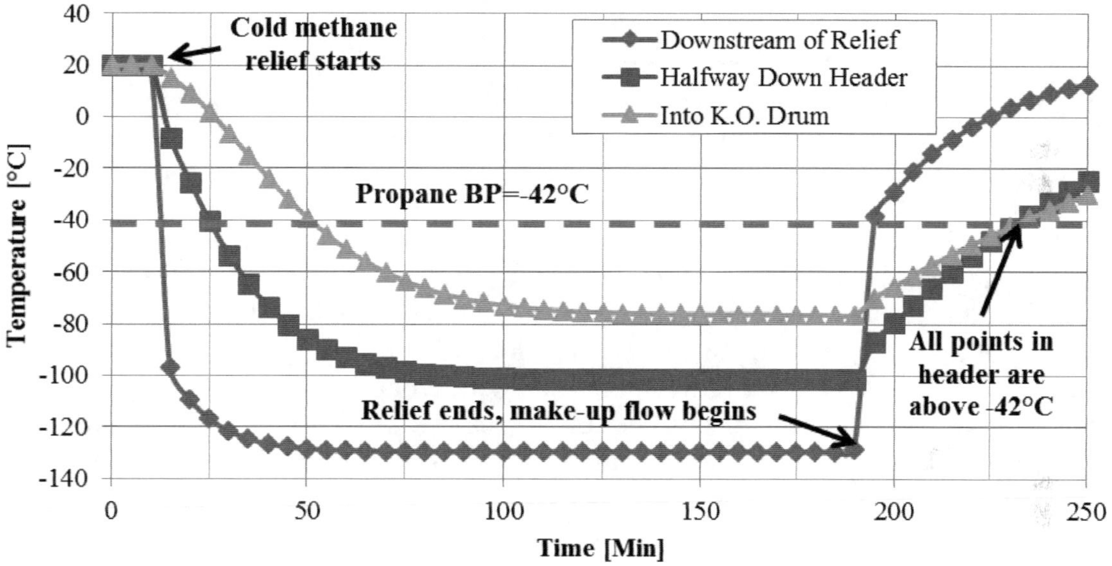

Figure 8: Temperature at three different locations along the header over time as the header is initially cooled by cryogenic methane for 3 hours, then subsequently warmed by make-up gas.

3.7 Potential Simplification of the Make-up Gas Addition Control Scheme

All the above scenarios were simulated with one pressure relief valve (PRV) relieving and a dedicated make-up gas scheme. An ideal make-up addition schematic is shown in Figure 9. In an ideal design, each PRV on the flare header system will have a dedicated sensor for relief and a make-up gas addition scheme that will add make-up gas only when a relief is sensed, with the addition rate and starting and stopping times tuned to that particular PRV. In this scheme, little or no make-up gas will be wasted.

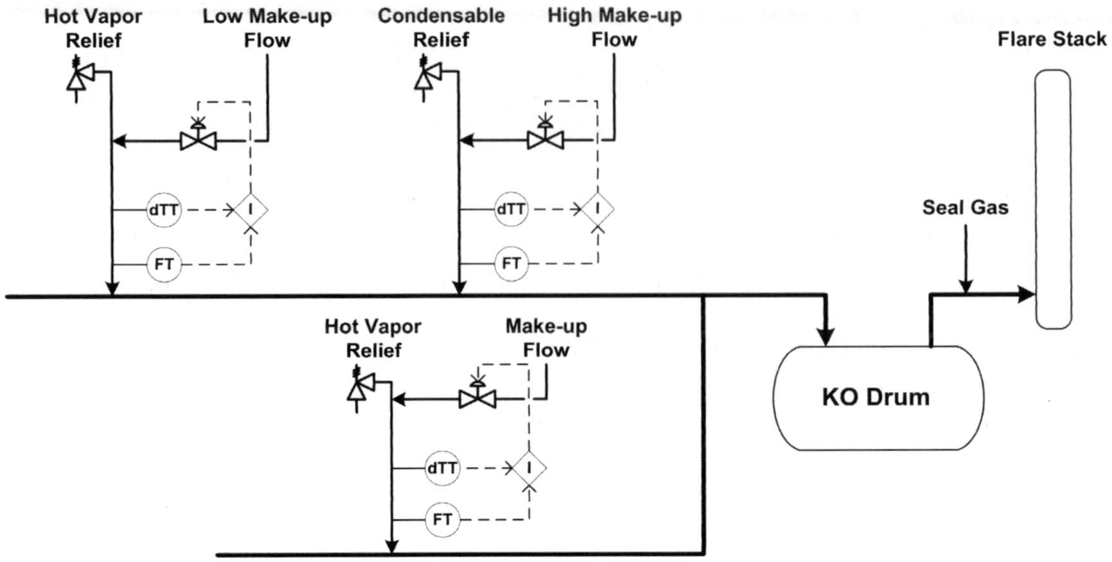

Figure 9: Ideal make-up gas addition scheme; dedicated temperature and flow sensors and separate make-up flows are provided for each PRV.

Obviously, such a scheme for each PRV on the flare header is generally not implemented because of economic and maintenance considerations. To reduce instrumentation, consideration could be made to group the PRVs that require low rates of make-up gas (essentially those where the relief would only be cooling) and the PRVs that require large rates of make-up gas addition (essentially those reliefs that may condense in the header). The location of the PRV laterals on the header and its branches would also be a consideration for such grouping. The temperature/differential temperature sensors and the flow meters should then be appropriately located to capture the start of relief to trigger make-up gas flow, as well as to measure the temperature subsidence in the header in order to stop the make-up gas flow at a safe point in time. Finally, if the flow meters are considered unreliable within the flow range required for reliable control purposes, they could be taken out of the control scheme altogether. In these circumstances the make-up gas flow duration would need to be increased in order to ensure early start and late stop.

A more typical schematic is shown in Figure 10. Note that in all schemes, the make-up gas addition point is as far upstream of all PRVs as possible so that it not only cools at the temperature sensor location, but effectively displaces the hot/condensable gases out of the header. Additionally, the differential temperature sensor is located downstream of all the PRVs, but as close as possible to the sources of relief in order to minimize any potential delays.

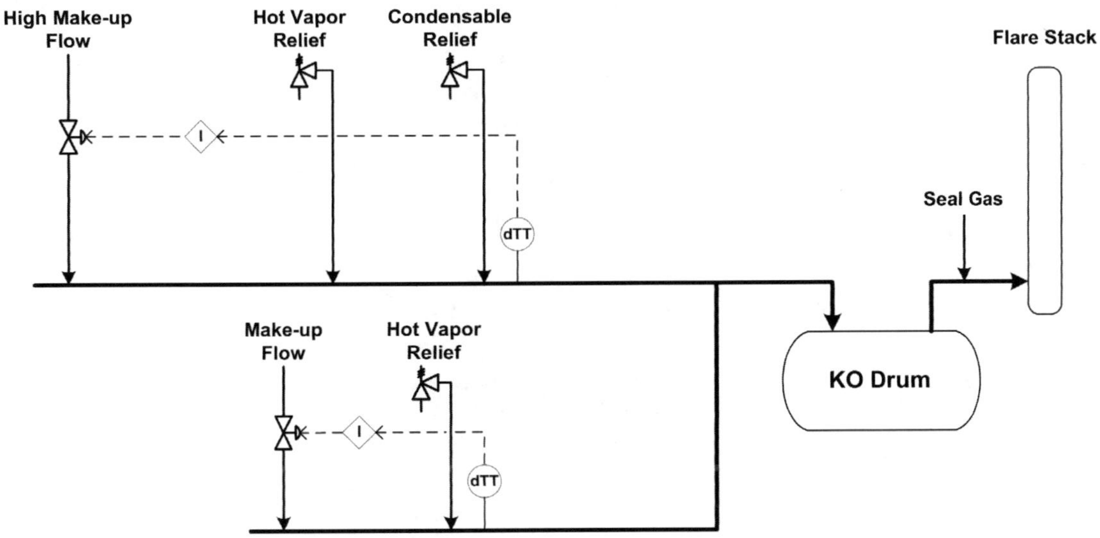

Figure 10: Simplified make-up gas addition scheme; a common temperature sensor and a single make-up flow rate cover all PRVs.

4. Conclusion

Designing an adequate and safe make-up gas injection system to protect a flare header system from various types of reliefs requires a thorough consideration of the physical properties of the relief gases, the locations of temperature and flow instrumentations, the seal purge rate, and the size of the flare header as well as prudent selection of when to start and stop make-up gas flow. The source of the make-up gas supply should be highly reliable. Simplification from a fully instrumented system is possible, albeit at the expense of potentially wasting make-up gas.

5. References

[1] Husa, H.W., "How to Compute Safe Purge Rates", Hydrocarbon and Petroleum Refiner, Vol. 43, No. 5, May 1961.

[2] Pressure-relieving and Depressuring Systems, API Standard 521, Sections 5.7.6 and 5.7.8, Sixth Ed., January 2014.

Additional references not cited:

[3] Reed, Robert, "What is Flare's Proper Purge Rate?" The Oil and Gas Journal, February 14, 1972.

[4] Bader, Adam; Baukal, Charles; Bussman, E., "Selecting the Proper Flare Systems", CEP, July 2011.

CSB INVESTIGATION OF THE WEST (TEXAS) FERTILIZER FIRE AND EXPLOSION

Presenter:
Johnnie A. Banks, CFEI, Supervisory Investigator,
U.S. Chemical Safety and Hazard Investigation Board, Washington, DC
johnnie.banks@csb.gov

Prepared for Presentation at
American Institute of Chemical Engineers
2016 Spring Meeting
12th Global Congress on Process Safety
Houston, Texas
April 10-14, 2016

CSB INVESTIGATION OF THE WEST (TEXAS) FERTILIZER FIRE AND EXPLOSION

Presenter:
Johnnie A. Banks, CFEI, Supervisory Investigator,
U.S. Chemical Safety and Hazard Investigation Board, Washington, DC
johnnie.banks@csb.gov

Keywords: *Fertilizer Grade Ammonium Nitrate (FGAN), Hazard Awareness, Emergency Planning and Response, FGAN Storage Practices, Land Use Planning*

Abstract

On April 17, 2013, a fire and explosion occurred at the West Fertilizer Company (WFC), a fertilizer blending, retail, and distribution facility in West, Texas. The violent detonation fatally injured 12 emergency responders and three members of the public. Local hospitals treated more than 260 injured victims, many of whom required hospital admission. The blast completely destroyed the WFC facility and caused widespread damage to more than 150 offsite buildings. The WFC explosion is one of the most destructive incidents ever investigated by the U.S. Chemical Safety and Hazard Investigation Board (CSB) as measured by the loss of life among emergency responders and civilians; the many injuries sustained by people both inside and outside the facility fenceline; and the extensive damage to residences, schools, and other structures.

The CSB conducted an exhaustive investigation that lasted more than two years and which yielded 19 recommendations directed to ten recipients. The tragic loss of life to the civilians and emergency responders, in addition to issues related to community-wide property damage were central in the development of these recommendations, and cover a range of elements which include:

- Regulatory oversight
- Insurance
- Emergency response
- Emergency planning
- Land use planning
- Technical conditions

This presentation will provide an overview of the incident, consider the key findings from the investigation, and illustrate the relationship of these findings to the recommendations. Click here for the final West Fertilizer Investigation Report and related videos and information.

Process Safety Incidents: Big Picture Revelations

Paul Baybutt
Primatech Inc.
Columbus, Ohio, USA
paulb@primatech.com

Prepared for Presentation at
American Institute of Chemical Engineers
2016 Spring Meeting
12th Global Congress on Process Safety
Houston, Texas
April 11-23, 2016

Process Safety Incidents: Big Picture Revelations

Paul Baybutt
Primatech Inc.
Columbus, Ohio, USA
paulb@primatech.com

Keywords: process safety instruments; process safety management; risk management programs

Abstract

Many incidents have helped to define and develop process safety. Each has provided valuable learning opportunities. However, it is even more important to identify insights that can be obtained from an analysis of a large set of incidents that represents those that typically occur. This larger picture illuminates trends and commonalities and provides learning opportunities that are even more important than the causes of any one individual incident.

The Chemical Safety Board has published the results of numerous investigations of process safety incidents and the Center for Chemical Process Safety has published information on incidents that have helped to define process safety. These data have been analyzed to identify commonalities and trends so that measures to help protect against future incidents can be developed. Recommendations are made to address key issues identified.

1.0 Introduction

Process safety regulations have been promulgated in various parts of the world to address protecting process plants against catastrophic incidents in response to the occurrence of a number of major incidents such as Flixborough, Seveso, and Bhopal [1]. However, despite the widespread adoption of process safety practices, major incidents continue to occur. The continued occurrence of such incidents is unacceptable and the issue of what can be done to protect against them must be addressed. This paper addresses the issue of why such incidents continue to occur.

Process safety regulations contain requirements for periodic audits for programs to identify any omissions or deficiencies and some process safety programs require periodic management reviews. However, while omissions and deficiencies must be identified and corrected, knowledge of what omissions and deficiencies have actually resulted in process safety incidents is essential to help prevent further incidents.

Compilations of incident descriptions have been published by various authors, for example, [1 - 7]. Usually, incidents are organized in various ways, such as by root cause, release type, equipment involved, operation being conducted, and process safety element involved. Descriptions of individual incidents also can be found, for example, in reports from the U.S. Chemical Safety Board (CSB) [8], and in journals such as the Loss Prevention Bulletin [9]. Such publications usually focus on providing a description of the incident, identifying its causes, and providing recommendations for corrective actions. Certainly, lessons can be learned from each incident. However, it is also important to take an overall view to identify commonalities and trends across incidents so that measures can be identified that will help to protect against future incidents. In particular, the identification of issues that have contributed to multiple incidents is especially important so that future incidents may be prevented. For example, the CSB has flagged reactive hazards [10], combustible dusts [11], and hot work [12] as significant issues across multiple incidents. This paper identifies additional issues that were important contributors in numerous incidents.

CSB investigation reports have been analyzed concerning management system failures [13], process hazard analysis (PHA) [14], and the hierarchy of controls [15]. The evolution of process safety standards and legislation following nine key process safety incidents has been addressed in a recent article [16].

Section 2 identifies the incidents reviewed and the analysis approach followed. Section 3 identifies and discusses the common contributors that were found for the incidents studied and provides recommendations to address them. Section 4 provides a discussion of the results. Section 5 provides some additional noteworthy observations that were made during the review of incident descriptions. Section 6 identifies three issues that underlie all of the other issues. Each underlying issue is discussed and recommendations are made for addressing them. Conclusions are presented in section 7.

2.0 Incidents Addressed and Analysis Approach

Incidents investigated by the Chemical Safety Board (CSB) [8] and incidents described by the Center For Chemical Process Safety (CCPS) [1] were reviewed. CCPS provides descriptions of over 100 incidents. Some date back to the early 20th century but most occurred within the past 40 years. They cover a variety of industries including chemicals, oil and gas production and refining, aviation, shipping, and nuclear power generation. The CSB has issued over 60 incident investigation reports, case studies, and safety bulletins dating back to 1998 when the CSB began investigating incidents. The CSB incidents cover various sectors of the process industries. The CSB currently is investigating one incident from 2010, three from 2013, three from 2014 and one from 2015. Sixty-eight CSB incidents and 12 CCPS incidents making a total of 80 incidents were studied (see Tables 1 and 2, respectively). About 40% of the CSB incidents occurred in processes covered by the Occupational Safety and Health Administration's (OSHA's) process safety management (PSM) regulations. The descriptions of incidents and their causes were taken at face value.

Table 1. CSB Incidents Reviewed.

No.	Incident	Incident Date	Report Date	PSM[1]	PHA[2]
1	Sonat Exploration Co. Catastrophic Vessel Overpressurization	03/04/1998	09/21/2000	N	N
2	Morton International Inc. Runaway Chemical Reaction, Explosion and Fire	04/08/1998	08/16/2000	N	Y
3	Concept Sciences Hydroxylamine Explosion	02/19/1999	02/01/2002	Y	Y
4	Tosco Avon Refinery Petroleum Naphtha Fire	02/23/1999	03/21/2001	Y	N
5	Bethlehem Steel Corporation Gas Condensate Fire	02/02/2001	12/06/2001	N	N
6	BP Amoco Thermal Decomposition Incident	03/13/2001	05/20/2002	N	Y
7	Motiva Enterprises Sulfuric Acid Tank Explosion	07/17/2001	08/28/2002	N	Y
8	Georgia-Pacific Corp. Hydrogen Sulfide Leak	01/16/2002	11/20/2002	N	N
9	Third Coast Industries Petroleum Products Facility Fire	05/01/2002	03/06/2003	N	U
10	DPC Enterprises Festus Chlorine Release	08/14/2002	05/01/2003	Y	U
11	First Chemical Corp. Reactive Chemical Explosion and Fire	10/13/2002	10/15/2003	N	N
12	Environmental Enterprises Hydrogen Sulfide Release	12/11/2002	09/17/2003	N	U
13	Catalyst Systems Inc. Reactive Chemical Explosion and Fire	01/02/2003	10/29/2003	N	N
14	BLSR Operating Ltd. Vapor Cloud Deflagration and Fire	01/13/2003	09/17/2003	N	U
15	West Pharmaceutical Services Dust Explosion and Fire	01/29/2003	09/23/2004	N	N
16	Technic Inc. Ventilation System Explosion and Fire	02/07/2003	08/20/2004	N	N
17	CTA Acoustics Dust Explosion and Fire	02/20/2003	02/15/2005	N	N
18	D.D. Williamson & Co. Catastrophic Vessel Failure	04/11/2003	03/12/2004	N	N
19	Honeywell Chemical Releases (3 incidents)	a. Chlorine 07/20/2003 b. Antimony pentachloride 07/29/2003 c. Hydrogen fluoride 08/13/2003	08/08/2005	Y N Y	Y N N
20	Isotec / Sigma Aldrich Nitric Oxide Explosion	09/21/2003	08/24/2004	Y	Y
21	Hayes Lemmerz Dust Explosions and Fire	10/29/2003	09/27/2005	N	N
22	DPC Enterprises Glendale Chlorine Release	11/17/2003	02/28/2007	Y	Y
23	Giant Industries Oil Refinery Explosion and Fire	04/8/2004	10/26/2005	Y	Y
24	MFG Chemical Inc. Toxic Gas Release	04/12/2004	04/11/2006	Y	N
25	Formosa Plastics Vinyl Chloride Explosion	04/23/2004	03/06/2007	Y	Y
26	Sterigenics Ethylene Oxide Explosion	08/19/2004	03/30/2006	Y	Y
27	Marcus Oil and Chemical Tank Explosion and Fire	12/03/2004	06/06/2006	N	N

No.	Incident	Incident Date	Report Date	PSM[1]	PHA[2]
28	Acetylene Service Company Gas Explosion	01/25/2005	01/26/2006	Y	Y
29	BP Texas City Refinery Explosion and Fire	03/23/2005	03/20/2007	Y	Y
30	Formosa Plastics Propylene Fire and Explosions	10/06/2005	07/20/2006	Y	Y
31	Bethune Point Wastewater Plant Explosion and Fire	01/11/2006	03/13/2007	N	N
32	Synthron Chemical Reaction and Vapor Cloud Explosion	01/31/2006	07/31/2007	Y	N
33	Partridge Raleigh Oilfield Explosion and Fire	06/05/2006	06/12/2007	N	N
34	Universal Form Clamp Co. Explosion and Fire	06/14/2006	04/10/2007	Y	N
35	EQ Hazardous Waste Plant Explosions and Fire	10/05/2006	04/16/2008	N	U
36	CAI / Arnel Chemical Plant Explosion	11/22/2006	05/13/2008	Y	N
37	Valero Refinery Propane Fire	02/16/2007	07/09/2008	Y	Y
38	Barton Solvents Explosions and Fire	07/17/2007	06/26/2008	N	U
39	Xcel Energy Company Hydroelectric Tunnel Fire	10/02/2007	08/25/2010	N	Y
40	Barton Solvents Flammable Liquid Explosion and Fire	10/29/2007	09/18/2008	N	U
41	T2 Laboratories Inc. Reactive Chemical Explosion	12/19/2007	09/15/2009	N	N
42	Imperial Sugar Company Dust Explosion and Fire	02/07/2008	09/24/2009	N	U
43	Goodyear Heat Exchanger Rupture and Ammonia Release	06/11/2008	01/27/2011	Y	U
44	Packaging Corporation Storage Tank Explosion	07/29/2008	03/04/2010	N	N
45	Bayer CropScience Pesticide Waste Tank Explosion	08/28/2008	01/20/2011	Y	Y
46	INDSPEC Chemical Corporation Oleum Release	10/11/2008	09/30/2009	Y	Y
47	Allied Terminals Fertilizer Tank Collapse	11/11/2008	05/26/2009	N	N
48	Veolia Environmental Services Flammable Vapor Explosion and Fire	05/04/2009	07/21/2010	Y	N
49	ConAgra Natural Gas Explosion	06/09/2009	02/04/2010	N	U
50	CITGO Refinery Fire and Hydrofluoric Acid Release	07/19/2009	12/09/2009	Y	Y
51	Caribbean Petroleum Refining Tank Explosion and Fire	10/23/2009	10/21/2015	N	N
52	Silver Eagle Refinery Explosion	11/04/2009	04/14/2014	Y	U
53	NDK Crystal Inc. Explosion	12/07/2009	11/14/2013	N	U
54	DuPont Corporation Toxic Chemical Releases (3 incidents)	a. Methyl chloride 01/22/2010 b. Oleum 01/23/2010 c. Phosgene 01/23/2010	09/20/2011	U U Y	U U Y
55	Kleen Energy Natural Gas Explosion	02/07/2010	06/28/2010	N	U
56	Tesoro Refinery Explosion and Fire	04/02/2010	05/01/2014	Y	Y
57	Horsehead Holding Company Explosion and Fire	07/22/2010	03/11/2015	N	U

No.	Incident	Incident Date	Report Date	PSM[1]	PHA[2]
58	Millard Refrigerated Services Ammonia Release	08/23/2010	01/15/2015	U	U
59	E. I. DuPont De Nemours Co. Hotwork Explosion	11/09/2010	04/19/2012	Y	Y
60	AL Solutions Dust Explosion	12/09/2010	07/16/2014	N	U
61	Carbide Industries Fire and Explosion	03/21/2011	02/07/2013	N	U
62	Hoeganaes Corporation Flash Fires and Explosion	05/27/2011	01/05/2012	U	U
63	Chevron Refinery Fire	08/06/2012	01/28/2015	Y	Y
64	US Ink Fire	10/09/2012	01/15/2015	U	U

[1] Covered by OSHA's PSM standard.
[2] PHA was performed
Y – Yes
N – No
U – Unknown

Table 2. CCPS Incidents Reviewed.

No.	Incident	Incident Date
1	ELF refinery, LPG tank farm, Feyzin, France	01/04/1966
2	DSM Nypro plant, Flixborough, UK	06/01/1974
3	ICMESA, chemical plant, Seveso, Italy	07/10/1976
4	BP hydrocracker unit, Grangemouth, UK	03/21/1987
5	Shell refinery, FCC unit, Norco, LA	05/05/1988
6	Occidental Petroleum, Piper Alpha Platform, North Sea, UK	07/06/1988
7	Phillips Chemical Company, polyethylene reactor, Pasadena, TX	10/23/1989
8	Total refinery, FCC unit, La Mede, France	11/09/1992
9	Texaco Oil Refinery, FCC Unit, Milford Haven	07/24/1994
10	Napp Technologies, Lodi, NJ	04/21/1995
11	Tosco Avon Refinery, hydrocracker unit, Martinez, CA	07/21/1997
12	Sonatrach LNG plant, Skikda, Algeria	01/19/2004

All incident descriptions were first reviewed to identify commonalities. A second reading of the incident descriptions was performed to confirm the initial findings and document characteristics of the incidents. A third reading of the incident descriptions was conducted to corroborate the findings and provide further detail.

3.0 Incidents Addressed and Analysis Approach

This section identifies common contributors to the incidents. Of course, each incident has its own individual characteristics but there is a remarkable degree of similarity in some deficiencies and omissions across the incidents. The focus in this paper is on issues common to many of the incidents.

Incidents where each issue was a contributor are identified, the percentage of CSB incidents where the issue was present is given, aspects of each issue are discussed, and recommendations are made to address the issues.

3.1 Process Design

Many incidents occurred in processes that arguably had deficient designs {1, 3, 6, 15, 16, 17, 18, 21, 25, 29, 31, 34, 37, 40, 41, 42, 57, 58, 64} and {*10, 11*}[3]. This issue was present in 28% of CSB incidents and 18% of incidents covered by OSHA's PSM standard. Deficiencies arose in the initial design and as a result of modifications.

Hazards were not recognized, designs created hazards, safety systems were inadequate, process instrumentation was insufficient, human factors issues were not addressed, hazards of scale-up were not recognized, and throughput was increased without review. In some cases deficient equipment design contributed to incidents {54a, 54c}. Possibly designers were unaware of the problems and/or design reviews were inadequate but these reasons are not acceptable.

Design deficiencies can be addressed by making process design and engineering an element in process safety programs [17]. Examples of topics that should be addressed by requirements for process design and engineering include design practices, design reviews, inherent safety, hazardous area classification, pressure relief, alarm management, damage mechanisms, human factors, facility siting, competency, and training for designers. Also, process design and engineering should be subject to the same management system requirements as other elements in a process safety program such as written policies and procedures, document management, assigning roles and responsibilities, auditing, management review, and continuous improvement. Process design and engineering should be an element of process safety management systems in order to provide a specific focus on the activities involved and to ensure the application of management system practices to it. Also, the competency of process designers must be addressed.

For new designs and process modifications, proof of safety should be required rather than proof of danger. This will require a change in culture.

[3] Braces or curly brackets, that is, { }, are used to enclose the number designations of incidents from Table 1, using normal font, and Table 2, using italic font.

3.2 Safeguards

Many incidents involved processes that were not adequately protected by safeguards {2, 7, 8, 9, 11, 12, 13, 17,18, 20, 21, 22, 24, 25, 26, 28, 29, 30, 32, 34, 35, 36, 37, 38, 40, 41, 42, 46, 48, 49, 50, 51, 54a, 54b, 54c, 60, 62, 64} and *{1, 3, 4, 5, 6, 9, 11}*. This issue was present in 56% of CSB incidents and 57% of incidents covered by OSHA's PSM standard. Needed safeguards were not present or those that were present were insufficient, unreliable, disabled, or lacked independence. Commonly, the hierarchy of controls was not applied (see Figure 1). Often reliance was placed on procedural safeguards which are at the lower end of the hierarchy (19a, 22, 25, 29, 32, 46, 54c}. Process risks should be matched with safeguards in the hierarchy of controls. Higher consequence scenarios, such as a chlorine release in a densely populated area, require safeguards higher in the hierarchy and likely multiple, independent safeguards. Procedural safeguards typically are viewed as less reliable, and, therefore, inadequate for protection against high-consequence events.

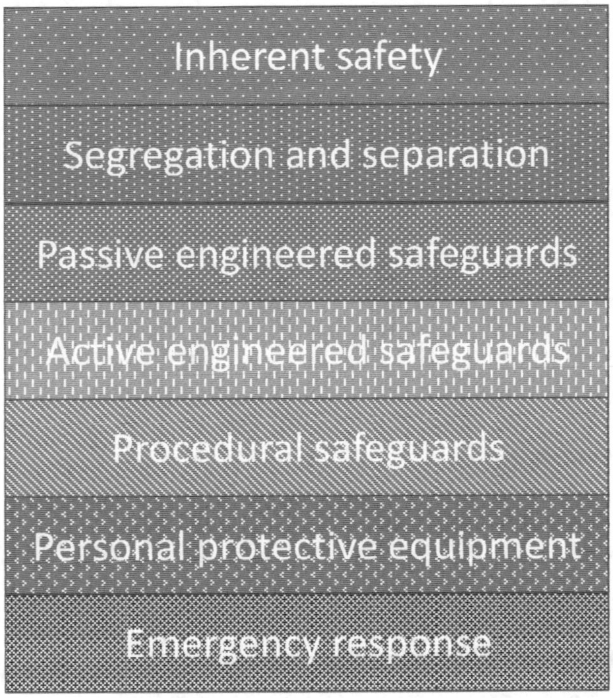

Figure 1. Hierarchy of Controls.

Inherently safer designs or practices could have prevented or mitigated some incidents {29, 39, 55, 56, 54b, 63}. Currently, there are no requirements to address inherently safer systems or the hierarchy of controls in Federal process safety regulations. Process safety regulations in New Jersey require an inherently safer technology review [18] and in Contra Costa County, California require that inherently safer systems be addressed [19]. Proposed amendments to California's process safety regulations for refineries contain a requirement to perform a hierarchy of controls analysis [20].

The operability of safeguards was an issue during several incidents. Explosions and fires disabled safeguards {15, 30, 42} and {*7*}, and toxic releases and fires prevented access to controls {19a, 30, 37, 50} and {*6*}.

Currently, there is no requirement in US process safety regulations to address the design basis for safeguards, evaluate or document the effectiveness of process safeguards, or show that safeguards adequately reduce risks. Proposed amendments to California's process safety regulations for refineries contain a requirement to perform a hierarchy of controls analysis and a safeguards protection analysis (SPA) [20]. The SPA should be performed before a PHA study and updated after the PHA study has been completed. The SPA should include the preparation of a Safety Requirements Specification (SRS) for each individual safeguard [21]. An SRS documents the functional and integrity requirements for a safeguard. Functional requirements describe what the safeguard does to prevent a hazardous state of the process and integrity requirements specify how well the safeguard should perform and what is needed to support that performance level. The SPA should validate that safeguards adequately protect against the hazards of the process, have a specified integrity level, have been installed correctly, are functional, are being maintained appropriately, etc.

Once the SPA has been performed, the PHA team can focus on the qualification of safeguards for specific hazard scenarios [21, 22]. Qualification of safeguards addresses whether safeguards are appropriate for each hazard scenario in a PHA study. Information developed in a PHA study may necessitate revisions to the SPA and SRSs. For example, it may be determined that a redundant power supply is needed to support the integrity required for a scrubber system. SPAs should be updated during management of change (MOC) reviews and when developing recommendations from incident investigations.

3.3 Operations and Maintenance

Problems with procedures were common {1, 2, 3, 5, 6, 11, 12, 13, 14, 18, 19b, 19c, 21, 22, 25, 28, 29, 32, 33, 36, 41, 42, 45, 46, 47, 51, 54c, 59, 61, 62} and {*4, 9*}. . This issue was present in 44% of CSB incidents and 43% of incidents covered by OSHA's PSM standard. In numerous incidents, procedures had not been developed, were not followed, were incomplete, or were otherwise deficient. Such situations are unacceptable. Properly written procedures must be developed following best practices for all process operations and activities before they occur and for all stages and modes of operation. Procedures are needed to deal with any situation where there is the potential for a process safety incident. Procedures must be workable in all circumstances where they apply and personnel need to be trained in the procedures and practice them regularly in drills. Facility personnel must be competent.

In six CSB investigations that identified problems with standard operating procedures (SOPs), the CSB noted that each incident involved SOP deviations that became "necessary violations" to get the job done {45}. Such situations cannot be tolerated.

Existing process safety regulations require written operating procedures. However, the regulations should be clarified to encompass all process tasks and activities and a mechanism is needed to ensure the regulatory requirements are followed.

Deficiencies in maintenance were present in a number of incidents {5, 12, 21, 29, 31, 32, 51, 54c}. Issues regarding the adequacy of inspections were common {4, 7, 47, 53, 56, 62} and {*2, 8*}. For some incidents, there was no or an inadequate preventive maintenance (PM) program {11, 13, 16, 41} and no or inadequate MI program for some incidents {10, 23, 29, 52, 61, 63}. Such issues were present in 34% of CSB incidents and 32% of incidents covered by OSHA's PSM standard. Existing process safety regulations address the integrity of equipment but a mechanism is needed to ensure the requirements are followed for all safety-critical equipment. Current US process safety regulations require that covered equipment comply with recognized and generally accepted good engineering practices (RAGAGEP) but the term is not defined. OSHA issued a Request for Information (RFI) in advance of rulemaking to revise the PSM standard [23]. Amongst other topics, the RFI sought comment on whether an explicit definition of RAGAGEP should be provided, and, if so, what definition should be used. The current OSHA PSM regulation provides a limited list of covered equipment [24] and the RFI sought comments on whether mechanical integrity requirements should apply to all equipment designated as safety-critical equipment.

The CCPS Risk Based Process Safety (RBPS) element, Conduct of Operations, provides requirements for operational discipline that tolerate no deviation from approved procedures, even if the outcome is inconsequential or desirable [26, 27]. Appropriate work procedures must be determined, documented, and enforced. The importance of following procedures must be emphasized to facility personnel. The adoption of an element on operational discipline would help in this regard coupled with an element that addresses process safety culture. The CCPS RBPS program contains both these elements but they are not included in existing US regulations. Additionally, deviations from approved procedures should be identified and investigated as process safety incidents.

An operational discipline element would encourage and enforce strict adherence to established procedures and imbue facility personnel with the belief that no operation or activity should be performed without a governing procedure. When people's lives are at risk it should not be acceptable to perform activities without proper preparation or by improvising.

Another deficiency present in several incidents was the acceptance of practices as normal and acceptable by facility personnel that outside observers would view as deviant and unacceptable {17, 19c, 21, 29, 42, 57, 62, 63}. This phenomenon has been called "normalization of deviance" [28]. It involves allowing operation of a process that deviates from established limits by relaxing performance standards rather than identifying and correcting the underlying causes of deviations so the deviations become the new, normal operating conditions. Hazards have been normalized in some cases {12, 27, 46, 56}. Normalization of deviance was present in 18% of CSB incidents and 18% of incidents covered by OSHA's PSM standard.

While the acceptance of a lesser performance standard in such cases may not immediately result in an incident, continued successful operation instills a sense of false security. It will be only a matter of time before uncorrected hazardous conditions eventually result in an incident.

Unfortunately, facility personnel can become so accustomed to exposure to hazardous situations that the risks are not recognized even when their lives are at stake. Just because a facility has not experienced a catastrophic incident, perhaps ever in its lifetime, does not mean one could not happen tomorrow.

Normalization of deviance can be addressed by ensuring that any deviations from established performance standards are identified and resolved as soon as they occur. Operational discipline will help in this regard.

3.4 Abnormal and Non-Routine Operations

Many incidents involved non-routine operations and situations {1, 3, 4, 5, 6, 10, 11, 16, 17, 19c, 20, 21, 23, 24, 26, 27, 29, 32, 33, 39, 41, 43, 45, 46, 47, 49, 52, 54a, 54c, 55, 56, 58, 59, 60, 61, 62, 63, 64} and {*1, 7, 9*}. In some cases, they were not recognized as non-routine operations and were not reviewed to ensure that work could proceed safely. Flaws in hot work {7, 31, 33, 44, 59} and other safe work practices {4, 5, 6, 11, 14, 23, 39, 43} and {*6*} also contributed to incidents. Lack of equipment isolation was a contributor to some incidents {11, 33, 59}. Abnormal and non-routine operations were involved in 62% of CSB incidents and 64% of incidents covered by OSHA's PSM standard.

People play an intimate role in non-routine operations. This both increases the likelihood of human error being a factor and places people closer to process hazards. Furthermore, a number of incidents occurred in processes that involved manual operations or were manually controlled {2, 12, 13, 16, 18}.

Current US process safety regulations contain requirements to develop and implement safe work practices and utilize hot work permits. However, the regulations do not specifically address non-routine operations. There are requirements for written procedures under the Operating Procedures element for specific phases of operation listed in the regulations (initial startup, normal operations, temporary operations, emergency shutdown, emergency operations, normal shutdown, and startup following a turnaround or after an emergency shutdown). However, other phases of operation are not specified including commissioning, standby, bypass, unattended operation, maintenance on live equipment, regeneration, troubleshooting, operation with disabled safety systems, cleaning, decommissioning, etc. Process safety programs must address all phases of operation that are or may be experienced by a process. Emphasis should be placed on non-steady state operations including abnormal operations because they are the process states in which most incidents have occurred.

Other elements of process safety programs besides operating procedures should include requirements to address non-routine operations, for example, mechanical integrity and PHA [29]. A separate element on non-routine operations may be merited to provide requirements in a single place. Absent specific requirements to address non-routine operations, many non-routine operations are likely not to be viewed as such and their hazards not recognized or addressed.

Procedures are needed for identifying and planning for non-routine work and abnormal situations, e.g. restart after emergency shutdown or an extended unplanned shutdown. Processes should be operated defensively and conservatively. Procedures should specify alternative courses of action if safety preconditions cannot be met. Furthermore, job hazard analysis [29] should be performed immediately prior to any non-routine activity.

3.5 Human and Organizational Factors

Human and organizational factors contributed to many incidents, for example, valve positions were not clear {23, 51}, valve operation was difficult {56}, lighting was insufficient {51}, operators were distracted {18}, operators were fatigued {29, 45}, operators had insufficient information {57}, procedures were complex and lengthy {45}, communications were problematic {25, 29, 43, 45}, layout was error prone {25}, staffing was not sufficient {25, 29, 45, 51}, operators became desensitized to an alarm due to its history of unreliability {54a}, the work environment encouraged operations personnel to deviate from procedures {29}, a computerized control system was poorly designed {29}, a new control system posed challenges {45}, and there were difficulties with organizational structure {25, 29}. Undoubtedly, problems with safety culture played an important role in many incidents, with BP Texas City being the best known {29}. Most CSB investigations have not explicitly identified contributions to incidents from human and organizational factors so it is not possible to provide meaningful statistics for their contributions to incidents. Undoubtedly, human and organizational factors played a role in most accidents.

Fundamentally, all of the incidents reviewed resulted from human failures ranging from design to operation to maintenance. Even when equipment failures such as corrosion were the basic cause of an incident, the underlying cause usually was human failure {63}.

Some incidents involved conflicts and tradeoffs between process safety and other imperatives {4, 7, 29, 39, 45, 51, 54c, 58, 63}, often financial but sometimes other issues such as noise attenuation {36}.

OSHA's PSM standard requires that "the process hazard analysis shall address human factors" [24]. However, it is not sufficient simply to require PHA studies to address human factors. Certainly, PHA teams must address the role of human factors in influencing human failures but this approach leads to insufficient consideration of human factors issues for processes. Commonly, human factors are addressed by using a checklist with questions that are asked about the entire process so that analysts can try and think of any problems. A much more detailed approach is needed. Otherwise, many aspects of human factors will be missed because of practical limits on the level of detail that can be addressed in PHA studies, limitations on the ability of PHA methods to address human factors, and time pressures on PHA teams. Also, human factors issues affect each of the elements of a process safety program including PHA itself [31]. It is not reasonable to expect a PHA team to address human factors issues that impact its own performance. The Contra Costa industrial safety ordinance requires the development and implementation of a written human factors program [19]. Proposed amendments to California's process safety regulations for refineries contain a requirement to develop, implement and maintain an effective written human factors program that goes beyond the requirements of Contra Costa's industrial safety ordinance [20].

All process safety management programs should have a distinct element that addresses human factors for processes with specific requirements to be addressed. Many aspects of human factors for process safety have been addressed by CCPS [32] and they should be covered by process safety programs.

For many incidents, it is difficult to believe that the conditions that resulted in the incident were allowed to occur. An independent observer easily can conclude that surely someone could have intervened at some stage in the sequence of events that occurred and taken action to prevent the incident. However, such a view does not take account of the psychological factors that influence decision making by people, particularly in groups and when under pressure to achieve competing goals. Factors such as group think, peer pressure, and other cognitive biases interfere with objective decision making [33]. Questionable decision making played a role in numerous incidents, for example {58, 63}. Facility personnel need to apply critical thinking in their daily work activities [34]. Critical thinking can be taught and key facility personnel should be so trained.

3.6 Process Hazard Analysis

Failures in PHA are a common finding in CSB incident investigation reports. Indeed, issues with PHA were present for all CSB incidents where a PHA study was performed. Of particular note is that CSB found that incidents were not identified by the PHA studies performed. However, many of the incidents involved deliberate operation of processes in ways that a PHA team would not consider credible. PHA teams look for ways in which a process may deviate from its design intent but usually the specified design intent identifies what should be done, not what should *not* be done. It is the latter that occurs in many incidents and, while PHA teams try to identify ways that a process may be misoperated, it is unreasonable to expect a PHA team to identify the ways in which a process may be maloperated. Moreover, there is essentially an infinite number of ways in which a process may be maloperated and it would be impractical to ask a PHA team to try and identify all of them. Maloperation must be addressed through operating discipline, proper change reviews, and designing robust processes.

In some cases, the CSB found that PHA studies did not adequately address safeguards because safeguards listed in PHA studies were non-specific or general {19a, 56, 63} and, in many cases, relied too heavily on administrative procedures. For example, safeguards were recorded as "design", "standard operating procedures", "testing and inspection", and "proper emergency response procedures". PHA studies must identify specific safety-critical equipment, for example, a specific relief valve, and specific safety-critical actions, for example, activation of an emergency quench system by an operator. PHA teams should consider the hierarchy of controls when evaluating existing safeguards and deciding on recommendations for further risk reduction. PHA team leaders and team members must be competent [35].

Some CSB incident investigation reports noted the failure to validate critical assumptions used in PHAs {45, 56, 29}. However, such validations should not be part of PHA. They should be performed before a PHA study is performed [21]. Such validations require a different skill set than that required for PHA team members. Moreover, if they were included in PHA studies they would add complexity to the studies and detract from their primary goal of scenario identification. Of course, the adequacy of safeguards for individual hazard scenarios must be addressed by the PHA team.

Some incidents have occurred when more than one activity was being performed simultaneously or sequentially and the activities conflicted {4, 7, 8, 25, 33, 43, 47, 48, 59}. Simultaneous and sequential operations (SIMOPS) reviews [36] should be performed as adjuncts to PHA studies to address such issues. However, they are not yet common and there are no regulatory requirements to perform them.

PHA studies were known to be performed for 22 of the 68 CSB incidents that were reviewed, although incident descriptions did not indicate whether or not a PHA study was performed for 21 of the incidents. However, it is likely that a PHA study was not performed for most of these incidents. Of the 22 PHA studies that were performed, only 4 studies were conducted for processes not covered by OSHA's PSM standard (see Table 1). This finding suggests that PHA studies are only performed when required by regulations.

3.7 Process Changes

Many incidents involved changes that occurred but were not reviewed for their impact on process safety {2, 4, 6, 7, 8, 16, 17, 18, 21, 25, 27, 29, 37, 45, 46, 54c, 59, 63} and {*2, 9, 11*}, and inadequate reviews were performed in some cases {7, 15, 19a, 29, 32, 45, 54a, 56, 63}. Issues with process change reviews were present in 34% of CSB incidents and 39% of incidents covered by OSHA's PSM standard. Omissions and deficiencies can arise because changes are not recognized as requiring MOC reviews, personnel lack awareness of MOC requirements, or MOC reviews are bypassed for expediency. These problems relate to the competency of facility personnel and operational discipline. If personnel are involved in unauthorized and unreviewed changes, they cannot be viewed as competent to work in the facility in the same way as if they maloperated a process or performed maintenance work incorrectly. The establishment of a competency management program should help in this regard. It should be reinforced by operational discipline.

Also, MOC reviews required by the OSHA PSM standard have a narrow definition of covered changes. They cover only changes to PSI, procedures, and facilities that affect a covered process [24]. These requirements are neither global nor sufficiently specific and they are open to interpretation. Organizational changes have played a role in several incidents {29, 45, 56, 64} but they are not currently covered by the regulations, although OSHA's RFI addresses this omission [23]. However, incidents have involved other types of changes not explicitly defined in US process safety regulations including changes in feedstock {4, 63}, addition or removal of equipment {16, 22, 37, 42, 59}, degraded equipment {59}, scale-up of recipes {32}, and prolonged operation {6}. Furthermore, MOC programs should encompass operational deviations and variances, and abnormal situations as well as preplanned changes [37]. Regulations and best practices should specify that any change that may impact on or affect a release of a highly hazardous chemical should be covered by MOC requirements.

Additionally, activities involved in making process changes can be hazardous and are not covered by current US process safety regulations. MOC reviews of changes should include the performance of job hazard analysis for the activities involved.

3.8 *Previous Incidents*

Information from previous incidents that could have helped to prevent incidents was not used in many cases {2, 5, 6, 11, 17, 19a, 20, 21, 25, 26, 29, 41, 42, 45, 51, 53, 54a, 54c, 60, 61, 63, 64} and {*4, 11*}. Issues with previous incidents were present for 32% of CSB incidents and 29% of incidents covered by OSHA's PSM standard. Previous incidents were not investigated properly, or at all, and near misses were ignored. The issue of retaining information from previous incidents in the corporate memory of companies is well known [38]. However, information cannot be retained if it has not first been developed.

Facility personnel may not have the time or inclination to study incident descriptions and they may not see or understand their relevance to their own facilities. Reliance cannot be placed on people to address lessons learned from previous incidents unless process safety programs contain specific requirements to do so. OSHA's PSM standard requires that previous incidents be addressed [24]. However, the regulatory requirement is part of the PHA element which simply states that the PHA shall address previous incidents. This weak requirement is likely to result in the inadequate consideration of how information from previous incidents can be used to prevent repeat or similar incidents. Process safety programs should have a separate element on previous incidents, including near misses, that requires companies to identify relevant previous incidents from within the company and from other companies in the process industries, compile pertinent information on them, analyze the information in the context of reducing risks in existing processes, and develop recommendations to do so and track them to implementation. The previous incidents review should be performed periodically, e.g. annually, and documented. It should address both the issue of ensuring that incidents are identified, reviewed, and lessons learned are implemented, and that information is retained.

3.9 *Facility and Stationary Source Siting*

The proximity of facilities to members of the public was an issue in multiple incidents {2, 3, 9, 10, 15, 16, 17, 18, 19a, 20, 22, 23, 24, 25, 26, 27, 29, 30, 32, 35, 36, 41, 45, 46, 48, 51, 52, 53, 63}. This issue was present for 43% of CSB incidents and 64% of incidents covered by OSHA's PSM standard. It raises concerns regarding land use planning in the US. Proximity of facility personnel to hazards in plants also was an issue in numerous incidents {11, 19a, 20, 21, 25, 26, 29, 32, 41, 48, 61} and {*2, 5, 8, 9, 12*]. This issue was present for 16% of CSB incidents and 25% of incidents covered by OSHA's PSM standard. The consequences of some incidents have been exacerbated by the presence of non-essential personnel {29, 49, 56}.

OSHA's PSM standard requires that "the process hazard analysis shall address facility siting" [24] and EPA's RMP rule requires that "the process hazard analysis shall address stationary source siting [39]. However, it is not sufficient simply to require PHA studies to address facility and stationary source siting. Certainly, PHA teams must address the impact of facility and stationary source siting on hazard scenarios but this approach likely leads to insufficient consideration of facility and stationary source siting issues for processes. Aspects of facility and stationary source siting will be missed because of practical limits on the level of detail that can be addressed in PHA studies, limitations on the ability of PHA methods to address facility and stationary source siting, and time pressures on PHA teams.

Process safety management programs should have a distinct element that addresses facility and stationary source siting for processes with specific requirements to be addressed. These requirements should include topics that often are overlooked including using designs that facilitate minimal operator presence; limiting the number of people exposed to hazards, particularly at more hazardous times, by moving unnecessary people away from non-routine operations; training and requiring people to leave when hazardous conditions or situations arise; and assessing the potential for domino events.

The problem of the proximity of people to hazards is compounded by poor emergency planning and response which was present for about 40% of these incidents (see next section).

3.10 Emergency Preparedness and Response

Deficiencies in emergency preparedness and response existed in many cases {5, 8, 10, 11, 16, 19a, 20, 22, 24, 25, 32, 34, 35, 39, 42, 43, 45, 51, 54a, 54b, 63, 64} and {*1, 3, 6, 10, 12*}. This issue was present for 32% of CSB incidents and 39% of incidents covered by OSHA's PSM standard. Deficiencies occurred in hazard communication, evacuation procedures and emergency egress, community notification, emergency preparedness, knowledge of responders, and coordination between responders. OSHA's RFI addresses requiring facilities to coordinate emergency planning with local emergency response authorities [23].

An effective way to help ensure that emergency response plans are in place and will work is to hold regular, periodic drills to test them. It is important to ensure that emergency preparedness and response correlates emergency response with actual hazard scenarios and addresses access to fire water monitors and HAZMAT equipment during incidents and provisions for containment of fire fighting water to prevent environmental pollution.

A hazard analysis should be performed for emergency response activities to identify issues before emergency response is needed. Such a hazard analysis should incorporate task analysis, job safety analysis, and a SIMOP study.

3.11 Audits and Inspections

CSB investigations found that audits by companies, regulators, insurance providers, and others often did not identify process safety omissions and deficiencies and that audits and inspections missed issues that contributed to incidents in several cases {17, 18, 22, 42, 56, 59, 64}. Inadequate audits were identified in several cases {4, 22, 29}. Issues with audits and inspections were present for 13% of CSB incidents and 18% of incidents covered by OSHA's PSM standard.

The CSB noted that in the case of a refinery incident {56} a refinery National Emphasis Program (NEP) inspection was carried out about 18 months before the incident occurred but the mechanical integrity issues that were at the heart of the incident were not identified. NEP audits are the most comprehensive ones that have been performed by regulators. This audit was conducted by a team of people that spent several weeks at the refinery and it focused on the unit that subsequently experienced the catastrophic incident investigated by the CSB.

OSHA's PSM standard requires that a company conduct a compliance audit every three years to "verify that the procedures and practices developed under the standard are adequate and are being followed" [24]. This terse requirement can make for perfunctory audits. Furthermore, process safety compliance audits do not focus on fundamental inadequacies in the design of a process safety program; rather they evaluate the design against regulatory requirements and their implementation. Audits may not properly address non-routine operations because usually they do not occur when audits are performed. Additionally, audits are conducted under time limitations and use sampling techniques.

More comprehensive audits should be performed. They should cover all aspects of a process safety program including safety culture, roles and responsibilities, Auditors must be competent and should be certified [40]. OSHA's RFI addresses requiring third-party audits, increasing the required frequency of compliance audits, and requiring specific time frames for responding to deficiencies found by audits. There is an international standard for auditing management systems [41]. Process safety regulations should require audits be conducted in compliance with the standard.

3.12 Equipment Damage Mechanisms

Equipment damage mechanisms played an important role in several incidents {4, 10, 19a, 20, 53, 56, 63}. Interestingly, all but one of these incidents occurred in a facility covered by OSHA's PSM standard. The CSB has noted the need to perform damage mechanism reviews (DMRs) {63}. Proposed amendments to California process safety regulations include requirements for conducting DMRs [20].

3.13 Process Safety Regulations

Incidents occurred in facilities both covered and not covered by US process safety regulations. Many facilities that are not covered by process safety regulations in the US pose serious process safety risks, for example, facilities handling reactive chemicals and dusts. These issues have been recognized by the CSB [10, 11].

In some cases, facilities that are covered by process safety regulations had not implemented a process safety program {24, 32, 34, 36, 48}.

3.14 Compliance with Industry and Company Standards and Practices and Regulations

Non-compliance with industry and company standards and practices were common in numerous incidents {1, 4, 8, 9, 13, 15, 16, 17, 21, 24, 27, 28, 29, 31, 33, 40, 42, 43, 47, 50, 51, 56, 54b, 54c, 62, 64} and {2, 7}. This issue was present for 38% of CSB incidents and 29% of incidents covered by OSHA's PSM standard. Also, there were failures to follow regulations for multiple incidents {8, 12, 16, 24, 25, 33, 34, 36, 39, 40, 41, 64}. This issue was present for 18% of CSB incidents and 14% of incidents covered by OSHA's PSM standard.

Non-compliance with standards, practices and regulations is a fundamental issue for process safety management. If standards, practices and regulations are ignored, the comprehensiveness of their content does not matter and incidents will continue to occur. Certainly, management review and continuous improvement may go some way to address this problem and the inclusion of this CCPS RBPS element [25] was addressed in OSHA's RFI [23]. The RFI also addressed the inclusion of the CCPS RBPS element on measurement and metrics [25] to track the effectiveness of process safety management programs. Regulators in California have proposed amendments to the state's process safety regulations for refineries that will require the use of process safety performance indicators [20]. Public disclosure of process safety performance would encourage management to pay close attention to ensuring those measures reflect favorably on the company. Thus, for example, it can be expected that there would be fewer action items that remain unclosed past their due dates. Of course, if such an approach were to be applied, it may still be ignored by some companies. Changes in regulatory enforcement may be needed.

A recent case in the UK contrasts markedly with US practices. Phosphine was released from a plant requiring 4,500 people to shelter in place for 2-3 hours and causing disruption to a motorway [42]. The U.K. Health and Safety Executive (HSE) investigated the incident and found that an equipment failure caused loss of containment. The company was fined a total of £333,000 and ordered to pay costs of £110,000 after pleading guilty to an offence under the U.K. Health and Safety at Work Act of 1974 having failed to properly assess and act upon the risk of the equipment failing. Such a penalty for an incident of relatively low consequences contrasts with smaller penalties for greater consequence incidents in the United States.

After the prosecution, an HSE inspector said, "This case should serve as a warning to other companies dealing with harmful substances that they need to get their processes absolutely right, in order to ensure the safety of the public, if they don't they will face the consequences." In virtually, every incident analyzed in this paper, the company did not get it right. There are very few, if any process safety accidents; they are preventable incidents.

3.15 Industry Standards and Guidelines

Numerous CSB incident investigation reports identified flaws and omissions in industry standards and recommended practices {9, 14, 16, 25, 26, 29, 36, 37, 48, 51, 53, 54c, 55, 56, 63} from organizations such as the American Petroleum Institute (API), the American Society of Mechanical Engineers (ASME), the Compressed Gas Association (CGA), and the National Fire Protection Association (NFPA). Issues with industry standards and guidelines were present for 22% of CSB incidents and 32% of incidents covered by OSHA's PSM standard.

Such standards and practices usually are subject to periodic review and updating as part of a process of continuous improvement so such omissions and deficiencies ultimately can be addressed. However, the CSB has criticized the permissive nature of some of these standards and recommended practices {56} which leave options open to companies not to follow some of the guidance. Also, the CSB has criticized the process by which standards are set by some organizations, for example API {56}.

This finding suggests that CSB should investigate a diversity of incidents to identify as many deficiencies and omissions as possible in industry standards and guidelines.

3.16 Product Stewardship

Some incidents involved the use of highly hazardous chemicals obtained by one company from another which did not recognize their hazards {14, 17}. In another case, a company was not informed by a supplier that process safety regulations would need to be implemented before receiving shipments {24}. While chemical supplier companies will not necessarily know the details of how purchasers intend to use the chemicals, they should at least communicate their known hazards to the purchaser. Process safety considerations should be a key part of product stewardship initiatives such as that in the ACC's Responsible Care Code of Management Practices [43]. Suppliers of highly hazardous chemicals have responsibilities in supplying such chemicals to other companies.

Some additional issues were identified for a few CSB incidents (see Table 3).

Table 3. Additional Issues Noted for CSB Incidents.

Issue	Examples
Process boundaries defined too narrowly	59
Inadequate workforce participation	56 / *24*
Inadequate process safety information	3, 6
Inaccurate or incomplete safety data sheets	2, 5, 15, 38
Loss of process design and safety information due to poor management of records	11
Lack of hazard recognition and understanding	14, 15, 21, 26, 31
Inadequate communication of plant hazards	5, 17
Need for positive materials identification during commissioning, modifications, and maintenance	10
Inadequate or no pre-startup safety review	3, 13, 29, 45
No or inadequate near miss reporting	54c
Inadequate or no incident investigation	12, 20
Inadequate contractor management	39, 59, 64
Outstanding action items	2, 4, 6, 7, 16, 25, 45, 54b, 54c, 63 / *27*
Inadequate oversight by management	4, 7, 10, 12, 29, 32

4.0 Discussion of the Analysis

A summary of the results of the analysis is provided in Table 4. Clearly the top issues are abnormal / non-routine work and safeguards which are about equally important for all incidents and for PSM incidents and stationary source siting for PSM incidents. These are followed by procedures which are about equally important for all incidents and for PSM incidents and stationary source for all incidents. Compliance with standards and practices, process changes, emergency preparedness and response, maintenance, and previous incidents come next.

Table 4. Summary of Analysis for CSB Incidents.

Issue	% Incidents	% PSM Incidents
Design	28	18
Safeguards	56	57
Procedures	44	43
Maintenance	34	32
Normalization of deviance	18	18
Abnormal and non-routine operations	62	64
Process changes	34	39
Previous incidents	32	29
Stationary source siting	43	64
Facility siting	16	25
Emergency preparedness and response	32	39
Audits and inspections	13	18
Compliance with industry and company standards and practices	38	29
Failure to follow regulations	18	14
Industry standards and guidelines	22	32

Notable differences between the issues for all incidents and PSM incidents occur in some cases. There are fewer issues for both design and compliance with standards and practices for PSM incidents than for all incidents, but there are more issues for PSM incidents than for all incidents for each of stationary source siting, facility siting, and industry standards and guidelines. The difference for stationary source siting is particularly significant.

The lowest percentage of incidents with any of the issues is 13% and the highest is 64%. This is poor process safety performance.

Time trends for the CSB incidents are provided in Table 5. The numbers of incidents displayed per year in the table largely reflects resources available and decisions made by the CSB. However, it appears that the trend for virtually all issues is for them to continue.

Table 5. Time Trends for CSB Incidents.

Issue	1998		1999		2000		2001		2002		2003		2004		2005		2006		2007		2008		2009		2010		2011		2012	
	N	P	N	P	N	P	N	P	N	P	N	P	N	P	N	P	N	P	N	P	N	P	N	P	N	P	N	P	N	P
Design	1	-	-	1	-	-	1	-	-	-	5	-	-	1	-	1	1	1	2	1	1	-	-	-	2	-	-	-	1	-
Safeguards	1	-	-	-	-	-	1	-	4	-	4	2	-	3	-	3	1	3	2	1	1	1	2	2	3	1	1	-	1	-
Procedures	2	-	-	1	-	-	1	-	3	-	5	2	-	1	-	2	1	1	1	-	2	2	1	-	-	2	2	-	-	-
Maintenance	-	-	-	1	-	-	2	-	2	1	3	-	-	1	-	1	1	1	1	-	1	-	2	1	-	2	2	-	-	1
Normalization of deviance	-	-	-	-	-	-	-	-	1	-	2	1	1	-	-	1	-	-	-	-	1	1	-	-	1	-	1	-	-	1
Abnormal and non-routine operations	1	-	-	2	-	-	3	-	1	1	4	2	1	3	-	1	2	1	2	-	2	3	1	1	4	3	2	-	1	1
Process changes	1	-	-	1	-	-	2	-	1	-	5	-	1	1	-	2	-	1	-	1	-	2	-	-	1	3	-	-	-	1
Previous incidents	1	-	-	-	-	-	2	-	1	-	2	2	-	2	-	1	-	-	1	-	1	1	2	-	2	1	1	-	1	1
Stationary source siting	1	-	-	1	-	-	-	-	1	1	4	3	1	4	-	2	1	2	1	-	-	2	2	2	-	-	-	-	-	1
Facility siting	-	-	-	-	-	-	-	-	1	-	1	2	-	2	-	1	-	1	1	-	-	-	-	1	-	-	1	-	-	-
Emergency preparedness and response	-	-	-	-	-	-	1	-	2	1	1	3	-	2	-	-	1	2	1	-	1	2	1	-	2	-	-	-	1	1
Audits and inspections	-	-	-	1	-	-	-	-	-	-	2	1	-	-	-	1	-	-	-	-	1	-	-	-	-	2	-	-	1	-
Compliance with industry and company standards and practices	1	-	-	1	-	-	-	-	2	-	5	-	1	1	-	2	2	-	1	-	2	1	1	1	1	2	1	-	1	-
Failure to follow regulations	-	-	-	-	-	-	-	-	2	-	1	-	-	2	-	-	1	2	3	-	-	-	-	-	-	-	-	-	1	-
Industry standards and guidelines	-	-	-	-	-	-	-	-	1	-	2	-	-	2	-	1	-	1	-	1	-	-	2	1	1	2	-	-	-	1

N – Non PSM Covered
P – Covered by PSM

There is an average of 5 issues per incident and as many as 13 issues. Some CSB investigations appear to be more thorough than others so the number of issues identified in the investigation reports does not necessarily reflect all of the issues that were present. Consequently, these numbers may be low. However, it is clear that most incidents involved multiple issues and this conclusion would not change. It is likely that the underlying reason for multiple issues being present for the majority of incidents is a systemic problem with the process safety management system and process safety culture.

5.0 Further Observations

This section provides some additional noteworthy observations that were made during the review of incident descriptions.

5.1 Age of Plants

Many process facilities in the US were constructed some time ago. Of course the longer a process operates the more exposure there is to risk. Plant aging factors need to be addressed by process safety programs.

Equipment deteriorates over time and this has played a role in some incidents {63}. Damage mechanism reviews can help to address this issue. Old equipment may be used, for example, a tank constructed in 1929 failed catastrophically in 2008 {47}. Use of old equipment must be subject to close review.

As time passes, industry standards and best practices evolve but the designs of processes are frozen in time. Thus, gaps between current industry standards and best practices and existing company standards and practices may exist. Use of dated standards and practices has played a role in several incidents {30}. Outdated technology also has played a role in some incidents {29}. Process safety programs should require periodic formal reviews to identify and evaluate the feasibility of adopting current standards, best practices, and technology retrospectively. Current standards, best practices, and technology should be addressed when developing recommendations to reduce risk, for example, in PHA studies, MOC reviews, and incident investigations. OSHA's PSM standard does not require employers to evaluate updates to applicable RAGAGEP or to examine new RAGAGEP after evaluating and documenting compliance with mechanical integrity requirements. OSHA's RFI sought comments on the best approach for requiring the evaluation of updates to applicable RAGAGEP. However, RAGAGEP is defined narrowly by CCPS which is the definition being advanced by OSHA. It covers generally approved ways to perform specific engineering, inspection, or mechanical integrity activities, such as fabricating a vessel, inspecting a storage tank, or servicing a relief valve. There are many more types of standards and practices that could be included in the definition of RAGAGEP.

Informal procedures may become institutionalized with the passage of time with operators passing them on verbally without others being aware {46}. Key tribal knowledge may be lost through departures of personnel including retirements {11, 21, 54}. An aging workforce may display diminishing physical and mental faculties. These are human factors issues that should be addressed by a human factors program.

Often process facilities have been built in rural areas away from the public but over time housing developments are built near the facility. This issue was present for several incidents. It must be addressed through land use planning.

5.2 Magnitude of Consequences

The consequences of some incidents could have been much worse. For example, releases occurred under favorable weather conditions {22}, with the wind blowing away from populated areas {10}, when projectiles created by explosions missed hitting adjacent processes {11}, and at a time when the number of people potentially exposed was at a minimum {3, 36}. In some cases, people exposed to serious hazards miraculously escaped harm {63} and avoided being caught in the collapse of structures {49}.

These situations were serendipitous. However, it does suggest a hazard control strategy in which riskier activities, such as non-routine work, should be timed for when the risk to receptors is at a minimum. Also, facility personnel in the vicinity of a riskier activity who are not part of the activity should be relocated for the duration of the activity. Such considerations should be an integral part of planning for non-routine work.

Some injuries occurred when people investigated process abnormalities {32, 45, 64} or tried to fix leaks {62, 63}. Such situations must be taken seriously and not considered part of normal operations. Appropriate precautions must be taken.

6.0 Underlying Issues

This section identifies three issues that underlie all of the other issues. Each underlying issue is discussed and recommendations are made for addressing it.

6.1 Competency of Facility Personnel

The CSB has not explicitly addressed the issue of competency in many incident investigations. However, it was mentioned as a contributor in at least one CSB incident {29} and two CCPS incidents {*2, 4*}. In reading incident descriptions it is hard not to reach the conclusion that the people responsible for the processes were not competent in one or more ways in many cases. This issue likely underlies most of the process safety incidents that have occurred.

The decisions made by personnel ranging from managers to operators can be questioned for many incidents. While facility personnel were not obviously incompetent, in numerous cases clearly they were not competent to perform the work required of them safely. In particular, facility personnel often lacked an awareness and/or understanding of one or more critical aspects of a process, such as its hazards {6, 8, 11, 24, 32, 36, 39, 41, 44, 54a, 54c, 62}. Inadequate operator and mechanic training was an issue in a number of incidents {1, 2, 3, 10, 11, 12, 13, 14, 18, 21, 22, 25, 28, 29, 32, 39, 42, 45, 64}.

OSHA's RFI sought comments on the addition of an element on process safety competency [23] as described by CCPS [25]. Certainly, facilities should implement a competency management program for process safety but the CCPS RBPS competency approach should be modified [44]. Critical personnel should be certified and periodically re-certified. For example, operators and mechanics should not be allowed to operate or maintain a process without demonstrating their competence, including such topics as knowledge and understanding of process hazards, the higher risks of non-steady state operations such as startup, previous incidents, the safe operating envelope for the process, handling abnormal and emergency situations, conditions under which the process must be shut down, use of permit-to-work systems, and the identification and management of changes. Responsibility for supervision of the operation of a process should be formally assigned to a single person and the responsibility formally transferred to another individual as necessary. There must be supervisory oversight of processes at all times.

In some incidents, the availability of technical expertise has been an issue {29, 32, 45}. When critical decisions are faced regarding process safety, someone must be present who understands mass and energy balances, physical and chemical phenomena that may occur, process controls and safeguards, hazards present, and the safe operating envelope, that is tolerable ranges of values of safety-critical parameters. Individuals should be designated as engineering and technical authorities who ensure the proper application of engineering standards and technical practices and adjudicate when deviations are proposed. They must be available for consultation at all times and they must have the authority to shut down processes without the need for consultation with managers and without recrimination. They must be free of production or commercial pressures and should report directly to upper management. A stop work authority that authorizes personnel to stop work in the face of imminent risk or dangerous activity and an ultimate work authority that requires designation of who has the authority for operational safety and decision making are addressed by OSHA's RFI [23].

6.2 Standards for Process Safety Elements

Process safety programs comprise multiple elements such as employee participation, PSI, PHA, etc. For example, OSHA's PSM standard has 14 elements [24] and CCPS's RBPS program has 20 elements [25]. However, it is possible for companies to address each of these elements in widely varying ways because their requirements are general in nature and do not address the effectiveness of their contribution to risk reduction. For example, PHA studies can be conducted with widely varying levels of quality. A solution to this problem is to establish minimum performance standards for each element. These standards should be established by an international organization such as ISO so they apply around the world. There is precedent for such standards in the IEC 61511 standard for safety instrumented systems [45 - 47] and the Hazard and Operability Study (HAZOP) standard [48], although the latter is outdated. These standards are still performance-based in nature, as should be new standards for other process safety elements.

One benefit of standards for process safety elements is that they can be updated more easily than regulations. Current US process safety regulations are static and unable to adapt to innovation and advances in the management of major hazard risks.

This problem could be overcome if the US process safety regulations simply established overall requirements for each process safety element and required companies to follow RAGAGEP for each element. There is already a precedent for doing so with the PSI and mechanical integrity PSM elements although RAGAGEP is interpreted more narrowly in that context. The proposal is to broaden the meaning of RAGAGEP, establish RAGAGEPs for all process safety elements, and require that RAGAGEPs be followed. The process safety regulations would effectively update automatically as RAGAGEPs evolve. Continuous improvement is an essential element of a management system for any surety such as quality, environmental protection, and occupational health and safety, and is part of CCPS's RBPS program [25].

6.3 Safety Goals and Risk Tolerance Criteria

In an least one incident, the CSB concluded that one company had knowledge of and acceptance of high risk {29}. Most companies do not really know what risk their process facilities pose because they do not evaluate risk as there are no requirements to do so under the current US process safety regulatory regime. Indeed, it is possible that similar facilities may pose quite different risks because the regulations do not contain any safety goals or risk tolerance criteria.

The goal of process safety programs is to ensure safety and control risks. Consequently, some form of safety goals or risk tolerance criteria are needed [49]. Absent some metric to judge achievement of process safety goals, resources may be allocated inappropriately. There is a current precedent for the use of numerical risk tolerance criteria provided by the New Jersey Toxic Catastrophe Prevention Act (TCPA) process safety regulations [18]. Although the particular approach used is erroneous [50], companies in the state have accepted the use of numerical criteria.

Many jurisdictions around the world employ the tolerability of risk framework developed by the UK HSE including the As Low As Reasonably Practicable (ALARP) principle [51]. Its use should be incorporated into US regulations. Although the ALARP principle usually incorporates the use of numerical criteria, it can be used without specifying such criteria but using its essential element that risk should be reduced until the incremental expenditures of resources is grossly disproportionate to the value of the incremental risk reduction achieved. This is a performance-based requirement that is consistent with the current philosophy of US process safety regulations. Use of the ALARP principle would help to accommodate improved process safety practices without the need for periodic modification of regulations.

One of the difficulties faced by PHA teams is the use of risk ranking in studies to determine if recommendations for risk reduction, that is, additional safeguards, are needed. The CSB has been critical of the subjectivity involved and the criticism is warranted [31]. The CSB has suggested that Layers of Protection Analysis (LOPA) be used for this purpose. Of course, LOPA requires the use of risk tolerance criteria that are allocated from facility criteria [52]. Proposed amendments to the California process safety regulations include requirements for such an analysis [20]. Simply put, the adequacy of safeguards cannot be determined without risk tolerance criteria.

Use of risk tolerance criteria with LOPA raises the issue of what type of risk is being addressed and should be addressed. Certainly, no loss of life is acceptable but in US society single fatalities are more tolerable than multiple fatalities as evidenced by the relative attention paid by the media to single versus multiple fatality incidents. Perhaps the focus in process safety should be on protecting groups of people, particularly members of the public.

All process safety elements would benefit from the use of appropriate performance indicators [25, 53, 54]. Taken together, their use helps to control overall facility and process risks.

7.0 Conclusions

This survey of incidents covered a range of company and facility sizes, and types of processes. Also, it included facilities covered and not covered by process safety regulations. The study results can be used by companies to focus on improving areas where other companies have fallen short.

Of the 80 incidents studied, only the explosion resulting from corrosion failure of piping at the Shell refinery in Norco, LA, that occurred in 1988 {5} was truly unexpected owing to a unique corrosion phenomenon that occurred under conditions that were expected to be less corrosively onerous and with pH within the tolerable range. However, the failure occurred as a result of process changes and it is possible that the incident could have been prevented. All other incidents were predictable and therefore preventable.

Process safety programs aim to identify possible incidents before they occur and implement practices to prevent them. Remarkably, many of the incidents reviewed were attributable to deficiencies or omissions in the design, operation, or maintenance of processes. Therefore, it should have come as no surprise that these processes experienced incidents.

The quote, "It is said that there is nothing new in safety - it has all happened before" was published by CCPS [1]. Virtually every incident or something like it had happened somewhere before. All the incidents involved some type of deficiency or omission in adhering to established process safety practices. In many cases there were multiple deficiencies and omissions. All of these incidents were predictable and preventable. Just because an event may be described as unthinkable does not necessarily mean it is unpredictable.

Several recommendations can be made to strengthen process safety programs and regulations.

7.1 Recommendations for Operating Companies

- Add process safety elements on:
    - Design and engineering

- Human factors review
- Facility and stationary source siting review
- Safeguards protection analysis
- Operational discipline
- Competency management
- Previous incidents review
- Best practices review

- Address non-routine operations throughout all process safety elements.
- Require the use of JSA for non-routine work.
- Expand the types of changes addressed under the MOC element to include any change that may impact on or affect a release of a highly hazardous chemical.
- Make compliance audits more comprehensive.
- Require annual emergency response drills for facilities.
- Require that process risks be managed using the ALARP principle.
- Require facilities to set safety goals and/or risk tolerance criteria and demonstrate compliance with them.

7.2 Recommendations for Standards Organizations

- Develop international standards for key process safety elements that can be referenced by process safety regulations.

7.3 Recommendations for the Chemical Safety Board

- Expand the incident attributes that are addressed in investigations to gain more insights into process safety performance, for example, explicitly address competency, human and organizational factors, safety culture, plant aging factors, and operational discipline.
- Provide an appendix to each investigation report that characterizes each attribute for the incident. Not only is information on deficiencies and omissions valuable but also knowledge of what was done correctly or that there were no problems in certain areas is valuable.
- Compile a data base of incidents and their attributes that is searchable.

- Provide periodic analyses of the full data base of incidents looking for commonalities and trends and make recommendations for improvements in process safety practices.
- Publish the criteria used for the selection of incidents that are investigated by the CSB and update as appropriate.
- Expand on the current documentation in incident investigation reports to explain how the criteria were applied in selecting the incident for investigation.

The CSB adopted a motto in its early days. It still applies:

Salus populi suprema lex esto.

(People's safety should be the highest law.)

8.0 References

[1] J. Atherton and F. Gil, Incidents That Define Process Safety, Wiley-AIChE/CCPS, New York, NY, 2008.

[2] K. Gugan, Unconfined Vapor Cloud Explosions, Gulf Publishing, Houston, TX, 1979.

[3] T. A. Kletz, Learning from Accidents, 3rd edition, Routledge, Abingdon, UK, 2001.

[4] T. A. Kletz, What Went Wrong?: Case Histories of Process Plant Disasters and How They Could Have Been Avoided, 5th edition, Elsevier, Burlington, MA, 2009.

[5] T. A. Kletz, Still Going Wrong!: Case Histories of Process Plant Disasters and How They Could Have Been Avoided, Butterworth-Heinemann, Burlington, MA, 2003.

[6] Roy E. Sanders, *Chemical Process Safety: Learning from Case Histories*, Butterworth-Heinemann; 4th edition, Butterworth-Heinemann, 5th edition, 2015.

[7] The 100 Largest Losses 1974-2013, Large Property Damage Losses in the Hydrocarbon Industry, 23rd Edition, Marsh and McLennan.

[8] http://www.csb.gov.

[9] http://www.icheme.org/lpb.aspx

[10] Hazard Investigation: Improving Reactive Hazard Management, U.S. Chemical Safety Board, Washington, DC, 2002.

[11] Combustible Dust Hazard Study, U.S. Chemical Safety Board, Washington, DC, 2006.

[12] Seven Key Lessons to Prevent Worker Deaths During Hot Work In and Around Tanks, Safety Bulletin, U.S. Chemical Safety Board, Washington, DC, 2010.

[13] A. S. Blair, Management system failures identified in incidents investigated by the U.S. Chemical Safety and Hazard Investigation Board, Process Safety Progress, Volume 23, Issue 4, pages 232–236, December 2004.

[14] M. Kaszniak, Oversights and omissions in process hazard analyses: Lessons learned from CSB investigations, Process Safety Progress, Volume 29, Issue 3, pages 264–269, September 2010.

[15] Amyotte, P. R., MacDonald, D. K. and Khan, F. I. (2011), An Analysis of CSB Investigation Reports Concerning the Hierarchy of Controls. Process Safety Progress, Volume 30, Issue 3, pages 261–265, September 2011.

[16] T. Kerin, The evolution of process safety standards and legislation following landmark events—what have we learnt?, Process Safety Progress, DOI: 10.1002/prs.11762.

[17] P. Baybutt, Process Design - A Missing Process Safety Element (letter to the editor), Process Safety Progress, Volume 35, Issue 1, March, 2016.

[18] New Jersey Department of Environmental Protection, Title 7, Chapter 31, Toxic Catastrophe Prevention Act Program, Consolidated Rule Document.

[19] Contra Costa County Ordinance, Chapter 450-8-Risk Management.

[20] Proposed General Industry Safety Order (GISO) §5189.1, Process Safety Management for Petroleum Refineries, State of California, Department of Industrial Relations, Division of Occupational Safety and Health, Version 5.0 - September 14, 2015.

[21] P. Baybutt, On the Validation of Safeguards for Process Hazard Analysis, Process Safety Progress, Volume 32, Issue 2, pps. 165–169, June 2013.

[22] P. Baybutt, "What risk reduction measures should be credited in process hazard analysis?", Process Safety Progress, Vol. 31, No. 4, pps 359–362, December, 2012.

[23] Process Safety Management and Prevention of Major Chemical Accidents, Request for Information, Federal Register Volume 78, Number 236, Monday, December 9, 2013.

[24] Process Safety Management (PSM) of Highly Hazardous Chemicals, 29 CFR 1910.119.

[25] Guidelines for Risk Based Process Safety, Wiley-AIChE/CCPS, New York, NY, 2007.

[26] Risk Management Program for Chemical Accident Prevention, 40 CFR Part 68.

[27] Conduct of Operations and Operational Discipline, Wiley-AIChE, New York, NY, 2011

[28] D. Vaughn, Risky Technology, Culture, and Deviance at NASA, 1st edition, University of Chicago Press, Chicago, IL, 1996.

[29] P. Baybutt, Process hazard analysis for phases of operation in the process life cycle, Process Safety Progress, Vol. 31, No. 3, pps 279–281, September, 2012.

[30] G. Swartz, Job Hazard Analysis: A Guide to Identifying Risks in the Workplace, Government Institutes, Rockville, MD, 2001.

[31] P. Baybutt, The Role Of People And Human Factors In Performing Process Hazard Analysis And Layers Of Protection Analysis, J. of Loss Prevention in the Process Industries, Vol. 26 pps.1352-1365, 2013.

[32] Human Factors Methods for Improving Performance in the Process Industries, Wiley-AIChE/CCPS, New York, NY, 2007.

[33] P. Baybutt, The Psychology of Decision Making in Process Hazard Analysis, 11th Global Congress on Process Safety, Austin, TX, April 26 - 30, 2015.

[34] B. N. Moore and R. Parker, Critical Thinking, Eleventh Edition, McGraw- Hill Education, New York, NY, 2015.

[35] P. Baybutt, Competency requirements for process hazard analysis (PHA) teams, Journal of Loss Prevention in the Process Industries, Volume 33, January 2015, Pages 151-158.

[36] Guidance on SIMOPS, IMCA M 203, International Marine Contractors Association, March, 2010.

[37] Management of Change, Safety Bulletin No. 2001-04-SB, U.S. Chemical Safety Board, 2001.

[38] T. A. Kletz, Lessons from Disaster: How organisations have no memory and accidents recur, Institute of Chemical Engineers, Rugby, UK, 1993.

[39] Accidental Release Prevention Requirements: Risk Management Programs Under Clean Air Act Section 112(r)(7), Federal Register / Vol. 61, No. 120 / Thursday, June 20, 1996.

[40] P. Baybutt, Competency requirements for process safety auditors, Process Safety Progress, Volume 34, Issue 2, pages 179–182, June 2015.

[41] Guidelines for auditing management systems, ISO 19011:2011, International Organization for Standardization, Geneva, Switzerland, 2011.

[42] http://press.hse.gov.uk/2016/chemical-company-fined-for-dangerous-gas-release/?ebul=hsegen&cr=8/08-feb-16, last accessed 2/8/16.

[43] Responsible Care® Process Safety Code, American Chemistry Council.

[44] P. Baybutt, The meaning and importance of process safety competency, Process Safety Progress, DOI: 10.1002/prs.11767.

[45] ANSI/ISA—84.00.01—2004 Part 1 (IEC 61511-1 Mod), Functional Safety: Safety Instrumented Systems for the Process Industry Sector — Part 1: Framework, Definitions, System, Hardware and Software Requirements.

[46] ANSI/ISA—84.00.01—2004 Part 2 (IEC 61511-2 Mod), Functional Safety: Safety Instrumented Systems for the Process Industry Sector — Part 2: Guidelines for the Application of ANSI/ISA—84.00.01—2004 Part 1 (IEC 61511-1 Mod).

[47] ANSI/ISA—84.00.01—2004 Part 3 (IEC 61511-3 Mod), Functional Safety: Safety Instrumented Systems for the Process Industry Sector — Part 3: Guidance for the Determination of the Required Safety Integrity Levels — Informative.

[48] Hazard and Operability Studies (HAZOP Studies) – Application Guide, IEC 61882:2001, International Electrotechnical Commission, Geneva, Switzerland, 2001.

[49] P. Baybutt, Letter to the Editor, Managing process risks, Process Safety Progress, Volume 34, Issue 2, page 106, June 2015.

[50] P. Baybutt, Numerical Risk Tolerance Criteria in the United States: a Critique of the Risk Criterion Used for the New Jersey Toxic Catastrophe Prevention Act, Journal of Loss Prevention in the Process Industries 32 (2014), Volume 32, November 2014, Pages 428-435.

[51] P. Baybutt, The ALARP Principle in Process Safety, Process Safety Progress, Vol. 33, Issue 1, Pages: 36–40, March 2014.

[52] P. Baybutt, Allocation of risk tolerance criteria, Process Safety Progress, Volume 33, Issue 3, pages 227–230, September 2014.

[53] Guidelines for Process Safety Metrics, Wiley-AIChE, New York, NY, 2009.

[54] Guidelines for Integrating Management Systems and Metrics to Improve Process Safety Performance, Wiley-AIChE, New York, NY, 2016.

Accidents during Turnarounds, Cleanings, and other Infrequent Operations

Michael C. Stern
Exponent, Inc.
Natick, Massachusetts
mstern@exponent.com

Vijay Somandepalli,[1] Noah Budiansky,[1] Ali Reza,[2] Timothy J. Myers[1]
Exponent, Inc.
[1]Natick, Massachusetts
[2]Los Angeles, California

Prepared for Presentation at
American Institute of Chemical Engineers
2016 Spring Meeting
12th Global Congress on Process Safety
Houston, Texas
April 11-13, 2016

Accidents during Turnarounds, Cleanings, and other Infrequent Operations

Michael C. Stern
Exponent, Inc.
Natick, Massachusetts
mstern@exponent.com

Vijay Somandepalli, Noah Budiansky, Ali Reza, Timothy J. Myers
Exponent, Inc.

Keywords: Process Safety, Turnarounds, Startup, Shutdown, Cleaning

Abstract

Infrequent operations, such as turnaround services and cleaning operations, pose unique hazards due to their lesser familiarity to operators and plant personnel. For example, cleaning chemicals and methods can have unforeseen reactions with scale material, the chemistry of which may not be well-understood. Turnaround services may create a false sense of security because the plant is not in operation, which eliminates many of the commonly encountered hazards. Turnaround services, however, can be hazardous, especially if proper precautions have not been taken when shutting down the plant. Avoidance of accidents during these infrequent operations requires similar hazard analyses to those for standard operations, but these may not be thoroughly performed if the operation is assumed safe. Multiple case studies will be presented on accidents that have occurred during infrequent operations and the lessons learned from those incidents will be discussed.

1 Introduction

Operations including startups, shutdowns, turnarounds, and cleanings can pose additional risk for facilities due to their infrequency and uniqueness. The infrequency of these types of operations provides less opportunities for operators to gain experience and understanding of the procedures. The uniqueness, such as replacing or modifying a part that has been relatively untouched for years, can lead to unintended changes in the process when operation resumes.

Variability is a common risk factor for cleaning and turnaround operations. The same cleaning procedure or chemicals that might have worked for a certain type of scale could be hazardous if used for to clean out a different contaminant. Turnarounds and cleanings may also involve the use of contractors who are less familiar with the facility and may require site specific training to safely operate.

In this paper, the heighted risks of different infrequent operations are discussed and relevant case studies are presented. The case studies, some previously published and some presented for the first time, span a wide range of industries from manufacturing facilities to chemical plants, but all experienced or observed failures during an infrequent and transient operations.

2 Cleaning Operations

Both scheduled and unscheduled cleaning operations are often required at chemical and manufacturing facilities to maintain the equipment functionality and minimize contaminants in the products. Cleaning with chemicals can pose hazards due to undesired chemical reactions between the scale material and the cleaner. In certain cases when multiple cleaners are used, the cleaners could react to form hazardous substances or gases. Table 1, shown below, provides a few examples of chemical mixtures that could form explosive, toxic, or flammable substances leading to hazardous conditions. As will be described in the Condea Vista case study below, even steam cleaning can lead to an explosion in a process vessel when contacted with an aluminum powder containing sludge.

Table 1. Potential reaction products and hazards for certain chemical mixtures

Chemical Type One	Chemical Type Two	Reaction Products	Hazards
Acetone	Peroxides	Acetone-Peroxides	Explosives
Salts with (Cl^-, F^-, CN^-) ions	Acids	Acid gases	Corrosive and toxic
Metal Powders	Water or Steam	Hydrogen	Flammable

When cleaning areas that are already hazardous locations, such as areas near equipment containing flammable liquids or combustible dusts, the cleaning process can be a potential ignition source if proper precautions are not taken. Routine maintenance or cleaning actions can also release combustible dust accumulations creating a hazardous environment in an area where ignition controls are not present, leading to a fire or explosion. The case studies of the Jahn Foundry and CTA Acoustics combustible dust explosions provide examples of how ignition of disturbed accumulations can lead to catastrophic explosions.

2.1 Condea Vista Chemical Plant Explosion[1]

On October 13, 1998, an explosion and fire occurred inside a 3000-gallon Hastelloy reactor (R-2) that was part of the linear alkyl benzene (LAB) process at the Condea Vista plant in Baltimore, Maryland. Process materials released from the various vessels and process piping by the explosion fueled a fire that took approximately 2 hours to extinguish.

The investigation revealed that prior to the explosion, the bottom of R-2 was plugged with approximately 180 gallons of a sludge-like mixture that likely consisted of a mixture of coarse aluminum powder, aluminum chloride and various hydrocarbons. The explosion occurred approximately 10 minutes after steam was introduced at the bottom of the reactor in an effort to break up the sludge accumulation.

Figure 1. Photographs showing the overview of the explosion (left) and top of the exploded R-2 reactor (right).

A series of small-scale chemical reactivity tests were performed to investigate the potential reactions between steam and manufactured sludge that was similar to the material accumulated inside the R-2 reactor. Rapid runaway chemical reactions were observed after steam was injected into this mixture, while pressures in excess of 700 psig and temperatures in excess of 400 F were recorded before the test vessels ruptured.

A review of possible chemical reactions between steam, aluminum and aluminum chloride confirms that the predominant reaction mechanism is the exothermic hydrolysis of aluminum chloride to produce HCl vapor. Significant amounts of hydrogen gas can also be generated as the aluminum reacts with the acid/steam mixture after it is stripped of its protective oxide coating. Exponent concluded that the root cause of this accident was direct and sustained contact between the steam and accumulated sludge at the bottom of the reactor.

2.2 Dust Explosion at CTA Acoustics[2,3]

At approximately 7:30 am on February 20, 2003, a fire in at the CTA Acoustics facility in Corbin, Kentucky, lead to multiple secondary explosions and fires resulting in the death of seven workers and 37 injuries.

CTA Acoustics manufactured thermal acoustic insulation. The insulation material consisted of fiberglass bound together by phenolic resin powder. The product was used in automotive applications including the insulation on the hood or floor board of a car. CTA had four fiberglass forming lines, 401, 402, 403 and 405. The CSB report on the incident stated that the explosion originated near line 405 where dust generated during cleaning was ignited by the oven on line 405.

The CSB investigation concluded that workers started a routine cleaning of the dust collection system on line 405 thirty minutes before the incident occurred. The cleaning was expected to take approximately one hour. The night shift had reported significant clogging on the line. During the cleaning, the operators observed that the transition leading to the baghouse was the source of the clog and proceeded to use compressed air to break up the clog. When the baghouse was restarted at approximately 7:30 am, an initial explosion occurred that spread to the rest of the facility through secondary explosions fueled by clouds of combustible dust that were dislodged during the initial explosion.

2.3 Dust Explosion at the Jahn Foundry[3]

Figure 2. Photographs showing dust deposits on exhaust hoods in Shell Mold department (left) and the structural damage to the foundry (right).

At approximately 3:04 PM, on February 25, 1999, a series of explosions and a subsequent fire caused extensive damage within the Shell Mold building at the Jahn Foundry Corp. in Springfield, MA. Three employees suffered fatal injuries and nine others were injured. The foundry produced gray iron cast products for various customers.

The casting process included the production of shell molds, into which the molten material was ultimately poured. The molds were produced from a 95/5 percent mixture of sand and phenolic resin, added as a thermosetting bonding agent to hold the mold together. Even though the resin was supplied in the form of a fine powder, a dust collection system was not installed in the Shell Mold building. However, an odor control system designed to remove volatile organic compounds emitted from the resin during heating of the shell molds was sufficiently powerful to entrain fine particulates. The odor control system consisted of hoods or canopies above each shell mold station that vented into duct manifolds that connected to a scrubber tower where volatile organic compounds were removed.

An engineering investigation of the incident concluded the explosions were initiated due to the ignition of resin dust dislodged from an overhead exhaust duct or inadvertently dispersed during routine maintenance. A subsequent deflagration (dust explosion) within the odor control system was fueled by excessive, resin rich deposits of dust within the exhaust ducts. A second, larger dust explosion in the upper half of the Shell Mold building caused the majority of the damage to the structure. This second explosion was fueled by resin rich dust deposited on various horizontal surfaces on the upper level of the Shell Mold building.

Personnel at Jahn Foundry were generally unaware of the specific dust explosion hazard associated with this product and the foundry did not comply with the requirements of NFPA 654, *Standard for the Prevention of Fire and Dust Explosions from the Manufacturing, Processing and Handling of Combustible Particulate Solids*

3 Startup and Shutdown Operations

Operator inexperience and unidentified operational differences can lead to additional hazards when starting up or shutting down a plant. For plants that only shut down for maintenance annually or less, the operators will have only limited experience with the tasks required for startup and shutdown due to the infrequency of the operation. As a result, lessons learned from previous experiences may be less likely to be remembered, recorded, and passed on to future operators.

Differences that occur in the process during startup can be difficult to predict and assess. In systems with dust generation, for example, the properties of the collected dust during startup can vary from the dust generated during steady-state operations. It can be difficult to quantify these differences because process sampling during startup is complicated by the inherent transience of the operation.

In processes that include recycle loops or process material recycled downstream, the feedstock during startup can vary in considerable but unidentified ways. Small process changes, equipment maintenance and replacement, and other alterations to the system can introduce small but significant changes into the performance of the system that could lead to equipment failures if the changes are not anticipated. The case of the ammoxidation reactor incident described below provides an example of a chemical reactor that was damaged during startup due to the changes that occurred while the reactor was offline.

3.1 Ammoxidation reactor failure during startup

Several startup attempts of an ammoxidation reactor led to unplanned shutdowns within hours of starting due to over-temperature alarms. The ammoxidation reactor was a packed-bed counter-current tube-in-shell reactor where the tubes were packed with catalyst and an organic heat transfer fluid was pumped through the shell to cool the tubes and remove heat from the exothermic reaction. After a two-day period involving several attempted startups that were followed with subsequent shutdowns due to over-temperature conditions, the reactor was taken offline. Inspection of the reactor revealed a significant leak of heat transfer fluid and severe localized damage to the top tubesheet. The reactor was repaired by replacement of tubesheet sections and a number of tubes.

The reactor was brought back online and was used to produce a different isomer and performed without issue. After two months, the reactor was switched back to producing the original isomer. After two unsuccessful startup attempts that led to shutdowns due to over-temperature conditions, the reactor was inspected and thermal damage was observed.

Investigation of the reactor incidents revealed that prior to the startup, new catalyst was installed to replace the old catalyst that had been in the reactor. The catalyst packing gradient had also been adjusted such that higher concentrations of catalyst were present near the beginning of the reactor tubes. The new catalyst packing gradient and the increased activity of the fresh catalyst both contributed to increased reaction rates within the reactor during operation. It was also discovered that the heat transfer fluid reacted exothermically when exposed to conditions inside the reactor tubes.

It was determined that the failure of the reactor was a direct result of thermal expansion of tubes that occurred as feed rates were increased in the reactor and over-temperature conditions occurred. The thermal expansion led to the mechanical failure of a tube-to-tubesheet connection, which allowed heat transfer fluid to enter the tube side of the reactor. Exothermic reaction of the heat transfer fluid in the tube side of the reactor during repeated startups of the reactor led to the melting and heat damage of the tubesheet.

4 Plant Turnarounds

Plant turnarounds that involve significant maintenance, cleaning, replacement of equipment, and plant modifications can pose additional hazards due to the uniqueness and unfamiliarity of the activities, the involvement of contractors who are not as intimately aware of the hazards at the plant, and the difficulty of entirely purging the process lines and equipment of hazardous chemicals. Corrosion or contaminant buildup that occurred during the operation of the process can alter the condition of the process equipment in ways unbeknownst to the operators until it is inspected during the turnaround. The case study presented below provides an example of equipment that became corroded during operation but was detected before it created hazardous conditions.

4.1 Corrosion in a Flue Gas Desulfurization Absorber

All equipment is subject to both periodic planned shutdowns to conduct routine maintenance and inspections and unplanned shutdowns to replace components or make repairs. During these shutdowns careful inspections of the equipment are important to prevent accumulating degradation of a system or fix other unforeseen issues.

During a routine shutdown to perform periodic cleanout of a flue gas desulfurization (FGD) absorber associated with an incineration plant, corrosion damage was observed in the corrosion resistant liner of the absorber. FGD systems are designed to remove sulfur dioxide and other aggressive gaseous species from power plant and incineration emissions by spraying the flue gas with a limestone slurry, producing a calcium sulfite (gypsum) or other reaction species solid that falls to the bottom of the tower. The FGD slurry environments can have pH values as low as 1 and chloride concentrations as high as 200,000 ppm. The resulting slurry can be very corrosive and highly alloyed metals (or non-metallic coatings) are required to prevent corrosion of the liner to prevent exposure of the underlying carbon steel support structure.

In this case, 2205 duplex stainless was selected as the absorber liner was selected to prevent corrosion based on the nominal operating conditions, but corrosion of the duplex stainless steel liner was observed around a cleanout door. The corrosion damage had penetrated through the stainless steel liner in the time frame since the last inspection and caused minor damage to the underlying structure. A corrosion failure analysis was conducted and it was determined that the isolated environment in the gap between the cleanout access door and the door sill was more aggressive than the nominal conditions in the absorber. Laboratory corrosion testing indicated that corrosion resistance of the stainless steel liner, that was highly resistant to corrosion under the nominal design conditions, was significantly reduced in this area due to accumulated ions resulting in a reduced pH. In this case, periodic inspection of the equipment resulting in detection of corrosion degradation that was ultimately repaired prevented more serious failures from occurring.

5 Discussion and Conclusions

A review many of the industrial failures that have been investigated by Exponent shows a significant number that occurred during infrequent and transient operations including startups, shutdowns, cleanings (scheduled and unscheduled), and chemical transfers. While the specific incidents vary substantially in their magnitude and cause, an unknown change is the system can be traced to the cause. In three cases it was accumulation (combustible dust or aluminum sludge) and in another cases it was a change in a process variable (catalyst packing) that occurred during a turnaround that caused a failure upon startup.

Preparation for these upsets require thorough process hazard analyses (PHAs) and an effective management of change (MoC) program to identify these key variables before they become an issue and develop strategies for operators to employ when these events occur. The case of the FGD unit highlights the importance of careful inspection during turnarounds to identify corrosion or other changes in equipment. In some cases, such as combustible dust incidents, significant industry and governmental focus has raised awareness of the hazards of combustible dust. An OSHA National Emphasis Program and updated NFPA standards have been developed to address these hazards. In other cases, the lessons learned from the incidents should be used in future PHAs to mitigate the likelihood of similar incidents in the future.

6 References

[1] A. Reza, A. Kemal and P. Markey. "Runaway Reactions in the Aluminum, Aluminum Chloride, HCl and Steam System. An Investigation of the 1998 CONDEA Vista Explosion in Maryland," 36th Annual Loss Prevention Symposium. AICHE Spring National Meeting, March 2002.

[2] CSB. *Combustible Dust Fire and Explosions: CTA Acoustics*. United States Chemical Safety and Hazard Investigation Board, Report No. 2003-09-I-KY, February, 2005.

[3] T. Myers and A. Ibarreta. "Investigation of the Jahn Foundry and CTA Acoustics dust explosions: similarities and differences," *Journal of Loss Prevention*, pp. 740-745, 2009.

12TH GLOBAL CONGRESS ON PROCESS SAFETY

Biomass Industry Explosion Case Study

Amy E. Theis, P.E.
Fauske & Associates, LLC
16W070 83rd Street, Burr Ridge, IL USA 60524-5802
Theis@Fauske.com

Timothy L. Cullina, M.S., P.E. and Kenneth N. Kurko
Fauske & Associates, LLC
16W070 83rd Street, Burr Ridge, IL USA 60524-5802
Cullina@Fauske.com and Kurko@Fauske.com

Prepared for Presentation at
American Institute of Chemical Engineers
2016 Spring Meeting
12th Global Congress on Process Safety
Houston, Texas
April 11-23, 2016

Biomass Industry Explosion Case Study

Amy E. Theis, P.E.
Fauske & Associates, LLC
16W070 83rd Street, Burr Ridge, IL USA 60524-5802
Theis@Fauske.com

Timothy L. Cullina, M.S., P.E. and Kenneth N. Kurko
Fauske & Associates, LLC

Keywords: Explosion, thermal stability, combustible dust, biomass, incident, dust hazard analysis (DHA), management of change (MOC)

Abstract

The focus of this case study is an explosion that occurred in a biomass processing facility. There were three pieces of equipment involved including a dryer, conveyor and a cyclone. Thankfully, there were no injuries or fatalities due to this incident. This event occurred at a biomass company, resulting in damaged equipment with one piece needing to be completely replaced. Other equipment was able to be repaired and returned to service.

Material handled in the process had been observed to smolder on prior occasions, but the cause of these near miss incidents was not investigated or identified. The material was a potential combustible dust and there was also the possibility for flammable gas to be generated due to decomposition. Through material characterization, several hazards were identified. Combustible dust and calorimetry testing were employed to evaluate the thermal stability, reactivity and combustibility of the material. In addition, off-gases were evaluated to determine if flammable gas was generated. This case study will outline the test design strategy to obtain key process safety information, identify the multiple hazards based on the test results, review existing explosion safeguards and determine a postulated series of events that could have led to the incident. Lessons learned from the incident will be reviewed including recommended risk reduction strategies that could be used to prevent future events. One example is the importance of obtaining appropriate process safety information to facilitate data driven decisions to be made regarding safe operating conditions.

1 Introduction

An explosion occurred at a biomass processing facility. Fortunately no one was injured or killed in this incident. It is believed that a fire, smoldering or other ignition source began in the conveyor. Due to a lack of explosion protection, this led to a subsequent explosion in a connected piece of equipment. Findings indicate that the source of the explosion was consistent with a combustible dust deflagration that appeared to have ignited in or near the transfer equipment and accelerated through the duct work to a cyclone.

Lessons learned presented in this paper are based on experimental test results provided by the customer as well as testing performed by Fauske & Associates, LLC after the incident. Deficiencies that yield opportunities for improvement at this facility include a lack of hazards identification for the process, lack of combustible dust management systems and lack of a risk assessment.

2 Background

2.1 Process Description

In the process, the biomass is fed to a dryer that heats the material to elevated temperatures. The biomass has some residence time in the dryer before it is transferred through the bottom to a conveyor. Forced air is fed to the dryer and then goes to a cyclone where the entrained biomass is collected downstream. See Figure 1 for a process flow diagram.

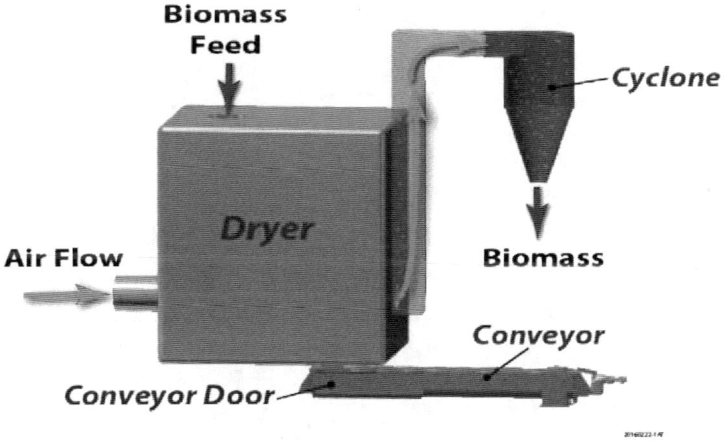

Figure 1 Biomass Process Flow Diagram

2.2 Explosion Event Description

On the day of the explosion event, a fire occurred earlier during the process. It is postulated that heated material accumulated in the conveyor, which led to self-decomposition at elevated temperatures, ultimately resulting in a fire. Employees used a water spray to suppress the flames, but perhaps this did not fully extinguish embers.

When sufficient oxygen was available, hot embers from the transfer equipment could have been drawn into the duct work and provided an ignition source for airborne biomass particulate solids. After ignition of dust, gases, or both, the rapidly expanding pressure wave and flame front moved into all of the available space in the first few milliseconds of the event.

The pressure wave blew the conveyor access door open and ejected burnt material. The flame front propagated to the dryer, fueled by airborne biomass then proceeded to the cyclone, ultimately leading to a rupture along the top of the cyclone. No personnel observed either explosive discharge. All witnesses reported hearing only one explosion. The access door opening and the cyclone failure could have occurred within less than 100 milliseconds of each other, producing a noise identifiable to witnesses as a single explosion. Pressure effects were observed downstream of the cyclone as well.

2.3 Previous Incidents

Based on history from processing similar biomass, personnel believed that smoking and fires, while not desirable, were typical and expected during the process. Fires had not only occurred during processing, but also during transportation of a by-product generated during the process.

In addition, biomass color and odor were not consistent during process runs. Operators believed there was a correlation between initial biomass color, intermediate color/odor, smoking events and fires. A lack of familiarity with combustible dust and gas hazards, especially the potential for deflagrations and explosions by both the management and operators contributed to the incident.

3 Hazard Evaluation

3.1 Material Properties

3.1.1 Biomass #1

Before the explosion, limited combustible dust and calorimetric data were available for this type of biomass. Previous combustible dust test results of a related biomass (Biomass #1) include a minimum explosive concentration (MEC) well above concentrations expected during the normal process conditions and a minimum ignition energy (MIE) value greater than 500 mJ. These results indicate that Biomass #1 was a combustible dust, but was not sensitive to ignition by electrostatic discharge.

Differential Scanning Calorimeter (DSC) testing of Biomass #1 identified one significant exotherm. This exotherm was detected above the normal operating temperature of the process and was confirmed by Carius tube testing. The Carius tube testing also indicated that non-condensable gas was formed during the event.

3.1.2 Biomass #2

After the incident, testing was performed on a sample obtained from biomass being processed at the time of the event (Biomass #2). First, DSC testing was conducted on Biomass #2 per ASTM E 537-12, *Standard Test Method for the Thermal Stability of Chemicals by Differential Scanning Calorimetry* [1] to determine the endothermic and exothermic activity of the material. A Differential Scanning Calorimeter was utilized for this test in conjunction with a 20 μl gold-plated stainless steel high pressure crucible. The DSC cell was heated at a preset rate while the differential heat flow, as a function of temperature, was measured and recorded. Five (5) DSC experiments were run on the sample at different heating rates: 1.0, 2.0, 4.0, 8.0, and 16.0°C/min.

The lowest heating rate DSC test showed that exothermic activity was detected at temperatures just less than the maximum temperature of the normal operating range for the process. Test results indicated that Biomass #2 yielded exothermic activity at lower temperatures compared to Biomass #1. The DSC results are summarized in Figure 2.

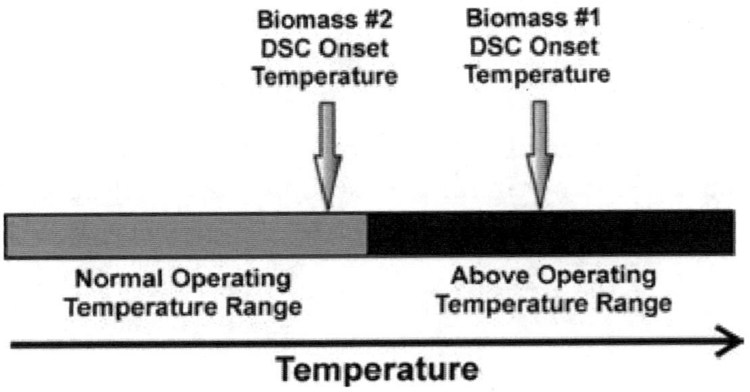

Figure 2 Thermal stability summary for biomass materials

The DSC data were evaluated to determine kinetic parameters for Biomass #2. These results showed that the material exhibited a self-accelerating decomposition temperature (SADT) value according to the United Nations Department of Transportation (UN DOT) H-1 test definition [2] that was within the normal operating temperature range. Similar evaluations were performed for smaller quantities of material (10 kg compared to 50 kg) and were also within the normal operating temperature range of the process.

Kinetic parameters were also used to determine adiabatic times to maximum rate for Biomass #2 as a function of temperature. These results indicated that the temperature that would yield a 24 hour adiabatic time to maximum rate would be in the middle of the normal operating temperature range of the process.

The next test performed on Biomass #2 was conducted using an Accelerating Rate Calorimeter (ARC), which is a high thermal inertia adiabatic calorimeter that is used to obtain data regarding the relationships between time, temperature, and pressure for exothermic reactions. The ARC utilized a 10 ml spherical test cell constructed of stainless steel with ¼" neck. In this experiment, the ARC was run in isothermal mode in order to monitor pressure generation. The instrument was programmed to track a reaction if exothermic activity was detected or maintain the desired temperature. ARC testing was conducted per ASTM E 1981-98, *"Standard Guide for Assessing Thermal Stability of Materials by Methods of Accelerating Rate Calorimetry"* [3].

The ARC isothermal test was performed on Biomass #2 at the maximum operating temperature range of the process over a time period of approximately 6 days. The temperature of the ARC test remained constant for the duration of the experiment. The evidence of non-condensable gas generated indicates that a reaction did occur. After the sample cooled, the non-condensable gas was subjected to a flame. No changes in the flame were observed, indicating that flammable gases were not present.

The last test performed on Biomass #2 was to determine the Minimum Explosible Concentration (MEC) using a 20-L Siwek chamber. The MEC is the minimum concentration of a combustible dust that is capable of propagating a deflagration through a uniform mixture of the dust in air, under the specified conditions of the test. Data obtained from this MEC test method can provide a relative measure of the concentration of a dust in the form of a uniform cloud necessary for an explosion. The test was performed per ASTM E1515, *Standard Test Method for Minimum Explosible Concentration of Combustible Dusts* [4].

Results from the MEC test on Biomass #2 showed that the MEC measured was within the range of what could be expected to occur within the process. If this concentration was present in the process, the material could be readily ignited by a burning/smoldering ember. A comparison of MEC values are shown in Figure 3.

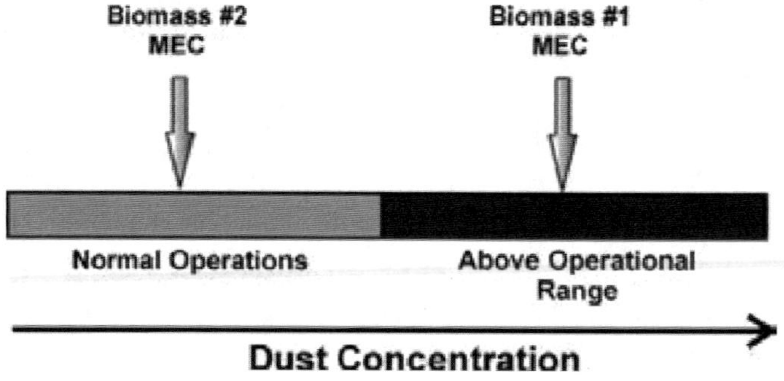

Figure 3 Minimum explosible concentrations summary for biomass materials

3.2 Incident Evaluation

It is postulated that heated material accumulated in the conveyor, which led to self-decomposition at elevated temperatures, ultimately resulting in a fire. Though fire-fighting measured were taken, it is possible that remaining material in the conveyor continued to heat or smolder. This could provide sufficient heat or an ember to serve as an ignition source for airborne particulate solids.

The combustible dust hazards of Biomass #2 handled in an existing process were not adequately identified or understood. It followed that safeguards were not in place to manage the risks associated with each piece of equipment in the process because the hazards and consequences were not characterized or evaluated. Since the hazards were not characterized at the time of the incident, employees were not made aware of the hazards associated with processing the material.

There was a concern whether flammable gas was present that could have created a hybrid mixture in addition to the combustible dust in the system. The presence of a small quantity of volatiles well below the lower flammability limit (LFL) of the gas can create a hybrid mixture with combustible dust that effectively reduces the minimum ignition energy and the minimum explosive concentration required to initiate a deflagration. Testing indicated that non-condensable gas formed during the isothermal experiment was not flammable. The gas tested may not be representative of gases formed during the deflagration event or even after the material was burned, so it is not possible to conclude whether flammable gases were present during the deflagration event. Additional testing is recommended to further evaluate this concern.

The lack of explosion prevention or mitigation devices on the equipment and absence of appropriate training are both a result of this inadequate hazard recognition. Installation of a properly designed explosion vent and installation of isolation technology could have mitigated damage to the cyclone. Figure 4 illustrates the pressure increase in a vessel both with and without the application of explosion mitigation by venting.

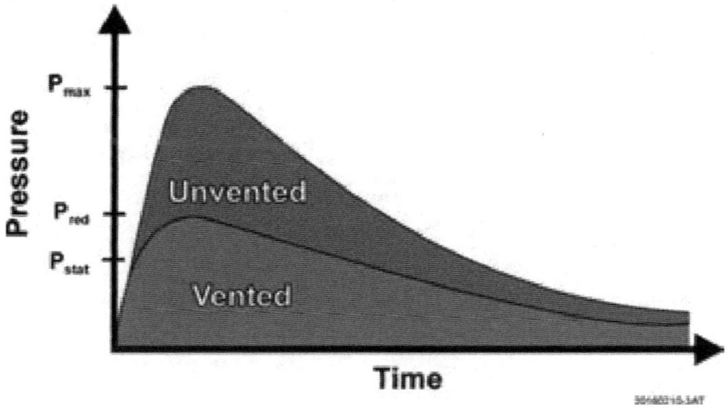

Figure 4 Progress of Explosion With and Without Explosion Protection

There was no record of a dust hazard analysis being performed for this process. Many of the required elements of a comprehensive combustible dust management program as outlined in National Fire Protection Association (NFPA) standards were not utilized to reduce the risk of dust fires and explosions.

3.3 Lessons Learned

This process included equipment covered by NFPA 654, *Standard for the Prevention of Fire and Dust Explosions from the Manufacturing, Processing, and Handling of Combustible Particulate Solids*. While processing Biomass #1, testing was performed to characterize combustible dust hazards and thermal stability for the material. Results indicated that the material was thermally stable in the normal operating temperature range and MEC values were above what would be expected during normal operations. At that time, no further protective measures were implemented. Data was gathered, but a risk basis of safety was not formally established.

When the raw material for the process changed from Biomass #1 to Biomass #2, this modification did not trigger additional testing to be performed. A management of change process, as required in NFPA 654, would have triggered the need to characterize Biomass #2 since it was different from Biomass #1. Even though the process equipment and technology remained the same, the facility changed raw materials which should have triggered Biomass #2 to be evaluated for combustible dust explosibility characteristics. This evaluation should have included thermal stability testing which were also performed previously for Biomass #1.

Fires were not uncommon in this process when processing Biomass #2 and the cause of these fires was not evaluated. It did not appear that an incident investigation process was in place at this facility, as required per NFPA 654 or the most recent NFPA 652, *Standard on the Fundamentals of Combustible Dust*. It appeared that management had allowed an acceptance of deviation to become the norm since fires were expected in the process. The response was for the operator to extinguish the fire and continue processing.

It was not evident if an inspection, cleaning and maintenance schedule was identified for pieces of equipment covered by NFPA 654 and 652. Material was allowed to accumulate in the conveyor which could have self-heated and led to fires observed in this equipment. It is not clear that inspection and access panels were in place to allow adequate cleaning of ductwork and the conveyor. It would be appropriate to establish and review cleaning and maintenance schedules per NFPA 654 and 652 requirements.

A dust hazard analysis (DHA) was not performed on this process. Given the operating temperatures for the dryer, the material should additionally be evaluated for thermal stability. This process should have triggered the need for testing to characterize the material for combustible dust hazards as well as for thermal stability. Once the hazards were identified, the equipment could then be evaluated to determine if additional protective strategies such as explosion protection and isolation were necessary.

After hazards have been identified, employees should be trained and provided guidance on required work practices in order to ensure their safety. Insufficient training contributed to the near miss of exposing employees to the deflagration. Without knowledge of the deflagration potential and the likely effects, employees were preparing to position themselves in high risk areas during the unintended explosion vent.

Data reviews after the incident revealed that employees did not understand the full intent of information recorded during the process. Design of the data forms likely contributed to the uncertainty regarding data collected. Employees did not have an awareness regarding the critical safety parameters and were not adequately trained on how to respond to deviations in the process.

Lessons learned from this case are summarized as follows:

- There was a lack of formal risk based safety documentation for the process.
- A management of change (MOC) process did not trigger data to be obtained when a change in the raw materials for the process were made.
- Sufficient data for the material/process were not obtained to establish safe operating conditions and identify potential upset scenarios.
- Performing a DHA could have identified the hazards, potential ignition sources, potential consequences and provided a framework to evaluate the adequacy of safeguards or mitigation strategies to determine if additional protection was needed.
- Proper hazard identification for the process is critical to ensure that adequate safeguards are in place and that employees receive adequate and appropriate training regarding normal operations and emergency response procedures.

- Routine fires during processing and transportation should serve as warning signs that trigger an investigation to identify preventive and corrective actions.
- The inspection, maintenance and cleaning procedures for dust handling equipment should be consistent with NFPA 654 and 652 requirements.

4 Conclusion

Testing performed after the incident indicated that the new material (Biomass #2) handled in the process was thermally unstable during normal operating conditions as well as being a combustible dust hazard.

In summarizing the lessons learned, it appears that several combustible dust management systems required by NFPA 654 and 652 did not exist, or if they existed, were not effective to prevent this incident. Key process safety information was not obtained to identify the hazards of the process. The company did not understand the warning signs for a potential combustible dust explosion in this process. Performing a DHA on this process could have identified the need to obtain relevant combustible dust and thermal stability data and understand that normal operating conditions could lead to an upset scenario. This information could then be used to propose changes to the process and determine if existing systems and equipment were adequately designed to protect against the consequences of the identified upset scenarios.

The following steps are recommended for identifying hazards associated with a process that contains equipment that could potentially handle combustible dust. These steps are consistent with the practices recommended in NFPA 652.

- Identify equipment in the process that is intended to handle particulate solids and the desired processing conditions (temperature and pressure operating range, equipment design requirements)
- Identify material to be handled in the particulate handling equipment
- Identify the hazards (safety data sheets, combustible dust hazard data, thermal stability data as warranted by process conditions, identify potential consequences due to upset conditions, perform a DHA)
- Understand the consequences in terms of severity and likelihood (determine if flash fire and/or explosion scenarios exist, determine if normal operating conditions or process deviations could lead to thermal decomposition, fire and/or explosion scenarios, identify if flammable gas could be generated)
- Determine process safety controls, safeguards and mitigation strategies (protect against overpressure situations, control and alarm set points, change in process design or procedures, explosion protection systems)
- Implement and maintain the safety measures (document safe operating ranges, perform preventive maintenance and inspections, train appropriate personnel, implement emergency response plans, employ other procedures as identified in NFPA 652)

5 References

[1] ASTM Standard E 537-12. *Standard Test Method for the Thermal Stability of Chemicals by Differential Scanning Calorimetry*, ASTM International, West Conshohocken, PA, 2012.

[2] United Nations. Recommendations on the Transport of Dangerous Goods: Manual of Tests and Criteria (6th ed.), New York and Geneva, 2015.

[3] ASTM Standard E 1981-98 e2. *Standard Guide for Assessing Thermal Stability of Materials by Methods of Accelerating Rate Calorimetry*, ASTM International, West Conshohocken, PA, 2012.

[4] ASTM E1515-14. *Standard Test Method for Minimum Explosible Concentration of Combustible Dusts*, ASTM International, West Conshohocken, PA, 2014.

[5] NFPA. NFPA 652, *Standard on the Fundamentals of Combustible Dust, 2016 Edition.* NFPA, Quincy, MA, September 2015.

[6] NFPA. NFPA 654, *Standard for the Prevention of Fire and Dust Explosions from the Manufacturing, Processing, and Handling of Combustible Particulate Solids, 2013 Edition*, Quincy, MA, June 2012.

[7] Ebadat, V. Testing to Assess Explosion Characteristics of Dust Clouds, NFPA Symposium on Dust Explosion Hazard Recognition and Control, 2009.

[8] U.S. Chemical Safety and Hazard Investigation Board. Investigation Report: Combustible Dust Hazard Study. Rep. No. 2006-H-1, 2009.

[9] Grossel, S.S. and R.G. Zalosh. Guidelines for Safe Handling of Powders and Bulk Solids. Center for Chemical Process Safety, AIChE, New York, 2005.

Atmospheric Tank Explosions: Mechanisms and an Unexpected Case Study

Michael S. Schmidt, P.E.
Principal, Bluefield Process Safety LLC
3101 Cherokee Street
Saint Louis, Missouri 63118, USA
bluefieldsafety@gmail.com

Prepared for Presentation at
American Institute of Chemical Engineers
2016 Spring Meeting
12th Global Congress on Process Safety
Houston, Texas
April 11-13, 2016

Atmospheric Tank Explosions: Mechanisms and an Unexpected Case Study

Michael S. Schmidt, P.E.
Principal, Bluefield Process Safety LLC
3101 Cherokee Street
Saint Louis, Missouri 63118, USA
bluefieldsafety@gmail.com

Keywords: storage tank, atmospheric tank, tank explosion, caustic

Abstract

Most of the literature on catastrophic failures of low pressure tanks focuses on the ignition of explosive mixtures in the vapor space of those tanks, and rightly so, because most catastrophic failures of atmospheric tanks involve explosions of such mixtures. Moreover, there is a general sense in the industry that other than explosions of flammable mixtures, the only other hazard associated with low pressure tanks is the occasional dramatic implosion of a vessel that is inadequately protected against vacuum.

In fact, there are a number of mechanisms that can lead to the catastrophic failure of a low pressure tank that have nothing to do with combustion or unprotected vacuum. Under certain circumstances, even a tank equipped with an atmospheric vent and containing nothing other than salt water can explode, with disastrous impacts.

This paper reviews the mechanisms for catastrophic failure of low pressure tanks, both implosion and explosion, and serves as a reminder for experienced process safety practitioners and as a tutorial for new process safety practitioners of what to look for during a hazard review. It also includes a case study of an atmospheric caustic tank explosion that resulted from a previously unreported mechanism that involved neither combustion nor vacuum.

1 Introduction

It is probably fair to say that atmospheric tanks are not very high on most people's list of hazards to worry about. Usually, they just sit there, storing liquids, doing what they are designed to do. Issues are typically the result of overfilling the tank and, on occasion, leaking components. Nothing sudden and catastrophic. When the words "sudden and catastrophic" are associated with atmospheric tanks, it is usually in relation to the flammable hazards of the liquids being stored. Chang and Lin [1] conducted a study of 242 storage tank incidents and found that 145 of them involved fire and another 61 involved explosion, mostly of flammable vapors from the liquids being stored. Outside of flammability issues, sudden and catastrophic incidents associated with atmospheric tanks are not what we typically worry about. Setting flammability issues aside, which have been addressed thoroughly elsewhere [2], [3], [4], [5] there are three types of atmospheric tank failures that can be sudden and catastrophic: collapses, implosions, and explosions.

Before continuing, however, it is important to first establish what is meant by the term, "atmospheric tank." The ASME boiler and pressure vessel code [6] divides vessels into two categories: pressure vessels, which operate at pressures greater than 15 psig and low pressure vessels, which operate at 15 psig or less. The BPVC applies only to pressure vessels. There are two API standards that apply to low pressure vessels. API 620 [7] applies to tanks that operate between 2½ psig and 15 psig, and at temperatures not greater than 250°F. API 650 [8] applies to tanks that operate at pressures up to 2½ psig, and at temperatures not greater than 200°F. OSHA defines an atmospheric tank as "a storage tank which has been designed to operate at pressures from atmospheric through 0.5 psig" [9].

All atmospheric tanks are low pressure tanks, but not all low pressure tanks are atmospheric tanks. There is not a consensus as to the limit—API 620 uses 2 ½ psig, OSHA uses 0.5 psig, others [10], [11] use a limit somewhere in between. For the purposes of this paper, an atmospheric tank is not intended to operate under vacuum and is not intended to operate at greater than 1 psig, or 30 inch water column. Human lungs have a greater operating pressure range.

2 Catastrophic Atmospheric Tank Failure

2.1 Collapse

One of the most famous catastrophic tank failures of the industrial era is the Great Molasses Flood of 1919 in Boston, Massachusetts [12]. The 50-foot tall, 90-foot diameter cast iron tank, built in a rush during World War I, ruptured and the wave of molasses killed 21 people. There was no fire. There was no explosion. The tank collapsed because it was under-designed for its application.

A more recent example is the failure of a liquid fertilizer storage tank at the Allied Terminals facility in Chesapeake, Virginia in 2008 [13]. The tank, originally designed for service with petroleum products, ruptured while being filled with a 32% solution of urea-ammonium nitrate (UAN). Two welders who were working on the tank as it was being filled for the first time after weld repairs were made were seriously injured when the tank collapsed.

2.1.1 Mechanisms for catastrophic failure

Atmospheric tanks fail for three reasons:

- The design or construction of the tank was inadequate.
- The tank is converted to a service that exceeds the capabilities of the initial design.
- The tank deteriorates over time.

Inadequate design or construction. While an atmospheric tank does not need to withstand high pressures, it does need to withstand the hydrostatic head of the liquid being stored. Atmospheric tanks collapse because the material of construction is not strong enough for the service, as was the case in the Great Molasses Flood, where the tank walls were 50% too thin, even for the relaxed standards of the 1910s [12]. Atmospheric tanks also collapse because the welds holding the tank together are inadequate. The Chemical Safety Board's report on their investigation into the Allied Terminals tank collapse includes an appendix of similar incidents, in which the listed cause of the incidents is "faulty welds" in more than 90% of the incidents where a cause is known [13]. Because of these issues, the CSB report notes that "tanks are more likely to fail when being filled to the maximum level for the first time."

Converted to a new service. Atmospheric storage tanks are designed for a specific service. Many tanks are designed to store petroleum products, for which a typical specific gravity is around 0.8. Simply converting the use from hydrocarbons to water increases the hydrostatic pressure on a tank by 25% when filled to the same level. Converting the use from hydrocarbons to 32% UAN, which has a specific gravity of 1.33, increases the hydrostatic pressure on a tank by two thirds. Converting the use from hydrocarbons to 50% caustic, which has a specific gravity of more than 1.5, almost doubles the hydrostatic pressure on a tank. Converting the service of atmospheric storage tanks to higher density liquids can lead to catastrophic tank failure by exceeding the design strength of the tank. Likewise, increasing the "safe fill level" has a similar impact on hydrostatic pressure, which can also lead to catastrophic tank failure.

Deterioration with time. Lastly, atmospheric storage tanks, like all equipment, deteriorate with time. Walls thin due to corrosion and erosion. Ideally, the thinning is uniform, steady, and slow. Catching this kind of deterioration during a tank inspection that includes thickness testing is straightforward. A tank with an adequate corrosion allowance will give lots of warning before the condition of the tank warrants a change in the operating conditions.

But not all deterioration is uniform. Tanks where nozzles, dip pipes, or eductors are aimed at the wall or floor of the tank are subject to localized thinning. Even a stream running down the interior wall of a tank can cause localized thinning. This weak spot then become the nucleus of a catastrophic failure. Stresses concentrate at a weak spot, the weak spot fails, and the stresses are immediately propagated through the rest of the material. The tank unzips and can collapse suddenly and catastrophically, often leading bystanders to believe there has been an explosion.

2.1.2 Protections against catastrophic failure

The causes of catastrophic tank failures also give a strong hint as to the protections that might be used against them.

Inadequate design or construction. This is best addressed with good equipment specifications according to recognized and generally accepted good engineer practice (RAGAGEP) and appropriate pre-startup safety reviews (PSSRs) to assure that the tank, as designed, constructed, and installed, meets those specifications. During commissioning, when many vulnerabilities in the tank are revealed, personnel should be kept out of harm's way, based on the assumption that the tank will fail.

Converted to a new service. Converting to a new service is a clear case for management of change, and the MOC review should consider changes in level, density, and corrosion rates, both general and local.

Deterioration with time. All equipment deteriorates, which is why a mechanical integrity program that adheres to RAGAGEP is essential. Tank inspections should test for thickness in enough spots to assure that general thinning is detected. Interior inspections should also look for specific evidence of localized thinning, especially at points where nozzles, dip pipes, or eductors are directed at the walls or floor of the tank.

2.2 Implosion

Atmospheric tanks are not designed to withstand vacuum, and they do not have to be exposed to very much vacuum to implode suddenly and catastrophically. The emptier the tank is when exposed to vacuum, the more dramatic the implosion. Tank implosions make such dramatic film that there are countless YouTube videos of tank implosions.

The lore that has been told and retold is of the maintenance team that was steaming out an atmospheric tank to prepare it for vessel entry, but did not finish before the end of the day. Rather than conduct the vessel entry at night, they closed up the tank and left. During the night, a storm front moved in, showering the tank with cold rain. The steam condensed, creating a vacuum, and when the crew returned in the morning they found the tank crumpled like an empty beer can. The crew from *Mythbusters* set out to prove or disprove this scenario using a DOT 111 tank car and decided that this story was "busted" [14]. *Mythbusters* notwithstanding, atmospheric storage tanks are not DOT 111 tank cars and the scenario should be considered credible.

Unlike explosions, implosions do not produce shock waves or flying shrapnel. Breaches may occur at the creases that form, but generally any loss of containment is not significant. The primary hazard associated with implosions are the potential injuries to personnel in the immediate area. The Center for Disease Control and Prevention reports of a worker killed while repairing a tank when it imploded [15]. Similarly, a worker in Spokane, Washington was killed in 2011 when the sewage digester he has checking imploded [16].

2.2.1 Mechanisms for catastrophic failure

Atmospheric tanks implode when they experience an internal pressure lower than they are capable of withstanding, i.e. a vacuum. Source of vacuum include the deliberate vacuum created by vacuum pumps and steam vacuum jets, as well as the incidental vacuum created by the pump suction of other pumps during pump out. Even in the absence of a pump, a gravity drain to empty a tank and other forms of emptying a tank of its liquid contents will create a vacuum. Finally, as described above, phase changes, particularly condensing vapors will create a vacuum.

2.2.2 Protections against catastrophic failure

In the absence of a source of vacuum, protections are unnecessary. However, all atmospheric tanks have a source of vacuum, if for no other reason than they are emptied. Protection include vacuum breakers, pressure/vacuum conservation vents, nitrogen pads, and most commonly, atmospheric vents. As long as these are properly sized and designed and remain fully operational, they are sufficient to protect against catastrophic implosion.

2.3 Explosion

Most atmospheric tank explosions involve the storage of flammable liquids, where it is the explosion of the flammable material that is the concern and which is outside the scope of this paper. There are also physical explosions that can occur, resulting from overpressure in the tank for reasons other than sudden combustion.

2.3.1 Mechanisms for catastrophic failure

Atmospheric tanks explode when they experience an internal pressure higher than they are capable of withstanding, which is not much. Sources of this high pressure include the pressure of inlets to the tank and heating within the tank.

Inlet line pressure can be of either gases or liquids. When the tank is open to atmosphere, the pressure drop at the inlet is complete, and the pressure in the inlet is not transferred to the interior of the tank. However, when the tank is blocked off from the atmosphere, the pressure in the tank will approach the pressure of the inlet line. Gases will flow into the tank until the pressure in the tank matches the supply pressure. Liquids will flow into the tank, compressing the headspace, until the headspace pressure matches the deadhead pressure of the liquid line. A tank does not need to be liquid full to reach the deadhead pressure of the liquid line.

Heating a vessel can result from either external heating or the heat of reaction, planned or unplanned, within the vessel. In either case, the liquid will expand and the vapor pressure will increase. Liquid expansion is only a concern when the tank is blocked off from atmosphere and the tank is completely full of liquid. In this case, since liquids are incompressible fluids, any additional expansion will sharply increase the pressure. However, atmospheric tanks are not absolutely rigid, so the effect will be to cause the tank to bulge, rather than to cause it to burst.

Increased vapor pressure is the real concern. Vapor pressure increases as temperature increases, and is independent of head space. An atmospheric tank blocked in while containing any liquid will experience an increase in pressure when heated, and atmospheric tanks are not designed to contain much increase. The vapor pressure of water, for instance, will increase by 1 psi when heated from 72°F to 113°F; it will increase by 5 psi when heated from 80°F to 165°F.

2.3.2 Protections against catastrophic failure

The key to protecting against physical explosion of atmospheric tanks is to assure that any source of pressure is released from the vessel before the pressure becomes too high. The obvious means is an atmospheric vent. Other protections include pressure/vacuum conservation vents and venting entryways or "trashcan lid."

To be useful, however, it is imperative that these protections be properly sized and designed and then remain fully operational. A vent that is too small or blocked is no protection. An atmospheric vent may be directly to atmosphere, or pass through a seal pot. Many creatures—

birds, wasps, rodents—find the outlet of an atmospheric vent a perfect place to build a nest. Bird screens, when in place, will keep out birds and rodents, but not nest-building insects. Likewise, a seal pot that has frozen solid will not vent. Vent lines that contain monomer vapor, whether directly to atmosphere or through a device can become plugged with polymer or simply glued shut.

Pressure vessels can be protected with rupture discs and pressure relief valves, but these require set pressures that are too high to be of use in atmospheric tanks. Likewise, pressure vessels are frequently protected from inlet pressure with regulators on the inlet. Unfortunately, atmospheric tanks operate at such low pressures that to regulate inlet pressures down to the operating pressure of the tank would reduce flow to unacceptably low rates.

Weak-seam roof tanks are frequently used to protect against the failure of atmospheric tank walls in the event of overpressure. They are designed so that the roof-to-shell joint will fail before any of the seams in the shell or the floor of the tank fail, thus preventing a release of the contents. However, this approach must be used with care. Accessories, such as railing or stairs, can make the roof-to-shell joint stronger than intended, as can the workmanship of an enthusiastic welder. (Welders are rarely scolded for making their welds too strong.) In general, no credit should be given for a weak-seam roof on a tank smaller than 10 meters (30 feet) in diameter because of the concern that the joint will be too strong [17]. When a weak-seam roof is used, the roof itself should be restrained with some kind of tether or hinge. Otherwise, the roof, ripped away from its weak seam, will be launched as a projectile.

3 A Case Study

While process air was blowing through the transfer line from a caustic storage tank to a caustic day tank, the caustic day tank failed catastrophically, launching the top head of the tank approximately 125 feet, where it came down through the roof of a nearby process building and triggered an impromptu evacuation in which one employee fell and had to receive medical treatment.

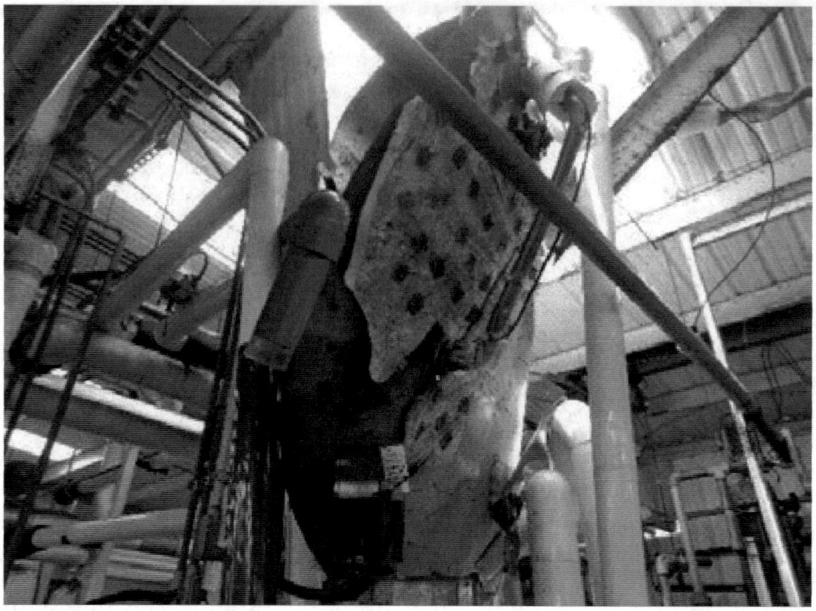

Figure 1. Caustic Day Tank head, after coming down through the roof of a nearby process building.

3.1 Process Background

The process consisted of a 3-inch transfer line constructed of 304 stainless steel that connected the discharge of a caustic transfer pump at the caustic storage tank to the top of a caustic day tank. A 50% solution of NaOH in water entered the caustic day tank through a dip pipe to the bottom of the tank. The transfer line was purged after each transfer of 50% caustic solution from the caustic storage tank to the caustic day tank by blowing out the line from the caustic storage tank to the caustic day tank for about 10 to 15 minutes with process air.

When the transfer line was being prepared for maintenance, there was a more extensive purge procedure. The normal procedure for purging the transfer line was to blow out the line with process air, blowing from the caustic storage tank to the caustic day tank for about 10 to 15 minutes. The next step was to flush the line with city water for 5 to 10 minutes, again flowing from the caustic storage tank to the caustic day tank. After the water flush, any residual water in the transfer line was blown out with process air, again blowing from the caustic storage tank to the caustic day tank for about 10 to 15 minutes.

The transfer line was routinely purged following transfers and was purged successfully for maintenance a month before the incident.

3.2 Incident Description

On a balmy, spring Friday morning, preparation began for maintenance on a leaking caustic transfer line from the caustic storage tank to the caustic day tank at about 9:30 am. The first step – purging the caustic from the transfer line with process air – was completed without incident.

The second step – purging residual caustic from the transfer line with city water – started at around 9:50 am. This step was also completed without incident, and at around 10 am the operator disconnected the water hose and left it to drain while he turned his attention to other matters in the unit.

At around 10:30 am, the operator returned to the caustic storage tank, disconnected the water hose, and reconnected the air line. He then opened the air supply valve in order to blow the water out of the transfer line.

At 11:10 am (the time shown on the time-stamp from the video monitoring the area), while the process air was still blowing through the transfer line from the caustic storage tank to the caustic day tank, the caustic day tank failed catastrophically. The pressure was sufficient to separate the conical head from the top of the tank and launch it approximately 125 feet, where it came down through the roof of a nearby process building to the east.

While the sides and bottom of the caustic day tank remained intact after the incident, the damage was enough to consider the caustic day tank a total loss. The piping attached to the top head of the caustic day tank was also damaged beyond repair. There were no fatalities or serious injuries associated with the incident, but one employee did require medical treatment for a sprained ankle following a fall that occurred during his evacuation from the immediate area of the incident.

3.3 Possible Issues

An explosion of a caustic tank can result from:

- Sources of pressure considerably higher than the design of the tank was capable of handling,
- Blockage in the equipment that lowered the pressure the tank was capable of handling, or
- Material failures in the tank that lowered the pressure the tank was capable of handling.

Guidance documents from suppliers of 50% caustic solution suggest the following as sources of high pressure, causes of blockage in caustic equipment, and causes of equipment material failure at pressures lower than design pressure [18], [19].

3.3.1 Possible sources of high pressure in a generic caustic tank

- Reaction of caustic solution with reactive metals (aluminum, magnesium, zinc, tin, chromium, brass, and bronzes made with zinc or tin), releasing hydrogen that forms a flammable mixture with air and subsequently ignites.
- Plant air or nitrogen used to transfer caustic solution by pressure.
- Steam used to clean out transfer lines.
- Direct steam heating of frozen caustic, exposing equipment to steam pressure.
- Pump deadhead pressure in unvented tank or line.
- Diluting strong caustic with water, rather than diluting strong caustic by adding it to water.
- Heating caustic solution while a crust of frozen caustic is formed over the top of the liquid

3.3.2 Possible causes of blockage in generic caustic equipment

- Block valves inadvertently closed.
- Lines plugged with frozen caustic solution.

3.3.3 Possible causes of generic equipment failure at lower than design pressure

- "Caustic embrittlement" – stress-corrosion failure which occurs at areas of mechanical stress or where fabricating operations produce high residual stress such as at bends in piping and at welds.
- Using iron and steel as materials of construction for equipment handling 50% caustic solutions at temperatures above 140°F. While iron and steel are the usual materials for equipment handling 50% caustic solutions below 140°F, nickel, nickel alloys, and stainless steel are required at higher temperatures and for higher concentrations.
- Using unsupported plastic pipe, which can fail because of the high density of 50% caustic solution, and using fiberglass reinforced plastic pipe except for specific conditions for which it is suited.
- Continued heating after caustic soda is melted, which may damage the special lining on the inside of railcars. The temperature of 50% caustic solution should never exceed 150°F (66°C) in lined cars.
- Wet or saturated steam above a maximum of 15 psig and 250°F. Steam at higher temperatures will damage rail car linings.

3.4 Conditions at Incident Scene

3.4.1 Noted from video

There were several dozen closed circuit video cameras positioned around the plant, three of which captured evidence of the catastrophic failure of the caustic day tank:

- The truck entrance camera, located about 600 feet southeast of the caustic day tank and aimed northwest.
- The tram gate camera, located about 275 feet southwest of the caustic day tank and aimed northeast.
- The main crosswalk camera, located about 175 feet west of the caustic day tank and aimed east.

The time stamps on the truck entrance camera and the tram gate camera coincide closely, and both showed the initial cloud release from the tank rupture at 11:10:19. The main crosswalk camera did not show the cloud release from the tank rupture, but shuddered at the time the cloud was released. Both the truck entrance camera and the main crosswalk camera captured an employee falling down and then getting up and walking away, hardhat in hand, at 14 seconds after the tank rupture.

Figure 2. Detail of screen capture from truck entrance camera at 11:10:19.647, showing launch of the caustic day tank top and associated cloud.

The main crosswalk camera, which was closest to the caustic day tank, shuddered for 0.4 sec, consistent with the passing of a shockwave. Neither of the more distant cameras showed the shudder. The more distant cameras, however, showed the initial cloud released. The cloud formed and was completely dispersed within 2 seconds. There was no evidence at all in those videos of combustion or of a secondary BLEVE. The shockwave confirms that there was an explosion. The absence of combustion or BLEVE and the quick dispersion of the cloud that was released confirm that the explosion resulted from simple overpressure, not from the ignition of a flammable mixture of hydrogen and air in the headspace or from the boil-up of the contents of the vessel.

The truck entrance camera was the only view that showed the launch of the top head. In the first 0.2 seconds, a cloud appeared above the west side of the tank. Then, 0.2 seconds after the initial release of the cloud, the conical head fully separated from the tank and began its 125-foot trajectory eastward, where it came down through the roof of a nearby process building. The video suggests that the separation began on the west side of the conical head and the break proceeded almost instantly around the tank head, propelling the conical head upward before the pressure inside the tank could be relieved.

3.4.2 Noted from field observations

The most relevant observation was that of the condition of the vent from the caustic day tank. A 3-inch gooseneck constructed from glued heavy-wall PVC pipe and elbows, the vent was attached to the top of the caustic day tank at a flanged nozzle 40 inches from the center peak of the conical head. The vent sheared off during the incident just above the flanged connection.

The internal surface of the vent (the surface that would have been visible from the outside while it was still attached to the functioning tank) was open and free of any obstruction. However, from the point the nozzle was attached to the top head, through the flange and up to the first elbow, the vent was encrusted with a white solid that tested positive for NaOH. The morphology of the solid was consistent with a slow accumulation over an extended period of time, rather than the result of an incident of overfilling the tank and then freezing in the vent. Moreover, there was no evidence of accumulation or discoloration on the head under the discharge of the vent, which would have been consistent with discharge through the vent onto the surface of the conical head.

The accumulation of solids within the vent choked it down to a diameter of about a half inch, and the rough, convoluted surface of the choked vent would have rendered it less useful than a clean half-inch pipe as a vent.

Figure 3. Vent nozzle, showing build-up of caustic.

Likewise, there were no solids built up on the interior walls of the vessel or on the interior surface of the head. This suggests that the buildup inside the vent line was the result of an entrained mist of 50% caustic solution being carried to the vent, where it collected and evaporated, leaving behind solid NaOH. The monthly vent inspections conducted on the tank since it was put into service in late 2003 would not have detected the gradual accumulation of solids.

Figure 4. View of interior of the caustic day tank after the incident. The thin white solid is caustic that dried there at the time of the incident. The interior shows some staining, but no build-up of the NaOH encrustation seen in the vent nozzle.

A thorough inspection of the vents of the two caustic tanks in the area showed no similar build-up of solids. However, one of the caustic storage tanks was put into service about 5 years before the incident and the other caustic storage tank was put into service about a year before the incident, so there would not have been the same opportunity for buildup.

The caustic day tank was specified to be constructed of 304 stainless steel. A field inspection of the vessel was consistent with that specification. The metal looked like stainless steel and was not magnetic; short of testing the metallurgy, this was considered enough to conclude that the tank was built as specified. Suppliers of 50% caustic recommend that carbon steel is a suitable material of construction for up to 140°F; over 140°F, they recommend 304 stainless steel to provide good corrosion resistance. This tank was operated at ambient temperatures. There was no reason, then, to believe that temperature or corrosion played any role in this incident.

The conical head separated completely. An inspection of the head showed that the separation was at the weld of the head to the 3"x3" angle that formed the top flange of the vertical cylinder of the tank. While the failure of an atmospheric tank is typically expected to be in the form of a split, where pressure is released without accompanying shockwave or shrapnel, an explosive separation of the head from the body of the vessel is not unheard of, e.g., Barton Solvents, Valley Center, Kansas, on July 17, 2007 [20].

Figure 5. The separation of the head from the body of the caustic day tank was at the weld of the head to the 3”x3” angle that formed the top flange of the vertical cylinder of the tank.

3.4.3 Noted from interviews

The pressure gauge on the process air was not read at the time of the incident. The production supervisor noted that the pressure should be set at around 35 to 40 psig, while the maintenance engineer stated that the process air pressure is normally around 80 psig, and that the regulator was normally wide open, with little motivation to crank it back closed. This is consistent with the pressure reading observed in the field, which indicated that the process air pressure was at 90 psig. The caustic day tank itself has a design pressure of atmospheric (0 psig) and was pneumatically tested at 5 psig at the time it was manufactured in 2003.

The supply pressure for the city water used to flush the transfer line is around 40 to 60 psig. This means that the line could have been successfully flushed into the caustic day tank, even if the pressure in the tank had risen to 35 psig during the initial purge. During the second process air purge, which lasted for around 45 minutes before the incident, there would have been more time for the pressure to build.

3.5 Contributing Factors

3.5.1 Non-contributing factors

The investigation was able to eliminate many of the generic issues identified by caustic solution suppliers.

In regard to possible sources of high pressure, the materials of construction – 304 stainless steel and PVC – make hydrogen formation as a result of reaction with reactive metals an extremely unlikely factor. This is further supported by the absence of any evidence of combustion during the incident. Likewise, neither steam nor a pump were in use, so steam pressure and pump deadhead pressure did not factor in. Temperatures were well above 65°F, the temperature at which frozen 50% caustic solution become a concern.

The purging procedure did include a step where water was added to strong caustic, a step that is normally discouraged. However, this did not create the high pressure that resulted in the incident, and overall, is a reasonable way to deal with the flush water.

In regard to possible causes of blockage in caustic equipment, there were no block valves to be inadvertently closed, and ambient temperatures were high enough to prevent lines from being plugged with frozen caustic solution.

In regard to possible causes of equipment failure at lower than design pressure, the materials of construction – 304 stainless steel and PVC – also eliminate the concerns with caustic embrittlement and with temperatures above 140°F. There was no unsupported plastic pipe to fail, and steam was not in use.

3.5.2 Direct causes

A "direct cause" is the uncontrolled release of energy or hazardous material that cannot be absorbed safely. In the case of this incident, the direct cause was high pressure in the caustic day tank, the source of which was the approximately 80 to 90 psig process air used to blow down the transfer line from the caustic storage tank to the caustic day tank.

3.5.3 Indirect causes

An "indirect cause" is the action or condition that results in direct causes. In the case of this incident, the indirect cause was the plugged tank vent on the caustic day tank. A contributing factor was probably the extended time that process air was allowed to blow into the tank. However, the vent was gradually becoming clogged and eventually, even shorter periods would have been sufficient to overpressure the vessel.

3.5.4 Basic causes

A "basic cause" is the policy, procedure, or decision that leads to indirect or direct causes. There were three basic causes. The first basic cause was the procedure that called for blowing fluids into the caustic day tank at pressures higher than the rated pressure of the tank. While pressures higher than atmospheric pressure, either as a result of pressurized fluids or pumping, are absolutely necessary to effect transfer into the tank, the higher pressure also made a functioning atmospheric vent an imperative element of normal operation.

The second basic cause was a tank design and operation that resulted in carrying entrained 50% caustic solution mist to the vent, where it could deposit and plug the vent.

The third basic cause was the vent inspection procedure. In hindsight, it is now clear that monthly visual inspection did not detect that the vent was becoming plugged. In the absence of this hindsight, there is nothing to fault the vent inspection procedure. Caustic solution is not volatile, and there is no general expectation in the industry that vents on caustic solution tanks are subject to plugging. However, now that the problem has been identified, it should be addressed.

3.6 Recommendations

3.6.1 Vent design to assure pressure relief is adequate

Dow Chemical has the following recommendation in regard to vents on caustic solution storage tanks: "In addition to a solution high-level alarm, the tank should have a liquid-level gauge and a vent. The vent should have a cross-sectional area at least four times that of the fill line (which is normally 2"), extending from the top of the tank to within about 3' of ground level. The vent, which also acts as an overflow, should be visible from the tank truck or tank car loading area so that personnel will see and/or hear caustic discharging from this overflow pipe in case the tank is overfilled. Valves should never be installed in the vent line. It must be open to the atmosphere at all times" [18].

In this particular case, the vent and fill are both 3-inch lines. The vent pipe is a gooseneck, rather than a pipe that extends to within 3 feet of ground level. Modifications to the vent piping may be helpful.

3.6.2 Vent inspection to catch any buildup

It was unexpected to discover that the vent was plugged with caustic. The likely cause of the plugging was the deposit of entrained mist in the vent. The mist, composed as it was of 50% caustic solution, would be the source of solids. The practice of blowing process air into the tank through the dip pipe assures that there for mist to be entrained after every transfer into the tank. The evidence is that process air carries the mist all the way to the vent. As a result, this is an issue that should be expected in any tank containing solutions of dissolved solids that can be entrained as mist.

In this particular case, the purpose of the vent inspection was to look for evidence of external blockage, primarily as the result of natural activity, e.g., bird nests, wasp nests, etc. In the case of tanks storing solutions of dissolved solids, a more thorough inspection of the vents is in order. This may involve the use of boroscopes and cameras, or as an alternative, the installation of inspection and clean-out ports in the vent line. The frequency of inspection should be adjusted as experience dictates, beginning with quarterly inspection, but probably settling on annual inspection.

3.6.3 Limit use of dip pipes for inlet piping

The practice of blowing process air or nitrogen into a tank through a sub-surface dip pipe assures that mist will be entrained. While the density of aqueous solutions such as 50% caustic is high enough to lead to the expectation that mist droplets will settle out before reaching a vent, especially if the level is low enough, droplets that are small enough will be carried to the vent, regardless of the level. Operating at lower liquid levels will slow down the rate of blockage, but not stop it, and the improvement would come at the cost of lost capacity.

A more effective solution is to eliminate the dip pipe so that process air is not blown through the caustic solution to create a mist in the first place. Neither Dow Chemical nor PPG Industries show dip pipes in their example configurations of storage tanks [18], [19]. Because aqueous solutions of solids are neither flammable nor non-conducting, the usual concerns regarding static discharge from free-falling liquid are simply not an issue.

In this particular case, the dip pipe also serves to avoid erosion of the head and vessel walls that result from feeding directly through a tank nozzle. Cutting the dip pipe off a few inches below the entrance to the head and fitting it with a slight elbow to direct the free-falling liquid away from the wall still avoids erosion problems but eliminates the mist that forms by sparging process air through the caustic solution.

3.6.4 Pressure regulation to reduce potential for overpressure

Atmospheric tanks are frequently connected to high pressure sources. Examples include fill lines, where the pressure can be as high as the deadhead pressure of the transfer pump; water supplies, which typically run around 40 to 60 psig; and plant air, which is often in the range of 80 to 120 psig. Any one of these can be a source of overpressure, but to regulate them down to atmospheric pressure would render them useless. While the source pressure must be above atmospheric pressure in order for these connections to work, they should not be any higher than necessary.

In this particular case, the plant air supply used to blow out the lines would have been just as effective at 20 psig as it was at 90 psig. Regulating the air supply pressure down creates no burden. The water pressure and the transfer pump pressure need to be where they are.

4 Conclusions

Atmospheric tanks are relatively fragile, especially when compared to vacuum vessels, low pressure vessels, and high pressure vessels. They are also widely used and considered relatively benign. However, because they are considered relatively benign, they are often overlooked as a hazard and neglected as a maintenance priority.

When atmospheric tanks contain flammable materials, the consequence of concern is most often a tank fire. Beyond that, the typical consequence of concern associated with atmospheric tanks is a leak. The impacts of these events are usually limited to injuries and low-consequence impacts to the environment, although there have been some notable exceptions [21], [22]. Rarely is the concern one of sudden, catastrophic failure. However, an atmospheric tank can fail suddenly and catastrophically, with life-threatening consequences. The catastrophic failure can result in a collapse, an implosion, or an explosion. Collapses result when a tank is badly installed or when a tank is misused. The best way to avoid collapses is to first make sure the tank is properly designed and installed, and then to use it within the limits of that design.

Explosions can result from the ignition of flammable vapors, but surprising to many, can also occur for purely physical reasons. Implosions result primarily when sources of vacuum are overlooked or neglected. Atmospheric tanks are called atmospheric tanks because they vent to atmosphere. When that vent becomes inoperable, it becomes all too easy for the pressure in an atmospheric tank to become lower or higher than the thin walls of the tank can withstand. So it is imperative for operators to understand what causes a vent to become inoperable, and to prevent that from happening. The response in a HazOp to "pressure-too high" or to "pressure-too low" cannot be, "That can't happen, we're open to atmosphere."

5 Recommendations

5.1 Regarding catastrophic tank collapse

To avoid catastrophic vessel collapse, apply rigorous management of change and pre-startup safety review when putting an atmospheric tank into new or different service. In particular, pay attention to the impact of using a tank for higher density liquids. When a tank is first put into service is the time when the tank is most vulnerable. Resist the temptation to fast-track atmospheric tank commissioning when it is being filled for the first time; act as though you know it will fail and deploy personnel accordingly.

Once an atmospheric tank has been successfully put into service, it should provide good service. The greatest concern will be wall thinning that compromises the strength of the vessel walls. Ideally, the thinning will be uniform and caught with a minimum of testing. However, uniform thinning is rarely the actual experience, so special attention needs to be paid to localized thinning due to erosion, which can only be detected by looking for it. Pay attention to where nozzles, dip pipes, and eductors are aimed and confirm that weak spots are not being created.

5.2 Regarding catastrophic tank implosion

Avoiding catastrophic tank implosion requires identifying and controlling sources of vacuum and assuring that measures to protect tanks from implosion remain operable. Obvious vacuum sources include vacuum pumps, pump out and other forms of emptying a tank of its liquid contents, and condensing vapors (typically, but not limited to, steam). When the presence of any of these is unavoidable, they should only be present with full awareness of their hazard and the total reliance on the protection device.

All of the devices to protect atmospheric tanks against implosion have mechanisms by which they can fail. Facilities that take credit for these devices – atmospheric vents, conservation vents, inerting pads, and vacuum breakers – need to periodically validate that they are working as designed. The means to validate the functionality of protective devices needs to be designed with the same care with which the device itself is designed.

5.3 Regarding catastrophic tank explosion

In addition to the obvious concerns of flammable vapor combustion, avoiding catastrophic tank explosion requires identifying and controlling sources of pressure. Obvious pressure sources include high inlet gas pressure, high inlet liquid pressure, and heating. Again, the presence of any of these may be unavoidable, so they should only be present with full awareness of their hazard and the importance of the protective devices.

As with any device, the devices to protect atmospheric tanks against overpressure and explosion have mechanisms by which they can fail. The term "fail-safe" does not mean they cannot fail; it means that in the absence of energy sources, they are supposed to go to the safe state. It does not mean that they cannot fail to a dangerous state. Facilities that take credit for these devices – typically atmospheric vents and conservation vents – need to periodically validate that they are working as designed. As with all protective devices, the means to validate the functionality of protective devices needs to be designed with the same care with which the device itself is designed.

For atmospheric vents in particular, the design needs to allow for easy inspection of the entire vent. While this may involve the use of boroscopes and cameras, the installation of inspection and clean-out ports in the vent line may be more cost effective and easier to implement. It is not enough to have inspection measures, however. They must be used and at a sufficient frequency to detect a developing problem in time to address it. The frequency of inspection may be adjusted as experience dictates, but two failed inspections in a row indicates that the test interval is too long.

5.4 Regarding use of dip pipes in atmospheric storage tanks

Dip pipes have two important features when used in atmospheric storage tanks. The most important is that when the liquid being stored is both non-polar and flammable, dip pipes prevent free-falling liquid, which is known to create static sparks, a source of ignition of the flammable vapors. The other feature is that dip pipes direct liquid away from areas of the tank vulnerable to erosion. When these features are not required, dip pipes should not be used.

This is of particular concern when the liquid being stored is a solution of dissolved solids. When the line and its dip pipe are blown clear, which is a fairly common practice, the gas used to blow the line clear will sparge through the liquid and carry some quantity of mist to the vent. At the vent, the droplets will collect, evaporate, and leave a deposit of solids. Unless inspections explicitly look for this deposit, it is likely to go unnoticed. When dip pipes are installed and their use with solutions of dissolved solids involves blowing them out, vent inspections for blockage should not be limited to looking for nests or other biological activity, but should also include inspection for solids build-up.

6 References

[1] Chang, J.I. and C-C Lin. "A study of storage tank accidents," *Journal of Loss Prevention*, 19 (2006) 51-59.

[2] Atherton, W. and J.W.Ash. *Review of Failures, Causes & Consequences in the Bulk Storage Industry*, National Lightning Safety Institute. 2006. http://lightningsafety.com/nlsi_lls/Causes-of-Failures-in-Bulk-Storage.pdf (accessed 02-Sep-2015).

[3] EPA. "Catastrophic Failure of Storage Tanks," *Chemical Safety Alert*, EPA 550-F-97-002b, Chemical Emergency Preparedness and Prevention Office, USEPA, Washington, DC. May 1997. www.nrt.org/production/nrt/nrtweb.nsf/AllAttachmentsByTitle/A-107tanks/$File/tanks.pdf?OpenElement (accessed 02-Sep-2015).

[4] Marsh. "Atmospheric Storage Tanks," *Risk Engineering Position Paper-01*, Rev 1.0, Marsh & McLennan Companies, February 2011. uk.marsh.com/Portals/18/Documents/Atmospheric%20Storage%20Tanks_lowres.pdf (accessed 02-Sep-2015).

[5] OGP. "Storage incident frequencies," Risk Assessment Data Directory, Report No. 434-3, International Association of Oil & Gas Producers, March 2010. http://www.ogp.org.uk/pubs/434-03.pdf (accessed 02-Sep-2015).

[6] ASME. *Boiler and Pressure Vessel Code, Section VIII-Rules for Construction of Pressure Vessels*, 2015 Ed. American Society of Mechanical Engineers, New York, NY. 2015.

[7] API. *API 620: Design and Construction of Large, Welded, Low-pressure Storage Tanks*, 12th Edition. American Petroleum Institute, Washington, DC. 2013.

[8] API. *API 650: Welded Tanks for Oil Storage*, 12th Edition. American Petroleum Institute, Washington, DC. 2013.

[9] OSHA. *29 CFR 1910.119, Process safety management of highly hazardous chemicals.*, Occupational Safety and Health Administration, U.S. Department of Labor, Washington, DC. Last revised 08-Feb-2013.

[10] API. *API 2000: Venting Atmospheric and Low-Pressure Storage Tanks*, 7th Edition. American Petroleum Institute, Washington, DC. 2014.

[11] Ennis, T. "Pressure Relief Considerations for Low-Pressure (Atmospheric) Storage Tanks," *IChemE Symposium Series No. 151*, Institution of Chemical Engineers, Rugby, UK. https://www.icheme.org/communities/subject_groups/safety%20and%20loss%20prevention/resources/hazards%20archive/~/media/Documents/Subject%20Groups/Safety_Loss_Prevention/Hazards%20Archive/XIX/XIX-Paper-63.pdf (accessed 02-Sep-2015).

[12] Schworm, P. "Nearly a century later, structural flaw in molasses tank revealed," The Boston Globe, January 14, 2015. www.bostonglobe.com/metro/2015/01/14/nearly-century-later-new-insight-into-cause-great-molasses-flood/CNqLYc0T58kNo3MxP872iM/story.html (accessed 15-Feb-2016).

[13] CSB. *Investigation Report: Allied Terminals, Inc.-Catastrophic Tank Collapse*. No. 2009-03-I-VA, U.S. Chemical Safety and Hazard Investigation Board, Washington, DC, May 2009. 49 pp.

[14] Savage, A. "On MythBusters' FIRST-EVER Implosion," *TESTED*. January 22, 2016. www.tested.com/art/makers/561346-mythbusters-first-ever-implosion/ (accessed 16-Feb-2016).

[15] FACE. "Worker Dies While Repairing a Vacuum Evaporator Tank in Virginia," FACE 8725. Fatal Accident Circumstances and Epidemiology Project, NIOSH, CDC, Atlanta, GA. October 15, 2014. www.cdc.gov/niosh/face/In-house/full8725.html (accessed 16-Feb-2016).

[16] Clouse, T. "Ruling upheld in sewage death: Supreme Court says CH2M Hill liable for $6.5 million," The Spokesman-Review, Spokane, WA, May 27, 2011. www.spokesman.com/stories/2011/may/27/ruling-upheld-in-sewage-death/ (accessed 16-Feb-2016).

[17] CCPS. *Guidelines for Engineering Design for Process Safety*, Center for Chemical Process Safety, American Institute of Chemical Engineers, New York, NY. 2010. p. 110.

[18] Dow. *Caustic Soda Solution Handbook*, Form Number 102-00011-0810XBBI, The Dow Chemical Company, Midland, Michigan. August 2010. 23 pp.

[19] PPG. *NaOH-Caustic Soda*, #0100 Rev 0406 pdf, PPG Industries, Inc., Monroeville, Pennsylvania. 2008. 63 pp.

[20] CSB. *Case Study: Barton Solvents*. No. 2007-06-I-KS, U.S. Chemical Safety and Hazard Investigation Board, Washington, D.C., June 26, 2008. 14 pp.

[21] DoJ. *Former Freedom Industries official sentenced for role in chemical spill.* U.S. Attorney's Office, Southern District of West Virginia, February 1, 2016. www.justice.gov/usao-sdwv/pr/former-freedom-industries-official-sentenced-role-chemical-spill (accessed 15-Feb-2016).

[22] Asahi. Police push for charges against TEPCO, execs over pollution problems after nuclear disaster. The Asahi Shimbun, Asia & Japan Watch. October 3, 2015. http://ajw.asahi.com/article/0311disaster/fukushima/AJ201510030052 (accessed 15-Feb-2016).

Explosions in SMPO Hydrogenation Unit

Presenter/Principal Author: Nick Gonzales
Affiliation: Shell Global Solutions
Address: 3333 Hwy 6 South, Houston TX 77082
Email address:nick.gonzales@shell.com

Prepared for Presentation at
American Institute of Chemical Engineers
2016 Spring Meeting
12th Global Congress on Process Safety
Houston, Texas
April 11-23, 2016

LEARNING FROM INCIDENTS
AWARENESS ALERT

For Industry-Wide Sharing **September 2015**

Explosions in SMPO Hydrogenation Unit

Figure 1 – Hydrogenator structure after the explosion.

Target audience for this alert

- Process Safety and Process Engineers
- Production staff in Refining/Chemical and hydrogenation processes

What happened

On June 3rd 2014, a Styrene Monomer and Propylene Oxide (SMPO) unit was in the pre start-up phase after the scheduled shutdown and replacement of catalyst in the hydrogenation reactors. The reaction system contained two fixed-bed adiabatic, trickle-flow reactors. Operations were heating up the hydrogenation unit in preparation for the catalyst reduction step. The heat-up was performed by the circulation of Ethyl Benzene (EB) through the reaction system under a nitrogen blanket and continuous nitrogen purge. During this heat up, one hydrogenation reactor and an additional upstream separation vessel suddenly exploded. Two contractor workers, who were in the immediate process area, were injured with burns from hot catalyst particles. The explosion caused extensive damage requiring over a year to repair.

Why it happened

The incident investigation concluded that both the reactor and vessel had failed because of overpressure, resulting from unexpected exothermic reactions between Ethyl Benzene (EB) and copper chromite oxide catalyst which produced an excessive amount of gas.

Prior to the incident, it was generally believed that under inert nitrogen conditions EB would not react with the passivated catalyst. However, it was subsequently determined and proven in the laboratory that two exothermic reactions actually occurred during the heat up:

- EB reacted with active oxygen present in the catalyst (mainly but not limited to copper chromate $CuCr^{VI}O_4$). The catalyst contained more copper chromate than the previous catalyst used at the site. The redox reaction started when the temperature exceeded approximately 90°C. The target temperature at the end of the heat up phase was 130°C, but due to the exothermic reaction the temperature increased further in local dry zones of the trickle bed reactors.
- Copper oxide ($Cu^{II}O$) was reduced by EB when localized temperatures reached 180°C. This redox reaction generated large amounts of heat and gaseous molecules (such as CO_2 and H_2O) leading to the overpressure.

AWARENESS ALERT

The early heat generation by the EB-chromate reaction, in combination with reduced cooling capacity in localized dry zones created localized temperatures in the reactor bed, high enough to initiate the EB-copper oxide reaction. Contributing factors include:

Low nitrogen flow: In a trickle flow reactor, sufficient nitrogen flow is required to achieve a proper liquid distribution from the distribution trays to wet the catalyst in the catalyst bed. The actual nitrogen flow was lower than needed, which led to non optimum liquid distribution in both trickle bed reactors causing dry zones in the top part of the catalyst beds.

Low EB flow: The EB flow to the hydrogenator was, at times, lower than needed for proper wetting of the catalyst and the flow variations led to preferential or rivulet flow and stagnant wet zones in the trickle flow reactor catalyst bed. These conditions led to non optimal heat removal and as a consequence high localized temperatures

Increasing Pressure: About thirty minutes before the explosion, the vent gas valve on the second separator tripped closed on high level and stayed closed. The system pressure increased up to the nitrogen supply pressure and caused the nitrogen flow to the reactor to decrease. The increasing pressure reduced the evaporative cooling of the EB from the catalyst.

Reactive hazard understanding: A Reactive Hazard Assessment (RHA) had been conducted in 2011 for this section of the SMPO unit. The RHA was based on process development work, design knowledge, literature search and application of a tool that was developed and used for identifying unintended reactions. The RHA only identified the reactions with hydrogen during the reduction phase. The reactions of EB with the catalyst in the absence of hydrogen were not identified in the RHA. This, together with the previous operational reduction experiences, reinforced the belief that EB was an inert medium in combination with the catalyst.

Lessons learned

- The passivated copper chromite hydrogenation catalyst contains bonded oxygen in the form of copper chromate and copper oxide prior to reduction. Where a hydrocarbon and oxidizing substance (i.e. both a "fuel" and reactive oxygen) are combined and in sufficient quantities, an exothermic reaction may occur and the only mitigation available may be the total amount of cooling provided.
- Heating up in EB, prior to the reduction step with hydrogen, was seen as normal and safe from a chemical reaction point of view. This influenced the choices made during the process design, process development work, the design of the trickle bed reactor, qualification of the catalyst and operational procedures.
- The purpose of a Reactive Hazard Assessment (RHA) is to identify consequences of the potential reactions, intended and unintended, between the components of the reaction streams. At the time of the Assessment, The composition of the reactants and the potential initiation temperatures were not understood adequately. Post incident investigation suggests that suitable analytical tools such as Thermogravimetric Analysis (TGA) may help to identify such exothermic reactions.
- A series of incremental and seemingly innocuous changes can allow unknown reactions to proceed and run out of control, resulting in a severe incident. Introduction of a new catalyst or catalyst supplier can introduce new components or changes even in small quantities of components that could lead to unidentified and unintended chemical reactions. The catalyst selection and subsequent change processes should address these with the focus on reactive hazards in both transient and steady state operations. Note that this may also apply to changes in process chemicals in other services (e.g. polymerization inhibitors, cleaning solutions, etc.).

AUTHOR INDEX

AUTHOR INDEX